Advance in Barley Sciences

Guoping Zhang • Chengdao Li • Xu Liu
Editors

Advance in Barley Sciences

Proceedings of 11th International Barley Genetics Symposium

ZHEJIANG UNIVERSITY PRESS
浙江大学出版社

Springer

Editors
Guoping Zhang
Department of Agronomy
Zhejiang University
Hangzhou, China

Chengdao Li
Department of Agriculture and Food,
government of Western Australia
Australia

Xu Liu
Department of Agriculture and Food,
government of Western Australia
Australia

ISBN 978-94-007-4681-7 ISBN 978-94-007-4682-4 (eBook)
DOI 10.1007/978-94-007-4682-4
Springer Dordrecht Heidelberg New York London

Jointly published with Zhejiang University Press
ISBN: 978-7-308-09790-1 Zhejiang University Press

Library of Congress Control Number: 2012954562

Printed on acid-free paper

Springer is part of Springer Science+Business Media (www.springer.com)

Preface

The last half century has seen extraordinary progress in barley genetics research. In the first International Barley Genetics Symposium in 1963, scientists just started to discuss a map of barley chromosomes. Today, the International Barley Genome Sequencing Consortium is at the dawn of the completion of the barley genome sequence. The regular International Barley Genetics Symposia provide an important platform for barley breeders and scientists to share their research results and understand the future trends of barley genetics research. The proceedings are not only the permanent record; they also provide key references for interested outsiders at the symposium and future barley research scientists. The organizing committee received around 150 abstracts and put them into a special volume for all symposium participants to have access, which will be helpful for improving the discussions at the poster sessions. Moreover, we have selected 38 full-length papers and published them as The Proceedings of 11th IBGS. Hopefully, they will provide in-depth content of current research and development in barley genetics and breeding.

This is the first time the symposium was held in China. Chinese scientists first participated in the Okayama Symposium in 1986, and increasing numbers of Chinese scientists joined the following symposia. However, the majority of Chinese barley research is still unknown to the international barley community except the fact that China is the world's largest beer producer and malting barley importer. Barley has been a major crop over thousands of years, and it is one of the most widely distributed crops in China. The Qing-Tibetan Plateau in south-west China is a unique agricultural region in the world with an average altitude of 4 km above sea level. Barley is still the major food crop for millions of people in this region. The harsh environments have created unique germplasm for low soil fertility, drought, frost and salinity tolerance. Tibetan barley also contains multiple functional components for human nutrition, e.g. high β-glucan and antioxidants. It also has unique enzyme activity and thermostability to determine barley malting quality. Tibetan barley germplasm is a highly valuable but underutilized component of the world barley gene pool. This region is arguably another centre of origin of cultivated barley. We hope that the symposium will unveil the secret of Tibetan barley, enhance understanding of China's barley research and promote collaboration between Chinese scientists and the international barley community.

The organizing committee gratefully acknowledges financial support from the Natural Science Foundation of China, the Natural Science Foundation of Zhejiang, Zhejiang University and the K. C. Wong Education Foundation, Hong Kong. We thank the international organizing committee members for their support and guidance for this symposium. We also appreciate support from the Zhejiang University Press and the Springer Press in publishing the proceedings.

The 11th International Barley Genetics Symposium will be held in the First World Hotel, Hangzhou, China, from 15 to 20 April 2012. Hangzhou is a core city of the Yangtze River Delta and has a registered population of 3.8 million people. The city is located on Hangzhou Bay, 180 km southwest of Shanghai. It has been one of the most renowned and prosperous cities of China for much of the last 1,000 years, due in part to its beautiful natural scenery. The city's West Lake is its best-known attraction, and Hangzhou is the oriental leisure capital.

Department of Agronomy, Zhejiang University — Guoping Zhang
Department of Agriculture and Food, government of Western Australia — Chengdao Li
Department of Agriculture and Food, government of Western Australia — Xu Liu

INTERNATIONAL COMMITTEE

LOCAL COMMITTEE

SPONSORS

Zhejiang University, China
National Science Foundation of China
Zhejiang Provincial Natural Science Foundation of China
Zhejiang Academy of Agricultural Sciences, China
Barley Professional Committee of Crop Science Society of China
The KC Wong Education Foundation, Hong Kong
Key Laboratory of Crop Germplasm Resource of Zhejiang Province, China

Contents

Chapter 1
Evolution of Wild Barley and Barley Improvement

Eviatar Nevo

Abstract Wild barley, Hordeum spontaneum, the progenitor of cultivated barley, Hordeum vulgare, originated 5.5 million years ago in southwest Asia, is distributed in the Eastern Mediterranean, Balkans, North Africa, Central Asia, and Tibet. H. vulgare, the fourth important world crop, used for animal feed, beer, and human food was domesticated polyphyletically by humans 10,000 years ago in the Neolithic revolution in at least three centers: Fertile Crescent, Central Asia, and Tibet. H. vulgare with thousands of land races and cultivars is widespread where other crops cannot adapt, yet it deteriorated genetically, especially due to pure breeding, and needs genetic reinforcement. H. spontaneum, the best hope for barley improvement, is a hardy ecological generalist, adapted to a wide range of extreme latitudes, altitudes, climates (warm and cold), and soils. Adaptations occur at all levels: genomically, proteomically, and phenomically both regionally and locally. It displays "archipelago" genetic structure, rich genetically, and harbors immense adaptive abiotic and biotic resistances precious to barley and cereal improvement. Sequencing the H. spontaneum genome will reveal huge, mostly untapped, genetic resources. The current global warming stresses H. spontaneum, and so it is imperative to conserve it in situ and ex situ to safeguard its future immense contribution to barley and cereal improvement, thereby helping to fight hunger.

Keywords *Hordeum spontaneum* • Polyphyletic domestication • Genetic resources

E. Nevo (✉)
Institute of Evolution, University of Haifa, Mount Carmel, Haifa 31905, Israel
e-mail: nevo@research.haifa.ac.il

G. Zhang et al. (eds.), *Advance in Barley Sciences: Proceedings of 11th International Barley Genetics Symposium*, DOI 10.1007/978-94-007-4682-4_1,

1.1 Introduction

1.1.1 Cultivated Barley, Hordeum vulgare L.: Domestication and Origin

Cultivated barley, *Hordeum vulgare* L., is one of the main cereals of the belt of Mediterranean agriculture, a founder crop of Old World Neolithic food production, and one of the earliest domesticated crops (Harlan and Zohary 1966; Zohary et al. 2012). It is an important crop, ranking fourth (at 136 million tons) in 2007, in world crop production in an area of 566,000 km^2 (http://faostat.fao.org/faostat). Barley is used for animal feed, brewing malts, and human food (in this order). In Mediterranean agriculture, barley is a companion of wheat but is regarded as an inferior staple and is known as the poor man's bread. Barley is a short-season, early-maturing grain with a high yield potential and grows in widely varying environments, including extreme latitudes where other crops cannot adapt (Harlan 1976). It extends far into the arctic, reaching the upper limit of cultivation in high mountains, desert oases, and desert fringes. It is more salt and drought resistant than other cereals. Barley is a cool-season crop. It can tolerate high and low temperatures if the humidity is low but avoids warm-humid climates. It grows in cold highlands such as Mexico, the Andes, East Africa, and Tibet. Major production areas of barley are Europe, the Mediterranean region, Ethiopia, the Near East, Russia, China, India, Canada, United States, and Australia.

1.2 Domestication

Barley first appeared in several preagriculture or incipient sites in southwest Asia. The remains are of brittle two-rowed forms, morphologically identical with present-day *H. spontaneum* wild barley and apparently collected in nature (Zohary et al. 2012 and their Fig. 16). The earliest records of such wild barley harvest come from ca. 50,000 years BP Kebara (Lev et al. 2005) and from ca. 23,000 years BP Ohallo II, a submerged early Epipaleolithic site on the south shore of the Lake of Galilee, Israel (Kislev et al. 1992; Weiss et al. 2004), as brittle two-rowed forms, morphologically identical with the progenitor of barley, *Hordeum spontaneum*, apparently collected in the wild. Other sites of *H. spontaneum* collection from the wild from 15.500 to 10.150 BP, including Jericho (Kislev 1997), appear in Zohary et al. (2012, p. 56). Like the domestication model advocated for wild emmer wheat (Feldman and Kislev 2007), domestication occurred independently in sites across the Levant. Moreover, recent proposals suggest Central Asia (Morrell and Clegg 2007) and Tibet (see discussion in Dai et al. 2012) as additional centers of wild barley domestications. According to this view, the linked genes for nonbrittleness (Bt_1 and Bt_2) were transferred to numerous wild barley genotypes through multiple spontaneous hybridizations, followed by human selection. The rich genetic variation of the progenitor,

H. spontaneum (Nevo 1992, and see later), as well as its superb thermogenesis (Nevo et al. 1992), has enabled it to tolerate biotic and abiotic stresses and succeed under cultivation in the warm-dry Near East and in cold-dry Tibet. These advantages of wild barley may explain the wider ecogeographic range of wild and cultivated barley as compared to those of wheat. Current archaeological finds show barley as a founder crop of the southwest Asian Neolithic agriculture and as a close companion of emmer and einkorn wheats (Zohary et al. 2012; but see Morrell and Clegg 2007 and Dai et al. 2012). Future studies will highlight the full domestication scenario of *H. spontaneum* in Asia.

1.3 Origin

The grass family Poaceae originated at the Upper Cretaceous (Prasad et al. 2005). The genera *Hordeum* and *Triticum* diverged about 13 Mya (million years ago) (Gaut 2002). The genus *Hordeum* evolved ~12 Mya in southwest Asia and spread into Europe and Central Asia. Multiple intercontinental dispersals shaped the distribution area of *Hordeum* (Blattner 2006). The divergence of *H. spontaneum* from the Near East and Tibet is around 5.5 Mya (Dai et al. 2012), whereas barley (*Hordeum vulgare*) was domesticated around 10,000 years ago (see earlier).

The progenitor of barley is wild barley, *Hordeum spontaneum* (Harlan and Zohary 1966; Zohary 1969). *H. spontaneum* is an annual brittle two-rowed diploid ($2n = 14$), predominantly self-pollinated (but see Brown et al. 1978, reporting on 0–9.6% outcrossing, averaging 1.6%) and a strong colonizer species penetrating Central Asia and Tibet (Fig. 1.1). A population-genetic analysis based on 795 loci, in 506 individuals of the progenitor *Hordeum spontaneum*, the cultivar *Hordeum vulgare*, and their hybrid *Hordeum agriocrithon*, concluded that barley cultivars form a distinct species, derived from the progenitor. Numerous cultivars and land races of barley have nonbrittle ears. Nonbrittle mutations survive primarily under domestication, and nonshattering ears, as well as *thresh-1*, the locus of threshability (Schmalenbach et al. 2011), are signatures of cultivation. In spite of these substantial differences between *H. spontaneum*, the progenitor, and *H. vulgare*, the human derivative, Zohary et al. (2012) concluded that splitting the two entities into separate species is genetically unjustified and that the main cultivated barley types represent races of a single crop species. Notably, Darwin (1859) considered domestication as a gigantic evolutionary experiment in adaptation and speciation, generating incipient species. Domestication was performed by humans primarily during the last 10,000 years mimicking speciation in nature (Wei et al. 2005). It leads to adaptive syndromes fitting human ecology (Harlan 1992). *H. spontaneum* and *H. vulgare* appear to be both reproductively (hybrids are selected against) and ecologically (occupying separate ecological niches) independent species. They conform to the biological species concept based on reproductive biology and ecological compatibility. If agriculture was to disappear, then, in all likelihood, cultivated barley would disappear also, along with man-made habitats, with *H. spontaneum* the only surviving

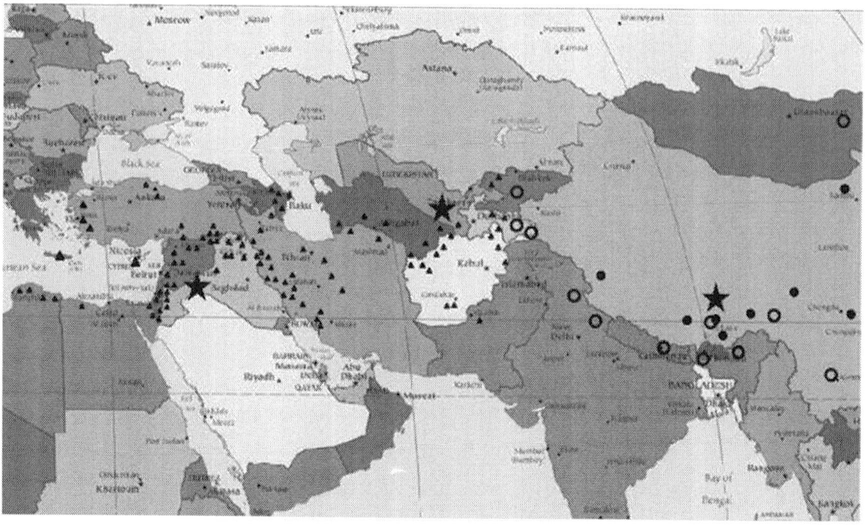

Fig. 1.1 Distribution of sites of wild barley (*Triangle*: wild barley in the Near East *(left star)* and Central Asia (*middle star*) (Harlan and Zohary 1966). *Dot*: wild barley in the Tibetan Plateau (*right star*) and its vicinity (Ma 2000). *Circle*: wild barley collected by Prof. Kazuhiro Sato (Okayama University, personal communication))

species (Wei et al. 2005). Even wild *H. spontaneum* shows signatures of phenotypic and genotypic stresses due to global warming (Nevo et al. 2012).

Hordeum spontaneum is distributed in the Eastern Mediterranean basin and the west Asiatic countries, penetrating into the Aegean region and North Africa to Morocco. It extends eastward to Central Asian areas (Turkmenia, Afghanistan, Ladakh, and Tibet) (Fig. 1.1). Wild barley occupies both primary and segetal habitats. Its center of origin and diversity was considered, until recently, in the Near East Fertile Crescent, displaying high genetic diversity in Israel, Golan Heights, and Jordan and extending across Asia to Tibet (Zohary et al. 2012). See the current distribution of *H. spontaneum* in Zohary et al. (2012), Morrell and Clegg (2007), and Dai et al. (2012) (Fig. 1.1). Extensive genetic diversity was found in *H. spontaneum* in Tibet using 1,300 markers across the genome (Dai et al. 2012, and their Fig. 1). Zohary (1999) argued that the fixation of independent mutations at nonallelic, nonbrittle ear loci in cultivated barley is strongly suggestive of at least two centers of domestications. Chloroplast DNA microsatellites support a polyphyletic origin for barley (Molina-Cano et al. 2005). Likewise, differences in haplotype frequency among geographic regions at multiple loci infer at least two domestications of barley: one within the Fertile Crescent (see Badr et al. 2000; Lev-Yadun et al. 2000; Zohary et al. 2012) and the second 1,500–3,000 km farther east (Morrell and Clegg 2007). Finally, another center of domestication of barley in Tibet was suggested, based on 1,300 DArT P/L (Diversity Arrays Technology Pty Ltd) (Dai et al. 2012). In the Near East, wild barley is adapted primarily to warm and dry climates and only rarely found above 1,500 m. However, in Tibet it thrives above 4,000 m and is adapted to cold and dry environments and may have developed after diverging from the Near East adaptive complexes to a cold and dry climatic regime (Dai et al. 2012).

1.4 Genetic Diversity in Wild Barley

Genetic diversity is the basis of evolutionary change (Nevo 1978, 1988, 1998a, b, 2004, 2005; Nevo and Beiles 2011). Wild barley, *H. spontaneum*, is rich in adaptive genetic diversity at the genetic (allozyme and DNA), genomic, proteomic, and phenomic levels (see Nevo, wild cereals at http://evolution.haifa.ac.il, and specifically Nevo et al. 1979, 1981, 1983, 1986a, b, c, d, 1997, 1998; Nevo 2004, 2005, 2009b; Chalmers et al. 1992; Baum et al. 1997; Pakniyat et al. 1997; Forster et al. 1997; Owuor et al. 1997; Li et al. 1998; Vicient et al. 1999; Turpeinen et al. 1999, 2003; Kalendar et al. 2000; Close et al. 2000; Gupta et al. 2002, 2004; Sharma et al. 2004; Ivandic et al. 2002, 2003; Huang et al. 2002; Baek et al. 2003; Owuor et al. 2003).

The regional and local allozyme studies in Israel, Turkey, and Iran highlight the following patterns. *H. spontaneum* in the Near East is very variable genetically. Genetic divergence of populations includes some clinal but primarily regional and local patterns, often displaying sharply geographic divergence over short distances at both single and multilocus genome organization. The average relative genetic differentiation (GST) was 54% within populations, 39% among populations (range 29–48%) within countries, and 8% among the three countries (Table 7 in Nevo et al. 1986c). Allele distribution is characterized by a high proportion of unique alleles (51%) and a high proportion of common alleles that are distributed either locally or sporadically, as well as displaying an "archipelago" genetic structure, where high-frequency allele levels can reside *near* low ones or none at all. Discriminant analysis by allele frequencies successfully clustered wild barley of each of the three countries (96% correct classification). A substantial portion of allozyme variation in nature is significantly correlated with the environment and is predictable ecologically, chiefly by a combination of humidity and temperature variables. Natural populations of wild barley are, on average, more variable than two composite crosses and landraces of cultivated barley (Nevo 2004). Genetic variation of wild barley is not only rich in the Near East but at least partly adaptive and predictable by ecology and allozyme markers (Nevo 1987). Consequently, *conservation* and *utilization* programs should optimize sampling strategies by following the ecological-genetic factors and molecular markers as effectively predictive guidelines (Nevo et al. 1986c; Nevo 1987; Chalmers et al. 1992; Volis et al. 2001, 2002).

DNA genetic diversity and divergence patterns parallel those of allozymes (Li et al. 2000). This conclusion suggests that climatic selection through aridity stress may be an important factor acting on both structural protein coding and presumably partly regulatory noncoding DNA regions resulting in adaptive patterns, for example, in intergenic and genic SSRs (Li et al. 2002, 2004). The population structure of *H. spontaneum* is strongly correlated with temperature and precipitation (Hubner et al. 2009). These and other multiple cases indicate that genetic diversity across the genome is driven, to a substantial yet unknown quantity, by natural selection. The latter overrides nonselective forces like gene flow and stochastic factors, revealing how plants respond to stressful environments (Cronin et al. 2007; Fitzgerald et al. 2011; Hubner et al. 2009; Nevo 1992, 2011a, b, c). Recent increases in the availability of expressed sequence tag (EST) data have facilitated the development of microsatellite (SSR) markers in plants, including cereals, enabling interspecific transferability and comparative mapping of barley EST-SSR markers in wheat, rye, and rice (Varshney et al. 2005). Development of new microsatellites in barley reinforces genetic mapping (Li et al. 2003). Genomic SSR markers displayed higher polymorphism than EST-SSRs. The latter, however, displayed clearer separation between wild and cultivated barley (Chabane et al. 2005). The EST-SSRs are applicable to barley genetic resources, providing direct estimates of functional biodiversity. EST-SNP are the best markers for typing gene bank accessions, and the AFLP and EST-SSR markers are more suitable for diversity analysis and fingerprinting (Varshney et al. 2007). Analysis of molecular diversity, population structure, and linkage disequilibrium was conducted in a worldwide survey of cultivated barleys (Malysheva-Otto et al. 2006). Low levels of linkage disequilibrium in wild barley were recorded despite the high rate of self-fertilization, ~98% (Morrell et al. 2005). High-resolution genotyping of wild barley and fine mapping facilitate QTL fine mapping and cloning (Schmalenbach et al. 2011). This enabled the fine mapping of the threshability locus *thresh-1* on chromosome 1H. *Thresh-1* controls grain threshability and played an important role in domestication.

1.5 Adaptive Complexes in the Near East

1.5.1 Phenotypic Adaptations

Israeli populations of *H. spontaneum* display dramatic variation in phenotypic traits across Israel in accordance with climatic and edaphic variations from robust mesic phenotypes to slender xeric genotypes. The genetic basis of this phenotypic variation in ten variables, including germination, earliness, biomass, and yield, was identified in common garden experiments in the mesic (Mount Carmel, Haifa) and xeric Avedat and Sde Boker in the northern Negev desert (Nevo et al. 1984, including several figures demonstrating the variation). Adaptive variation patterns of germination and desiccation of mesic and xeric phenotypes include longer seed dormancy, roots, and desiccation tolerance in xeric plants (Chen et al. 2002, 2004a, b). Likewise, small and dark kernels characterize xeric and high-solar-exposed populations (Chen et al. 2004c). The genetic basis of wild barley caryopsis dormancy and seedling desiccation tolerance at germination was described by Zhang et al. (2002, 2005). Xeric phenotypes had deeper dormancy but less seedling salt tolerance (Yan et al. 2008). Fifteen agronomic, morphological, developmental, and fertility traits differentiated at the

100-m Tabigha microsite, subdivided into 50 m of wetter basalt and 50 m of drier terra rossa soil (Ivandic et al. 2000). Terra rossa genotypes had better resistance to drought than basalt genotypes.

Edaphic natural selection strongly diverges phenotypes and genotypes at microscales as was also shown in the microclimatic divergent microsites of Newe Ya'ar (Nevo et al. 1986a) and "Evolution Canyon" (Nevo et al. 1997). Growth characteristics diverge distinctly in wild barley from different habitats (van Rijn et al. 2000) associated with AFLP markers (van Rijn et al. 2001; Vanhala et al. 2004), growth rates (Verhoeven et al. 2004a, b), and seedling desiccation tolerance (Zhang et al. 2002, 2005; Yan et al. 2008, 2011).

1.6 Genotypic Adaptations

1.6.1 Abiotic Genetic Resources of Drought and Salinity Resistances in Hordeum spontaneum

Drought and salinity are the major abiotic stresses that threaten food supplies around the world. Wild relatives of wheat and particularly barley harbor immense potential for drought and salt tolerance. *Triticum dicoccoides* (Gustafson et al. 2009; Nevo 2011a), but particularly *Hordeum spontaneum* (Chen 2005), the progenitors of cultivated wheat and barley, respectively, developed rich genetic diversities for drought and salt tolerance (Nevo and Chen 2010 and their references) with great potential in plant breeding for stress environments (Blum 1988). Drought resistance in wild barley from Israel, including physiology, gene identification, and QTL mapping, was extensively studied by Chen (2005) and Chen et al. (2002, 2004a, b, c, 2009, 2010, 2011a, b). Multilevel regulation and signaling processes associated with adaptation to terminal drought in wild emmer wheat were analyzed by Krugman et al. (2010 and references therein), followed by transcriptomic and metabolomic profiles in drought adaptation mechanisms in wild emmer (Krugman et al. 2011). A total of 5,892 transcripts were identified in this study between drought-resistant and drought-susceptible genotypes. Two hundred and twenty-one well-studied genes involved 26% regulatory genes including transcriptional regulation, RNA binding, kinase activity, and calcium and abscisic acid signal affecting stomatal closure. Additional adaptive genes were involved in wall adjustment, cuticular wax deposition, lignification, osmoregulation, redox homeostasis, dehydration protection, and drought-induced senescence. Tolerant genes, gene networks, and QTLs within a multidisciplinary context will play an increasing role in crop breeding programs to develop drought- and saline-tolerant crops, especially with the ongoing global warming associated with drought (IPCC 2010). A huge amount of literature, impractical to cite here, describes patterns and mechanisms of genetic resources from wild relatives as candidate genes for crop improvement. I will briefly overview three in-depth studies involving dehydrins and two novel genes, *Hsdr4* and *Eibi 1*, studied at the Institute of Evolution.

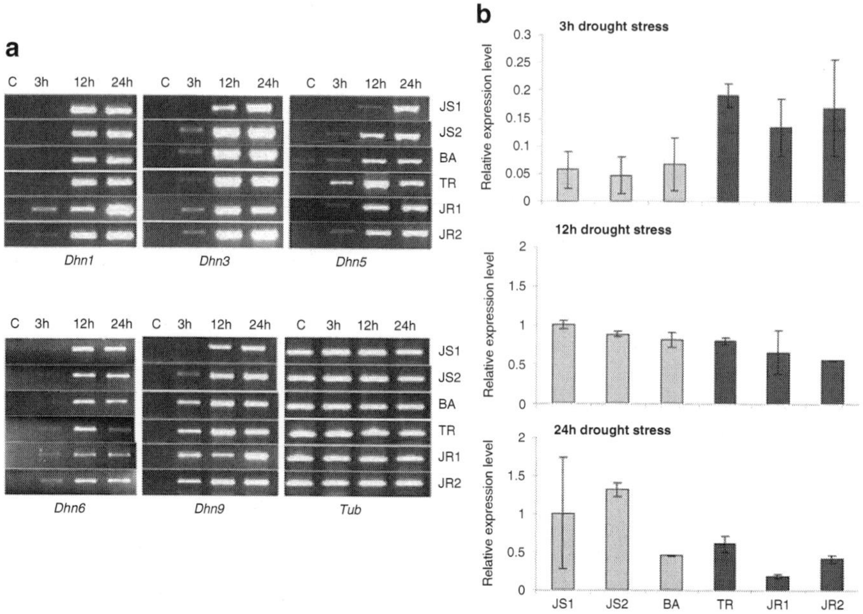

Fig. 1.2 (**a**) Differential expression patterns of *Dhn*1, 3, 5, 6, and 9 detected by RT-PCR. The RT-PCR was carried out with gene-specific primers, using cDNA obtained from six wild barley genotypes (JS1, JS2, BA, JR1, JR2, TR) after 0 (control, C), 3, 12, and 24 h of dehydration. As a control for relative amount of DNA, RT-PCR with gene-specific primers for α-*tubulin (Tub)* was performed. (**b**) Expression of Dhn6 detected by quantitative real-time PCR. Real-time PCR was carried out with cDNA obtained from six barley genotypes (JS1, JS2, BA, TR, JR1, and JR2) after 3, 12, and 24 h of dehydration. Quantification is based on C_t values that were normalized using the *C1* value corresponding to a barley (housekeeping) α-*tubulin* gene. Two independent plant samples for each genotype were examined in triplicate. Each value is the mean ± SE ($n=2$) (From Suprunova et al. 2004)

1.6.2 Differential Expression of Dehydrins in Wild Barley at Regional and Local Scales

Dehydrins (DHNs; Lea D-11) are water-soluble lipid vesicle-associated proteins involved in adaptive responses of plants to drought, low temperature, and salinity (Close et al. 2000). The assembly of several domains into consistent permutations resulted in DHN polypeptide lengths from 82 to 575 amino acid residues. Allelic variation in *Dhn* genes provides a rich repertoire for drought-stress tolerance in barley and other Triticeae species. Regionally, tolerant and sensitive genotypes were identified from Israeli and Jordanian wild barley *H. spontaneum* populations in dehydrin genes (*Dhn* 1, 3, 5, 6, and 9) (Suprunova et al. 2004) (Fig. 1.2). The five *Dhn* genes were upregulated by dehydration in resistant and sensitive wild barley genotypes and remarkably so in *Dhn1* and *Dhn6* genes, depending on the duration of dehydration stress. *Dhn1* reacted earlier, after 3 h, and displayed higher resistance (at 12 and 24 h) in resistant compared to sensitive genotypes. The expression level of *Dhn6* was significantly higher in the resistant genotypes at earlier stages of stress, but after 12 and 24 h, *Dhn6* expression was relatively higher in sensitive genotypes. These results indicate adaptive responses of these genes in dehydration tolerance *regionally* in wild barley (Fig. 1.2). We continued to test *Dhn* genes *locally* at "Evolution Canyon."

"Evolution Canyon" I, at lower Nahal Oren, Mount Carmel, Israel, is a natural microscale model for studying *evolution in action* highlighting biodiversity evolution, adaptive radiation, and incipient sympatric speciation across life (Nevo list of "Evolution Canyon" at http://evolution.haifa.ac.il and reviewed in Nevo 1995, 2006a, 2009a, 2011b, c, 2012). Wild barley, *Hordeum spontaneum,* is a major model organism at ECI, displaying interslope adaptive molecular-genetic divergence (Nevo et al. 1997; Owuor et al. 1997) and incipient sympatric speciation (Parnas 2006; Nevo 2006a). The adaptive divergence occurs between the "African" xeric, tropical south-facing slope (AS = SFS), and the "European," mesic, temperate north-facing slope (ES = NFS), separated, on average, by 200 m. *Dhn1* of wild barley was examined in 47 genotypes at "Evolution Canyon" I, 4–10 individuals in each of seven stations (populations) in an area of 7,000 m^2. The analysis was conducted on sequence diversity at the 5' upstream flanking region of the *Dhn1* gene. Rich diversity was found in 29 haplotypes, derived from 45 SNPs in a total of 708 bp sites. Most haplotypes, 25 of 29 (86.2%), were represented by one genotype, that is, unique to one population. Only a single haplotype was common to both slopes. Nucleotide diversity was higher on the AS (Fig. 1.3a) as in 64% of other model organisms tested at ECI (Nevo 2009a). Haplotype diversity was higher on the ES. *Inter*slope divergence was significantly higher than *intra*slope divergence, and SNP neutrality was rejected by the Tajima test. *Dhn1* expression under dehydration displayed interslope divergent expression between AS and ES genotypes (Fig. 1.3b), unfolding the adaptive nature of Dhn1 drought resistance. Microclimatic natural selection appears to be the most likely evolutionary driving force causing adaptive interslope *Dhn1* divergent evolution at ECI.

We also examined the genetic pattern of *Dhn6* in 48 genotypes of wild barley at ECI (Yang et al. 2012) because it is also strictly related to drought resistance in barley. A recent insertion of 342 bp in 5'UT primarily at the upper more xeric stations of the opposite slopes, AS and ES, was associated with earlier upregulation of *Dhn6* after dehydration. Both coding SNP nucleotide and haplotype diversity (see Fig. 2 in Yang et al. 2012) were higher on the AS than on the ES, and the applied Tajima D and Fu Li tests rejected neutrality of SNP diversity. Differential expression patterns of *Dhn6* were detected after different hours of dehydration. The interslope genetic divergence of amino acid sequences indicated significant positive selection of *Dhn6*. Clearly, *Dhn6* diversity was subjected to microclimatic divergent natural selection and was adaptively associated with drought resistance of wild barley at "Evolution Canyon" I, paralleling *Dhn1* (Yang et al. 2012).

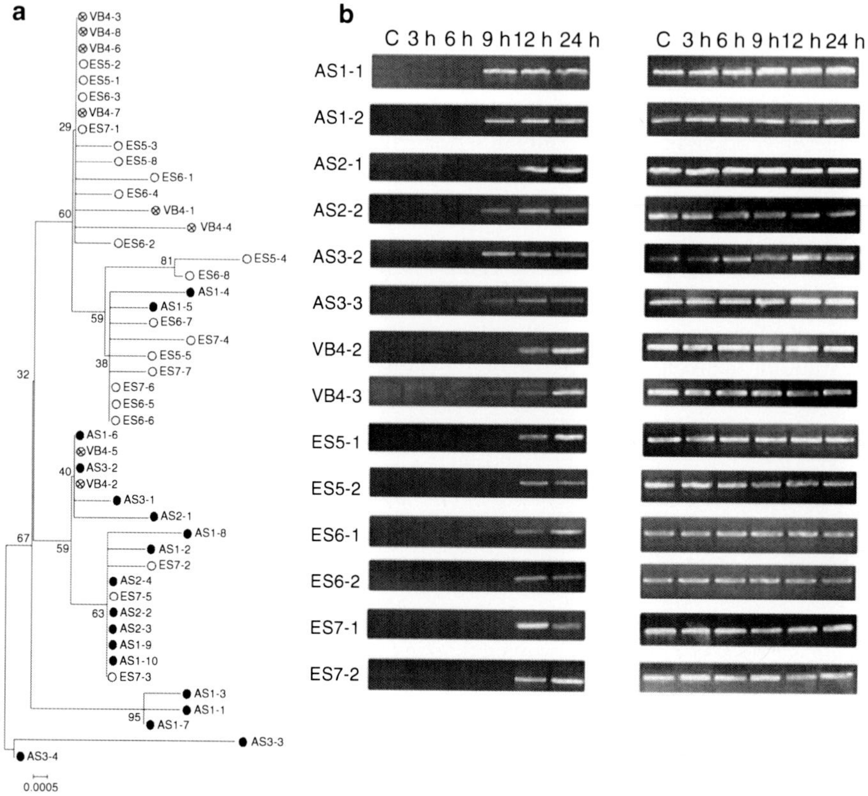

Fig. 1.3 (**a**) Dendrogram of the genetic relationships of 47 genotypes of wild barley, *H. spontaneum*, from seven populations representing the AS (*black circles*), ES (*blank circles*), and VB (*x-circles*) in "Evolution Canyon" I, Israel. These genotypes were obtained from the dehydrin 1 alignment sequence, based on Nei's calculated nucleotide diversity (*p*-distance) values (see scale), using the neighbor-joining method. Numbers on branches are percentage values from bootstrap analysis (1,000 replicates).(**b**) Differential expression patterns of Dhn1 detected by RT-PCR. The RT-PCR of *Dhn*1 was using the cDNA from the 14 genotypes (two from each station) after 0 (control, C), 3, 6, 8, 12, and 24 h after dehydration with α-tubulin (*Tub*) as a control (From Yan et al. 2008, Figs. 4 and 5)

1.6.3 Hsdr4 Novel Gene Involved in Water-Stress Tolerance in Wild Barley

In search of drought-resistant genes in wild barley, we identified a novel gene, *Hsdr4* (Suprunova et al. 2007). Gene expression profiles of water-stress-tolerant versus water-stress-sensitive wild barley genotypes were compared under severe dehydration stress applied at the seedling stage using cDNA AFLP analysis. Seventy out of 1,100 transcript-derived fragments (TDFs) displayed differential expression between control and stress conditions. Eleven of them showed clear up- or down-regulation differences between tolerant and susceptible genotypes. These TDFs were isolated, sequenced, and tested by RT-PCR. TDF-4 was selected as a promising candidate gene for water-stress tolerance. The corresponding gene, designated *Hsdr4* (*Hordeum spontaneum* dehydration responsive), was sequenced. The deduced amino acids were similar to the rice Rho-GTPase-activating protein-like with a Sec 14 p-like lipid-binding domain. Analysis of the *Hsdr4* promoter region revealed a new putative miniature inverted-repeat transposable element (MITE) and several potentially stress-related binding sites for transcription factors (MYC, MYB, LTRE, and GT-1), suggesting a role of the *Hsdr*4 gene in plant tolerance of dehydration stress. The *Hsdr4* was mapped to the long arm of chromosome 3H within a region that was previously shown to affect osmotic adaptation in barley. Transgenic work will validate the role of *Hsdr4* in resisting dehydration stress. This study indicates the great potential for identifying novel candidate genes related to water-stress tolerance in wild barley.

1.6.4 Wild Barley eibi1 Mutation Identified a Gene Essential for Leaf Water Conservation

The colonization of land by water plants necessitated the evolution of a cuticle, a cutin matrix embedded with and covered by wax, sealing the aerial organ's surface, thus protecting the plant from uncontrolled water loss. The incidental discovery by Guoxiong Chen of a spontaneous wilty mutant (*eibi1*), hypersensitive to drought in a desert wild barley in Israel, led to the identification of a major gene contributing to the generation of cutin and enabling land life (Chen 2005; Chen et al. 2004a, b, c, 2009, 2010, 2011a, b). *eibi1* showed the highest relative water-loss rate among the known wilty mutants, indicating that it is one of the most drought-sensitive mutants. *eibi1* had the same abscisic acid (ABA) level, the same ability to accumulate stress-induced ABA, and the same stomatal movement in response to light, dark, drought, and exogenous ABA as the wild type. Thus, *eibi1* was neither an ABA-deficient nor an ABA-insensitive mutant. The transpiration rate of *eibi1* was closer to the chlorophyll efflux rate than to stomatal density, demonstrating that the cuticle of *eibi1* was cutin defective (Figs. 1.4 and 1.5). A fine-scale genetic mapping of the *eibi1* locus on chromosome 3H is perfectly collinear with the equivalent region on rice chromosome 1. Gene prediction revealed that this rice segment harbors 16 genes. Some of them were proposed as candidate genes of *eibi1* (Chen et al. 2009). Map-based cloning revealed that *eibi1* encodes a HvABCG31 full transporter. The gene is highly expressed in the elongation zone of a growing leaf (the site of cutin synthesis), and its gene product also was localized in developing, but not in mature, tissue (Chen et al. 2011a). A *de novo* wild barley mutant, named "eibi1.c," along with two transposon insertion lines of rice, mutated in the ortholog of HvABCG31, was also unable to restrict water loss from detached leaves. HvABCG51 was hypothesized to function as a transporter involved in cutin formation, and the *eibi1* mutant resulted in the loss of function of HvABCG31, which is an ABCG full transporter. The sequence of the wild-type and mutant alleles revealed a single-nucleotide difference in exon 14, predicted to alter a tryptophan codon into a TAG stop codon (Fig. 1.5), and thereby induced a probable loss-of-function mutation. The full-length wild-type *eibi1* sequence consisted of 11,695 bp arranged in 24 exons, producing a transcript length of 4.799 (Fig. 1.5) with an ORF of 4.293 BP (1.430 residues) (Fig. S1 in Chen et al. 2011a, b). *eibi1* epidermal cells contain lipid-like droplets consisting presumably of cutin monomers that have not been transported out of the cells (Chen et al. 2011b). Homologs of HvABCG31 were found in green algae, moss, and lycopods indicating that this full transporter is highly conserved in the evolution of land plants (Chen et al. 2011a). The *eibi1* full story appears in Figs. 1.4 and the model in Fig. 1.5.

1.6.5 Salt Tolerance in *Hordeum spontaneum*

T. dicoccoides and *H. spontaneum* display salt tolerances in salty ecologies primarily on the eastern mountains of Samaria and Judea (wild emmer) and the coastal plain (wild barley) (Nevo 2007). Both cereals have successful genotypes ripening under dry and salty desert conditions (Nevo et al. 1984). Both *T. dicoccoides* and *H. spontaneum* have proven to be potential donors of salt-tolerant genes, which could be transferred to their respective cultivars by classical and modern techniques. Wild barley has an advantage in salt-stressed environments. When compared with wild emmer wheat, superior genotypes ripen at 350 Mm NaCl (60% sea water), whereas wild emmer ripened only at 250 mM of NaCl (40% sea water). Wild barley genotypes from more saline areas are able to maintain lower values of relative shoot $^{22}Na^{+}$. Associations of AFLP markers in wild barley in the Fertile Crescent were identified with salt tolerance measurements of shoot $^{22}Na^{+}$ content and shoot delta^{13}C (carbon isotope composition) after 4 weeks of treatment (100 mol m^{-3} NaCl) (Pakniyat et al. 1997). Shoot delta^{15}N correlates with salt and stress in barley (Handley et al. 1997a). Correlating stable isotopes with physiological expression in *Hordeum* is part of an integrated physiology-molecular genetics effort (Ellis et al. 1997; Forster et al. 1997). Seedling salt tolerance showed significant differences for salt tolerance under specific criteria (Yan et al. 2008). Higher seedling salt tolerance was revealed in the mesic versus xeric accessions. Molecular-based improvement of wheat and barley and transgenic plants is reviewed in Nevo and Chen (2010). Future studies should focus on networks of molecular drought- and salt-tolerant traits, regulation of fitness components at both the *coding* and *noncoding* genomic regions, and the development of new breeding strategies.

1.6.6 Biological Trace Elements in Wild Barley: Selenium Grain Concentration

Selenium is an essential trace mineral for human and animals with antioxidant, anticancer, antiarthropathy, and antiviral effects. The grain Se concentration (GSeC) was studied in 92 *H. spontaneum* genotypes from nine populations representing

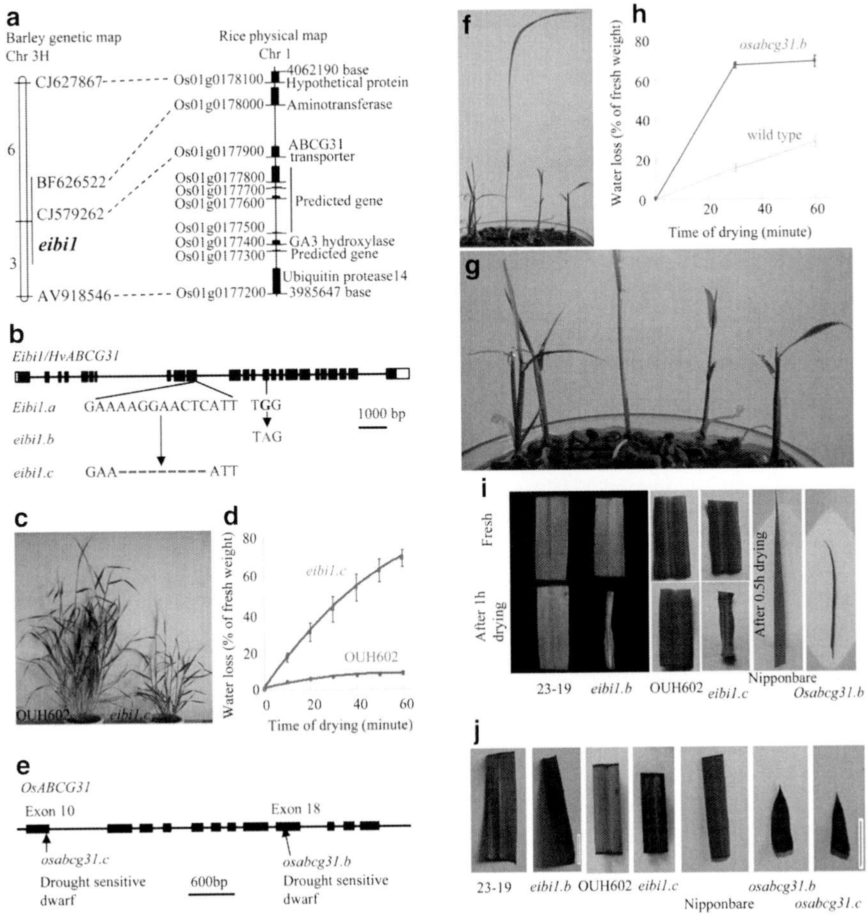

Fig. 1.4 Positional cloning of *eibi1* and its functional assignment. (**a**) The *Eibi1* locus maps to barley chromosome 3H and its ortholog to rice chromosome 1. Numbers to the left of the barley map indicate the number of observed recombination events in a population of 9,070 gametes. Rice and barley orthologs are connected by dashed lines. (**b**) Exon/intron structure of barley *Eibi1*. The single-nucleotide difference between the wild-type and the *eibi1.b* mutant sequence is indicated. The red dashed line indicates the 9-bp deletion in *eibi1.c* exon 10. Untranslated regions are indicated by an empty box at each end. (**c**) The γ irradiation-induced eibi1.c mutant and its wild type at the flowering stage. (**d**) Water-loss test of detached leaves from *eibi1.c* and OUH602. (**e**) Exon/intron structure of rice OsABCG31 indicating Tos17 insertions (*arrows*). (**f**) Wild-type (one tall seedling) and osabcg31.b mutants (four dwarf seedlings) at the three-leaf stage. (**g**) Enlarged view of dwarf seedlings in f. (**h**) Water loss in the detached leaf of the wild type and the mutant. (**i**) Mutant and wild-type leaves of barley air dried for 1 h and rice air dried for 0.5 h. (**j**) Toluidine blue staining of leaf segments of mutants and wild types (scale bar: 20 mm for wild barley; 5 mm for rice) (From Chen et al. 2011a)

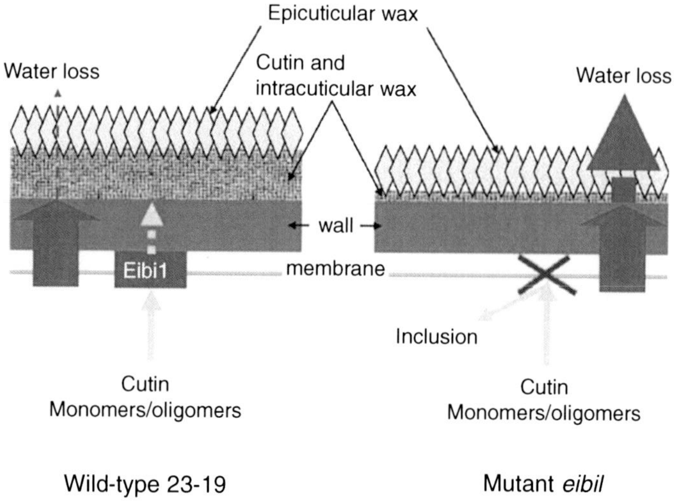

Fig. 1.5 Putative model of *Eibi1* function in barley (From Chen et al. 2011b)

different mesic and xeric habitats in Israel (Yan et al. 2011). Remarkable variation in GSeC was found between and within *H. spontaneum* populations, ranging from 0 to 0.387 mg kg^{-1} with an average of 0.047 mg kg^{-1}. The highest level was found in a genotype from the xeric Sde Boker population. Significant correlations were found between GSeC and 12 ecogeographical factors, including soil type, out of 14 factors studied. The wild barley exhibited wider GSeC ranges and greater diversity than cultivated barley, indicating that the progenitor has higher ability for Se uptake and accumulation.

Wild barley unfolds an important resource of a biological trace element that is significant nutritionally and in barley improvement.

1.6.7 Biotic Genetic Resources of Multiple Disease Resistance in Hordeum spontaneum

Hordeum spontaneum is known to be a rich resource of disease resistance genes (Moseman et al. 1983; and their references). Overall, a very high level of macroscale and microscale diversity for disease genetic resistance was found in 116 accessions of *H. spontaneum* evaluated at the seedling stage (Fetch et al. 2003). Genetic heterozygosity for resistance loci was common in wild barley. The frequency of resistance in genotypes from Jordan and Israel was high for *Septoria* speckled leaf blotch (77 and 98%, respectively), leaf rust (70 and 90%), net blotch (72 and 68%), and powdery mildew (58 and 70%); intermediate for spot blotch (53 and 46%); and low for stem rust (2 and 26%). The level of disease resistance in *H. spontaneum* was not strongly correlated with any of the climatic variables (temperature, precipitation, and humidity) monitored near the collection sites. However, in general, resistance was more often found in genotypes from mesic (e.g., Mediterranean coast) than in xeric (e.g., Negev desert) regions. Two *H. spontaneum* genotypes (from Shekhem and Damon) were resistant to all six pathogens and may be useful parents in programs breeding barley for multiple resistance. This study unfolded high levels of diversity and heterozygosity for disease reaction indicating that *H. spontaneum* is an extraordinarily rich and largely untapped source of unique disease resistance alleles for cultivated barley improvement. Patterns of diversity distribution in the *Isa* defense locus in wild barley populations suggest adaptive selection at this locus (Fitzgerald et al. 2011, and references therein). The *Isa* gene codes for a bifunctional amylase subtilisin inhibitor (BASI), which provides defense against bacterial and fungal pathogens.

1.6.8 Genetic Resources in Hordeum spontaneum for Barley Improvement

Wild cereals generally, and *H. spontaneum* particularly, harbor rich genetic resources and are the best hope for cereal improvement. The desirable traits that can be transferred from wild barley to cultivated barley, wheat, and other cereals include the following: (1) resistance to a variety of abiotic (e.g., drought, cold, heat, salt, and low mineral tolerances) and biotic (viral, bacterial, fungal, and herbicide resistances) stresses; dormancy, seedling revival after drought (Zhang et al. 2002, 2005; Yan et al. 2008, 2011; Chen et al. 2010); and vitamin E (Shen et al. 2011); (2) high-quantity and high-quality storage proteins; (3) differential richness of amino acids; and (4) amylases and photosynthetic yield (Nevo 1992, 2004). Genes for most of these and other potential traits available in wild barley and wild emmer wheat are still largely untapped and provide potential precious sources for cereal improvement.

Extensive QTL has been mapped in both cereals, wild emmer (Peng et al. 2003) and wild barley (Chen et al. 2004a, b, c, 2010). These include brittle rachis, *thresh-1*, heading date, plant height, grain size, yield, and yield components. The domestication and drought resistance QTLs are *nonrandomly* distributed among and along chromosomes. In the domestication-related QTLs in wild emmer wheat, we identified QTL clustering and association with gene-rich regions. The cryptic beneficial alleles at specific QTLs derived from wild cereals may contribute to cereal improvement. They could be introduced into cultivated cereals by using the strategy of marker-assisted selection and simultaneously eliminating undesirable alleles. Wild cereals, particularly wild barley, harbor very valuable wild germplasm resources for future cereal improvement. Recently, we identified the effects of global warming on the wild progenitors of barley and wheat in Israel from 1980 till 2008 (Nevo et al. 2012). Phenotypic effects across Israel caused a universal earliness, on average of 10 days, in flowering time, presumably escaping the increasing heat. Likewise, we identified a general depletion of SSR genetic diversity, associated in some populations with novel alleles, presumably adapted to drought resistance. These changes, most likely caused by global warming, dictate a punctuate strategy of *in situ* and *ex situ* conservation of these precious populations, which are the best hope for crop improvement.

1.7 Conclusions and Prospects

Domesticated barley, *Hordeum vulgare*, a human domesticate around 10,000 years ago, is associated with emmer wheat and a founder crop of the southwest Asian Neolithic and Bronze ages in human food production. Though inferior to wheat as human food, it is the source of beer and animal feed. However, barley copes with harsher ecology, drought, salinity, and poor soil better than wheat, inherited from its progenitor. Wild barley, *Hordeum spontaneum*, is the indisputable progenitor of cultivated barley, and it contributed its multiple resistances against *abiotic* and *biotic* stresses to the cultivar. *H. spontaneum* originated in southwest Asia, diverged some 5.5 million years ago from Tibetan wild barley, and migrated eastward. It is currently distributed in the Fertile Crescent, Central Asia, and Tibet, having been domesticated, apparently independently, in all three major centers. However, while it is adapted to drought stress across its current range, it is coupled with warm-stress adaptations in the Near East and cold-stress adaptations in Tibet. It displays both *micro-* and *macro-*geographic genetic adaptations to environmental stresses. Domestication entailed the fixation of *nonbrittle* mutations, *threshability*, and subsequently the emergence of six rowed-hulled and naked types. The evolutionary biology of *H. spontaneum* and the history of its domestication centers call for necessary future research.

What is next? Biologically, the full genome sequencing of *H. spontaneum* will unfold immense, mostly untapped genetic resources at both the *coding* and *noncoding* levels, partly regulatory genomes, for barley and cereal improvement. The rich *adaptive* and *nonrandom* genetic resources for resisting *abiotic* and *biotic* stresses discovered to date make *H. spontaneum* a major source for future cereal improvement. Its generalist distribution adapted to diverse and harsh ecological conditions from the warm-dry Near East to the cold-dry Tibet harbor wide precious resources for future barley and cereal improvement. Nevertheless, wild barley together with wild emmer wheat was affected *phenotypically* and *genotypically* by the current ongoing global warming (Nevo et al. 2012). Hence, it is of utmost importance to unfold and conserve its rich genetic resources both *in situ* and *ex situ* for utilization in human and animal economies. This is particularly crucial in a world where human populations and hunger are on the rise.

Acknowledgments I thank all of my colleagues and students from the Institute of Evolution, the International Graduate Center of Evolution, University of Haifa, and worldwide for their excellent collaboration in our research project of the evolution of wild cereals in the Near East Fertile Crescent that started in the early 1970s. I am deeply grateful to Robin Permut for the editing and to the Ancell-Teicher Research Foundation for Genetics and Molecular Evolution, which provides continuous support of my research.

References

Badr, A., Muller, K., Schafer-Pregl, R., Rabey, H. E. L., Effgen, S., Ibrahim, H. H., Pozzi, C., Rohde, W., & Salamini, F. (2000). On the origin and domestication history of barley (*Hordeum vulgare*). *Molecular Biology and Evolution, 17*, 499–510.

Baek, H. J., Beharav, A., & Nevo, E. (2003). Ecological-genomic diversity of microsatellites in wild barley, *Hordeum spontaneum*, populations in Jordan. *Theoretical and Applied Genetics, 106*, 397–410.

Baum, B. R., Nevo, E., Johnson, D. A., & Beiles, A. (1997). Genetic diversity in wild barley (*Hordeum spontaneum* Koch) in the Near East: A molecular analysis using random amplified polymorphic DNA (RAPD). *Genetic Resources and Crop Evolution, 44*, 147–157.

Blattner, F. (2006). Multiple intercontinental dispersals shaped the distribution area of *Hordeum* (Poaceae). *The New Phytologist, 169*, 603–614.

Blum, A. (1988). *Plant breeding for stress environments*. Boca Raton: CRC Press.

Brown, A. H. D., Zohary, D., & Nevo, E. (1978). Outcrossing rates and heterozygosity in natural populations of *Hordeum spontaneum* Koch in Israel. *Heredity, 41*, 49–62.

Chabane, K., Ablett, G. A., Cordeiro, G. M., Valkoun, J., & Henry, F. J. (2005). EST versus genomic derived microsatellite markers for genotyping wild and cultivated barley. *Genetic Resources and Crop Evolution, 52*, 903–909.

Chalmers, K. J., Waugh, R., Waters, J., Forster, B. P., Nevo, E., & Powell, W. (1992). Grain isozyme and ribosomal DNA variability in *Hordeum spontaneum* populations from Israel. *Theoretical and Applied Genetics, 84*, 313–322.

Chen, G. (2005, February). *Drought resistance in wild barley,* Hordeum spontaneum, *from Israel: physiology, gene identification, and QTL mapping*. PhD thesis, University of Haifa, Haifa, Israel.

Chen, G., Krugman, T., Fahima, T., Korol, A. B., & Nevo, E. (2002). Comparative study of morphological and physiological traits related to drought resistance between xeric and mesic *Hordeum spontaneum* lines in Israel. *Barley Genetics Newsletter, 32*, 22–33.

Chen, G., Sagi, M., Weining, S., Krugman, T., Fahima, T., Korol, A. B., & Nevo, E. (2004a). Wild barley *Eibi1* mutation identifies a gene essential for leaf water conservation. *Planta, 219*, 684–693.

Chen, G., Krugman, T., Fahima, T., Zhang, Z., Korol, A. B., & Nevo, E. (2004b). Differential patterns of germination and desiccation tolerance of mesic and xeric wild barley (*Hordeum spontaneum*) in Israel. *Journal of Arid Environments, 56*, 95–105.

Chen, G., Suprunova, T., Krugman, T., Fahima, T., & Nevo, E. (2004c). Ecogeographic and genetic determinants of kernel weight and color of wild barley (*Hordeum spontaneum*) populations in Israel. *Seed Science Research, 14*, 137–146.

Chen, G., Pourkheirandish, M., Sameri, M., Wang, N., Nair, S., Shi, Y., Li, C., Nevo, E., & Komatsuda, T. (2009). Genetic targeting of candidate genes for drought sensitive gene eibi1 of wild barley (*Hordeum spontaneum*). *Breeding Science, 59*, 637–644.

Chen, G., Krugman, T., Fahima, T., Chen, K., Hu, Y., Röder, M., Nevo, E., & Korol, A. B. (2010). Chromosomal regions controlling seedling drought resistance in Israeli wild barley *Hordeum spontatneum* C. Koch. *Genetic Resources and Crop Evolution, 57*, 85–99.

Chen, G., Komatsuda, T., Ma, J. F., Nawrath, C., Pourkheirandish, M., Tagiri, A., Hu, Y.-G., Sameri, M., Li, X., Zhao, X., Liu, Y., Li, C., Ma, X., Wang, A., Nair, S., Wang, N., Miyao, A., Sakuma, S., Yamaji, N., Zheng, X., & Nevo, E. (2011a). An ATP-binding cassette subfamily G full transporter is essential for the retention of leaf water in both wild barley and rice. *Proceedings of the National Academy of Sciences of the United States of America, 108*(30), 12354–12359.

Chen, G., Komatsuda, T., Ma, J. F., Li, C., Yamaji, N., & Nevo, E. (2011b). A functional cutin matrix is required for plant protection against water loss. *Plant Signaling & Behavior, 6*, 1297–1299.

Close, T. J., Choi, D. W., Venegas, M., Salvi, S., Tuberosa, R., Ryabushkina, N., Turuspekov, Y., & Nevo, E. (2000, October 22–27). *Allelic variation in wild and cultivated barley at the* Dhn4 *locus, which encodes a major drought-induced and seed protein, DHN4*. In 8th International Barley Genetics Symposium, Adelaide, Australia.

Cronin, J. K., Bundock, P. C., Henry, R. J., & Nevo, E. (2007). Adaptive climatic molecular evolution in wild barley at the *Isa* defense locus. *Proceedings of the National Academy of Sciences of the United States of America, 104*, 2773–2778.

Dai, F., Nevo, E., Zhou, M., Wu, D., Chen, Z., Beiles, A., Chen, G., & Zhang, G. (2012). *Tibet: One of the centers of origin of cultivated barley* (Submitted).

Darwin, C. (1859). *On the origin of species by means of natural selection*. London: Murray.

Ellis, R. P., Forster, B. P., Waugh, R., Bonar, N., Handley, L. I., Robinson, D., Gordon, D. C., & Powell, W. (1997). Mapping physiological traits in barley. *The New Phytologist, 137*, 149–157.

Feldman, M., & Kislev, M. E. (2007). Domestication of emmer wheat and evolution of free-threshing tetraploid wheat. *Israel Journal of Plant Sciences, 55*, 207–221.

Fetch, T. B., Steffenson, B. J., & Nevo, E. (2003). Diversity and sources of multiple disease resistance in *Hordeum spontaneum*. *Plant Disease, 87*, 1439–1448.

Fitzgerald, T. L., Shapter, F. M., McDonald, S., Waters, D. L. F., Chivers, I. H., Drenth, A., Nevo, E., & Henry, R. J. (2011). Genome diversity in wild grasses under environmental stress. *Proceedings of the National Academy of Sciences of the United States of America, 108*, 21140–21145.

Forster, B. P., Russell, J. R., Ellis, R. P., Handley, L. L., Robinson, D., Hackett, C. A., Nevo, E., Waugh, R., Gordon, D. C., Keith, R., & Powell, W. (1997). Locating genotypes and genes for abiotic stress in barley, a strategy using maps, markers, and the wild species. *The New Phytologist, 137*, 141–147.

Gaut, B. S. (2002). Evolutionary dynamics of grass genomes. *The New Phytologist, 154*, 15–28.

Gupta, P. K., Sharma, P. K., Balyan, H. S., Roy, J. K., Sharma, S., Beharav, A., & Nevo, E. (2002). Polymorphism at rDNA loci in barley and its relation with climatic variables. *Theoretical and Applied Genetics, 104*, 473–481.

Gupta, P. K., Sharma, S., Kumar, S., Balyan, H. S., Beharav, A., & Nevo, E. (2004). Adaptive ribosomal DNA polymorphism in wild barley at a mosaic microsite, Newe Ya'ar, in Israel. *Plant Science, 166*, 1555–1563.

Gustafson, P., Raskina, O., Ma, X., & Nevo, E. (2009). Wheat evolution, domestication, and improvement. In B. F. Carver (Ed.), *Wheat: Science and trade* (pp. 5–30). Ames: Wiley-Blackwell.

Handley, L. L., Robinson, D., Forster, B. P., Ellis, R. P., Scrimgeour, C. M., Gordon, D. C., Nevo, E., & Raven, J. A. (1997a). Shoot d15N correlates with genotype and salt stress in barley. *Planta, 201*, 100–102.

Harlan, J. R. (1976). Barley. In N. W. Simmonds (Ed.), *Evolution of crop plants* (pp. 93–98). London: Longman.

Harlan, J. R. (1992). *Crop and man* (2nd ed.). Madison: American Society for Agronomy.

Harlan, J. R., & Zohary, D. (1966). Distribution of wild wheats and barley. *Science, 153*, 1074–1080.

Huang, Q., Beharav, A., Li, Y. C., Kirzhner, V., & Nevo, E. (2002). Mosaic microecological differential stress causes adaptive microsatellite divergence in wild barley, *Hordeum spontaneum*, at Neve Yaar, Israel. *Genome, 45*, 1216–1229.

Hubner, S., Hoffken, M., Oren, E., Haseneyer, G., Stein, N., Graner, A., Schmid, K., & Fridman, E. (2009). Strong correlation of wild barley (*Hordeum spontaneum*) population structure with temperature and precipitation variation. *Molecular Ecology, 18*, 1523–1536.

IPCC, Israel's Second National Communication on Climatic Change. (2010). Submitted under the United Nations Framework Convention on Climate Change. Ministry of Environmental Protection, State of Israel, Jerusalem. Available at www.environment.gov.il. Accessed 2 Feb 2012.

Ivandic, V., Hackett, C. A., Zhang, Z., Staub, J. E., Nevo, E., Thomas, T. B., & Forster, B. P. (2000). Phenotypic responses of wild barley to experimentally imposed water stress. *Journal of Experimental Botany, 51*, 2021–2029.

Ivandic, V., Hackett, C. A., Nevo, E., Keith, R., Thomas, W. T. B., & Forster, B. P. (2002). Analysis of simple sequence repeats (SSRs) in wild barley from the Fertile Crescent: associations with ecology, geography and flowering time. *Plant Molecular Biology, 48*, 511–527.

Ivandic, V., Thomas, W. T. B., Nevo, E., Zhang, Z., & Forster, B. P. (2003). Associations of simple sequence repeats with quantitative trait variation including biotic and abiotic stress tolerance in *Hordeum spontaneum*. *Plant Breeding, 122*, 300–304.

Kalendar, R., Tanskanen, J., Immonen, S., Nevo, E., & Schulman, A. H. (2000). Genome evolution of wild barley (*Hordeum spontaneum*) by BARE-1 retrotransposon dynamics in response to sharp microclimatic divergence. *Proceedings of the National Academy of Sciences of the United States of America, 97*, 6603–6607.

Kislev, M. E. (1997). Early agriculture and palaeoecology of Netiv Hagdud. In O. Bar-Yosef & A. Gopher (Eds.), *An early Neolithic village in the Jordan Valley. Part 1: The archaeology of Netiv Hagdud*. Cambridge: Peabody Museum of Archaeology and Ethnology, Harvard University.

Kislev, M. E., Nadel, D., & Carmi, I. (1992). Epi-palaeolithic (19,000 BP) cereal and fruit diet at Ohalo II, Sea of Galilee, Israel. *Review of Palaeobotany and Palynology, 73*, 161–166.

Krugman, T., Chague, V., Peleg, Z., Balzergue, S., Just, J., Korol, A. B., Nevo, E., Saranga, Y., Chalhoub, B., & Fahima, T. (2010). Multilevel regulation and signaling processes associated with adaptation to terminal drought in wild emmer wheat. *Functional & Integrative Genomics, 10*, 167–186.

Krugman, T., Peleg, Z., Quansah, L., Chagué, V., Korol, A. B., Nevo, E., Saranga, Y., Fait, A., Chalhoub, B., & Fahima, T. (2011). Alteration in expression of hormone-related genes in wild emmer wheat roots associated with drought adaptation mechanisms. *Functional & Integrative Genomics, 11*, 565–583.

Lev, B., Kislev, M. E., & Bar-Yosef, O. (2005). Mousterian vegetal food in Kebara Cave, Mt. Carmel. *Journal of Archaeological Science, 32*, 475–484.

Lev-Yadun, S., Gopher, A., & Abbo, S. (2000). The cradle of Agriculture. *Science, 288*, 1602–1603.

Li, Y. C., Krugman, T., Fahima, T., Beiles, A., & Nevo, E. (1998). Genetic diversity of alcohol dehydrogenase 3 in wild barley population at the "Evolution Canyon" microsite, Nahal Oren, Mt. Carmel, Israel. *Barley Genetics Newsletter, 28*, 58–60.

Li, Y. C., Fahima, T., Krugman, T., Beiles, A., Röder, M. S., Korol, A. B., & Nevo, E. (2000). Parallel microgeographic patterns of genetic diversity and divergence revealed by allozyme, RAPD, and microsatellites in *Triticum dicoccoides* at Ammiad, Israel. *Conservation Genetics, 1*, 191–207.

Li, Y. C., Korol, A. B., Fahima, T., Beiles, A., & Nevo, E. (2002). Microsatellites: Genomic distribution, putative functions and mutational mechanisms: A review. *Molecular Ecology, 11*, 2453–2465.

Li, J. Z., Sjakste, T. G., Röder, M. S., & Ganal, M. W. (2003). Development and genetic mapping of 127 new microsatellite markers in barley. *Theoretical and Applied Genetics, 107*, 1021–1027.

Li, Y. C., Korol, A. B., Fahima, T., & Nevo, E. (2004). Microsatellites within genes: Structure, function and evolution. *Molecular Biology and Evolution, 21*, 991–1007.

Ma, D. Q. (2000). *Genetic resources of Tibetan barley in China*. Beijing: China Agriculture Press.

Malysheva-Otto, L. V., Ganal, M. W., & Röder, M. S. (2006). Analysis of molecular diversity, population structure and linkage disequilibrium in a worldwide survey of cultivated barley germplasm (*Hordeum vulgare* L.). *BMC Genetics, 7*, 6.

Molina-Cano, J., Russell, J. R., Moralejo, M. A., Pscacena, J. I., Arias, G., & Powell, W. (2005). Chloroplast DNA microsatellite analysis supports a polyphyletic origin for barley. *Theoretical and Applied Genetics, 110*, 613–619.

Morrell, P. L., & Clegg, M. (2007). Genetic evidence for a second domestication of barley (*Hordeum vulgare*) east of the Fertile-Crescent. *Proceedings of the National Academy of Sciences of the United States of America, 104*, 3289–3294.

Morrell, P. L., Toleno, D. M., Lundy, K. E., & Clegg, M. T. (2005). Low levels of linkage disequilibrium in wild barley (*Hordeum vulgare* ssp. *spontaneum*) despite high rates of self-fertilization. *Proceedings of the National Academy of Sciences of the United States of America, 102*, 2442–2447.

Moseman, J. G., Nevo, E., & Zohary, D. (1983). Resistance of *Hordeum spontaneum* collected in Israel to infection with *Erysiphe graminis hordei*. *Crop Science, 23*, 1115–1119.

Nevo, E. (1978). Genetic variation in natural populations: patterns and theory. *Theoretical Population Biology, 13*, 121–177.

Nevo, E. (1987). Plant genetic resources: Prediction by isozyme markers and ecology. In M. C. Rattazi, J. G. Scandalios, & G. S. Whitt (Eds.), *Isozymes: Current topics in biological research* (Agriculture, physiology and medicine, Vol. 16, pp. 247–267). New York: Alan R. Liss.

Nevo, E. (1988). Genetic diversity in nature: Patterns and theory. *Evolutionary Biology, 23*, 217–247.

Nevo, E. (1992). Origin, evolution, population genetics and resources for breeding of wild barley, *Hordeum spontaneum*, in the Fertile Crescent. In P. Shewry (Ed.), *Barley: Genetics, molecular biology and biotechnology* (pp. 19–43). Wallingford: C.A.B International.

Nevo, E. (1995). Asian, African and European biota meet at "Evolution Canyon", Israel: Local tests of global biodiversity and genetic diversity patterns. *Proceedings of the Royal Society of London, Series B, 262*, 149–155.

Nevo, E. (1998a). Molecular evolution and ecological stress at global, regional and local scales: The Israeli perspective. *The Journal of Experimental Zoology, 282*, 95–119.

Nevo, E. (1998b). Genetic diversity in wild cereals: Regional and local studies and their bearing on conservation *ex situ* and *in situ*. *Genetic Resources and Crop Evolution, 45*, 355–370.

Nevo, E. (2004). Population genetic structure of wild barley and wheat in the Near East Fertile Crescent: Regional and local adaptive. In P. K. Gupta & R. K. Varshney (Eds.), *Cereal genomics* (pp. 135–163). Dordrecht: Kluwer Academic.

Nevo, E. (2005). Genomic diversity in nature and domestication. In R. J. Henry (Ed.), *Genotypic and phenotypic variation in higher plants* (pp. 287–316). Wallingford: CAB International.

Nevo, E. (2006a). "Evolution Canyon": A microcosm of life's evolution focusing on adaptation and speciation. *Israel Journal of Ecology & Evolution, 52*(3–4), 485–506.

Nevo, E. (2006b). Genome evolution of wild cereal diversity and prospects for crop improvement. *Plant Genetic Resources, 4*, 36–46.

Nevo, E. (2007). Evolution of wild wheat and barley and crop improvement: Studies at the Institute of Evolution. *Israel Journal of Plant Sciences, 55*, 251–262.

Nevo, E. (2009a). Evolution in action across life at "Evolution Canyon", Israel. *Trends in Evolutionary Biology, 1*, e3.

Nevo, E. (2009b). Ecological genomics of natural plant populations: The Israeli perspective. In D. J. Somers, P. Langridge, & J. P. Gustafson (Eds.), *Methods in molecular biology* (Plant genomics, Vol. 513, pp. 321–344). New York: Human Press.

Nevo, E. (2011a). *Triticum*. In C. Kole (Ed.), *Wild crop relatives: Genomic and breeding resources, cereals* (pp. 407–456). Berlin/Heidelberg: Springer.

Nevo, E. (2011b). Evolution under environmental stress at macro-and microscales. *Genome Biology and Evolution, 2*, 1039–1052.

Nevo, E. (2011c). Selection overrules gene flow at "Evolution Canyons, Israel". In K. V. Urbano (Ed.), *Advance in genetics research* (Vol. 5, Chap. 2, pp. 67–89). New York: Nova Science Publishers Inc.

Nevo, E. (2012). "Evolution Canyon" a potential microscale of global warming across life. *Proceedings of the National Academy of Sciences of the United States of America, 109*, 2960–2965.

Nevo, E., & Beiles, A. (2011). Genetic variation in nature. *Scholarpedia, 6*, 8821.

Nevo, E., & Chen, G. (2010). Drought and salt tolerances in wild relatives for wheat and barley improvement. *Plant, Cell & Environment, 33*, 670–685.

Nevo, E., Zohary, D., Brown, A. H. D., & Haber, M. (1979). Genetic diversity and environmental associations of wild barley, *Hordeum spontaneum*, in Israel. *Evolution, 33*, 815–833.

Nevo, E., Brown, A. H. D., Zohary, D., Storch, N., & Beiles, A. (1981). Microgeographic edaphic differentiation in allozyme polymorphisms of wild barley (*Hordeum spontaneum*, Poaceae). *Plant Systematics and Evolution, 138*, 287–292.

Nevo, E., Beiles, A., Storch, N., Doll, H., & Andersen, B. (1983). Microgeographic edaphic differentiation in hordein polymorphisms of wild barley. *Theoretical and Applied Genetics, 64*, 123–132.

Nevo, E., Beiles, A., Gutterman, Y., Storch, N., & Kaplan, D. (1984). Genetic resources of wild cereals in Israel and vicinity: II. Phenotypic variation within and between populations of wild barley, *Hordeum spontaneum*. *Euphytica, 33*, 737–756.

Nevo, E., Beiles, A., Kaplan, D., Golenberg, E. M., Olsvig-Whittaker, L., & Naveh, Z. (1986a). Natural selection of allozyme polymorphisms: A microsite test revealing ecological genetic differentiation in wild barley. *Evolution, 40*, 13–20.

Nevo, E., Beiles, A., Kaplan, D., Storch, N., & Zohary, D. (1986b). Genetic diversity and environmental associations of wild barley, *Hordeum spontaneum* (Poaceae), in Iran. *Plant Systematics and Evolution, 153*, 141–164.

Nevo, E., Beiles, A., & Zohary, D. (1986c). Genetic resources of wild barley in the Near East: Structure, evolution and application in breeding. *Biological Journal of the Linnean Society, 27*, 355–380.

Nevo, E., Zohary, D., Beiles, A., Kaplan, D., & Storch, N. (1986d). Genetic diversity and environmental associations of wild barley, *Hordeum spontaneum*, in Turkey. *Genetica, 68*, 203–213.

Nevo, E., Ordentlich, A., Beiles, A., & Raskin, I. (1992). Genetic divergence of heat production within and between the wild progenitors of wheat and barley: Evolutionary and agronomical implications. *Theoretical and Applied Genetics, 84*, 958–962.

Nevo, E., Apelbaum-Elkaher, I., Garty, J., & Beiles, A. (1997). Natural selection causes microscale allozyme diversity in wild barley and a lichen at "Evolution Canyon" Mt. Carmel, Israel. *Heredity, 78*, 373–382.

Nevo, E., Fu, Y. B., Pavlicek, T., Khalifa, S., Tavasi, M., & Beiles, A. (2012). Evolution of wild cereals during 28 years of global warming in Israel. *Proceedings of the National Academy of Sciences of the United States of America, 109*, 3412–3415.

Owuor, E. D., Fahima, T., Beiles, A., Korol, A. B., & Nevo, E. (1997). Population genetics response to microsite ecological stress in wild barley *Hordeum spontaneum*. *Molecular Ecology, 6*, 1177–1187.

Owuor, E. D., Beharav, A., Fahima, T., Kirzhner, V. M., Korol, A. B., & Nevo, E. (2003). Microscale ecological stress causes RAPD molecular selection in wild barley, Neve Yaar microsite, Israel. *Genetic Resources and Crop Evolution, 50*, 213–224.

Pakniyat, H., Powell, W., Baird, E., Handley, L. L., Robinson, D., Sorimgeour, C. M., Nevo, E., Hackett, C. A., Caligari, P. D. S., & Forster, B. P. (1997). AFLP variation in wild barley (*Hordeum spontaneum* C. Koch) with reference to salt tolerance and associated ecogeography. *Genome, 40*, 332–341.

Parnas, T. (2006, November). *Evidence for incipient sympatric speciation in wild barley,* Hordeum spontaneum, *at "Evolution Canyon", Mount Carmel, Israel, based on hybridization and physiological and genetic diversity estimates*. Thesis submitted in partial fulfillment of the requirements for master's degree, Institute of Evolution, University of Haifa, Israel.

Peng, J. H., Ronin, Y. I., Fahima, T., Roder, M. S., Li, Y. C., Nevo, E., & Korol, A. B. (2003). Domestication quantitative trait loci in *Triticum dicoccoides*, the progenitor of wheat. *Proceedings of the National Academy of Sciences of the United States of America, 100*, 2489–2494.

Prasad, V., Stromberg, C. A. E., Alimohammadian, H., & Sahni, A. (2005). Dinosaur coprolites and the early evolution of grasses and grazers. *Science, 310*, 117–1180.

Schmalenbach, I., March, T. J., Bringezu, T., Waugh, R., & Pillen, K. (2011). High-resolution genotyping of wild barley introgression lines and fine-mapping of the threshability locus *thresh-1* using the illumina Golden Gate assay. *Genomes, 1*, 187.

Sharma, S., Beharav, A., Balyan, H. S., Nevo, E., & Gupta, P. K. (2004). Ribosomal DNA polymorphism and its association with geographical and climatic variables in 27 wild barley populations from Jordan. *Plant Science, 166*, 467–477.

Shen, Y., Lebold, K., Lansky, E. P., Traber, M., & Nevo, E. (2011). "Tocol-Omic" diversity in wild barley. *Chemistry & Biodiversity, 8*, 2322–2330.

Suprunova, T., Krugman, T., Fahima, T., Chen, G., Shams, I., Korol, A. B., & Nevo, E. (2004). Differential expression of dehydrin *(Dhn)* in response to water stress in resistant and sensitive wild barley *(Hordeum spontaneum)*. *Plant, Cell & Environment, 27*, 1297–1308.

Suprunova, T., Krugman, T., Distelfeld, A., Fahima, T., Nevo, E., & Korol, A. B. (2007). Identification of a novel gene (*Hsdr4*) involved in water-stress tolerance in wild barley. *Plant Molecular Biology, 64*, 17–34.

Turpeinen, T., Kulmala, J., & Nevo, E. (1999). Genome size variation in *Hordeum spontaneum* populations. *Genome, 42*, 1094–1099.

Turpeinen, T., Vanhala, V., Nevo, E., & Nissila, E. (2003). AFLP genetic polymorphism in wild barley (*Hordeum spontaneum*) populations in Israel. *Theoretical and Applied Genetics, 106*, 1333–1339.

van Rijn, C. P. E., Heersch, I., Van Berkel, Y. E. M., Nevo, E., Lambers, H., & Poorter, H. (2000). Growth characteristics in *Hordeum spontaneum* populations from different habitats. *The New Phytologist, 146*, 471–481.

van Rijn, C. P. E., Vanhal, T. K., Nevo, E., Stam, P., van Eeuwijk, F. A., & Poorter, H. (2001). Association of AFLP markers with growth-related traits in *Hordeum spontaneum*. In C. van Rijn, Ph.D. Thesis: a physiological and genetic analysis of growth characteristics in *Hordeum spontaneum*. Universiteit Utrecht, Faculteit Biologie, Utrecht, pp. 75–93.

Vanhala, T., van Rijn, C. P. E., Buntjer, J., Stam, P., Nevo, E., Poorter, H., & van Eeuwijk, F. A. (2004). Environmental, phenotypic and genetic variation of wild barley (*Hordeum spontaneum*) from Israel. *Euphytica, 137*, 297–309.

Varshney, R. K., Raif, S., Borner, A., Korzun, V., Stein, N., Sorrelis, M. E., Langridge, P., & Graner, A. (2005). Interspecific transferability and comparative mapping of barley EST-SSR markers in wheat, rye, and rice. *Plant Science, 168*, 195–202.

Varshney, R. K., Chabane, K., Hendre, P. S., Aggarwal, R. K., & Graner, A. (2007). Comparative assessment of EST-SSR EST-SNP, and AFLP markers for evaluation of genetic diversity and conservation of genetic resources using wild, cultivated and elite barleys. *Plant Science, 173*, 638–649.

Verhoeven, K. J. E., Biere, A., Nevo, E., & van Damme, J. M. M. (2004a). Differential selection of growth rate-related traits in wild barley, *Hordeum spontaneum*, in contrasting greenhouse nutrient environments. *Journal of Evolutionary Biology, 17*, 184–196.

Verhoeven, K. J. E., Biere, A., Nevo, E., & van Damme, J. M. M. (2004b). Can a genetic correlation with seed mass constrain adaptive evolution of seedling desiccation tolerance in wild barley? *International Journal of Plant Sciences, 165*, 281–288.

Vicient, C. M., Suoniemi, A., Anamthawat-Jonsson, K., Tanskanen, J., Beharav, A., Nevo, E., & Schulman, A. H. (1999). Retrotransposon BARE-1 and its role in genome evolution in the genus *Hordeum*. *The Plant Cell, 11*, 1769–1784.

Volis, S., Yakubov, B., Shulgina, E., Ward, D., Zur, V., & Mendlinger, S. (2001). Test for adaptive RAPD variation in population genetic structure if wild barley, *Hordeum spontaneum* Koch. *Biological Journal of the Linnean Society, 74*, 289–303.

Volis, S., Mendlinger, S., & Ward, D. (2002). Differentiation in populations of *Hordeum spontaneum* along a gradient of environmental productivity and predictability: Life history and local adaptation. *Biological Journal of the Linnean Society, 77*, 479–490.

Wei, Y. M., Baum, B. R., Nevo, E., & Zheng, Y. L. (2005). Does domestication mimic speciation? 1. A population-genetic analysis of Hordeum spontaneum and Hordeum vulgare based on AFLP and evolutionary considerations. *Canadian Journal of Botany, 83*, 1496–1512 (Published on the NRC Research Press Web site on 1 February 2006).

Weiss, B., Wetterstrom, W., Nadel, D., & Bar-Yosef, O. (2004). The broad spectrum revisited: Evidence from plant remains. *Proceedings of the National Academy of Sciences of the United States of America, 101*, 9551–9555.

Yan, J., Yan, J., Chen, G., Cheng, J., Nevo, E., & Gutterman, Y. (2008). Phenotypic variation in caryopsis dormancy and seedling salt tolerance in wild barley, *Hordeum spontaneum*, from different habitats in Israel. *Genetic Resources and Crop Evolution, 55*, 995–1005.

Yan, J., Wang, F., Qin, H., Chen, G., Nevo, E., Fahima, T., & Cheng, J. (2011). Natural variation in grain selenium concentration of wild barley, *Hordeum spontaneum*, populations from Israel. *Biological Trace Element Research, 142*, 773–786.

Yang, Z., Zhang, T., Ki, G., & Nevo, E. (2012). Adaptive microclimatic evolution of the dehydrin 6 gene in wild barley at "Evolution Canyon", Israel. *Genetica (Online version)*. doi:10.1007/s10709-012-29641-1.

Zhang, F., Gutterman, Y., Krugman, T., Fahima, T., & Nevo, E. (2002). Differences in primary dormancy and seedling revival ability for some *Hordeum spontaneum* genotypes of Israel. *Israel Journal of Plant Sciences, 50*, 271–276.

Zhang, F., Chen, G., Huang, Q., Orion, O., Krugman, T., Fahima, T., Korol, A. B., Nevo, E., & Gutterman, Y. (2005). Genetic basis of barley caryopsis dormancy and seedling desiccation tolerance at the germination stage. *Theoretical and Applied Genetics, 110*, 445–453.

Zohary, D. (1969). The progenitors of wheat and barley in relation to domestication and agriculture dispersal in the Old World. In P. J. Ucko & G. W. Dimbleby (Eds.), *The domestication and exploitation of plants and animals* (pp. 47–66). London: Duckworth.

Zohary, D. (1999). Monophyletic vs. polyphyletic origin of the crops on which agriculture was founded in the Near East. *Genetic Resources and Crop Evolution, 46*, 133–142.

Zohary, D., Hopf, M., & Weiss, E. (2012). *Domestication of plants in the Old World: The origin and spread of domesticated plants in Southwest Asia, Europe, and the Mediterranean Basin* (4th ed.). Oxford: Oxford University Press.

Chapter 2
Genetic Diversity in Latvian Spring Barley Association Mapping Population

Ieva Mezaka, Linda Legzdina, Robbie Waugh, Timothy J. Close, and Nils Rostoks

Abstract Certified organic crop area is continuously increasing in European Union and in Latvia (Eurostat data), despite somewhat lower yield and higher potential for disease damage in organic farming. It is increasingly recognized that breeding varieties for organic farming requires focus on specific traits that may be less important under conventional agriculture. Molecular markers are becoming essential tools for plant breeding allowing reducing time and cost of development of new varieties by early selection of progeny with desired traits. However, there is lack of information on molecular markers for traits that may be important for organic farming, such as plant morphological traits ensuring competitive ability with weeds, yield and yield stability under organic growing conditions, nutrient use efficiency, and resistance to diseases. We have selected 145 Latvian varieties and breeding lines along with 46 foreign accessions for association mapping panel and genotyped those with 1,536 single-nucleotide polymorphism (SNP) markers using Illumina GoldenGate platform and barley oligo pooled array 1. In parallel to genotyping, 154 of the 191 spring barley genotypes contrasting for traits that are important for organic farming are currently in field trials under conventional and organic management. The success of association mapping in structured natural

I. Mezaka • L. Legzdina
State Priekuli Plant Breeding Institute, Zinatnes street 1a, Priekuli 4126, Latvia

R. Waugh
The James Hutton Institute, Invergowrie, Dundee DD2 5DA, Scotland, UK

T.J. Close
University of California, Riverside, CA 92521, USA

N. Rostoks (✉)
Faculty of Biology, University of Latvia, 4 Kronvalda Blvd., Riga 1586, Latvia
e-mail: nils.rostoks@lu.lv

G. Zhang et al. (eds.), *Advance in Barley Sciences: Proceedings of 11th International Barley Genetics Symposium*, DOI 10.1007/978-94-007-4682-4_2,

populations depends on the extent of linkage disequilibrium (LD) and ability to control for the population structure during statistical analyses. Preliminary results based on principal component and phylogenetic analyses of 1,003 SNP markers with average polymorphism information content (PIC) of 0.394 suggested that the set of germplasm is relatively uniform with the exception of a few six-row varieties. STRUCTURE analysis based on the ΔK value suggested that the population could be partitioned into two clusters. The mean LD (r^2>0.1) extended over 10-cM distance suggesting that the available marker density may be sufficient for association mapping. Plots of pairwise LD along the chromosomes indicated uneven distribution of LD blocks in barley genome.

Keywords Barley • Single-nucleotide polymorphism • Linkage disequilibrium • Association mapping • Population structure • Genetic diversity

2.1 Introduction

Genetic diversity present in a species is key factor determining its ability to adapt to specific environments. Barley (*Hordeum vulgare* L.) is adapted to a wide range of environments ranging from Mediterranean dry areas to polar circle and to Andean and Himalayan highlands. Barley breeders therefore need to breed not only for yield and quality traits but also for adaptability traits, such as tolerance to biotic and abiotic stresses characteristic to specific target geographic region. Moreover, recent trends toward organic agriculture and novel uses of conventional crops introduce additional challenges, such as necessity for improved disease resistance and nutrient use efficiency or suitability for biofuel production (Sarath et al. 2008; Sticklen 2008; Wolfe et al. 2008). In order to meet these challenges, key traits for the specific breeding purpose need to be identified along with the development of efficient tools to facilitate the breeding (Varshney et al. 2005). Molecular markers can be used for cloning of important genes and QTL as well as to facilitate transfer of useful alleles from exotic germplasm to locally adapted elite varieties (Feuillet et al. 2008; Hajjar and Hodgkin 2007). Characterization of the local barley germplasm and the basis of its adaptability to local environments are critical for successful marker-assisted selection, because of the interactions of the introgressed alleles with the genetic background of adapted germplasm.

Local germplasm has been affected by different factors, such as limited number of founder genotypes, selection by breeders, and introgression of novel alleles from exotic germplasm. Thus, it is important to study not only the overall genetic diversity of a germplasm set but the distribution of genetic diversity in the genome which may allow to identify genome regions under the selection.

Ultimately, for marker-assisted selection (MAS) to be successful, it is important to identify genes that are responsible for agronomically important traits. Once the genes are known, the molecular mechanisms affecting the trait can be elucidated, and molecular markers based on the specific gene can be designed. Moreover, the allelic diversity in the gene of interest can be studied allowing to estimate the effects of different alleles on the expression of trait. Association mapping in a selection of plant accessions provides alternative to traditionally used biparental mapping populations. The approach takes advantage of linkage disequilibrium (LD) between the marker locus and the trait of interest (Ardlie et al. 2002; Slatkin 2008). The limiting factors in association mapping are number of markers and size of population to be typed with markers. The number of markers depends on the size of LD blocks – the larger the block size, the smaller number of markers needs to be deployed; however, the resolution of the map will also be lower (Kruglyak 2005). The size of LD blocks depends on the material in study. In recently established populations, the extent of LD will be higher because of the limited number of recombination. Cultivated plant varieties represent a special case, as breeding practices have created unique populations with different amount of LD, and the varieties are often related (Ball 2007).

One of the most abundant types of molecular markers is single-nucleotide polymorphisms (SNPs) (Rafalski 2002). Even though majority of them are biallelic, they can be very useful as markers (Kruglyak 1997), in particular, when SNP haplotypes rather than individual SNPs are considered. Utilization of SNP markers is facilitated by development of high-throughput genotyping methods that allow typing of thousands of SNP in large sets of germplasm. Recently, barley EST data was used for resequencing-based SNP discovery (Rostoks et al. 2005; Stein et al. 2007) and for electronic SNP discovery (Close et al. 2009). SNP data was used for development of genotyping platform based on Illumina GoldenGate technology (Rostoks et al. 2006), which also allowed construction of a high-resolution consensus SNP linkage map of barley (Close et al. 2009) and paved the way for whole genome association mapping in barley (Waugh et al. 2009). Association mapping was successfully used to map and identify the *INT-C* gene (Ramsay et al. 2011) as well as to identify candidate gene for anthocyanin pigmentation (Cockram et al. 2010). Latvian barley varieties are mostly bred from Scandinavian, German, and Moravian barley varieties (Gaike 1992). Genome-wide diversity in several of these varieties has been studied using microsatellites (Sjakste et al. 2003) and DArT markers (Kokina and Rostoks 2008). Recently, high-throughput SNP genotyping platform was used for genotyping 95 Latvian barley varieties and breeding lines (Rostoks 2008), which revealed relatively low level of population structure with exception of six-row varieties and linkage disequilibrium extending over the distances suitable for association mapping. Currently, an association mapping project is in progress supported by the European Social Fund with the aim of identifying molecular markers for traits important for low input and organic barley cultivation. As a first step in this project, we report a preliminary characterization of the diversity and the extent of linkage disequilibrium in our association mapping panel.

2.2 Materials and Methods

2.2.1 Plant Material and DNA Extractions

The set of germplasm consisted of 25 Latvian barley varieties, 120 Latvian barley breeding lines, and 46 foreign varieties and breeding lines (list of varieties is available upon request). Only spring two-row-type varieties were included, with the exception of six Latvian six-row varieties. DNA for genotyping was extracted from leaves of a single plant using DNeasy Plant Mini Kit (Qiagen, Hilden, Germany).

2.2.2 High-Throughput SNP Genotyping and Data Quality Control

Illumina high-throughput genotyping was done as described (Rostoks et al. 2006). Currently, two barley oligo pooled assay, BOPA1 and BOPA2, each containing 1,536 SNP are available (Close et al. 2009). One set of 95 Latvian varieties and breeding lines was genotyped only with BOPA1, while the second set of 96 Latvian and foreign barley accessions was genotyped with BOPA1 and BOPA2. Current study is based only on BOPA1 data in the total set of 191 accessions. SNP genotyping data was controlled for presence of excessive number of heterozygous loci and missing data points. Barley consensus linkage map based on the SNP markers (Close et al. 2009) was used throughout the study.

2.2.3 Data Analyses

Principal component analysis (PCo) was done using DARwin5 (http://darwin.cirad.fr/darwin). Population structure was studied using Structure 2.2 software using a selection of SNP loci and admixture linkage model (Falush et al. 2003). Forty-one SNPs (five to six per chromosome) were selected based on criteria that they are at least 10 cM apart and have minor allele frequency (MAF) > 0.30. Hypothesis of 1–20 clusters (*K*) was tested using burn-in of 100,000, run length of 200,000, and admixture model in 15 iterations. Accessions were assigned to the most probable number of clusters according to a published method (Evanno et al. 2005). Largest value of an *ad hoc* statistic ΔK, which is based on the rate of change in the log probability of data between successive *K* values, was used as an indicator of the true number of clusters. Polymorphism information content (PIC) was calculated as described (Kota et al. 2008). Linkage disequilibrium between pairs of SNP loci was calculated in TASSEL 3.0 (Bradbury et al. 2007). One thousand and three mapped SNPs with MAF > 0.1

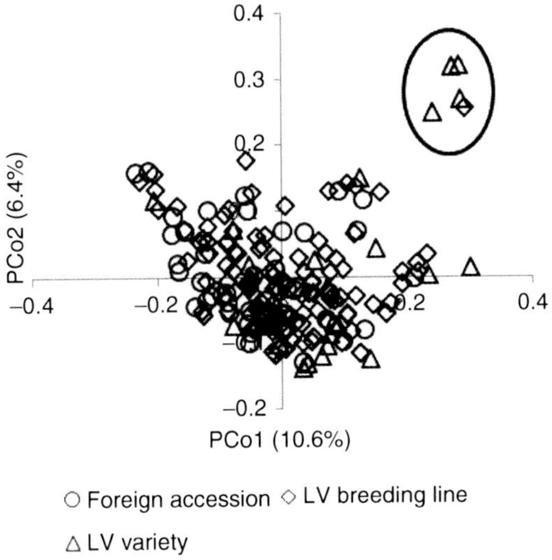

Fig. 2.1 Principal component analysis of SNP genotype data. Foreign barley accessions and Latvian breeding lines and varieties form a relatively uniform cluster with the exception of several Latvian six-row accessions (*circled*). *X*- and *Y*-axes show percent variation explained by the principal components 1 and 2, respectively

were used for calculations in the whole set of 191 accessions and after removal of the six six-row barley accessions. Once the foreign accessions were removed, MAF was recalculated, and 963 SNPs were retained for subsequent analyses.

2.3 Results and Discussion

One hundred and ninety-one Latvian and foreign barley accessions were genotyped with the Illumina oligo pooled array BOPA1 consisting of 1,536 SNPs (Close et al. 2009). SNP genotype data were controlled for the presence of excessive number of heterozygotes and missing data (1,463 SNPs retained), and of those, 1,084 were with minor allele frequency of over 10%. The 1,003 of these SNPs were positioned on the barley consensus linkage map (Close et al. 2009) and were used for all subsequent analyses in this study.

Principal component analysis of the genotype data revealed a relatively compact group of the accessions, while the PCo1 and PCo2 explained only 10.6 and 6.4 % of variation, respectively (Fig. 2.1). The only exception was a group of old Latvian six-row accessions, which were separated from the main cluster. Clustering according to row type is consistent with historical origins of barley (Pourkheirandish and Komatsuda 2007) and with previous studies that analyzed the genetic diversity in populations consisting of both six-row and two-row genotypes (Hamblin et al. 2010; Zhang et al. 2009). Thus, it appears that Latvian barley varieties show little population substructure and high degree of relatedness with foreign barley accessions. This is consistent with the pedigrees of Latvian barley varieties involving early German, Moravian, and Scandinavian varieties and with the recent breeding practices

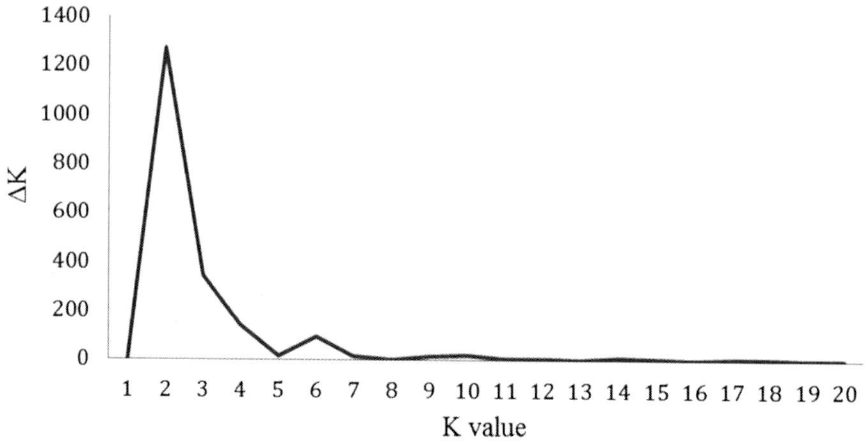

Fig. 2.2 Detection of number of clusters (*K*) by estimating Δ*K* over 15 iterations for each *K* value with STRUCTURE. Δ*K* was expressed as a mean of the absolute values of ratio of change of the likelihood function with respect to *K*. The highest value of Δ*K* indicates the strength of the signal

that attempt to combine agronomic and quality traits of foreign germplasm with the adaptive traits of the Latvian varieties.

Neighbor-joining dendrogram (results not shown) revealed several major clusters with low bootstrap support and generally good separation of accessions in terminal nodes often consistent with the known pedigrees. No separation according to breeder (Latvian or foreign) was apparent, which again was consistent with the exchange of breeding material between Latvian barley breeders and extensive use of foreign germplasm in Latvian barley breeding programs.

Population structure is known to affect association mapping in various crop species including barley (Hamblin et al. 2010), wheat (Le Couviour et al. 2011), rice (Jin et al. 2010), and oat (Newell et al. 2011). Population structure may result in finding spurious associations between markers and traits; therefore, population structure has to be assessed prior the association mapping (Pritchard et al. 2000). STRUCTURE was used to test highest probability of a number of clusters (*K*) among all barley accessions. The highest value of ΔK was observed with two clusters (Fig. 2.2) suggesting that the accessions most likely separate into two subpopulations, which is in contrast to the ten populations found in the set of 500 UK cultivars (Cockram et al. 2010) and seven populations in the set of 1,816 breeding lines from US breeding programs (Hamblin et al. 2010). This may be explained by more diverse set of genotypes with respect to row number (only six six-row genotypes included in our study) and growth habit (only spring type in our study), because clustering in US and UK genotypes was observed mostly based on row number and growth habit and their combinations. No clustering according to origin of genotypes was observed in our study, unlike the finding that Western European elite wheat varieties cluster according to country of origin and breeding history (Le Couviour et al. 2011).

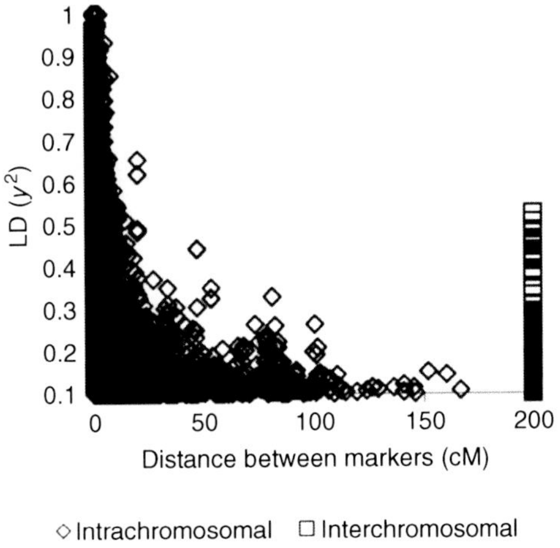

Fig. 2.3 Extent of the pairwise LD between 1,003 high-quality mapped SNP markers in the whole data set of 191 accessions. LD between pairs of SNP loci on different chromosomes (interchromosomal LD) is shown at a fixed distance of 200 cM

The extent of pairwise LD between all SNP markers was explored in Tassel 3.0 using the genetic distances from Close et al. (2009) (Fig. 2.3). The extent of pairwise LD significantly varied among the barley chromosomes, suggesting that this factor may affect the resolution of association mapping for QTL located on different chromosomes (Table 2.1). Median extent of LD was 2.0 cM on chromosome 6H, while chromosome 2H exhibited much more extensive LD (median 9.3 cM). Moreover, there were some differences in the subsets of germplasm, e.g., more extensive LD was observed on chromosome 1H in a set of Latvian accessions (15 cM) than in the whole set (10 cM). However, on all chromosomes, average distance between markers was smaller than the extent of pairwise LD suggesting that marker coverage could be sufficient for association mapping (Table 2.1).

Distribution of the LD along the chromosomes varied extensively (Fig. 2.4). While the chromosome 2H that showed the highest mean and median pairwise extent of LD also exhibited a relatively uniform distribution along the chromosome, the chromosomes 3H and 6H showed extensive LD mostly in the centromeric regions similar to results by Comadran et al. (2011). Interchromosomal and intrachromosomal variation of LD has been observed in various autogamous crop species, and regions where LD extends up to 50 cM (Hamblin et al. 2010) and 60 cM (Rostoks et al. 2006) in barley, 50 cM in rice (Jin et al. 2010), and 5 cM in oat (Newell et al. 2011) have been detected. Neumann et al. suggested that high LD in elite crop germplasm could be formed during breeding history resulting in assembled blocks of chromosomes containing genes for agronomic fitness (Neumann et al. 2011).

Whole genome association mapping in small and highly structured sets of germplasm has been shown to generate a large number of false-positive associations (Cuesta-Marcos et al. 2010). Recently, we successfully used whole genome association mapping to identify the *NUD* gene as a major determinant of β-glucan content in barley but failed to identify other QTL for β-glucan content in the set of 95 Latvian barley accessions (Mezaka et al. in press). Apparently, simple traits controlled

Table 2.1 Comparison of polymorphism information content (PIC) and linkage disequilibrium (LD, $r^2>0.1$) in the barley genome

Chromosome	Number of markers/average distance between markers in cM	PIC			Extent of pairwise LD $r^2>0.1$ (average/median in cM)		
		Whole set	Only two-row	Only LV two-row	Whole set	Only two-row	Only LV two-row
1 H	111/1.27	0.356	0.349	0.351	10.0/5.7	10.5/5.9	15.0/5.9
2 H	157/0.96	0.404	0.396	0.395	17.2/9.3	14.0/7.1	15.8/9.0
3 H	153/1.14	0.403	0.396	0.394	7.3/3.9	6.1/3.5	6.7/3.5
4 H	117/1.05	0.395	0.388	0.383	10.1/4.2	7.9/3.6	8.7/3.2
5 H	202/0.97	0.394	0.387	0.381	9.7/5.0	7.5/3.9	8.8/4.3
6 H	147/0.89	0.399	0.394	0.391	4.8/2.0	4.5/1.7	6.8/2.4
7 H	117/1.44	0.395	0.386	0.386	12.8/5.3	10.4/5.3	12.0/4.8
Genome-wide	1,003/1.08	0.394	0.387	0.384	10.3/4.7	8.5/4.1	10.2/4.5

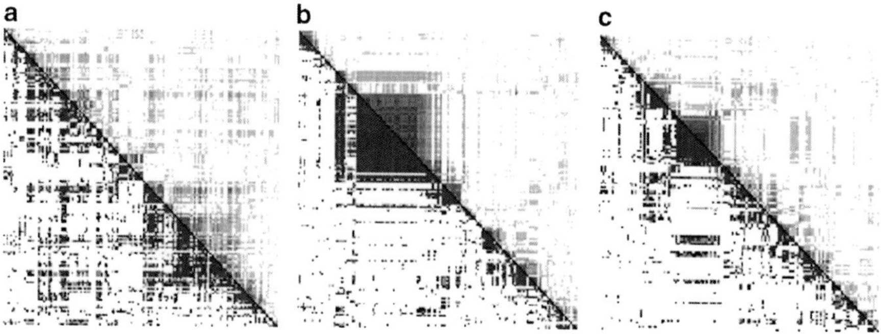

Fig. 2.4 Extent of the LD along selected chromosomes. (**a**) chromosome 2H; (**b**) chromosome 3H; (**c**) chromosome 6H. *Upper diagonal* in each panel shows LD (r^2); *lower diagonal* shows the corresponding *p* values

by a single gene may be successfully mapped in relatively small association mapping panels, provided that population structure is accounted for (Cockram et al. 2010; Comadran et al. 2011). However, other traits controlled by many small-effect QTL may be much more difficult to map. Careful selection of subsets of germplasm for a particular trait and statistical control of population structure may improve association mapping (Cockram et al. 2010), along with other approaches, such as use of SNP haplotypes rather than single SNPs (Lorenz et al. 2010). However, in the summary, the preliminary results reported in this chapter suggest a low degree of population substructure and a suitable extent of LD for association studies in our association mapping population.

Acknowledgements The study is funded by the European Social Fund cofinanced project 2009/0218/1DP/1.1.1.2.0/09/APIA/VIAA/099 and the Latvian Council of Science grant Z-956-090.

References

Ardlie, K. G., Kruglyak, L., & Seielstad, M. (2002). Patterns of linkage disequilibrium in the human genome. *Nature Reviews. Genetics, 3*, 299–309.

Ball, R. D. (2007). Statistical analysis and experimental design. In N. C. Oraguzie, E. H. A. Rikkerink, S. Gardiner, & H. N. De Silva (Eds.), *Association mapping in plants* (pp. 133–196). New York: Springer.

Bradbury, P. J., Zhang, Z., Kroon, D. E., Casstevens, T. M., Ramdoss, Y., & Buckler, E. S. (2007). TASSEL: Software for association mapping of complex traits in diverse samples. *Bioinformatics, 23*, 2633–2635.

Close, T., Bhat, P., Lonardi, S., Wu, Y., Rostoks, N., Ramsay, L., Druka, A., Stein, N., Svensson, J., Wanamaker, S., Bozdag, S., Roose, M., Moscou, M., Chao, S., Varshney, R., Szucs, P., Sato, K., Hayes, P., Matthews, D., Kleinhofs, A., Muehlbauer, G., DeYoung, J., Marshall, D., Madishetty, K., Fenton, R., Condamine, P., Graner, A., & Waugh, R. (2009). Development and implementation of high-throughput SNP genotyping in barley. *BMC Genomics, 10*, 582.

Cockram, J., White, J., Zuluaga, D. L., Smith, D., Comadran, J., Macaulay, M., Luo, Z., Kearsey, M. J., Werner, P., Harrap, D., Tapsell, C., Liu, H., Hedley, P. E., Stein, N., Schulte, D., Steuernagel, B., Marshall, D. F., Thomas, W. T., Ramsay, L., Mackay, I., Balding, D. J., Consortium, T. A., Waugh, R., & O'Sullivan, D. M. (2010). Genome-wide association mapping to candidate polymorphism resolution in the unsequenced barley genome. *Proceedings of the National Academy of Sciences of the United States of America, 107*, 21611–21616.

Comadran, J., Ramsay, L., MacKenzie, K., Hayes, P., Close, T. J., Muehlbauer, G., Stein, N., & Waugh, R. (2011). Patterns of polymorphism and linkage disequilibrium in cultivated barley. *Theoretical and Applied Genetics, 122*, 523–531.

Cuesta-Marcos, A., Szucs, P., Close, T., Filichkin, T., Muehlbauer, G., Smith, K., & Hayes, P. (2010). Genome-wide SNPs and re-sequencing of growth habit and inflorescence genes in barley: implications for association mapping in germplasm arrays varying in size and structure. *BMC Genomics, 11*, 707.

Evanno, G., Regnaut, S., & Goudet, J. (2005). Detecting the number of clusters of individuals using the software structure: A simulation study. *Molecular Ecology, 14*, 2611–2620.

Falush, D., Stephens, M., & Pritchard, J. K. (2003). Inference of population structure using multilocus genotype data: Linked loci and correlated allele frequencies. *Genetics, 164*, 1567–1587.

Feuillet, C., Langridge, P., & Waugh, R. (2008). Cereal breeding takes a walk on the wild side. *Trends in Genetics, 24*, 24–32.

Gaike, M. (1992). Spring barley. In I. Holms (Ed.), *Field crop breeding in Latvia (in Latvian)* (pp. 53–63). Riga: Avots.

Hajjar, R., & Hodgkin, T. (2007). The use of wild relatives in crop improvement: a survey of developments over the last 20 years. *Euphytica, 156*, 1–13.

Hamblin, M. T., Close, T. J., Bhat, P. R., Chao, S., Kling, J. G., Abraham, K. J., Blake, T., Brooks, W. S., Cooper, B., Griffey, C. A., Hayes, P. M., Hole, D. J., Horsley, R. D., Obert, D. E., Smith, K. P., Ullrich, S. E., Muehlbauer, G. J., & Jannink, J. L. (2010). Population structure and linkage disequilibrium in U.S. barley germplasm: implications for association mapping. *Crop Science, 50*, 556–566.

Jin, L., Lu, Y., Xiao, P., Sun, M., Corke, H., & Bao, J. (2010). Genetic diversity and population structure of a diverse set of rice germplasm for association mapping. *Theoretical and Applied Genetics, 121*, 475–487.

Kokina, A., & Rostoks, N. (2008). Genome-wide and Mla locus-specific characterization of Latvian barley varieties. *Proceedings of the Latvian Academy of Sciences, 62*, 103–109.

Kota, R., Varshney, R., Prasad, M., Zhang, H., Stein, N., & Graner, A. (2008). EST-derived single nucleotide polymorphism markers for assembling genetic and physical maps of the barley genome. *Functional & Integrative Genomics, 8*, 223–233.

Kruglyak, L. (1997). The use of a genetic map of biallelic markers in linkage studies. *Nature Genetics, 17*, 21–24.

Kruglyak, L. (2005). Power tools for human genetics. *Nature Genetics, 37*, 1299–1300.

Le Couviour, F., Faure, S., Poupard, B., Flodrops, Y., Dubreuil, P., & Praud, S. (2011). Analysis of genetic structure in a panel of elite wheat varieties and relevance for association mapping. *Theoretical and Applied Genetics, 123*, 715–727.

Lorenz, A. J., Hamblin, M. T., & Jannink, J. L. (2010). Performance of single nucleotide polymorphisms versus haplotypes for genome-wide association analysis in barley. *PLoS One, 5*, e14079.

Mezaka, I., Bleidere, M., Legzdina, L., & Rostoks, N. (in press). Whole genome association mapping identifies naked grain locus *NUD* as determinant of β-glucan content in barley. *Zemdirbyste – Agriculture*.

Neumann, K., Kobiljski, B., Denčić, S., Varshney, R., & Börner, A. (2011). Genome-wide association mapping: a case study in bread wheat (*Triticum aestivum* L.). *Molecular Breeding, 27*, 37–58.

Newell, M. A., Cook, D., Tinker, N. A., & Jannink, J. L. (2011). Population structure and linkage disequilibrium in oat (*Avena sativa* L.): implications for genome-wide association studies. *Theoretical and Applied Genetics, 122*, 623.

Pourkheirandish, M., & Komatsuda, T. (2007). The importance of barley genetics and domestication in a global perspective. *Annals of Botany London, 100*, 999–1008.

Pritchard, J. K., Stephens, M., Rosenberg, N. A., & Donnelly, P. (2000). Association mapping in structured populations. *The American Journal of Human Genetics, 67*, 170–181.

Rafalski, A. (2002). Applications of single nucleotide polymorphisms in crop genetics. *Current Opinion in Plant Biology, 5*, 94–100.

Ramsay, L., Comadran, J., Druka, A., Marshall, D. F., Thomas, W. T., Macaulay, M., MacKenzie, K., Simpson, C., Fuller, J., Bonar, N., Hayes, P. M., Lundqvist, U., Franckowiak, J. D., Close, T. J., Muehlbauer, G. J., & Waugh, R. (2011). *INTERMEDIUM-C*, a modifier of lateral spikelet fertility in barley, is an ortholog of the maize domestication gene *TEOSINTE BRANCHED 1*. *Nature Genetics, 43*, 169–172.

Rostoks, N. (2008, September 9–12) *High throughput genotyping for characterization of barley germplasm*. Proceedings of the EUCARPIA 18th General Congress, Valencia, Spain.

Rostoks, N., Mudie, S., Cardle, L., Russell, J., Ramsay, L., Booth, A., Svensson, J. T., Wanamaker, S. I., Walia, H., Rodriguez, E. M., Hedley, P. E., Liu, H., Morris, J., Close, T. J., Marshall, D. F., & Waugh, R. (2005). Genome-wide SNP discovery and linkage analysis in barley based on genes responsive to abiotic stress. *Molecular Genetics and Genomics, 274*, 515–527.

Rostoks, N., Ramsay, L., MacKenzie, K., Cardle, L., Bhat, P. R., Roose, M. L., Svensson, J. T., Stein, N., Varshney, R. K., Marshall, D. F., Graner, A., Close, T. J., & Waugh, R. (2006). Recent history of artificial outcrossing facilitates whole-genome association mapping in elite inbred crop varieties. *Proceedings of the National Academy of Sciences of the United States of America, 103*, 18656–18661.

Sarath, G., Mitchell, R. B., Sattler, S. E., Funnell, D., Pedersen, J. F., Graybosch, R. A., & Vogel, K. P. (2008). Opportunities and roadblocks in utilizing forages and small grains for liquid fuels. *Journal of Industrial Microbiology & Biotechnology, 35*, 343–354.

Sjakste, T. G., Rashal, I., & Roder, M. S. (2003). Inheritance of microsatellite alleles in pedigrees of Latvian barley varieties and related European ancestors. *Theoretical and Applied Genetics, 106*, 539–549.

Slatkin, M. (2008). Linkage disequilibrium – Understanding the evolutionary past and mapping the medical future. *Nature Reviews. Genetics, 9*, 477–485.

Stein, N., Prasad, M., Scholz, U., Thiel, T., Zhang, H., Wolf, M., Kota, R., Varshney, R. K., Perovic, D., Grosse, I., & Graner, A. (2007). A 1,000-loci transcript map of the barley genome: New anchoring points for integrative grass genomics. *Theoretical and Applied Genetics, 114*, 823–839.

Sticklen, M. B. (2008). Plant genetic engineering for biofuel production: Towards affordable cellulosic ethanol. *Nature Reviews. Genetics, 9*, 433–443.

Varshney, R. K., Graner, A., & Sorrells, M. E. (2005). Genomics-assisted breeding for crop improvement. *Trends in Plant Science, 10*, 621–630.

Waugh, R., Jannink, J. L., Muehlbauer, G. J., & Ramsay, L. (2009). The emergence of whole genome association scans in barley. *Current Opinion in Plant Biology, 12*, 1–5.

Wolfe, M., Baresel, J., Desclaux, D., Goldringer, I., Hoad, S., Kovacs, G., Loeschenberger, F., Miedaner, T., Ostergard, H., & Lammerts van Bueren, E. (2008). Developments in breeding cereals for organic agriculture. *Euphytica, 163*, 323–346.

Zhang, L. Y., Marchand, S., Tinker, N. A., & Belzile, F. (2009). Population structure and linkage disequilibrium in barley assessed by DArT markers. *Theoretical and Applied Genetics, 119*, 43–52.

Chapter 3
Genome-Wide Association Mapping of Malting Quality Traits in Relevant Barley Germplasm in Uruguay

Ariel Castro, Lorena Cammarota, Blanca Gomez, Lucia Gutierrez, Patrick M. Hayes, Andres Locatelli, Lucia Motta, and Sergio Pieroni

Abstract Knowledge about the genetic components of major malting quality traits is needed for efficient barley breeding, and although some of these traits have been mapped, little information about germplasm-specific QTL is known for South American germplasm. The aim of this study was to determine by genome-wide association mapping the key genetic basis of malting quality traits in a population of 76 different genotypes consisting of historical varieties, commercial cultivars, and advanced lines representative of barley breeding in Uruguay. Samples obtained in five contrasting environments were micromalted in order to obtain a phenotypic database. The population was genotyped with 1,033 polymorphic SNPs using the Illumina BOPA1. Marker-trait associations were detected through linkage disequilibrium mapping using a mixed linear model (MLM) Q + K containing a structure matrix (PCA) and a kinship matrix (K). Preliminary results showed QTL effects

A. Castro (✉) • A. Locatelli
Departamento de Producción Vegetal Est. Exp. "Dr. Mario A. Cassinoni", Universidad de la República, Ruta 3 Km 373, Paysandú 60000, Uruguay
e-mail: vontruch@fagro.edu.uy

L. Cammarota • S. Pieroni
Malteria Uruguay S.A., Ombúes de Lavalle, Uruguay

B. Gomez
Laboratorio Técnológico del Uruguay, Montevideo, Uruguay

L. Gutierrez
Departamento de Biometría, Estadística y Computación, Facultad de Agronomía, Montevideo, Uruguay

P.M. Hayes
Department of Crop and Soil Science, Oregon State University, Corvallis, OR 97331-3002, USA

L. Motta
Malteria Oriental S.A., Montevideo, Uruguay

G. Zhang et al. (eds.), *Advance in Barley Sciences: Proceedings of 11th International Barley Genetics Symposium*, DOI 10.1007/978-94-007-4682-4_3,

detected for all traits, with some genomic regions showing a high concentration of significant associations. Most QTL were environment specific. We are presently studying the relationship between malting quality traits and linkage disequilibrium blocks found in the population. The results provide some of the first data regarding genetic basis of malting quality relevant traits in the germplasm used in the region.

Keywords Barley • Malting quality • Association mapping • Linkage disequilibrium

3.1 Introduction

Barley is produced in Uruguay mainly as a source of malt for brewing. In order to provide farmers with an economical alternative of production, the development of improved genotypes with increasing malting quality is essential, and the knowledge of the genetic basis of these traits is relevant.

The malt quality phenotype is difficult and expensive to measure because there are many interrelated component traits that lead to malt of high quality (Hayes and Jones 2000). Breeding for malting quality is challenging, and most breeding programs resort to making relatively few "good quality" by "good quality" crosses. As a consequence, there have been modest genetic gains for malting quality (Han et al. 1997; Muñoz-Amatriaín et al. 2010). In order to accelerate the process of developing superior malting barley, other breeding methods are needed. One approach is to discover the genes determining the components of malting quality and to use this information for marker-assisted selection (MAS) or genomic selection (GS).

Biparental quantitative trait locus (QTL) mapping studies have been conducted to identify chromosome regions responsible for superior malting quality (see Szűcs et al. 2009 for a review). The results of these studies have been difficult to directly apply to cultivar development for various reasons. Most of the mapping populations did not reflect the germplasm used on cultivar development, favorable alleles at malting quality may already be fixed in relevant germplasm, and some biparental QTL mapping studies may have low accuracy due to small population size (Dekkers and Hospital 2002; Yu and Buckler 2006).

Genome-wide association mapping (GWA) may overcome many of the limitations of the biparental approach (Jannink et al. 2001). The advantages of GWA include simultaneous assessment of broad diversity, higher resolution for fine mapping, effective use of historical data, and immediate applicability to cultivar development because the genetic background in which QTL are estimated is directly relevant for plant breeding (Kraakman et al. 2004). GWA has been proposed as a promising tool for barley breeding (Hayes and Szücs 2006; Stracke et al. 2009; Waugh et al. 2009; Roy et al. 2010; Bradbury et al. 2011; von Zitzewitz et al. 2011).

The objective of this research is to determine the genetic factors affecting malting quality in relevant germplasm for spring malting barley breeding in Uruguay.

With that aim, we used a germplasm collection representative of barley breeding in Uruguay, a throughout genotypic and phenotypic characterization, and appropriate GWA analysis tools.

3.2 Materials and Methods

We used a population of 76 spring barley genotypes, consisting of varieties commercially used in Uruguay during the last century, advanced lines from breeding programs, and ancestors present in the pedigree of several varieties used in the country. The population is listed elsewhere (Locatelli et al., in these proceedings).

The population was phenotyped in five field experiments, planted at two locations in Uruguay. Three experiments were planted at the "Dr. Mario A. Cassinoni" Experimental Station, Paysandú (58° 03′W, 32° 55'S), on July 16, 2007 (Pay07), June 24, 2008 (Pay08a), and August 4, 2008 (Pay08b). The other two experiments were planted at Ombúes de Lavalle (57° 50′W, 33° 55′S) on July 9, 2007 (Col07), and July 10, 2008 (Col08). Samples from each experiment were micromalted at the facilities of Maltería Uruguay S.A. (Pay07, Col08), Maltería Oriental S.A. (Pay08b), and Laboratorio Tecnológico del Uruguay (Col07 and Pay08a). Malting procedures followed the standards used in Uruguay and are available upon request. In this work, we present results for malt extract (ME), malt protein (MP), diastatic power (DP), soluble nitrogen (SN), Kolbach index (KI), friability (F), and beta-glucans (BG). Measurements were taken following EBC protocols.

The population was genotyped using the barley oligonucleotide pool assay (BOPA 1), developed for the Coordinated Agricultural Project (www.barleycap.org), containing allele-specific oligos for a set of 1,536 SNPs. The estimated positions of the SNPs are based on the consensus map developed by Close et al. (2009). The SNPs were referred to the bins described by Kleinhofs and Graner (2001).

The structure of the population was estimated by developing a matrix of population structure (Q matrix) using the program STRUCTURE version 2.2 (Pritchard et al. 2000), in order to establish subpopulations (groups) within the population. From pedigree data, kinship relationships among individuals were estimated using coancestry coefficient (f) (Malecot 1949), with the procedure developed by Condón (2006). Coancestry coefficients were used for a cluster analysis (Ward's method), and the number of groups was estimated pseudo-F function of SAS 9.0 (SAS 2004).

To determine the associations between molecular markers and quantitative traits, a linear mixed model was used ($y=X\beta+Qv+Ku+e$) where X is the matrix of molecular markers, β is the vector of parameters related to the simple regression of the markers on the phenotype, Q is the structure matrix, v is a vector of effects of each group or subpopulation, K is the identity matrix, u is the vector of background polygenic effects, and e is the vector of random errors. The analysis was performed using TASSEL, version 2.0.1 (Bradbury et al. 2007). To consider a marker-trait association as significant, a false discovery rate (FDR) (Benjamini and Hochberg 1995) of $p \leq 0.05$ was used.

3.3 Results

As described elsewhere (Locatelli et al., these proceedings), a total of 1,033 SNP markers out of 1,536 were finally used for the analysis. The study of the population genetic structure indicated an optimum number of four groups. Based on coancestry coefficients, five major groups were identified, and the composition of the groups was consistent with the results from the structure analysis (for more details, see Locatelli et al., in these proceedings).

Average values for the analyzed traits (Table 3.1) indicated that the general quality was below industry standards, a somewhat expected fact considering that the population was relatively diverse, including genotypes with very low malting quality. Conditions in some experiments also favored high protein levels, particularly in Pay08b and Col08.

We find significant associations for all the traits analyzed. A total of 46 significant marker-trait associations were detected for the traits analyzed (Annex 1). In order to reach this number, we only included associations significant after an FDR test, plus associations rejected by that criteria but repeated in more than one environment and/or trait. Most of the associations were detected in only one experiment (37 out of 46). F was the trait with most associations (12) although only one was detected in more than one experiment. SN, by contrast, has only three significant associations detected.

Two experiments (Col07 and Pay08a) showed the highest number of significant associations (15 and 17, respectively). Pay07 and Col08, by contrast, had the lowest number (8 and 5, respectively). We are presently analyzing the relationships between environmental conditions and quality traits in these experiments. In any case, it is obvious that the certain particular conditions allowed the detection of significant associations. For example, most of the associations for MP were detected at Pay08a (5 out of 6), while no effects were detected in two experiments (Col07 and Col08), and the remaining two experiments showed only one significant association each.

Although, as mentioned above, most of the associations were experiment specific, some markers showed coincidence in their effects in different environments. The marker 1111258 (3 H), in particular, was significant for ME in three environments and in two environments for F (the only marker significant in more than one environment) and also was associated with IK. Also we found, genotyping more recent advanced lines from Uruguayan breeding programs (data not presented), that the favorable allele was fixed in modern germplasm.

Some genomic regions showed coincidence of significant associations for different traits (Fig. 3.1 and Annex 1). BIN 6 of 1H, BIN 5 of 2H, BIN 4 of 3H, BIN 5 of 4H, BIN 2 of 5H, and BIN 1 of 7H showed associations for three or more traits. These are the most interesting regions from a breeding point of view, and we are studying their coincidence with previous reports of QTL and genes associated with malting quality traits. Most of them (in particular, the ones in 1H, 2H, 3H, and 7H) are located within important LD blocks (see Locatelli et al., in these proceedings) and may have played a role in the historical breeding process.

Table 3.1 Average values of seven malting quality variables for each experiment (ME: %, MP: %, DP: uWK, SN: %, F: %, BG: ppm)

Experiment	ME	MP	DP	SN	KI	F	BG
Pay07	78.9	13.4	384	841	39.9	55.9	459
Col07	79.6	10.5	275	623	37.7	68.2	430
Pay08a	79.0	11.6	370	737	39.8	57.7	477
Pay08b	77.0	15.5	454	862	34.3	41.9	646
Col08	75.6	14.4	266	668	29.2	35.7	641

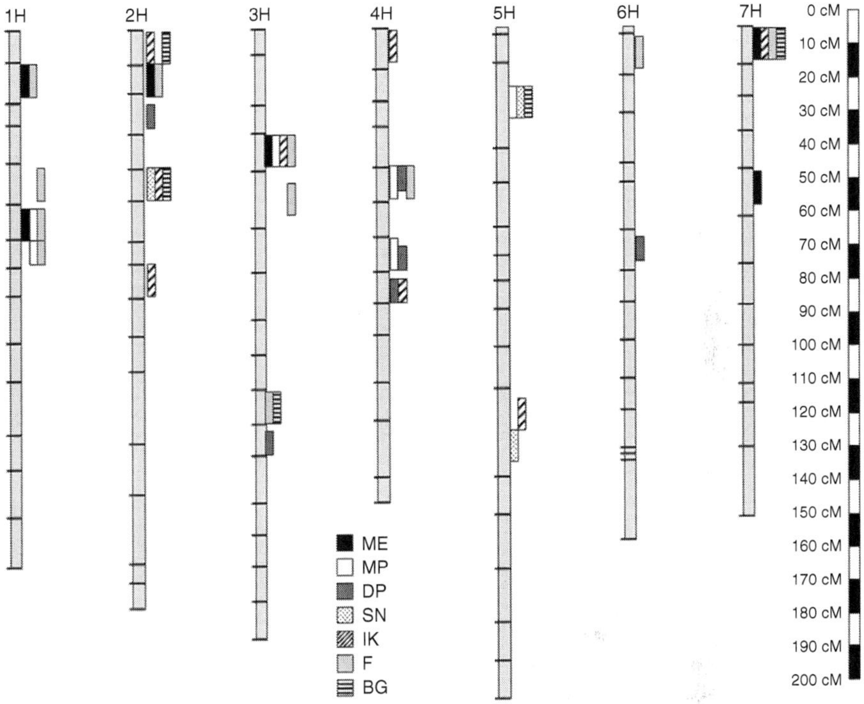

Fig. 3.1 Location in the barley BIN map (Kleinhofs and Graner 2001) of the significant marker-trait associations reported in Annex 1

At the present stage of the study, we have detected no repulsion linkage effects, meaning that alleles associated with favorable values for one trait were not associated with negative values for another one. Although these preliminary results do not preclude completely that possibility, they allow for a rather optimistic perspective.

Another issue of particular interest was to study the relationships between markers associated with malting quality traits and the ones associated with agronomic traits. Most of the markers associated with quality did not show associations with agronomic traits, with some exceptions as 111258, associated with grain size, phenology, harvest index, and biomass production (Locatelli et al., elsewhere in this publication). But if we considered the LD blocks associated with agronomic traits (analyzed by Locatelli et al. elsewhere in these proceedings), some of the genomic regions showing associations with more than one quality trait did have associations with those traits. BIN 5 of 2H and BIN4 of 3H are examples of this situation.

Nevertheless, we have to emphasize that the presented results are preliminary. We are refining the analysis, studying coincidences with previous reports and relationships between the association effects detected and the LD structure of the population. In any case, our results are an important advance toward the knowledge of the genetic basis of the most relevant phenotypes for barley breeding in the germplasm used in our geographical region.

Annex 1 Significant marker-trait associations detected for seven malting quality traits by experiment, including significance level (p: **0.001, *** 0.0001, ..., ***** 0.00001, non significant) and allele substitution effects (ASE). Marker position is indicated in cM (Dist.) and BIN location (Kleinhofs and Graner 2001). Markers significant with an FDR of 5 % are in bold. The rest of the markers include cases with coincident effects for several environments and/or several traits. When several tightly linked markers were significant only one was included in the table.

					Col07		Pay07		Col08		Pay08a		Pay08b	
Trait	Cr.	SNP	Dist.	Bin	p	ASE	p	ASE	p	ASE	p	ASE	p	ASE
ME	1 H	**1121226**	8.8	2	NS		NS		NS		NS		**	0.713
	1 H	1110516	65.5	6	NS		NS		NS		**	0.751	NS	
	2 H	**1120562**	10.9	2	NS		NS		NS		NS		**	0.900
	3 H	**1111258**	52.5	4	**	0.836	NS		NS		***	1.184	***	1.089
	7 H	**1121443**	0.6	1	**	0.295	**	0.705	***	0.914	NS		**	0.574
	7 H	**1120750**	68.5	5	NS		NS		NS		****	1.242	NS	
MP	1 H	**1110516**	65.5	6	NS		NS		NS		**	0.503	NS	
	1 H	**1120229**	71.4	7	NS		NS		NS		**	0.515	NS	
	3 H	**1110620**	56.4	4	NS		NS		NS		****	0.682	**	0.625
	4 H	**1121087**	62.1	5	NS		NS		NS		**	0.642	NS	
	4 H	1110334	103.1	7	NS		**	0.704	NS		NS		NS	
	5 H	**1120958**	51.0	2	NS		NS		NS		**	0.777	NS	
DP	2 H	**1110987**	31.0	3	NS		NS		***	29.7	NS		**	30.2
	3 H	**1121381**	102.2	6	***	44.7	NS		NS		NS		NS	
	4 H	**1110568**	55.6	5	**	30.4	NS		NS		NS		NS	
	4 H	**1120974**	106.0	7	***	28.2	NS		NS		NS		NS	
	4 H	**1110387**	119.8	8	***	25.8	**	36.6	NS		NS		NS	
	6 H	**1120972**	94.7	6	NS		NS		***	38.8	NS		NS	
SN	2 H	**1110733**	55.0	5	NS		***	83.6	NS		NS		NS	
	5 H	1121318	53.2	2	NS		NS		NS		NS		**	61.3
	5 H	**1120402**	195.4	10	NS		***	103.4	NS		NS		NS	
IK	2 H	**1121377**	8.6	1	NS		NS		NS		NS		***	2.36

(continued)

Annex 1 (continued)

					Col07		Pay07		Col08		Pay08a		Pay08b	
Trait	Cr.	SNP	Dist.	Bin	p	ASE	p	ASE	p	ASE	p	ASE	p	ASE
	2 H	1110733	55.0	5	NS		**	3.27	NS		NS		NS	
	2 H	**1111118**	113.5	8	NS		*****	2.12	NS		NS		**	2.85
	3 H	1111258	52.5	4	NS		NS		NS		**	2.67	NS	
	4 H	**1121228**	3.7	1	**	1.93	NS		NS		NS		**	1.68
	4 H	1110611	114.7	8	NS		NS		NS		**	1.59	NS	
	5 H	**1120897**	182.9	10	**	2.01	NS		NS		NS		NS	
	7 H	1110682	0.6	1	**	2.41	NS		NS		NS		NS	
F	1 H	**1121226**	8.8	2	NS		NS		NS		***	8.16	NS	
	1 H	1110438	50.0	5	NS		NS		**	7.56	NS		NS	
	1 H	1110516	65.5	6	NS		NS		NS		**	6.67	NS	
	1 H	**1120229**	71.4	7	NS		NS		NS		***	8.54	NS	
	2 H	1120562	10.9	2	NS		NS		NS		NS		**	6.69
	3 H	**1111258**	52.5	4	**	8.25	NS		NS		****	11.66	NS	
	3 H	**1110620**	56.4	4	NS		NS		NS		****	9.25	NS	
	3 H	**1111401**	58.0	5	NS		NS		NS		***	8.19	NS	
	3 H	1121008	162.1	10	**	6.03	NS		NS		NS		NS	
	4 H	1120453	62.8	5	**	7.57	NS		NS		NS		NS	
	6 H	**1120262**	8.1	1	NS		NS		NS		**	7.57	NS	
	7 H	1110682	0.6	1	**	6.66	NS		NS		NS		NS	
BG	2 H	1121377	8.6	1	NS		NS		**	102.3	NS		**	119.9
	2 H	1110733	55.0	5	NS		**	77.5	NS		NS		NS	
	3 H	1110646	162.1	10	**	78.3	NS		NS		NS		NS	
	5 H	1121318	53.2	2	NS		NS		**	112.7	NS		NS	
	7 H	**1110682**	0.6	1	***	99.1	NS		NS		NS		NS	

References

Benjamini, Y., & Hochberg, Y. (1995). Controlling the false discovery rate: a practical and powerful approach to multiple testing. *Journal of the Royal Statistical Society: Series B, 57*, 289–300.

Bradbury, P. J., Zhang, Z., Kroon, D. E., Casstevens, T. M., Ramdoss, Y., & Buckler, E. S. (2007). Tassel: Software for association mapping of complex traits in diverse samples. *Bioinformatics Applications Note, 23*, 2633–2635.

Bradbury, P., Parker, T., Hamblin, T., & Jannink, J. L. (2011). Assessment of power and false discovery rate in genome-wide association studies using the Barley CAP germplasm. *Crop Science, 51*, 52–59.

Close, T. J., Prasanna, R. B., Bhat, P. R., Wu, Y., Rostoks, N., Ramsay, L., Druka, A., Stein, N., Svensson, J. T., Wanamaker, S., Bozdag, S., Rose, M. L., Moscou, M. J., Chao, S., Varshney, R. K., Szücs, P., Sato, K., Hayes, P. M., Matthews, D. E., Kleinhofs, A., Muehlbauer, G. J., DeYoung, J., Marshall, D. F., Madishetty, K., Fenton, R. D., Condamine, P., Graner, A., & Waugh, R. (2009). Development and implementation of high-throughput SNP genotyping in barley. *BMC Genomics, 10*, 582.

Condón F. (2006) *Genetic grain, diversity, and marker-trait association in Minnesota barley germplasm*. Ph.D. dissertation, University of Minnesota, St. Paul.

Dekkers, J. C. M., & Hospital, F. (2002). The use of molecular genetics in improvement of agricultural populations. *Nature Reviews Genetics, 3*, 22–32.

Han, F., Romagosa, I., Ullrich, S. E., Jones, B. L., Hayes, P. M., & Wesenberg, D. M. (1997). Molecular marker-assisted selection for malting quality traits in barley. *Molecular Breeding, 3*, 427–437.

Hayes, P. M., & Jones, B. L. (2000, October) Malting quality from a QTL perspective. In *8th International Barley Genetics Symposium* (Vol 8, pp. 99–105). Adelaide: Adelaide Convention Centre.

Hayes, P., & Szücs, P. (2006). Disequilibrium and association in barley: Thinking outside the glass. *Proceedings of the National Academy of Sciences of the United States of America, 103*, 18385–18386.

Jannink, J.-L., Bink, M. C. A. M., & Jansen, R. C. (2001). Using complex plant pedigrees to map valuable genes. *Trends in Plant Science, 6*, 337–342.

Kleinhofs, A., & Graner, A. (2001). An integrated of the barley genome. In R. L. Philips & I. K. Vasil (Eds.), *DNA-based markers in plants* (2nd ed., pp. 187–200). Dordrecht: Kluwer.

Kraakman, A. T. W., Niks, R. E., Van den Berg, M. M. M., Stam, P., & van Eeuwijk, F. A. (2004). Linkage disequilibrium mapping of yield and yield stability in modern spring barley cultivars. *Genetics, 168*, 435–446.

Malecot, A. (1949). *Les mathématiques de l'heredite*. Paris: Masson & Cie.

Muñoz-Amatriaín, M., Xiong, Y., Schmitt, M. R., Bilgic, H., Budde, A. D., Chao, S., Smith, K. P., & Muehlbauer, G. J. (2010). Transcriptome analysis of a barley breeding program examines gene expression diversity and reveals target genes for malting quality improvement. *BMC Genomics, 11*, 653.

Pritchard, J. K., Stephens, M., & Donnelly, P. (2000). Inference of population structure using multilocus genotype data. *Genetics, 155*, 945–959.

Roy, J. K., Smith, K. P., Muelbahuer, G. J., Chao, S., Close, T. J., & Steffenson, B. J. (2010). Association mapping of spot blotch resistance in wild barley. *Molecular Breeding, 26*, 243–256.

SAS. (2004). *Statistical analysis system online documentation*. Cary: SAS.

Stracke, S., Haseneyer, G., Veyrieras, J. B., Geiger, H. H., Sauer, S., Graner, A., & Piepho, H. P. (2009). Association mapping reveals gene action and interactions in the determination of flowering time in barley. *Theoretical and Applied Genetics, 118*, 259–273.

Szűcs, P., Blake, V. C., Bhat, P. R., Close, T. J., Cuesta-Marcos, A., Muehlbauer, G. J., Ramsay, L. V., Waugh, R., & Hayes, P. M. (2009). An integrated resource for barley linkage map and malting quality QTL alignment. *The Plant Genome Journal, 2*, 134–140.

Von Zitzewitz, J., Cuesta-Marcos, A., Condon, F., Castro, A. J., Chao, S., Coorey, A., Filichkin, T., Fisk, S. P., Gutierrez, L., Haggard, K., Karsai, I., Muelhbauer, G. J., Smith, K. P., Veisz, O., & Hayes, P. M. (2011). The genetics of winterhardiness in barley: Perspectives from genome-wide association mapping. *The Plant Genome Journal, 4*, 76–91.

Waugh, R., Jannink, J. L., Muehlbauer, G. J., & Ramsay, L. (2009). The emergence of whole genome association scans in barley. *Current Opinion in Plant Biology, 12*, 218–222.

Yu, J. M., & Buckler, E. S. (2006). Genetic association mapping and genome organization of maize. *Current Opinion in Biotechnology, 17*, 1–6.

Chapter 4
The "Italian" Barley Genetic Mutant Collection: Conservation, Development of New Mutants and Use

Antonio Michele Stanca, Giorgio Tumino, Donata Pagani, Fulvia Rizza, Renzo Alberici, Udda Lundqvist, Caterina Morcia, Alessandro Tondelli, and Valeria Terzi

Abstract During the last 30 years, a collection of morphological barley mutants has been developed in Fiorenzuola d'Arda, Italy, as part of the international effort directed to the conservation of genetic stocks, to ensure the global availability and the correct maintenance of these genetic materials. The collection is comprehensive of stem, leaf, ear, flower, awn and grain mutants. Near isogenic lines (NILs) useful for genetic analysis and to study the effect of mutation on agronomic performance have been developed. New different hooded mutants have been obtained in rare outcrosses of the flower in the hood. The development of double and triple mutants obtained by intercrossing simple mutants has been done and will continue. The collection is continuously implemented by including new mutants through a collaborative work with the other germplasm banks and represents a unique source of alleles for better understanding the role of genes involved in plant architecture.

Keywords Mutants • *Leafy lemma* • *Rococò* • *Seeded in hood*

A.M. Stanca (✉) • G. Tumino
Department of Agricultural and Food Science, University of Modena and Reggio Emilia, Reggio Emilia, Italy
e-mail: michele@stanca.it

D. Pagani • F. Rizza • R. Alberici • C. Morcia • A. Tondelli • V. Terzi
CRA – GPG, Genomics Research Centre, Fiorenzuola d'Arda, Italy

U. Lundqvist
NordGen Plants, P.O. Box 41, SE-230 53, Alnarp, Sweden

G. Zhang et al. (eds.), *Advance in Barley Sciences: Proceedings of 11th International Barley Genetics Symposium*, DOI 10.1007/978-94-007-4682-4_4,

4.1 Introduction

Barley (*Hordeum vulgare*) is one of the best investigated crop plants and a model species for the *Triticeae*, due to the high degree of natural variation and wide adaptability, the diploid genome structure and the strongly inbreeding-based mating system, as well as to the availability of a wide array of genomic resources (Saisho and Takeda 2011). Among them, large collections of natural and induced mutants have been developed since the 1920s, with the aim of understanding developmental and physiological processes and exploiting mutation breeding in crop improvement (Castiglioni et al. 1998; Lundqvist and Franckowiak 2003; Stanca et al. 2004).

In recent years, the integration of the most advanced genomic technologies with the historical mutation-genetics research helped in the isolation and validation of morphologically and developmentally important barley genes. These include (1) genes controlling spike and spikelet characteristics such as *hooded* (*K*; Müller et al. 1995; Roig et al. 2004), *six-rowed spike* (*vrs1*; Komatsuda et al. 2007), *intermedium-C* (*int-c*), *naked caryopsis* (*nud*; Taketa et al. 2008) and *cleistogamy 1* (*cly1*) (2) genes regulating barley vegetative growth such as the plant height genes *slender 1* (*sln1*; Chandler et al. 2002) and *uzu dwarf* (*uzu*; Chono et al. 2003). Other interesting mutants such as *low number of tillers-1* (*lnt1*; Dabbert et al. 2010), *Dense spike-ar* (*dsp.ar*; Shahinnia et al. 2012) and *brittle rachis* (*btr1*; Azhaguvel et al. 2006) have been fine mapped. An overview of the genes responsible of the domestication syndrome is provided by Sakuma et al. (2011). Moreover, as primary barley mutants were induced or discovered from various genetic backgrounds, an extensive backcrossing programme has started with the final aim of introgressing hundreds of mutated loci into the genetic background of a common recurrent parent, thus generating pairs of near isogenic lines (NILs) for each mutant locus (Franckowiak et al. 1985). The extensive genotyping of these lines allowed to map more than 420 mutant alleles on barley chromosomes and will significantly increase the success of cloning the responsible genes (Druka et al. 2011).

Further international efforts are today directed to the conservation of these genetic stocks to ensure their global availability and correct maintenance. In this view, here we describe a collection of morphological barley mutants that has been developed in Fiorenzuola d'Arda (Italy) in the last 30 years. The collection is comprehensive of stem, leaf, ear, flower, awn and grain mutants, and includes new hooded mutants obtained in rare outcrosses of the extra flower in the hood, double and triple mutants obtained by intercrossing simple mutants, as well as near isogenic lines (NILs) useful for deep genetic analyses and to study the effect of mutations on barley agronomic performance.

4.2 Materials and Methods

The "Italian" barley genetic mutant collection nursery is grown every year in pure stock by using two sowing data, winter and spring. The majority of the mutants, for each organ, are reported in Tables 4.1, 4.2, 4.3, 4.4, 4.5. Moreover, double mutants

Table 4.1 Spike mutants

Character	Symbol
Absent lower laterals 1	*abs1*
Accordion rachis 1	*acr1*
Branched ear	*brc*
Deficiens 1	*Vrs1.t*
Dense spike 10	*dsp10*
Eceriferum-b	*cer-b.4*
Elongated outer glume 1	*eog1*
Extra floret-b	*flo-b.3*
Hairy leaf sheath 1/yellow head 1	*Hsh1/yhd1*
Hexastichon-v.3 = Six-rowed spike 1	*hex-v.3 (vrs1)*
Intermedium-a = Six-rowed spike 3	*int-a.1 (vrs3)*
Lax spike	*Dsp10*
Laxatum-a = lax spike with 5 anthers	*lax-a.01*
Many glumes on the laterals	*mgl*
Multiflorus 2	*mul2*
Opposite spikelets 1	*ops1*
Third outer glume 1	*trd1*
Two-rowed spike 1	*Vrs1*

Table 4.2 Awn mutants

Character	Symbol
Awned palea 1	*adp1*
Awnless 1	*Lks1*
Calcaroides-a	*cal-a.3*
Elevated hood (hooded lemma)	*Kap1.e*
Hooded lemma 1	*Kap1*
Hooded/branched	*mk*
Lateral lemma appendix reduced	*vrs1.c*
Leafy lemma 1	*lel1*
Long awn 2	*Lks2*
Macrolepis = elongated outer glume 1	*lep-e.1 (eog1)*
Rococò spike	*rcc*
Rough awn/long rachilla hair 1	*(Raw) (Srh1)*
Short awn 2	*lks2*
Smooth awn 1	*raw1*
Triaristatum.20	*tri.20*
Triple awned lemma 1	*trp1*

Table 4.3 Seed mutants

Character	Symbol
Black caryopsis	- -
Blue aleurone xenia1	*Blx1*
Covered caryopsis	*Nud1*
Globosum-b	*glo-b.1*
High lysine	*lys*
Nigrinudum (black lemma and pericarp 1)	*Blp1*
Naked caryopsis 1	*nud1*
Waxy endosperm 1	*Wax1*

Table 4.4 Stem mutants

Character	Symbol
Brachytic 2	*brh2*
Densinodosum.3	*den.3*
Low numbers of tillers 1	*lnt1*
Single internode dwarf 1	*sid1*
Uniculm 2	*cul2 (V)*
Uzu (semi-brachytic) 1	*uzu1*
Viviparoides-1	*viv.1*

Table 4.5 Leaf mutants

Character	Symbol
Albino seedling 1	*abo1*
Chlorina seedling 2	*fch2*
Liguleless 1	*lig1*
Mottled leaf 2	*mtt2*
Virescent seedling 2	*yvs2*
White streak 2	*wst2*
Yellow (xantha) virescent seedling 1	*yvs1*

Table 4.6 Hooded double mutants

Characters	Symbol
Hooded lemma1/accordion rachis 1	*Kap1/acr1*
Hooded lemma 1/auricleless-a	*Kap1/aur-a.1*
Hooded lemma 1/awnless 1	*Kap1/Lks1*
Hooded lemma 1/brachytic 2	*Kap1/brh2*
Hooded lemma 1/deficiens 1	*Kap1/Vrs1.t*
Hooded lemma 1/many noded dwarf 6	*Kap1/mnd6.6*
Hooded lemma 1/gigas.4	*Kap1/gig.4*
Hooded lemma 1/hairy leaf sheath 1/yellow head 1	*Kap1/Hsh1-yhd1*
Hooded lemma 1/laxatum-a.01	*Kap1/lax-a.01*
Hooded lemma 1/liguleless 1	*Kap1/lig1*
Hooded lemma 1/low number of tillers 1	*Kap1/lnt1*
Hooded lemma 1/short awn 2	*Kap1/lks2*
Hooded lemma 1/single internode dwarf 1	*Kap1/sid1*
Hooded lemma 1/third outer glume 1	*Kap1/trd1*
Hooded lemma 1/triaristatum.20	*Kap1/tri.20*
Hooded lemma 1/uniculm 2	*Kap1/cul2(V)*
Hooded lemma 1/viviparoides.4	*Kap1/viv.4*
Hooded lemma 1/elongated outer glume 1	*Kap1/eog1*
Hooded lemma 1/yellow head 1	*Kap1/yhd1*

in a common genetic background have been developed by using *Hooded* (Table 4.6) and *third outer glume* (Table 4.7) (Fig. 4.1). In Table 4.8, other double and triple mutants are reported.

Table 4.7 Third outer glume double mutants

Characters	Symbol
Absent lower laterals 1/third outer glume 1	*als1/trd1*
Accordion rachis.3/third outer glume 1	*acr.3/trd1*
Auricleless-a/third outer glume 1	*aur-a.1/trd1*
Awnless 1/third outer glume 1	*Lks1/trd1*
Black leaves lax spike/third outer glume 1	*--/trd1*
Brachytic 1/third outer glume 1	*brh1/trd1*
Bracteatum-a/third outer glume 1	*bra-a.1/trd1*
Bracteatum-d/many noded dwarf 6	*bra-d.2/mnd6.6*
Branched/third outer glume 1	*brc/trd1*
Branched/third outer glume 1	*mk/trd1*
Calcaroides-b/third outer glume 1	*cal-b.13/trd1*
Deficiens 1/third outer glume 1	*Vrs1.t/trd1*
Densinodosum.9/third outer glume 1	*den.9/trd1*
Elevated hood 1 (hooded lemma)/third outer glume 1	*Kap1.e/trd1*
Elongated outer glume/third outer glume 1	*lep-e.2/trd1*
Elongated outer glume 1/third outer glume 1	*eog1/trd1*
Gigas (giant plants) 1/third outer glume 1	*gig1/trd1*
Lateral lemma appendix reduced/third outer glume 1	*vrs1.c/trd1*
Lax spike 10/third outer glume 1	*Dsp10/trd1*
Laxatum-a (lax spike with 5 anthers)/third outer glume 1	*lax-a.01/trd1*
Liguleless 1/third outer glume 1	*lig1/trd1*
Long basal rachis internode 1/third outer glume 1	*lbi1/trd1*
Low numbers of tillers 1/third outer glume 1	*lnt1/trd1*
Many glumes on the laterals/third outer glume 1	*mgl/trd1*
Naked caryopsis 1/third outer glume 1	*nud1/trd1*
Opposite spikelets 2/third outer glume 1	*ops 2/trd1*
Short awn 2/third outer glume 1	*lks2/trd1*
Short basal rachis internode/third outer glume 1	*Lbi/trd1*
Six-rowed spike 1 (hexastichon-v.3)/third outer glume 1	*hex-v.3/trd1*
Six-rowed spike 5/third outer glume1	*vrs5/trd1*
Subjacient hood 1/third outer glume 1	*sbk1/trd1*
Triple awned lemma 1/third outer glume 1	*trp1/trd1*
Two-rowed spike/third outer glume 1	*Vrs1.d/trd1*
Uniculm 2/third outer glume 1	*cul2 (V)/trd1*
Uniculm 4/third outer glume 1	*cul4.15/trd1*
Uzu (semi-brachytic) 1/third outer glume 1	*uzu1/trd1*
Viviparoides.1/third outer glume 1	*viv.1/trd1*

Due to the outcrossing frequency in the nursery, all the offtypes discovered have been kept, introduced into a temporary genetic collection and, after stability and uniformity evaluations, eventually transferred in the general collection as new mutants or variant of previous ones. Allelism tests are used to discover new alleles for the traits under control.

For specific mutants, the progenies are grown in glasshouse and the spikes at heading are routinely bagged.

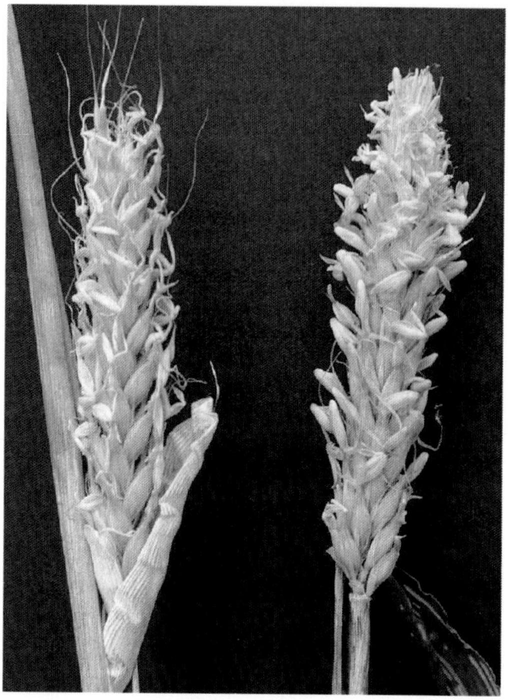

Fig. 4.1 Double mutant "hooded/third outer glume" (*left*). "Hooded" (*right*)

Table 4.8 Other double and triple mutants

Characters	Symbol
Brachytic 1/many noded dwarf 6	*brh1/mnd6.6*
Bracteatum-d/many noded dwarf 6	*bra-d.2/mnd6.6*
Branched/densinodosum.9	*branched/den.9*
Dense spike 10/many noded dwarf 6	*dsp10/mnd6.6*
Elongated outer glume 1/many glumes on the laterals	*eog1/mgl*
Elongated outer glume 1/triple awned lemma 1	*eog1/trp1*
Hairy leaf sheath 1 yellow head 1/third outer glume 1	*Hsh1/yhd1/trd1*
Lateral lemma appendix reduced/densinodosum.9	*vrs1.c/den.9*
Long awn 2/densinodosum.9	*Lks2/den.9*
Many glumes on the lateral/triple awned lemma 1	*mgl/trp1*
Opposited spikelets 1/many noded dwarf 6	*ops1.3/mnd6.6*
Triple awned lemma 1/elongated outer glume 1	*trp1/eog1*

The seeds are stored in a short-term chamber at 9 °C and 40 % of humidity.

For the mutants of potential agronomic interest, molecular analysis is used for genotyping. Phenotyping for photosynthetic efficiency, cold resistance (Fv/Fm), disease resistance, spring/winter growth habit tests and electrolytic leakage is routinely performed (Rizza et al. 2001).

4.3 Results

In a period of several decades, a collection of hundreds of morphological and physiological mutants has been implemented and maintained. A specific work is in progress, mainly focused on three mutants: *leafy lemma*, *rococò* and *seeded in hood*. The morphological characteristics and the genetic analysis of *leafy lemma*, *rococò* and *seeded in hood* are described.

4.3.1 Leafy Lemma

The *leafy lemma* phenotype was isolated in 1990 at the Istituto Sperimentale per la Cerealicoltura (Fiorenzuola d'Arda, Italy) in a plot in which the recessive mutant short awn (*lks2*) was grown. In the mutant, the lemma is transformed in a leaf-like structure. Particularly the transition zone of lemma is similar to the ligule-auricle region. Genetic analysis was carried out by crossing the mutant with several wild type genotypes, and the segregation ratio was according to 15:1 (Pozzi et al. 2000). A backcross programme has been developed by using the recurrent parent Kaskade with the aim to increase the seed size in the two-rowed barley. The physiological characteristics in terms of photosynthetic activity of *lel* by using LI-COR 6400, compared with WT, have been performed on 26 sister lines. Different classes on the basis of photosynthetic parameters and morphology of the mutated structure have been identified. DArT and SNPs markers have been developed and used to characterise the different classes. Moreover, the cross between *lel* and *hooded* has been done: the F1 is *elevated hooded*, probably for different expressivity of the trait. The progenies are now under study.

4.3.2 Rococò

Rococò was isolated in 2004 in a sodium azide mutagenesis programme in a *Hordeum spontaneum* accession. The mutant has brittle rachis, according to its origin and awns of different curly shape and in some case loop formation (Fig. 4.2). Several crosses have been performed with different normal varieties and the segregation ratio was 3:1. The *rcc* gene is therefore recessive. "*Rococò (rcc)* × curly spike (cud2)" cross has been performed to define the allelic status of the genotypes. The position of *rococò* on chromosomes is under study.

4.3.3 Seeded in Hood

By intercrossing single mutants a set of double mutants has been produced, most of them are in a *Hooded* common background. *Seeded in hood* is a spontaneous

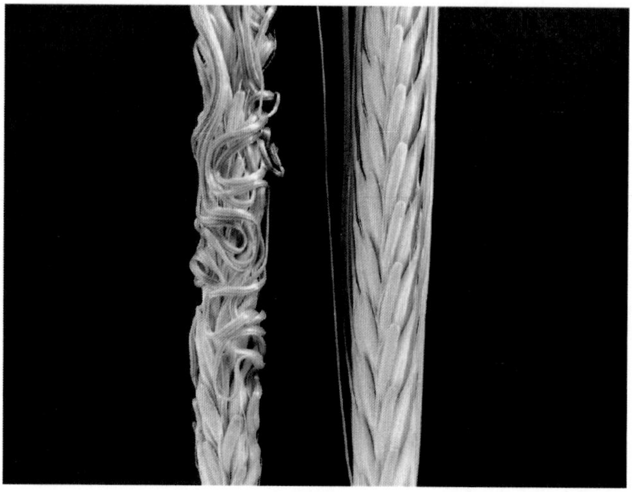

Fig. 4.2 *Rococò* spike (*left*) in comparison with wt

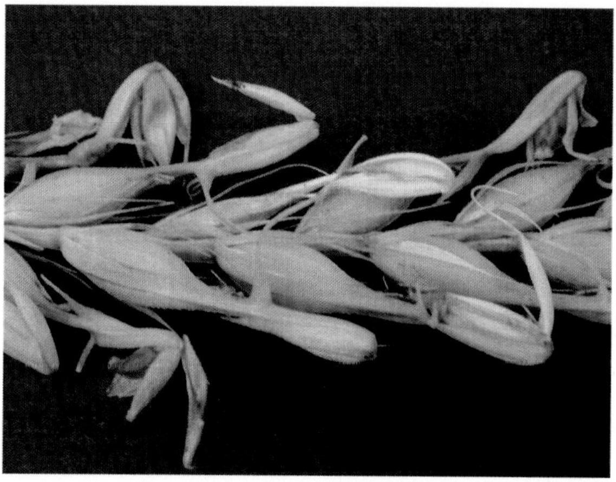

Fig. 4.3 Two small seeds in the hood *seeded in hood* in the double mutant *deficiens/hooded*

phenotype that occurred in a plot of the double mutant "deficiens(Vrs1.t) × Hooded (K)". In the *deficiens* background the size of the hood is bigger than in the other *Hooded* mutants and, probably due to this characteristic, some fertile florets can be developed in the hood. This fact is evidenciated by the presence of small seed in the hood (Fig. 4.3).

These small seeds have been grown in glasshouse and the F1 plants grown normally as *hooded* phenotype. The F2 seeds derived from the F1 plants, grown in glasshouse, produced different *hooded* spikes, including *elevated hooded, branched hooded* and the first stage of hood development (*trifurcatum*) (Fig. 4.4).

This complex segregation is under study through progenies derived from a cross between different form of hood and a wild type two-rowed variety Cometa.

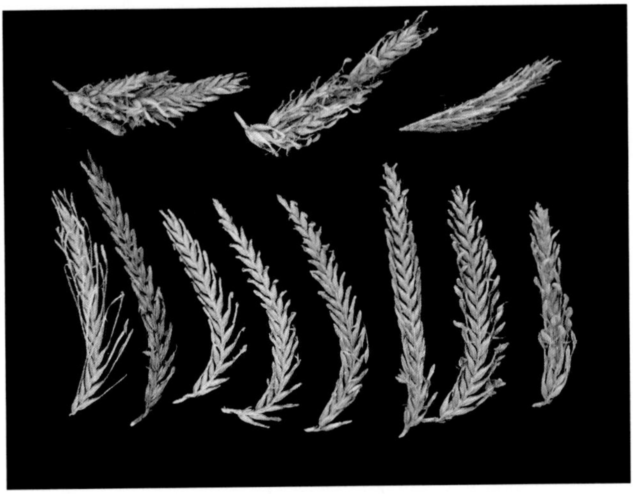

Fig. 4.4 F2 progenies derived from the segregation of the *seeded in hood* caryopsis.

4.4 Discussion

Behind the classical genetic studies of mutants, molecular analysis by using high-throughput platforms will be applied for positioning the genes on chromosomes. For the mutants obtained from *seeded in hood,* the complex segregation of seed in hood mutant will be more deeply studied genetically, and for this, F1 plants are available in a cross between *seeded in hood* and wild type Cometa. The pre-breeding work by using new *lel* genotypes will be developed with the final aim to increase the seed size and grain yield. Fine mapping of *lel* is in progress.

Acknowledgments This work was supported by "Agronanotech" and "FAO-RGV" MiPAAF projects.

References

Azhaguvel, P., Vidya-Saraswathi, D., & Komatsuda, T. (2006). High-resolution linkage mapping for the non-brittle rachis locus *btr1* in cultivated X wild barley (*Hordeum vulgare*). *Plant Science, 170*, 1087–1094.

Castiglioni, P., Pozzi, C., Heun, M., Terzi, V., Müller, K. J., Rohde, W., & Salamini, F. (1998). An AFLP-based procedure for the efficient mapping of mutants and DNA probes in barley. *Genetics, 149*, 2039–2056.

Chandler, P. M., Marion-Poll, A., Ellis, M., & Gubler, F. (2002). Mutants at the *Slender1* locus of barley cv Himalaya: molecular and physiological characterization. *Plant Physiology, 129*, 181–190.

Chono, M., Honda, I., Zeniya, H., Yoneyama, K., Saisho, D., Takeda, K., Takatsuto, S., Hoshino, T., & Watanabe, Y. (2003). A semidwarf phenotype of barley *uzu* results from a nucleotide substitution in the gene encoding a putative brassinosteroid receptor. *Plant Physiology, 133*, 1209–1219.

Dabbert, T., Okagaki, R. J., Cho, S., Heinen, S., Boddu, J., & Muehlbauer, G. J. (2010). The genetics of barley low-tillering mutants: *low number of tillers-1* (*lnt1*). *Theoretical and Applied Genetics, 121*, 70–715.

Druka, A., Franckowiak, J., Lundqvist, U., Bonar, N., Alexander, J., Houston, K., et al. (2011). Genetic dissection of barley morphology and development. *Plant Physiology, 155*, 617–627.

Franckowiak, J. D., Foster, A. E., Pederson, V. D., & Pyler, R. E. (1985). Registration of 'Bowman' barley. *Crop Science, 25*, 883.

Komatsuda, T., Pourkheirandish, M., He, C., Azhaguvel, P., Kanamori, H., Perovic, D., Stein, N., Graner, A., Wicker, T., Tagiri, A., et al. (2007). Six-rowed barley originated from a mutation in a homeodomain-leucine zipper I-class homeobox gene. *Proceedings of the National Academy of Sciences of the United States of America, 104*, 1424–1429.

Lundqvist, U., & Franckowiak, J. D. (2003). Diversity of barley mutants. In R. von Bothmer et al. (Eds.), *Diversity in barley (Hordeum vulgare)* (pp. 77–96). Amsterdam: Elsevier.

Müller, K. J., Romano, N., Gerstner, O., Garcia-Maroto, F., Pozzi, C., Salamini, F., & Rohde, W. (1995). The barley *Hooded* mutation caused by a duplication in a homeobox gene intron. *Nature, 374*, 727–730.

Pozzi, C., Faccioli, P., Terzi, V., Stanca, A. M., Cerioli, S., Castiglioni, Fink R., Capone, R., Müller, K. J., Bossinger, G., Rohde, W., & Salamini, F. (2000). Genetics of mutations affecting the development of a barley floral bract. *Genetics, 154*(3), 1335–1346.

Rizza, F., Pagani, D., Stanca, A. M., & Cattivelli, L. (2001). Use of chlorophyll fluorescence to evaluate the cold acclimation and freezing tolerance of winter and spring oats. *Plant Breeding, 120*, 389–396.

Roig, C., Pozzi, C., Santi, L., Müller, J., Wang, Y., Stile, M. R., Rossini, L., Stanca, A. M., & Salamini, F. (2004). Genetics of barley hooded suppression. *Genetics, 167*, 439–448.

Saisho, D., & Takeda, K. (2011). Barley: emergence as a new research material of crop science. *Plant & Cell Physiology, 52*, 724–727.

Sakuma, S., Salomon, B., & Komatsuda, T. (2011). The domestication syndrome genes responsible for the major changes in plant form in the triticeae crops. *Plant & Cell Physiology, 52*, 738–749.

Shahinnia, F., Druka, A., Franckowiak, J., Morgante, M., Waugh, R., & Stein, N. (2012). High resolution mapping of *Dense spike-ar* (*dsp.ar*) to the genetic centromere of barley chromosome 7H. *Theoretical and Applied Genetics, 124*, 373–384.

Stanca, A. M., Romagosa, I., Takeda, K., Lundborg, T., Sato, K., Terzi, V., & Cattivelli, L. (2004). Diversity in abiotic stresses. In R. von Bothmer, T. van Hintum, H. Knupffer, & K. Sato (Eds.), *Diversity in barley (Hordeum vulgare)* (pp. 179–199). Amsterdam/Boston: Elsevier.

Taketa, S., Amano, S., Tsujino, Y., Sato, T., Saisho, D., Kakeda, K., Nomura, M., Suzuki, T., Matsumoto, T., Sato, K., et al. (2008). Barley grain with adhering hulls is controlled by an ERF family transcription factor gene regulating a lipid biosynthesis pathway. *Proceedings of the National Academy of Sciences of the United States of America, 105*, 4062–4067.

Chapter 5
Alpha-Amylase Allelic Variation in Domesticated and Wild Barley

Suong Cu, Sophia Roumeliotis, and Jason Eglinton

Abstract Barley germplasm including breeding lines, wild barley accessions and mapping population parents were screened for variation in α-amylase using isoelectric focusing (IEF) in conjunction with activity staining. IEF screening found extensive variation in α-amylase IEF banding patterns with 14 and 49 different α-amylase isoenzymes identified for low- and high-pI groups, respectively. Alpha-amylase enzyme activity assayed at a range of temperatures (40–75 °C) revealed significant variation in the levels of enzyme activity and thermostability between varieties belonging to different IEF groups. The relationship between fermentability and α-amylase activity and thermostability was studied in 30 elite breeding lines and variation in thermostability between IEF groups was found to have a significant impact on fermentability.

Keywords Alpha-amylase • Allelic variation • Enzyme thermostability • Fermentability • Genetic mapping • *Hordeum vulgare* • *Hordeum vulgare* spp. *spontaneum*

5.1 Introduction

Barley α-amylase (α-(1,4)-D-glucanohydrolase; EC 3.2.1.1) is an endohydrolase that catalyses the cleavage of internal α-(1,4)-glucosyl linkages of amylose and amylopectin and is a key enzyme in the degradation of starch during brewing (Evans et al. 2005). This enzyme is produced during germination in response to hormone gibberellins (Filner and Varner 1967).

S. Cu (✉) • S. Roumeliotis • J. Eglinton
School of Agriculture Food and Wine, The University of Adelaide,
Waite Campus, Glen Osmond, Adelaide, South Australia 5064, Australia
e-mail: suong.cu@adelaide.edu.au

G. Zhang et al. (eds.), *Advance in Barley Sciences: Proceedings of 11th International Barley Genetics Symposium*, DOI 10.1007/978-94-007-4682-4_5,

Alpha-amylase family consists of two groups differentiated by their isoelectric points (pIs): the low-pI group AMY1 and the high-pI group AMY2 (Frydenberg and Nielsen 1965; Jacobsen and Higgins 1982). AMY1 and AMY2 show nearly 80 % sequence identity, but differ in enzymatic and stability properties with AMY2 having higher catalytic turnover but lower substrate affinity than AMY1 (Ajandouz et al. 1992; Bush et al. 1989; Jensen et al. 2003; Mundy et al. 1983; Vallée et al. 1998). In contrast to AMY1, AMY2 binds to an endogenous barley α-amylase/subtilisin inhibitor (BASI) (Vallée et al. 1998) and requires Ca^{2+} for maximum activity (Bush et al. 1989). AMY2 is also far more abundant accounting for up to 80–90 % of total α-amylase activity.

The gene families coding for AMY1 and AMY2 are *amy2* and *amy1,* located on chromosome 7H and 6H, respectively (Knox et al. 1987; Khursheed and Rogers 1988; Muthukrishnan et al. 1984). To date, the exact copy number for each gene family is still unknown.

Amongst all starch-hydrolysing enzymes, α-amylase is the most thermostable with less than 18 % of its activity lost during kilning. However, thermostability of this enzyme can vary quite considerably at 72.5 °C with the levels of activity retained after 10-min treatment ranging from 25 to 67 % (Evans et al. 2003). The relationship between variation in α–amylase thermostability and fermentability has not been established to date.

In this study, natural variation in α-amylase IEF isoforms was investigated in germplasm consisting of both domesticated and wild barley. The relationship between α-amylase IEF isoforms and enzyme thermostability to fermentability and other malt traits were examined.

5.2 Materials and Methods

5.2.1 Barley

Initial screening for the variation in α-amylase isoforms was conducted on germplasm consisting of 192 breeding lines from Intergrain Pty Ltd and the University of Adelaide, mapping population parents, and 68H. *spontaneum* lines originated from different sites in Israel. A subset of 30 elite breeding lines was then selected for the analysis of α-amylase thermostability and its relationship to other malt traits including fermentability. These lines were grown in experiment plots in 2012 at Brentwood, South Australia.

5.2.2 Petri Dish Germination

A hundred seeds of each barley line were germinated in petri dishes for IEF analysis and α-amylase enzyme assay. Germination was carried out in a time-course of 3, 4 and 5 days using the method of the European Brewing Convention (EBC) method 3.6.2 (EBC 1997) to examine any temporal variation in banding patterns or relative thermostability. A single germination period (4 days) was then selected for the mapping population. After germination, the green malts were freeze-dried and ground to flour for further analysis.

5.2.3 *Micromalting and Malt Quality Analysis*

Two hundred grams of each barley variety were micromalted in duplicate in an automated micromalting unit (Joe White systems, Joe White Maltings, Adelaide, Australia) employing a standard malting programme. Green malt samples of 30 elite breeding lines (10 g for each line) were removed just before kilning for the analysis of α-amylase activity and thermostability. All other malt quality parameters including diastatic power (DP), β-amylase, malt protein, grain protein, Kolbach index (KI), hot water extract (HWE) and fermentability (AAL) were measured using small-scale versions of the EBC Official Methods (Roumeliotis and Tansing 2004).

5.2.4 *Isoelectric Focusing (IEF)*

After germination, the kernels were extracted in 1 % glycine (0.2 ml buffer per seed) and heat-treated, as described by Frydenberg et al. (1969). IEF was performed following Evans et al. (1997) using a non-linear gradient of pH 4–7. Alpha-amylase isoenzyme patterns were detected by starch staining (Henson and Stone 1988), and pI values were estimated relative to IEF marker (Bio-rad, California, US).

5.2.5 *Alpha-Amylase Assay Using Ceralpha Assay Kit (Megazyme)*

Alpha-amylase activity was measured using the Ceralpha method (Megazyme). Green and kilned malts were extracted in triplicate and diluted in 0.1 M NaCl, 0.02 % $CaCl_2$ and 0.02 % sodium azide and heated at 70 °C for 15 min then cooled on ice. Enzyme activity was measured as described in the Ceralpha protocol. One unit of α-amylase activity is defined as the amount of enzyme in the presence of excess thermostable α-glucosidase, required to release 1 μmole of *p*-nitrophenol from BPNPG7 per min under the defined assay conditions, and is termed a Ceralpha unit.

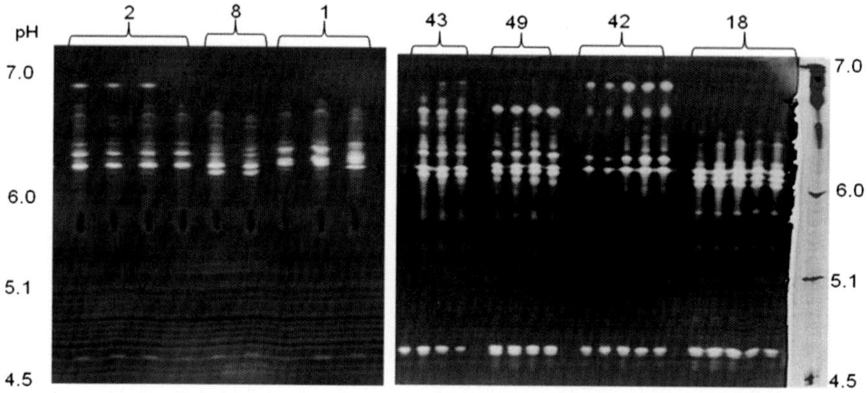

Fig. 5.1 Variation in α-amylase protein banding patterns in some IEF groups of cultivated barley (**a**) and wild barley (**b**)

5.2.6 Alpha-Amylase Assay at Elevated Temperature Using Starch Incubation and PAHBAH Reagent

Extracts were prepared and heat-treated as above. 0.1 ml extract was added to 4 ml of 2 % starch solution and incubated over a range of temperatures (40, 50, 55, 60, 65, 70 and 75 °C) for exactly 10 min. The reaction was stopped by the addition of 0.5 ml 0.5 M NaOH. The concentration of reducing sugar produced in the reaction was then determined using *p*-hydroxybenzoic acid hydrazide (PAHBAH) reagent as described in by Henry (1984). One unit of α-amylase activity is defined as the amount of enzyme required to release 1 μmole of maltose per min under defined assay conditions and is termed a PAHBAH unit.

5.3 Results

5.3.1 Examination of α-Amylase Polymorphism Using IEF

Based on the differences in IEF banding patterns, 49 high-pI and 14 low-pI α-amylase isoforms were identified within the selected germplasm (260 lines). This study focused only on the high-pI groups. Domesticated barleys (192 lines) are distributed amongst groups 1–17; however, the majority of them belong to group 1 (60 lines), 2 (28 lines), 3 (25 lines) and 8 (24 lines). In contrast to the skewed distribution of domesticated barley, 68 wild barley lines are evenly distributed over 34 IEF groups, sharing only 2 groups with domesticated barley (group 2 and 10). Thirty-two groups (18–49) were identified only in wild barley species. The difference in α-amylase banding patterns is generally quite subtle in cultivated barley varieties and far more apparent in the wild barley varieties (Fig. 5.1). With regard to temporal effects, germination time (3, 4 and 5 days) did not affect IEF banding patterns of α-amylase enzyme qualitatively, but it had a large effect on the intensity of some α-amylase protein bands (Table 5.1).

5.3.2 *Examination of α-Amylase Activity and Thermostability in Green and Kilned Malts*

5.3.2.1 Initial Screening from 49 IEF Groups

Representatives from the 49 IEF groups at 3-, 4- and 5-day germination were analysed for their α-amylase enzyme activity and at a range of temperatures from 25 to 75 °C. The means of α-amylase activity for each group (taking all three germination times into account) ranged from 174 to 773 U/g with the highest activities found in wild barley accession numbers CPI77140-36 and CPI77128-41 (group 47 and 35, respectively). The lowest enzyme activities were from OBW92 and Galleon (group 16 and 9, respectively). The frequency distribution of the means for enzyme activity at 25 °C followed a normal distribution (Fig. 5.2). With regard to temporal effects, α-amylase activity generally increased with increasing germination time with the means of enzyme activity for 49 representatives at 3-, 4- and 5-day germination being 299, 387 and 417 U/g, respectively. Statistical analysis also shows strong correlation between germination time and enzyme activity ($P<0.001$). However, there were exceptions to the general trend, for example, the enzyme activity of Hindmarsh (group 8) peaked at 4 days then decreased at 5 days of germination.

Alpha-amylase enzyme was assayed at elevated temperatures from 25 to 75 °C to examine the thermostability of the enzyme. In general, enzyme activity decreased as the temperature increased from 40 to 75 °C, and the difference in enzyme thermostability was significant between alpha-amylase groups. At 70 °C, the percentage of retained enzyme activity ranged from 10 % to over 100 % with the highest thermostability observed in CPI77145-44 (group 34) and the lowest in OBW92 (group 16). The temporal effect on thermostability of the enzyme, however, was found to be far more complicated compared to the initial activity as it differed significantly between varieties and between temperatures.

5.3.2.2 Variation Within Elite Breeding Lines

Thirty elite breeding lines and commercial varieties (group 1: 4 lines, group 2: 9 lines, group 4: 2 lines, group 5: 4 lines, group 8: 9 lines, group 16: 2 lines) were selected for examining the levels of α-amylase activity in green malts assayed at elevated temperatures 40–75 °C, and the levels of residual α-amylase activity retained after 15 min of heat treatment at 70 °C and being cooled down on ice. The differences in initial α-amylase activity levels in green malts assayed at 40 °C were not statistically significant between six groups at 5 % level (one-way ANOVA).

Table 5.1 Frequency distribution of α-amylase high-pI IEF groups

Group number	Number of domesticated barley lines	Number of wild barley accessions
1	60	0
2	28	1
3	25	0
4	5	0
5	3	0
6	15	0
7	11	0
8	24	0
9	3	0
10	5	1
11	1	0
12	2	0
13	1	0
14	1	0
15	3	0
16	2	0
17	3	0
18	0	2
19	0	4
20	0	1
21	0	1
22	0	1
23	0	2
24	0	2
25	0	1
26	0	2
27	0	4
28	0	4
29	0	1
30	0	2
31	0	1
32	0	2
33	0	3
34	0	3
35	0	2
36	0	2
37	0	1
38	0	6
39	0	1
40	0	2
41	0	2
42	0	2
43	0	2
44	0	1
45	0	1
46	0	2
47	0	1
48	0	3
49	0	2

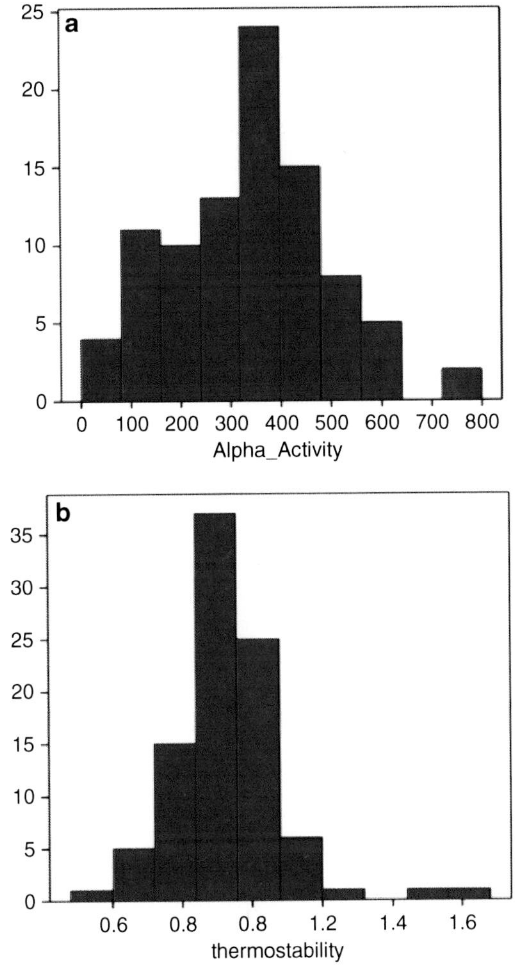

Fig. 5.2 Frequency distribution for (**a**) α-amylase activity in green malts from 49 IEF groups assayed at 40 °C and (**b**) thermostability determined by the percentage of enzyme activity assayed at 70 °C over that at 40 °C

However, significant differences between groups were detected for α-amylase activity in kilned malts ($P<0.001$) with the mean activity ranging from 40.4 U/g (group 4) to 64.8 (U/g) (group 16) (Fig. 5.3a) and residual α-amylase activity after heat treatment at 70 °C for 15 min ($P<0.05$) with the lowest residual activity observed in group 8 and the highest in group 4 (Fig. 5.3b). Significant differences were also found in the levels of enzyme activity assayed at 60 and 65 °C but not in other temperatures (data not shown).

5.3.2.3 Relationship Between α-Amylase Activity, Thermostability and Malt Quality Traits

Both diastatic power (DP) and fermentability (AAL) were significantly different between six examined groups ($P<0.05$ and $P<0.01$, respectively). Groups 1 and 2

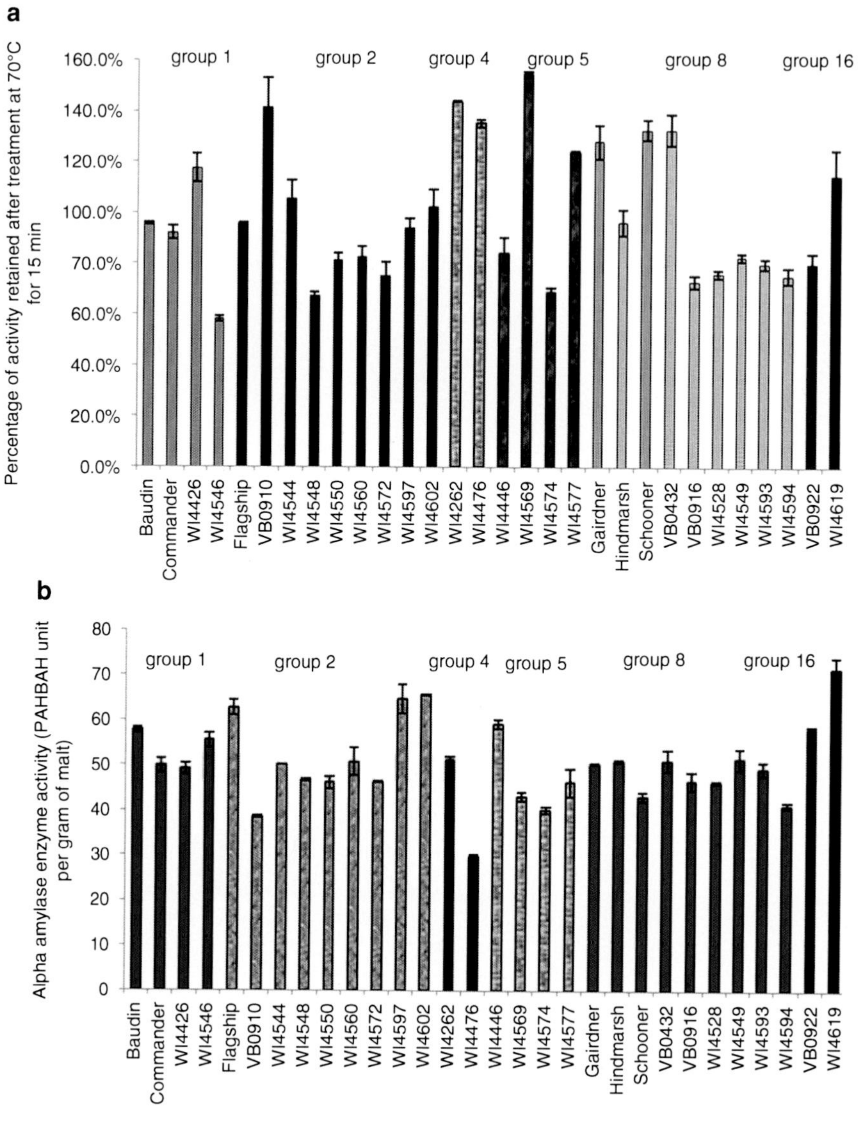

Fig. 5.3 (**a**) Alpha-amylase activity in kilned malts. Data bars represent the mean of two replicates. Error bars represent the standard error of the mean. (**b**) Residual α-amylase activity after incubation at 70 °C for 15 min, expressed as a percentage of initial activity. Data bars represent the means of three replicates. Error bars represent standard error of the mean

had the highest levels for DP and AAL, groups 4 and 5 had the lowest, and groups 8 and 16 had intermediate levels for both traits (Table 5.2).

With regard to the relationship between α-amylase activity, thermostability and other malt traits, our data shows that the initial level of α-amylase in green malts did

Table 5.2 Fermentability (AAL%) and DP across six examined groups. Analyses were performed in duplicate

	Group 1	Group 2	Group 4	Group 5	Group 8	Group 16
Replicate	10	18	4	8	16	4
Mean AAL%	84.1	83.5	81.0	81.4	82.8	82.0
S.E.	0.57	0.42	0.89	0.63	0.45	0.89
Mean DP	519.2	522.6	462.8	393.7	487.7	466.2
S.E.	28.00	20.87	44.27	31.30	22.13	44.27

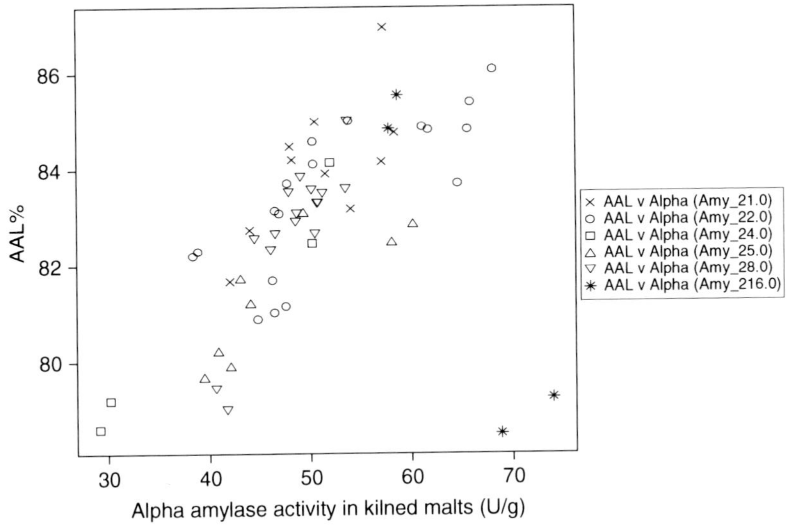

Fig. 5.4 Relationship between fermentability (AAL%) and α-amylase activity in kilned malts. × – group 1, o – group 2, □ – group 4, Δ – group 5, ∇ – group 8 and * – group 16

not significantly correlate with either DP or fermentability. However, α-amylase levels in kilned malts and in green malts assayed at elevated temperatures from 50 to 70 °C were all significantly correlated to DP and fermentability ($P<0.05$). The relationship between AAL and α-amylase activity in kilned malts is plotted in Fig. 5.4 ($r^2=0.57$). The main outliner was WI4619 (group 16) which yielded the lowest AAL % while having relatively high α-amylase enzyme activity.

5.4 Discussion

In the present study, extensive variation in the IEF isoforms of α-amylase has been identified. With 49 high-pI and 14 low-pI isoenzymes found within the selected germplasm, this variation is much larger than previously published data of 15 AMY2 and 5 AMY1 isoforms (Brown and Jacobsen 1982). This wide variation obtained may be due to a large number of wild barley accessions used in this study. A total of 68 wild barley lines from different geographical regions in Israel (coastal areas, mountains, Jordan valley and desert) were examined, and they exhibited far more diverse IEF banding patterns compared to their domesticated counterparts. The 68 lines were distributed over 34 IEF groups, of which 32 groups are novel to domesticated barley. These results reflect the richness in genetic variation of *H. spontaneum* in those regions which has been well documented in previous studies (Ahokas and Naskali 1990; Nevo et al. 1979) and confirm that *H. spontaneum* accessions are a good source of novel alleles for α-amylase.

Interestingly, the highest levels for both α-amylase activity and thermostability were detected in wild barley lines (CPI77140-36 and 77128-41, respectively), and these two lines are from similar mountainous regions in Israel. It is possible that the novel α-amylase alleles in these wild barley lines are the factors contributing to their high enzyme activity or thermostability. Similar findings have been reported for β-amylase with novel allele *Bmy1-Sd3* in wild barley conferring thermostable characteristics to the enzyme (Eglinton et al. 1998). However, further studies in mapping populations generated from these lines might be needed to elucidate this for α-amylase.

The variation in α-amylase IEF groups between domesticated barley, although less extensive than the wild counterparts, is still quite abundant with 17 distinguished isoforms, compared to only 4 isoforms identified for β-amylase in domesticated barley (Eglinton et al. 1998; Kihara et al. 1998). Within the subset samples of 30 elite breeding lines, variation in IEF isoforms shows significant impact on α-amylase activity in kilned malts, α-amylase thermostability at 60–70 °C, DP and fermentability (Fig. 5.3 and Table 5.2). Furthermore, α-amylase activity in kilned malts and α-amylase thermostability were positively correlated to DP and fermentability (Fig. 5.4). These results reconfirm the important role of α-amylase in brewing processes.

In contrast to α-amylase activity in kilned malts and α-amylase in green malts assayed at elevated temperature, initial α-amylase activity in green malts of six examined groups was not significantly influenced by variation in IEF groups. This result suggests that initial enzyme activity level and enzyme thermostability might be independently controlled. It has been well established that α-amylase activity can be confounded by many factors such as growth conditions, germination, KI, gibberellin level or the presence/absence of endogenous α-amylase inhibitors (Bertoft et al. 1984; Mundy et al. 1983; Mundy 1984). Enzyme thermostability, on the other hand, might not be influenced by those confounding factors to the same extent. The observation that the initial enzyme level did not significantly correlate with DP or fermentability implies the combination of activity and thermostability may be more important.

5.5 Conclusion

In summary, we have identified natural variation for α-amylase IEF isoforms and demonstrated the impact of this variation on enzyme thermostability and fermentability in a subset of samples. The extensive allelic variation for α-amylase together with their key role in brewing suggests that further studies in the biochemistry and genetics of this enzyme may provide opportunities to develop new malting varieties with specific quality profiles.

Acknowledgement The authors would like to acknowledge GRDC for funding this project (UA00108).

References

Ahokas, H., & Naskali, L. (1990). Geographic variation of a-amylase, b-amylase, b-glucanase, pullulanase and chitinase activity in germinating *Hordeum spontaneum* barley from Israel and Jordan. *Genetica, 82*, 73–78.

Ajandouz, E. L. H., Abe, J. I., Svensson, B., & Marchis-Mouren, G. (1992). Barley malt-a-amylase. Purification, action pattern, and subsite mapping of isozyme 1 and two members of the isozyme 2 subfamily using p-nitrophenylated maltooligosaccharide substrates, Biochimica et biophysica acta. *Protein Structure and Molecular Enzymology, 2*, 193–202.

Bertoft, E., Andtfolk, C., & Kulp, S. (1984). Effects of pH, temperature, and calcium ions on barley malt alpha-amylase isoenzymes. *The Journal of The Institute of Brewing & Distilling, 5*, 298–302.

Brown, A. H. D., & Jacobsen, J. V. (1982). Genetic basis and natural variation of alpha-amylase isozymes in barley. *Genetical Research, 3*, 315–324.

Bush, D., Sticher, L., van Huystee, R., Wagner, D., & Jones, R. (1989). The calcium requirement for stability and enzymatic activity of two isoforms of barley aleurone alpha-amylase. *The Journal of Biological Chemistry, 264*, 19392–19398.

EBC. (1997). *Analytica-EBC (European Brewery Convention)*. Nürnberg: Fachverlag Hans Karl.

Eglinton, J. K., Langridge, P., & Evans, D. E. (1998). Thermostability variation in alleles of barley beta-amylase. *Journal of Cereal Science, 28*, 301–309.

Evans, D. E., MacLeod, L. C., Eglinton, J. K., Gibson, C. E., Zhang, X., Wallace, W., Skerritt, J. H., & Lance, R. C. M. (1997). Measurement of beta-amylase in malting barley (*Hordeum vulgare* L.). I. Development of a quantitative ELISA for beta-amylase. *Journal of Cereal Science, 26*, 229–239.

Evans, E., Van Wegen, B., Ma, Y., & Eglinton, J. (2003). The impact of the thermostability of alpha-amylase, beta-amylase, and limit dextrinase on potential wort fermentability. *Journal of the American Society of Brewing Chemists, 61*, 210–218.

Evans, D. E., Collins, H. M., Eglinton, J. K., & Wilhelmson, A. (2005). Assessing the impact of the level of diastatic power enzymes and their thermostability on the hydrolysis of starch during wort production to predict malt fermentability. *Journal of the American Society of Brewing Chemists, 63*, 195–198.

Filner, F., & Varner, J. E. (1967). A test for the de novo synthesis of enzyme: density labeling with H_20^{18} of barley alpha-amylase induced by gibberellic acid. *Proceedings of the National Academy of Sciences of the United States of America, 58*, 1520–1526.

Frydenberg, O., & Nielsen, G. (1965). Amylase isozymes in germinating barley seeds. *Hereditas, 54*, 123–139.

Frydenberg, O., Nielsen, G., & Sandfaer, J. (1969). The inheritance and distribution of alpha-amylase types and DDT responses in barley. *Journal of Plant Breeding, 61*, 201–215.

Henry, R. (1984). A rapid method for the determination of diastatic power. *Journal of the Institute of Brewing, 90*, 37–39.

Henson, C., & Stone, J. (1988). Variation in alpha-amylase and alpha-amylase inhibitor activities in barley malts. *Journal of Cereal Science, 8*, 39–46.

Jacobsen, J. V., & Higgins, T. J. V. (1982). Characterization of the {alpha}-amylases synthesized by aleurone layers of Himalaya barley in response to gibberellic acid. *Plant Physiology, 70*, 1647–1653.

Jensen, M. T., Gottschalk, T. E., & Svensson, B. (2003). Differences in conformational stability of barley alpha-amylase isozymes 1 and 2. Role of charged groups and isozyme 2 specific salt-bridges. *Journal of Cereal Science, 38*, 289–300.

Khursheed, B., & Rogers, J. (1988). Barley alpha-amylase genes. Quantitative comparison of steady state mRNA levels from individual members of the two different families expressed in aleurone cells. *The Journal of Biological Chemistry, 263*, 18953–18960.

Kihara, M., Kaneko, T., & Ito, K. (1998). Genetic variation of beta-amylase thermostability among varieties of barley, *Hordeum vulgare* L., and relation to malting quality. *Plant Breeding, 117*, 425–428.

Knox, C., Sonthayanon, B., Chandra, G., & Muthukrishnan, S. (1987). Structure and organization of two divergent aamylase genes from barley. *Plant Molecular Biology, 9*, 3–17.

Mundy, J. (1984). Hormonal regulation of α-amylase inhibitor synthesis in germinating barley. *Carlsberg Research Communications, 49*, 439–444.

Mundy, J., Svendsen, I., & Hejgaard, J. (1983). Barley alpha-amylase/subtilisin inhibitor I. Isolation and characterization. *Carlsberg Research Communications, 48*, 81–90.

Muthukrishnan, S., Gill, B., Swegle, M., & Chandra, C. (1984). Structural genes for alpha-amylases are located on barley chromosomes 1 and 6. *The Journal of Biological Chemistry, 259*, 13637–13639.

Nevo, E., Zohary, D., Brown, A. H. D., & Haber, M. (1979). Genetic diversity and environmental associations of wild barley, *Hordeum spontaneum*, in Israel. *Evolution, 3*, 815–833.

Roumeliotis, S., & Tansing, P. (2004). *SA barley improvement program barley quality report 2001 and 2002 seasons*. Adelaide: Waite Barley Quality Evaluation Laboratory, The University of Adelaide.

Vallée, F., Kadziola, A., Bourne, Y., Juy, M., Rodenburg, K. W., Svensson, B., & Haser, R. (1998). Barley alpha-amylase bound to its endogenous protein inhibitor BASI: crystal structure of the complex at 1.9 å resolution. *Structure, 6*, 649–659.

Chapter 6
Novel Genes from Wild Barley *Hordeum spontaneum* for Barley Improvement

Xue Gong, Chengdao Li, Guoping Zhang, Guijun Yan, Reg Lance, and Dongfa Sun

Abstract Narrowing genetic basis is the bottleneck for modern plant improvement. Genetic variation in wild barley *Hordeum spontaneum* is much greater than that of either cultivated or landrace *H. vulgare* gene pool. It represents a valuable but under-utilised gene pool for barley improvement as no biological isolation barriers exist

X. Gong
College of Plant Science and Technology, Huazhong Agricultural University, Wuhan 430070, China

Department of agriculture and food, western Australia, 3 Baron-Hay Court, South Perth, Perth, WA 6155, Australia

School of Plant Biology, Faculty of Natural and Agricultural Sciences, The University of Western Australia, Crawley, Perth, WA 6009, Australia

C. Li
Department of agriculture and food, western Australia, 3 Baron-Hay Court, South Perth, Perth, WA 6155, Australia

Western Australian State Agricultural Biotechnology Centre, Murdoch University, Perth, Australia

G. Zhang
College of Agriculture and Biotechnology, Zhejiang University, Huajiachi Campus, Hangzhou 310029, China

G. Yan
School of Plant Biology, Faculty of Natural and Agricultural Sciences, The University of Western Australia, Crawley, Perth, WA 6009, Australia

R. Lance
Department of agriculture and food, western Australia, 3 Baron-Hay Court, South Perth, Perth, WA 6155, Australia

D. Sun (✉)
College of Plant Science and Technology, Huazhong Agricultural University, Wuhan 430070, China
e-mail: sundongfa@mail.hzau.edu.cn

G. Zhang et al. (eds.), *Advance in Barley Sciences: Proceedings of 11th International Barley Genetics Symposium*, DOI 10.1007/978-94-007-4682-4_6,

between *H. spontaneum* and cultivated barley. Novel sources of new genes were identified from *H. spontaneum* for yield, quality, disease resistance and abiotic tolerance. Quantitative trait loci (QTLs) were mapped to all barley chromosomes. A QTL on chromosome 4H from the wild barley consistently increased yield by 7.7% across six test environments. Wild barley *H. spontaneum* was demonstrated as key genetic resource for drought and salinity tolerance. Two QTLs on chromosomes 2H and 5H increased grain yield by 12–22% under drought conditions. Several QTL clusters were present on chromosomes 1H, 2H, 4H, 6H and 7H from *H. spontaneum* for drought and salinity tolerance. Numerous candidate genes were identified to associate with tolerance to drought or salinity, and some of the candidate genes co-located with the QTLs for drought tolerance. QTLs/genes for resistance to powdery mildew, leaf rust and scald were mapped to all chromosomes. Scald resistance was found in at least five chromosome locations (1HS, 3H, 6HS, 7HL and 7HS) from *H. spontaneum*, and simple molecular markers were developed to accelerate transferring of these genes into cultivated barley. Novel beta-amylase allele from *H. spontaneum* was used to improve barley malting quality. Advanced backcross QTL provides an efficiency approach to transfer novel genes from *H. spontaneum* to cultivated barley.

Keywords Abiotic stress tolerance • Disease resistance • Quality • Yield • AB-QTL • *Hordeum spontaneum*

6.1 Introduction

Barley is the fourth largest cereal crop after wheat, maize and rice in the world (Able et al. 2007). It is also the most widely adapted crop with a growing scale from the Arctic Circle and high mountainous regions to the desert fringes (Harlan 1976). To satisfy the burgeoning need for beverages like beer and whisky, breeders have made great efforts to improve productivity and quality of barley varieties. Progress of barley improvement relies on the ability to generate genetic variation and select for individuals with improved characteristics. Modern crop improvement efforts have heavily relied on the intensive use of favourable alleles present in cultivated germplasm collections, thereby contributing to the narrow genetic base of elite breeding germplasm (Matus and Hayes 2002). Wild barley *Hordeum spontaneum* is the direct ancestor of cultivated barley and contains valuable novel genes for barley improvement. Breeders have never stopped exploiting the value of *H. spontaneum*, which is well known as a rich source of disease resistance. Because there are no biological isolation barriers with the cultivated barley, hybrid between *H. spontaneum and H. vulgare* was possible in nature (Harlan 1976), and it makes *H. spontaneum* accessible for immediate use in barley breeding and allows for improved agronomic characteristics of cultivated varieties such as disease resistance, drought and salt tolerance.

One major limitation for utilising *H. spontaneum* for barley breeding is the genetic drags associated with the wild background. Development of molecular marker provided an efficient tool to tag and to transfer the novel genes from *H. spontaneum*. QTL mapping can not only determine and compare the roles of specific loci in a parent but also have clear knowledge about the relative size of their effects, the parent contributing to the allele for target traits and the interaction between target genes and the environment (Diab 2006; Qi et al. 1998; Young 1996). It can also be used as a tool for marker-assisted selection of complex disease-resistance characters and the positional cloning of partial resistance genes (Young 1996). Advanced backcross QTL (AB-QTL) is becoming a valuable approach to develop barley lines with introgressed novel genes from *H. spontaneum*. In this review, we mainly focus on novel genes identified from *H. spontaneum* for potential barley improvement.

6.2 Genes Affecting Barley Yield

There are many factors affecting barley productivity. Human being activity plays an important role in modifying species' genetic structure, which in turn alters the phenotypic characters. Natural selection may decide evolution direction, while human being selection helps evolution towards very direction to benefit production. Yield is controlled by many different genes, but the combination of available target genes within a specific breeding programme is limited. This may be largely caused by the long history of domestication process, during which the genetic diversity becomes narrower and narrower (Able et al. 2007). Barley breeders have explored germplasm collections and wild species as sources of favourable alleles for continued yield improvement. Studies demonstrated that the wild barley species *H. spontaneum* is a rich source of alleles for resistance to biotic and abiotic stresses and thus has potential to improve barley yield. These genes are seldom found in the cultivated barley germplasm (Ellis et al. 2000; Fetch et al. 2003; Nevo 1992; Williams 2003).

AB-QTL analysis has the potential to improve yield-related and quality-related agronomic traits. It is also a confirmation tool of classic QTL analysis. Pillen et al. (2003) use a BC_2F_2 population to find some agronomic-related QTLs on barley chromosomes. Day until heading, QTLs were mainly on chromosomes 1H, 2H, 4H, 5H and 7H; QTLs for plant height were located on four chromosomes 1H, 4H, 5H and 7H; QTLs for lodging tolerance at flowering were located on chromosome 5H; lodging-tolerance QTL at harvest was located on chromosome 1H; one QTL for kernels per ear was on chromosome 1H, and one QTL for above-ground biomass was on chromosome 7H; two QTLs were located for tenderness on chromosomes 5H and 7H; three QTLs were located for protein content on chromosomes 4H and 5H; five QTLs were located on chromosomes 4H, 5H and 7H for harvest index. A total of 12 QTLs were located for thousand grain weight on four chromosomes 2H, 4H, 5H and 7H; six QTLs were located for water absorption on chromosomes 4H, 5H and 7H; QTLs for yield were located on all barley chromosomes except for chromosome 6H. The maximum yield increase was due to exotic *H. spontaneum* donor. The increase ranged from 2.7 to 10.9%, with an average of 7.7% under six environments.

6.3 Disease Affecting Productivity

Disease can affect the profitability of barley production. It either reduces final yield or lowers grain quality. Both of them result in a relative lower financial return to growers. In the early 1980s, disease was seen as the key limitation on yield and was the major focus of breeding programmes. At least 30 diseases and pests are reported to affect barley (Matre 1982). The major ones included leaf scald, leaf rust and spot form and net form of net blotches (Wallwork 2000a, b).

Considerable effort has been made to identify and localise major disease-resistance genes and QTLs from cultivated and wild barleys. Major QTLs for resistance to each disease in wild barley were identified, and most resistance QTLs were from *H. spontaneum* (Yun et al. 2005). Currently, leaf scald- and powdery mildew-resistance QTLs were mapped on all chromosomes (Garvin et al. 1997, 2000; Genger et al. 2003a; Grønnerød et al. 2002; Backes et al. 2003; Falak et al. 1999; Heun 1992; Saghai Maroof et al. 1994). Five QTLs for spot blotch resistance were localised on chromosomes 1H, 2H, 3H, 5H and 7H (Bilgic et al. 2005; Steffenson et al. 1996). Four QTLs for SSLB resistance were mapped on chromosomes 2H, 4H, 1H and 6H (Toubia-Rahme et al. 2003; Yun et al. 2005). Loci conferring resistance to NTNB were mapped to every barley chromosome (Richter et al. 1998; Steffenson et al. 1996; Yun et al. 2005).

Exotic genes play a crucial role in genetic improvement. von Korff et al. (2005) used advanced backcross doubled haploid (BC_2DH) populations to detect nine QTLs for powdery mildew, six QTLs for leaf rust resistance and three QTLs for scald resistance. The presence of the exotic QTL alleles reduced disease symptoms by a maximum of 51.5, 37.6 and 16.5% for powdery mildew, leaf rust and scald, respectively.

6.3.1 Resistance to Powdery Mildew

Powdery mildew (*Blumeria graminis* DC. f. sp. *hor dei* Em. Marchal) is one of the most common diseases on barley. Yield reduction caused by powdery mildew can reach 30%, with an average of 5–10% across all the regions. The disease can also deteriorate quality characteristics (Czembor 2000). At least two genetically separable pathways control resistance to powdery mildew in barley (Jørgensen 1994; Peterhänsel et al. 1997). In the first pathway, resistance is mediated by recessive alleles at the *mlo* locus. Twenty-five alleles were identified so far for the *mlo* resistance but none of them from *H. spontaneum*. The second resistance pathway can be induced by a number of race-specific resistance genes (R genes) like *Mla*, *Mlg* and *Mlk* (Jørgensen 1994) and is almost invariably associated with the activation of rapid host cell death at attempted infection sites (Freialdenhoven et al. 1994). Mapping studies have localised these genes on chromosomes 1H and 4H (Jørgensen 1993).

Mla locus for mildew resistance on barley chromosome 1HS is known to be complex. At present, 31 genes/alleles are found in the *Mla* locus (Kintzios et al. 1995).

Multi-copy RFLP, SCAR, gene-specific SNP and SSR markers are closely linked to this locus (Schiiller et al. 1992). There are growing numbers of new alleles for powdery mildew-resistance genes for the *Mla* locus, mainly originating from *H. spontaneum* lines from Israel (Jahoor and Fischbeck 1993; Kintzios et al. 1995). Two cloned genes of the *Mla* locus have recently been shown to belong to the type of resistance genes containing a coiled-coil domain, a nucleotide binding site and a leucine-rich repeat domain (Halterman et al. 2001; Zhou et al. 2001).

Some of the powdery mildew-resistant genes originating from the *H. spontaneum* lines, neither allelic nor linked to the *Mla* locus, showed independent segregations. The resistance genes derive from *H. spontaneum* showed different resistant levels to powdery mildew: *mlt* with three alleles showed a recessive mode of inheritance on chromosome 7HS. *Mlf* and *Mlj* localised on chromosomes 7HL and 5HL were semi-dominant resistance genes (Schönfeld et al. 1996).

Powdery mildew resistance QTLs were mapped to all chromosomes. Three QTLs (*Rbgq1*, *Rbgq2* and *Rbgq3*) in the wild barley for resistance against powdery mildew were detected on chromosomes 2H, 3H and 5H, respectively (Shtaya et al. 2006).

6.3.2 Resistance to Leaf Scald Pathogen

Barley leaf scald was caused by the fungal pathogen *Rhynchosporium secalis*. The pathogen is highly variable among barley populations and can easily overcome newly developed resistance genes (Zhang et al. 1987; McDonald et al. 1999). Therefore, it is very difficult to find a host that can permanently resist this pathogen. The potential in finding some resistance from its wild progeny is huge and proved to be true.

Most genes for resistance to barley leaf scald were mapped either to the *Rrs1* locus on the long arm of chromosome 3H or the *Rrs2* locus on the short arm of chromosome 7H (Dyck and Schaller 1961; Schweizer et al. 1995). The major resistant QTLs to leaf scald pathogen in cultivated barley were mainly mapped to chromosome 3H. In contrast, the resistance in wild barley populations was identified at five chromosome locations (1HS, 3H, 6HS, 7HL and 7HS). SCAR markers were developed for selection of the scald-resistance genes located in the centromeric region of barley chromosome 3H (Genger et al. 2003a).

Currently, there are 16 scald resistance genes in barley: 11 scald resistance genes were reported in cultivated barley (Søgaard and von Wettstein-Knowles 1987; Goodwin et al. 1990; Bjørnstad et al. 2002), and five were found in wild barley. The five scald resistance genes from wild barley were mapped to four chromosome arms. *Rrs12* was mapped to the short arm of chromosome 7H and may be allelic to *Rh2* in cultivated barley (Abbott et al. 1992; Schweizer et al. 1995; Genger et al. 2003b). *Rrs13* was mapped to the short arm of chromosome 6H (Abbott et al. 1995). *Rrs14* was mapped to the short arm of chromosome 1H (Garvin et al. 1997, 2000), and *Rrs15* was allocated on chromosome 7HL (Genger et al. 2003b). The locus conferring scald resistance on 4HS was introgressed from *H. bulbosum*, designated as *Rrs16*^{Hb} (Pickering et al. 2006). Another gene mapped to chromosome 3HL may be allelic to *Rrs1* in cultivated barley (Genger et al. 2003b).

Scald resistance gene *Rrs14* derived from a wild population of *H. spontaneum* near Mehran of Iran showed a tight linkage in molecular marker *Hor2* (the seed storage protein hordein loci); thus, *Hor2* locus is an ideal codominant molecular marker for *Rrs14*. The tight linkage permits simple indirect selection of *Rrs14* in barley scald-resistance breeding programmes (Garvin et al. 2000). It also stimulates further study of hordatine in barley and the reaction mechanism between hordatine and the barley leaf fungal when it happens to powdery mildew infection (Smith and Best 1978).

The concentration of hordatine directly affects the efficiency of inhabiting the growth of fungal pathogens. The following experiment may explain why *H. spontaneum* species are more tolerant than cultivated barley. Fifty accessions of *H. spontaneum* from different habitats in Israel were used to determine the variation in the accumulation of hordatines. Noticeable different hordatine concentrations were found: Maximum hordatine concentrations in wild barley accessions were three to four times higher than in cultivated barley, and the decrease of the hordatine concentration between days 3 and 10 was much more pronounced and faster in cultivated barley than in *H. spontaneum* accessions (Batchu et al. 2006).

Up to present, four QTLs (*Rrsq1, Rrsq2, Rrsq3* and *Rrsq4*) for resistance against scald were detected on chromosomes 2H, 3H, 4H and 6H; one QTL was identified from a BC_2DH S42 (a cross between spring cultivated barley 'Scarlett' and a *spontaneum* accession 'ISR42-8') (von Korff et al. 2005). Both the resistant and the susceptible parents contributed alleles for resistance against scald (Shtaya et al. 2006).

6.3.3 Resistance to Leaf Rust

Leaf rust caused by *Puccinia hordei G.* Otth is another important disease of barley in many regions of the world. It caused yield losses up to 32% in susceptible cultivars (Griffey et al. 1994). There are two pathways of resistance against *P. hordei* in barley: partial resistance and race-specific resistance (Clifford 1985). Controlled by several or many genes, partial resistance is generally more durable than the race-specific resistance (Qi et al. 2000; Kicherer et al. 2000).

To date, there are 16 identified leaf rust-resistant genes on all chromosomes except 4H. Resistance in cultivated barley was shown to be very restricted and mainly limited to the genes *Rph3* and *Rph7* (Walther and Lehmann 1980; Jin et al. 1995). Four resistance genes (*Rph10, Rph11, Rph15* and *Rph16*) were found in *H. spontaneum*, and a large variability was found to exist in wild progenitor *H. spontaneum* from Israel. This confirmed the gene-to-gene hypothesis and furthermore demonstrated that the Near East represents a major centre for the evolution of resistance to *P. hordei Otth* (Manisterski et al. 1986; Moseman et al. 1990).

Rph1 and *Rph16* were mapped to chromosome 2H (Tuleen and McDaniel 1971; Tan 1978; Ivandic et al. 1998). *Rph2, Rph9, Rph12* and *Rph13* were located on chromosome 5H with a linkage of *Rph9* and *Rph13* (Franckowiak et al. 1997;

Borovkova et al. 1997, 1998; Tuleen and McDaniel 1971; Tan 1978). *Rph3* was detected on chromosome 7HL and found to have three alleles (Jin et al. 1993). *Rph4* was identified on chromosome 1HS and linked to the *Mla* mildew-resistance locus (McDaniel and Hathcock 1969; Tuleen and McDaniel 1971; Tan 1978); Chromosome 3H is the concentrate region of *Rph* genes including *Rph5*, *Rph6*, *Rph7* and *Rph10*. *Rph5* showed linkage to *Rph7* on chromosome 3H (Tuleen and McDaniel 1971; Tan 1978; Steffenson et al. 1993; Brunner et al. 2000; Graner et al. 2000; Feuerstein et al. 1990; Mammadov et al. 2003). *Rph11* was designated to 6H (Feuerstein et al. 1990). *Rph13*, *Rph14* and *Rph15* were effective at the adult plant stage.

QTLs for partial resistance were identified in the population L94×Vada. Three QTLs, *Rphq1* (designated on 1H), *Rphq2* and *Rphq3* (locate on chromosomes 2H and 6H, respectively), were effective at the seedling stage and contributed approximately 56% to the phenotypic variance. Four QTLs, *Rph2*, *Rphq3*, *Rphq4* (7H) and *Rphq5* (4H), contributed approximately 63% of the phenotypic variance and were effective at the adult plant stage. *Rphq6* (2H) which located the same QTL region as *Dh2* and *Ph2* was found to affect the latent period only at the adult plant stage. One significant interaction between *Rphq1* and *Rphq2* was detected in this population.

6.4 Abiotic Stress Affects Productivity

From the breeding perspective, stress tolerance can be described as the ability to maintain stable yield under stress environments. The wide ecological range of wild barley differs in water availability, temperature, soil type, altitude and vegetation generating a high potential of adaptive diversity to abiotic stresses (Forster et al. 1997). A significant proportion of diversity at the molecular level of stress responsive genes appears to be adaptive rather than neutral (Père de la Vega 1996; Favatier et al. 1997; Nevo et al. 1998; Liviero et al. 2002). Abiotic stress tolerance is a quantitative trait and controlled by genetic factors, and a lot of genes for abiotic tolerance have been lost during domestication (Cattivelli et al. 2002).

One of the most successful strategies to improve stress resistance is based on increased osmolyte (mainly glycine betaine or proline) concentration through transformation with genes controlling osmolyte biosynthesis (reviewed by Nuccio et al. 1999), while only few results have been obtained by over-expressing stress-related genes. However, the most effective way to improve productivity of barley grown in drought conditions is to use locally adapted germplasm and select in the target environment(s) (Ceccarelli et al. 1998).

6.4.1 Drought Resistance

Drought is a major abiotic stress that limits plant growth and crop productivity. It is a complex trait, which cannot be analysed genetically in the same way as monogenic resistance. When the crop is under stress, it can respond alternatively to environmental changes. Wild barley *H. spontaneum* was demonstrated as key genetic source for drought and salinity tolerance.

Phenotypic responses to water stress within a population of wild barley from Tabigha, Israel, were examined. Ten out of fifteen agronomic, morphological, developmental and fertility-related traits were significantly affected by the drought treatment. Soil types have different impact on formation of gene structure. Plants growing on terra rossa (TR) experienced more intense drought than plants growing on basalt (B). TR genotypes which showed accelerated development under water deficit conditions were significantly less affected by the imposed water stress than B genotypes, and moreover, TR genotypes exhibited yield stability which holds potential drought-resistant breeding programmes. This may be due to the water-holding capacity of the two soil types: Basalt soil type showed a greater water-holding capacity than terra rossa, while the resistant genotypes from TR soil were similar to the origin of the resistant genotype from the Tabigha microsite in Israel (Ivandic et al. 2000).

In *Triticeae*, the highest concentration of QTLs and major loci controlling plant's adaptation to the environment (heading date, frost and salt tolerance) were found on chromosome 5H. In addition, a conserved region with a major role in drought tolerance was localised to chromosome 7H (Cattivelli et al. 2002).

QTLs for agronomic traits related to drought resistance were detected in the cross 'Arta'×*Hordeum spontaneum* 41–1. QTLs for the most important character 'plant height under drought stress' were localised on chromosomes 2H, 3H and 7H. The 'plant height' QTLs, especially the one on chromosome 3H, showed pleiotropic effects on traits such as days to heading, grain yield and biological yield (Baum et al. 2003). A QTL on the long arm of chromosome 7H was identified as controlling winter hardness traits.

Carbon isotope discrimination as a selection criterion for drought-tolerance improvement has been used in cereals. It has been long time argued that molecular markers linked to genetic factors controlling CID could enhance selection in breeding programmes. CID has been associated with drought tolerance in terms of water-use efficiency and yield stability in drought-prone environments (Teulat-Merah et al. 2000). Variation for CID has been found within an *H. spontaneum* population in Tabigha, and the Tabigha varieties exhibit both rich phenotypic and genotypic diversities. It may provide valuable resources for important agronomic traits. Ten QTLs for CID have been found in maturing grain of Tadmor×Er/Apm in Mediterranean region, including the first QTL for grain yield measured in the Mediterranean field conditions related to drought stress (Teulat et al. 2002).

Water deficit has an immediate effect on plant growth and yield (Araus et al. 2002). Dehydrins (*Dhns*), peripheral membrane proteins which functionally protect the cell from water deficit or temperature change, are among the most frequently observed proteins in plants under water stress (Suprunova et al. 2004). A total of 13 *Dhn* genes were found on four barley chromosomes (Choi et al. 1999; Choi and Close 2000; Rodríguez et al. 2005). *Dhn1*, *Dhn2* and *Dhn9* (previously reported for *Dhn4a*) were mapped to the long arm of chromosome 5H; *Dhn3*, *Dhn4*, *Dhn5*,*Dhn7*,*Dhn8* and *Dhn12* were allocated to the long arm of chromosome 6H;

Dhn6 and *Dhn13* were allocated to chromosome 4SH; *Dhn10* and *Dhn11* were identified on chromosome 3HL. *Dhn1* and *Dhn2* are completely linked, together with *Dhn9*, and are located in the same QTL region of salt tolerance, freezing tolerance and ABA accumulation.

The drought-tolerant or drought-resistant mechanism of these genes is different, and the expression of some stress-related genes was shown to be linked to stress-tolerant QTLs (Cattivelli et al. 2002). This may due to differential expression patterns and furthermore indicates that each member in this family has a specific function in the process of plant response to drought. *Dhn* genes (*Dhn* 1, 3, 5, 6 and 9) were also found in wild barley (*H. spontaneum*), and these genes were not expressed in well-watered plants. *Dhn1* showed earlier expression (after 3 h dehydration) and higher levels (after 12 and 24 h) in resistant plants when compared with sensitive plants. *Dhn3* gene was induced by drought stress, ABA and salt tolerance (Choi and Close 2000), and expression of *Dhn3* and *Dhn9* was detected in sensitive and resistant genotypes during 3–12 h. However, after 12 and 24 h, they showed no clear differences. *Dhn5* was expressed after 3 h of dehydration stress in all genotypes and increased after 12 h of dehydration, and a slightly higher level was found in the resistant genotypes than in the sensitive types; low level of *Dhn5* was detected in TR and BA genotypes. It is interesting that *Dhn6* was expressed higher in resistant genotypes after 3 h dehydration; after 12 and 24 h of dehydration, the sensitive plants showed higher expression level as compared to the resistant plants (Suprunova et al. 2004).

The wild relatives of barley gene pool may provide an interesting source of new loci for drought tolerance. In wild barley, the role of *Dhn1* in drought tolerance is supported by several reports on co-localization of such QTLs with *Dhn* genes, e.g. QTLs for RWC (relative water content) (Teulat et al. 2003) and winter hardiness (Pan et al. 1994; van Zee et al. 1995) overlapping with a cluster of *Dhn* genes on chromosome 5H. Allelic variations in a stress-related gene can significantly alter plant stress-tolerance ability (Cattivelli et al. 2002). Allelic variations, potentially related to environmental adaptation, were found for the barley cold-regulated COR14b protein in the collection of *H. spontaneum* accessions (Crosatti et al. 1996). Wide allelic variation was found at the *Dhn4* locus in *H. spontaneum* germplasm from Israel (Close et al. 2000). High polymorphism with no geographic structure was found in *Dhn5* in a collection of wild barley from the Mediterranean across the Zagros Mountains and into Southwest Asia, and moderate polymorphism associated with geographic structure was found in *Dhn9* locus (Morrell et al. 2003).

6.4.2 Salt Tolerance

Besides drought tolerance, the gene pool of wild relatives may represent a valuable source of new loci for salt tolerance. In wild barley *H. spontaneum*, salt tolerance is associated with drought environments (Pakniyat et al. 1997).

A study of tolerance to salt at germination and seedling stages in wild *Hordeum* species was conducted (Mano and Takeda 1998). The results showed that seed germination tolerance in wild *Hordeum* species was generally lower than those found in cultivated barley (Mano et al. 1996). Among *Hordeum* species, *H. agriocrithon, H. spontaneum* and other wild *Hordeum* species showed high, middle and low levels of NaCl tolerance, respectively. The higher resistance level of *H. agriocrithon* probably originated from natural hybridization between *H. spontaneum* and cultivated barley (Briggs 1978). During the seedling stage, the NaCl-tolerant level in wild *Hordeum* species was generally higher than those found in cultivated barley (Mano and Takeda 1995). Thus, wild *Hordeum* species as a donor will benefit salt-resistant breeding at the seedling stage.

A number of QTLs affecting salt tolerance were detected on all chromosomes in a cross between *H. spontaneum* and *H. vulgare*, although several QTL clusters were present on chromosomes 1H, 4H, 6H and 7H (Ellis et al. 1997). In cultivated barley, QTLs controlling salt tolerance were mapped to chromosomes 1H, 4H, 5H and 6H at germination and to chromosomes 1H, 2H, 5H and 6H at the seedling stage (Mano and Takeda 1997). In a cross of Derkado × B83-12/21/5, the largest individual effects to salt tolerance were associated with the chromosomal regions around the two dwarfing genes *sdw1* (3H) and *ari-e.*GP (5H). The *sdw1* gene resulted in an overall yield increase but was only detected as a secondary QTL (Ellis et al. 2002). Loci involved in salt tolerance have been identified on 1Hch, 4Hch and 5Hch of *H. chilense* (Forster et al. 1990).

6.5 Malting Quality

Fermentability is a crucial quality parameter for brewing, particularly in systems using starch-based adjuncts. Barley ß-amylase (1,4-α-glucan maltohydrolase; EC 3.2.1.2) catalyses the liberation of ß-maltose from the non-reducing ends of 1,4-α-glucans and is a key enzyme in the degradation of starch during brewing. During the brewing process, the yield of degradation hydrolysate fermentable sugars directly affects the level of alcohol produced. The temperatures in excess of 63 °C are employed to gelatinise starch granules, which facilitate rapid and complete starch degradation. Native barley ß-amylase retains maximum activity up to 55 °C, but its stability decreases rapidly as temperature increases above 55 °C (Thacker et al. 1992). While in the breeding process, after a germination period of about 120 h, the β-amylase activity has reached its maximum level, and the bound activity has practically been released (Evans et al. 1997; Grime and Briggs 1996).

Bmy1 encode ß-amylase hydrolytic enzyme in the grain of malting (Zhang et al. 2007). The *Bmy1* allele has been located on the long arm of chromosome 4H and found to be tightly linked to a QTL for height (Hackett et al. 1992) and also a major spring habit gene, *sh* (*syn*, *Vrn1*, Chojecki et al. 1989; Hackett et al. 1992; Laurie et al. 1995).

Currently, 23 *Bmy1* allele types were identified, of which 16 were detected in cultivated barley, with seven major allele types and nine rare and unique allele types. The other seven allele types were devoted by *H. spontaneum* accessions, which turned to be unique. Noticeably, almost all the major allele types in cultivated barley were observed in wild barley except the B-I-8 allele, which was predominant in North European cultivated barley (Zhang et al. 2007). Nearly the same result was found in Chalmers' experiment: Wild barley from Israel possesses seven isoforms of ß-amylase enzyme, of which only two have been found in European cultivars. While one isoform was restricted to desert locations, 78% of the variation in the frequency of this isoform could be explained by site-of-origin rainfall and temperature (Chalmers et al. 1992).

Three discrete ß-amylase alleles (*Bmy*1-Sd2L, -Sd1 and -Sd2H) were identified in cultivated barley. They showed low, intermediate and high levels of thermostability, respectively. Analysis of the relationship between ß-amylase thermostability and fermentability in 42 commercial malt samples indicated that increased thermostability resulted in more efficient starch degradation (Eglinton et al. 1998).

Except these three ß-amylase alleles found in cultivated barley, another three alleles (*Bmy*1-Sd3, -Sd4 and -Sd5) were identified in a 154 accessions of *H. spontaneum.* The corresponding *Sd4* and *Sd5* enzymes showed middle levels of thermostability, similar to the *Sd1* ß-amylase in cultivated barley, while the *Sd3* ß-amylase exhibited greater thermostability than the other five allelic forms of ß-amylase. *Sd3* ß-amylase can sustain a stable state at T_{50} temperatures of 60.8 °C, while *Sd2L* enzyme and *Sd1* enzyme show sustainable temperature of 56.8 and 58.5 °C, respectively. Examination of the gene structure of the *Bmy1*-Sd3 allele revealed a 126 bp deletion in intron III, consistent with elevated gene expression (Eglinton et al. 2001). ß-amylase expression is thought to be modulated by intron-based gene regulation (Errkilä et al. 1998). The *Bmy1-Sd3* allele from *H. spontaneum* was used to develop commercial malting barley varieties through marker-assisted selection (Li et al. 2004).

The mean activities of α-amylase, β-amylase and β-glucanase showed significantly lower standard in the 257H. *spontaneum* accessions when compared with 32 unimproved Finnish landraces and five standard global barleys. These wild barley accessions also displayed less variation in β-amylase. Finnish landraces are useful gene resources of high β-amylase activity in barley (Ahokas et al. 1996). Landraces have adapted in many ways to the local environment and are morphologically closer to the desired types than wild barley. The yield data also suggest that the locally adapted genotypes can have a high yield potential. Compared with the variation obtained with wild barley (*H. spontaneum*) grown in Finland (Ahokas and Naskali 1990), the landraces have significantly higher mean activities for all the three enzymes assayed and showed superior maxima for α-amylase and β-amylase. The variance of β-amylase is highly significantly wider in the Finnish landraces than in the wild barleys (Ahokas and Poukkula 1999).

Beta-enzyme-less mutant lines had a naked seed character and various colours and could give a new taste and flavour to beer and other liquors. Utilisation of the mutant is also expected to extend to enzyme technology. The β-enzyme-less trait will be introduced into elite malting varieties by backcrossing. An enzyme assay showed that there were eight β-enzyme-less mutant lines in the Tibetan landrace barley. These mutants result from the same insert mutant in the β-enzyme structure gene of Tibetan barley. There was one structure gene in the mutant lines; both RFLP and PCR strongly suggest the insertion was suspected to exist between 2071 and 3101 bp from the initiation codon, and the structure gene causes the deficiency of enzyme activity (Kaneko et al. 2000). The α-glucosidase activity level of the mutant line was high in the initial period of germination, and the glucose ratio in the sugar confirmation was noticeable high in the germinated seed of mutant line. RFLP analysis showed a segregation ration of the enzyme activity of the normal to β-enzyme-less 30:10 (=3:1).

6.6 Conclusion

Numerous researches demonstrated that *Hordeum spontaneum* is a valuable gene pool for barley improvement. A large numbers of novel QTLs/genes were identified for abiotic and biotic stress tolerance and for improvement of productivity and sustainability. AB-QTL approach will provide an effective way to use the novel genes from *H. spontaneum*. Search for new alleles for malting quality has just started, and *H. spontaneum* may provide novel genes for improvement of malting quality.

Acknowledgements This project is supported by the National Natural Science Foundation of China (30630047) and Department of Agriculture and Food Western Australia.

References

Abbott, D. C., Brown, A. H. D., & Burdon, J. J. (1992). Genes for scald-resistance from wild barley (*Hordeum vulgare* ssp. *spontaneum*) and their linkage to isozyme markers. *Euphytica, 61*, 225–231.

Abbott, D. C., Lagudah, E. S., & Brown, A. H. D. (1995). Identification of RFLPs flanking a scald resistance gene on barley chromosome 6. *The Journal of Heredity, 86*, 152–154.

Able, J. A., Langridge, P., & Milligan, A. S. (2007). Capturing diversity in the cereals: many options but little promiscuity. *Trends in Plant Science, 12*, 71–79.

Ahokas, H., & Naskali, L. (1990). Geographic variation of α-amylase, β-amylase, β-glucanase, pullulanase and chitinase activity in germinating *Hordeum spontaneum* barley from Israel and Jordan. *Genetica, 82*, 73–78.

Ahokas, H., & Poukkula, M. (1999). Malting enzyme activities, grain protein variation and yield potentials in the displaced genetic resources of barley landraces of Finland. *Genetic Resources and Crop Evolution, 46*, 251–260.

Ahokas, H., Uutela, P., Erkkilä, M. J., & Vähämiko, S. (1996). Another source of genes with high beta-amylase activity in barley grain: Finnish landraces. *Barley Genetics Newsletter, 25*, 36–40.

Araus, J. L., Slafer, G. A., Reynolds, M. P., & Royo, C. (2002). Plant breeding and drought in C_3 cereals: what should we breed for? *Annals of Botany, 89*, 925–940.

Backes, G., Madsen, L. H., Jaiser, H., Stougaard, J., Herz, M., Mohler, V., & Jahoor, A. (2003). Localization of genes for resistance against *Blumeria graminis* f. *sp. hordei* and *Puccinia graminis* in a cross between a barley cultivar and a wild barley (*Hordeum vulgare* ssp.*spontaneum*) line. *Theoretical and Applied Genetics, 106*, 353–363.

Batchu, A. K., Zimmemann, D., Schulze-Lefert, P., & Koprek, T. (2006). Correlation between hordatine accumulation, environmental factors and genetic diversity in wild barley (*Hordeum spontaneum* C. Koch). *Genetica, 127*, 87–99.

Baum, M., Grando, S., Backes, G., Jahoor, A., Sabbagh, A., & Ceccarelli, S. (2003). QTLs for agronomic traits in the Mediterranean environment identified in recombinant inbred lines of the cross 'Arta'×*H. spontaneum* 41-1. *Theoretical and Applied Genetics, 107*(7), 1215–1225.

Bilgic, H., Steffenson, B. J., & Hayes, P. (2005). Comprehensive genetic analyses reveal differential expression of spot blotch resistance in four populations of barley. *Theoretical and Applied Genetics, 111*(7), 1238–1250.

Bjørnstad, A., Patil, V., Tekauz, A., Maroy, G., Skinnes, H., Jensen, A., Magnus, H., & Mackey, J. (2002). Resistance to scald (*Rhynchosporium secalis*) in barley (*Hordeum vulgare*) studied by near-isogenic lines: I. Markers and differential isolates. *Phytopathology, 92*, 710–720.

Borovkova, I. G., Jin, Y., Steffenson, B. J., Kilian, A., Blake, T. K., & Kleinhofs, A. (1997). Identification and mapping of leaf rust resistance gene in barley line Q21861. *Genome, 40*, 236–241.

Borovkova, I. G., Jin, Y., & Steffenson, B. J. (1998). Chromosomal location and genetic relationship of the leaf rust resistance genes *Rph9* and *Rph12* in barley. *Phytopathology, 88*, 76–80.

Briggs, D. E. (1978). The origin and classification of barleys. In D. E. Briggs (Ed.), *Barley* (pp. 76–88). London: Chapman and Hall.

Brunner, S., Keller, B., & Feuillet, C. (2000). Molecular mapping of the *Rph7* leaf rust resistance gene in barley (*Hordeum vulgare* L.). *Theoretical and Applied Genetics, 101*, 783–788.

Cattivelli, L., Baldi, P., Crosatti, C., Fonzo, N. D., Faccioli, P., Grossi, M., Am, Mastrangelo, Pecchioni, N., & Stanca, A. M. (2002). Chromosome regions and stress-related sequences involved in resistance to abiotic stress in Triticeae. *Plant Molecular Biology, 48*(5–6), 649–665.

Ceccarelli, S., Grando, S., & Impiglia, A. (1998). Choice of selection strategy in breeding barley for stress environments. *Euphytica, 103*, 307–318.

Chalmers, K. J., Waugh, R., Watters, J., Forster, B. P., Nevo, E., Abbott, R. J., & Powell, W. (1992). Grain isozyme and ribosomal DNA variability in Hordeum spontaneum populations from Israel. *Theoretical and Applied Genetics, 84*, 313–322.

Choi, D. W., & Close, T. J. (2000). A newly identified barley gene, *Dhn12*, encoding YSK2 *DHN*, is on chromosome 6H and has embryo-specific expression. *Theoretical and Applied Genetics, 100*, 1274–1278.

Choi, D. W., Zhu, B., & Close, T. J. (1999). The barley (*Hordeum vulgare* L.) dehydrin multigene family: sequences, allelic types, chromosome assignments, and expression characteristics of 11 *Dhn* genes of *cv* Dicktoo. *Theoretical and Applied Genetics, 98*, 1234–1247.

Chojecki, J., Barnes, S., & Dunlop, A. (1989). A molecular marker for vernalization requirement in barley. In T. Helentjaris & B. Burr (Eds.), *Development and application of molecular markers to problems in plant genetics* (pp. 145–148). Cold Spring Harbour: Cold Spring Harbour Laboratory.

Clifford, B. C. (1985). Barley leaf rust. In A. P. Roelfs & W. R. Bushnell (Eds.), *Cereal rust. Diseases, distribution, epidemiology, and control* (Vol. 2, pp. 173–205). New York: Academic.

Close, T. J., Choi, D. W., Venegas, M., Salvi, S., Tuberosa, R., Ryabushkina, N., Turuspekov, Y., & Nevo, E. (2000, October 22–27). *Allelic variation in wild and cultivated barley at the Dhn4 locus, which encodes a major drought-induced and seed protein, DHN 4.* In 8th International Barley Genetics Symposium, Adelaide, SA, South Australia.

Crosatti, C., Nevo, E., Stanca, A. M., & Cattivelli, L. (1996). Genetic analysis of the accumulation of COR 14 proteins in wild (*Hordeum spontaneum*) and cultivated (*Hordeum vulgare*) barley. *Theoretical and Applied Genetics, 93*, 975–981.

Czembor, J. H. (2000). Resistance to powdery mildew in populations of barley landraces from Morocco. *Australasian Plant Pathology, 29*, 137–148.

Diab, A. A. (2006). Construction of barley consensus map showing chromosomal regions associated with economically important traits. *African Journal of Biotechnology, 5*, 235–248.

Dyck, P. L., & Schaller, C. W. (1961). Association of two genes for scald resistance with a specific barley chromosome. *Canadian Journal of Genetics and Cytology, 3*, 165–169.

Eglinton, J. K., Langridge, P., & Evans, D. E. (1998). Thermostability variation in alleles of barley *Beta*-amylase. *Journal of Cereal Science, 28*, 301–309.

Eglinton, J. K., Evans, D. E., Brown, A. H. D., Langridge, P., McDonald, G., Jefferies, S. P., & Barr, A. R. (2001, September 16–20). *The use of wild barley (Hordeum vulgare ssp. spontaneum) in breeding for quality and adaptation.* In Proceedings of the 10th Australian Barley Technical Symposium, Canberra, ACT, Australia.

Ellis, R. P., Forster, B. P., Waugh, R., Bonar, N., Handley, L. L., Robinson, D., Gordon, D. C., & Powell, W. (1997). Mapping physiological traits in barley. *The New Phytologist, 137*, 149–157.

Ellis, R. P., Foster, B. P., Robinson, D., Handley, L. L., Gordon, D. C., Russell, J. R., & Powell, W. (2000). Wild barley: A source of genes for crop improvement in the 21st century? *Journal of Experimental Botany, 51*, 9–17.

Ellis, R. P., Forster, B. P., Gordon, D. C., Handley, L. L., Keith, R., Lawrence, P., Meyer, R. C., Powell, W., Robinson, D., Scrimgeour, C. M., Young, G. R., & Thomas, W. T. B. (2002). Phenotype/genotype associations of yield and salt tolerance in a barley mapping population segregating for two dwar.ng genes. *Journal of Experimental Botany, 53*, 1163–1176.

Errkilä, M. J., Leah, R., Ahokas, H., & Cameron-Mills, V. (1998). Allele-dependent grain ß-amylase activity. *Plant Physiology, 117*, 679–685.

Evans, D. E., Wallace, W., Lance, R. C. M., & MacLead, L. C. (1997). Measurement of beta-amylase in malting barley (*Hordeum vulgare* L.)II. The effect of germination and kilning. *Journal of Cereal Science, 26*, 241–250.

Falak, I., Falk, D. E., Tinker, N. A., & Mather, D. E. (1999). Resistance to powdery mildew in a doubled haploid barley population and its association with marker loci. *Euphytica, 107*, 185–192.

Favatier, F., Bornman, L., Hightower, L. E., Gunther, E., & Polla, B. S. (1997). Variation in *hsp* gene expression and *Hsp* polymorphism: do they contribute to differential disease susceptibility and stress tolerance? *Cell Stress Chaperones, 2*, 141–155.

Fetch, T. G., Steffenson, B. J., Jr., & Nevo, E. (2003). Diversity and sources of multiple disease resistance in *Hordeum spontaneum*. *Plant Disease, 87*, 1439–1448.

Feuerstein, U., Brown, A. H. D., & Burdon, J. J. (1990). Linkage of rust resistance genes from wild barley (*Hordeum spontaneum*) with isozyme markers. *Plant Breeding, 104*, 318–324.

Forster, B. P., Phillips, M. S., Miller, T. E., Baird, E., & Powell, W. (1990). Chromosome location of genes controlling tolerance to salt (NaCl) and vigour in *Hordeum vulgare* and *H. chilense*. *Heredity, 65*, 99–107.

Forster, B. P., Russel, J. R., Ellis, R. P., Handley, L. L., Hackett, C. A., Nevo, E., Waugh, R., Gordon, D. C., Keith, R., & Powell, W. (1997). Locating genotypes and genes for abiotic stress tolerance in barley: A strategy using maps, markers and the wild species. *New Phytologist, 137*, 141–147.

Franckowiak, J. D., Jin, Y., & Steffenson, B. J. (1997). Recommended allele symbols for leaf rust resistance genes in barley. *Barley Genetics Newsletter, 27*, 36–44.

Freialdenhoven, A., Scherag, B., Hollricher, K., Collinge, D. B., Thordal-Christensen, H., & Schulze-Lefert, P. (1994). *Nar-1* and *Nar-2*, two loci required for *Mla12*-specified race-specific resistance to powdery mildew in barley. *The Plant Cell, 6*, 983–994.

Garvin, D. F., Brown, A. H. D., & Burdon, J. J. (1997). Inheritance and chromosome locations of scald-resistance genes derived from Iranian and Turkish wild barleys. *Theoretical and Applied Genetics, 94*, 1086–1091.

Garvin, D. F., Brown, A. H. D., Raman, H., & Read, B. J. (2000). Genetic mapping of the barley *Rrs14* scald resistance gene with RFLP, isozyme and seed storage protein markers. *Plant Breeding, 119*, 193–196.

Genger, R. K., Brown, A. H. D., Knogge, W., Nesbitt, K., & Burdon, J. J. (2003a). Development of SCAR markers linked to a scald resistance gene derived from wild barley. *Euphytica, 134*, 149–159.

Genger, R. K., Williams, K. J., Raman, H., Read, B. J., Wallwork, H., Burdon, J. J., & Brown, A. H. D. (2003b). Leaf scald resistance genes in *Hordeum vulgare* and *Hordeum vulgare ssp spontaneum*: parallels between cultivated and wild barley. *Australian Journal of Agricultural Research, 54*, 1335–1342.

Goodwin, S. B., Allard, R. W., & Webster, R. K. (1990). A nomenclature for *Rhynchosporium secalis* pathotypes. *Phytopathology, 80*, 1330–1336.

Graner, A., Streng, S., Drescher, A., Jin, Y., Borovkova, I., & Steffenson, B. J. (2000). Molecular mapping of the leaf rust resistance gene *Rph7* in barley. *Plant Breeding, 119*, 389–392.

Griffey, C. A., Das, M. K., Baldwin, R. E., & Waldenmaier, C. M. (1994). Yield losses in winter barley resulting from a new race of *Puccinia hordei* in North America. *Plant Disease, 78*, 256–260.

Grime, K. H., & Briggs, D. E. (1996). The release of bound β-amylase by macromolecules. *The Journal of the Institute of Brewing & Distilling, 102*, 261–270.

Grønnerød, S., Marøy, A. G., MacKey, J., Tekauz, A., Penner, G. A., & Bjørnstad, A. (2002). Genetic analysis of resistance to barley scald (*Rhynchosporium secalis*) in the Ethiopian line 'Abyssinian' (CI668). *Euphytica, 126*, 235–250.

Hackett, C. A., Ellis, R. P., Forster, B. P., McNicol, J. W., & Macaulay, M. (1992). Statistical analysis of a linkage experiment in barley involving quantitative trait loci for height and ear-emergence time and two genetic markers on chromosome 4. *Theoretical and Applied Genetics, 85*, 120–126.

Halterman, D., Zhou, F. S., Wei, F., Wise, R. P., & Schulze-Lefert, P. (2001). The *Mla6* coiled-coil, NBS-LRR protein confers AvrMla6-dependent resistance specificity to *Blumeria graminis* f. sp. *hordei* in barley and wheat. *The Plant Journal, 25*, 335–348.

Harlan, J. R. (1976). Barley. In N. W. Simmonds (Ed.), *Evolution of crop plants* (Plant sciences, Vol. 13, pp. 97–119). London: Longman.

Heun, M. (1992). Mapping quantitative powdery mildew resistance of barley using a restriction fragment length polymorphism map. *Genome, 35*, 1019–1025.

Ivandic, V., Walther, U., & Graner, A. (1998). Molecular mapping of a new gene in wild barley conferring complete resistance to leaf rust (*Puccinia hordei Otth*). *Theoretical and Applied Genetics, 97*, 1235–1239.

Ivandic, V., Hackett, C. A., Zhang, Z. J., Staub, J. E., Nevo, E., Thomas, W. T. B., & Forster, B. P. (2000). Phenotypic responses of wild barley to experimentally imposed water stress. *Journal of Experimental Botany, 51*, 2021–2029.

Jahoor, A., & Fischbeck, G. (1993). Identification of new genes for mildew resistance of barley at the Mla locus in lines derived from *Hordeum spontaneum*. *Plant Breeding, 110*, 116–122.

Jin, Y., Statler, G. D., Franckowiak, J. D., & Steffenson, B. J. (1993). Linkage between leaf rust resistance genes and morphological markers in barley. *Phytopathology, 83*, 230–233.

Jin, Y., Steffenson, B. J., & Bockelman, H. E. (1995). Evaluation of cultivated and wild barley for resistance to pathotypes of *Puccinia hordei* with wide virulence. *Genetic Resources and Crop Evolution, 42*, 1–6.

Jørgensen, J. H. (1993). Durability of resistance in the pathosystems: barley-powdery mildew. In T. H. Jacobs & J. E. Parlevliet (Eds.), *Durability of disease resistance* (pp. 159–176). Dordrecht: Kluwer Academic.

Jørgensen, J. H. (1994). Genetic of powdery mildew resistance in barley. *Critical Reviews in Plant Sciences, 13*, 97–119.

Kaneko, T., Kihara, M., Ito, K., & Takeda, K. (2000). Molecular and chemical analysis of β-amylase-less mutant barley in Tibet. *Plant Breeding, 119*, 383–387.

Kicherer, S., Backes, G., Walther, U., & Jahoor, A. (2000). Localizing QTLs for leaf rust resistance and agronomic traits in barley (*Hordeum vulgare* L.). *Theoretical and Applied Genetics, 100*, 881–888.

Kintzios, S., Jahoor, A., & Fischbeck, G. (1995). Powdery mildew resistance genes *Mla29* and *Mla32* in *H. spontaneum* derived winter barley lines. *Plant Breeding, 114*, 265–266.

Laurie, D. A., Pratchett, N., Bezant, J. H., & Snape, J. W. (1995). RFLP mapping of five major genes and eight quantitative trait loci controlling flowering time in a winter x spring barley cross. *Genome, 38*, 575–585.

Li, C. D., Lance, R., Tarr, A., Broughton, S., Harasymow, S., Appels, R., & Jones, M. (2004, June). *Improvement of barley malting quality using a gene from* Hordeum spontaneum. In VI International Barley Genetic Symposium, Brno, Czech Republic.

Liviero, L., Maestri, E., Gulli, M., Nevo, E., & Marmiroli, N. (2002). Ecogeographic adaptation and genetic variation in wild barley: Application of molecular markers targeted to environmentally regulated genes. *Genetic Resources and Crop Evolution, 49*, 133–144.

Mammadov, J. A., Zwonitzer, J. C., Biyashev, R. M., Griffey, C. A., Jin, Y., Steffenson, B. J., & Saghai Maroof, M. A. (2003). Molecular mapping of leaf rust resistance gene *Rph5* in barley. *Crop Science, 43*, 388–393.

Manisterski, J., Treeful, L., Tomerlin, J. R., Anikster, Y., Moseman, J. G., Wahl, I., & Wilcoxson, R. D. (1986). Resistance of wild barley accessions from Israel to leaf rust collected in the USA and Israel. *Crop Science, 26*, 727–730.

Mano, Y., & Takeda, K. (1995). Varietal variation and effects of some major genes on salt tolerance in barley seedlings. *Bulletin of the Research Institute for Bioresources Okayama University, 3*, 71–81.

Mano, Y., & Takeda, K. (1997). Mapping quantitative trait loci for salt tolerance at germination and the seedling stage in barley (*Hordeum vulgare* L.). *Euphytica, 94*, 263–272.

Mano, Y., & Takeda, K. (1998). Genetic resources of salt tolerance in wild *Hordeum species*. *Euphytica, 103*, 137–141.

Mano, Y., Nakazumi, H., & Takeda, K. (1996). Varietal variation in and effects of some major genes on salt tolerance at the germination stage in barley. *Breeding Science, 46*, 227–233.

Matre, D. E. (1982). *Compendium of barley diseases*. St Paul: American Phytopathological Society Press.

Matus, I. A., & Hayes, P. M. (2002). Genetic diversity in three groups of barley germplasm assessed by simple sequence repeats. *Genome, 45*, 1095–1106.

McDaniel, M. E., & Hathcock, B. R. (1969). Linkage of the *Pa4* and *Mla* loci in barley. *Crop Science, 9*, 822.

McDonald, B. A., Zhan, J., & Burdon, J. J. (1999). Genetic structure of *Rhynchosporium secalis* in Australia. *Phytopathology, 89*, 639–645.

Morrell, P. L., Lundy, K. E., & Clegg, M. T. (2003). Distinct geographic patterns of genetic diversity is maintained in wild barley (*Hordeum vulgare* ssp. *spontaneum*) despite migration. *Proceedings of the National Academy of Sciences, 100*, 10812–10817.

Moseman, J. G., Nevo, E., & El-Morshidy, M. A. (1990). Reactions of *Hordeum spontaneum* to infection with two cultures of Puccinia hordei from Israel and United States. *Euphytica, 49*, 169–175.

Nevo, E. (1992). Origin, evolution, population genetics and resources for breeding of wild barley, *Hordeum spontaneum* in the fertile crescent. In P. R. Shewry (Ed.), *Barley genetics, biochemistry, molecular biology and biotechnology* (pp. 19–43). Wallingford: CAB International.

Nevo, E., Baum, B., Beiles, A., & Johnson, D. A. (1998). Ecological correlates of RAPD DNA diversity of wild barley, *Hordeum spontaneum* in the Fertile Crescent. *Genetic Resources and Crop Evolution, 45*, 151–159.

Nuccio, M. L., Rhodes, D., McNeil, S. D., & Hanson, A. D. (1999). Metabolic engineering of plants for osmotic stress resistance. *Current Opinion in Plant Biology, 2*, 128–134.

Pakniyat, H., Powell, W., Baird, E., Handly, L. L., Robinson, D., Scrimgeour, C. M., Nevo, E., Hackett, C. A., Caligari, P. D. S., & Forster, B. P. (1997). AFLP variation in wild barley (*Hordeum spontaneum* C. Koch) with reference to salt tolerance and associated ecogeography. *Genome, 40*, 332–341.

Pan, A., Hayes, P. M., Chen, F., Chen, T. H. H., Blake, T., Wright, S., Karsai, I., & Bedo, Z. (1994). Genetic analysis of the components of winter hardiness in barley (*Hordeum vulgare* L.). *Theoretical and Applied Genetics, 89*, 900–910.

Père de la Vega, M. (1996). Plant genetic adaptedness to climatic and edaphic environment. *Euphytica, 92*, 27–38.

Peterhänsel, C., Freialdenhoven, A., Kurth, J., Kolsch, R., & Schulze-Lefert, P. (1997). Interaction analyses of genes required for resistance responses to powdery mildew in barley reveal distinct pathways leading to leaf cell death. *The Plant Cell, 9*, 1397–1409.

Pickering, R., Ruge-Wehling, B., Johnston, P. A., Schweizer, G., Ackermann, P., & Wehling, P. (2006). The transfer of a gene conferring resistance to scald (*Rhynchosporium secalis*) from *Hordeum bulbosum* into *H. vulgare* chromosome 4HS. *Plant Breeding, 125*(6), 576–579.

Pillen, K., Zacharias, A., & Leon, J. (2003). Advanced backcross QTL analysis in barley (*Hordeum vulgare* L). *Theoretical and Applied Genetics, 107*, 340–352.

Qi, X., Niks, R. E., Stam, P., & Lindhout, P. (1998). Identification of QTLs for partial resistance to leaf rust (*Puccinia hordei*) in barley. *Theoretical and Applied Genetics, 96*, 1205–1215.

Qi, X., Fekadu, F., Sijtsma, D., Niks, R. E., Lindhout, P., & Stam, P. (2000). The evidence for abundance of QTLs for partial resistance to *Puccinia hordei* on the barley genome. *Molecular Breeding, 6*, 1–9.

Richter, K., Schondelmaier, J., & Jung, C. (1998). Mapping of quantitative traits loci affecting *Drechslera teres* resistance in barley with molecular markers. *Theoretical and Applied Genetics, 97*, 1225–1234.

Rodríguez, E. M., Svensson, J. T., Malatrasi, M., Choi, D. W., & Close, T. J. (2005). Barley *Dhn13* encodes a KS-type dehydrin with constitutive and stress responsive expression. *Theoretical and Applied Genetics, 110*, 852–858.

Saghai Maroof, M. A., Zhang, Q., & Biyashev, R. M. (1994). Molecular marker analysis of powdery mildew resistance in barley. *Theoretical and Applied Genetics, 88*, 733–740.

Schiiller, C., Backes, G., Fischbeck, G., & Jahoor, A. (1992). RFLP markers to identify the alleles on the *Mla* locus conferring powdery mildew resistance in barley. *Theoretical and Applied Genetics, 84*, 330–338.

Schönfeld, M., Ragni, A., Fischbeck, G., & Jahoor, A. (1996). RFLP mapping of three new loci for resistance genes to powdery mildew (*Erysiphe graminis* f. sp. *hordei*) in barley. *Theoretical and Applied Genetics, 93*, 48–56.

Schweizer, G. F., Baumer, M., Daniel, G., Rugel, H., & Röder, M. S. (1995). RFLP markers linked to scald (*Rhynchosporium secalis*) resistance gene *Rh2* in barley. *Theoretical and Applied Genetics, 90*, 920–924.

Shtaya, M. J. Y., Marcel, T. C., Sillero, J. C., Niks, R. E., & Rubiales, D. (2006). Identification of QTLs for powdery mildew and scald resistance in barley. *Euphytica, 151*, 421–429.

Smith, T. A., & Best, G. R. (1978). Distribution of the hordatines in barley. *Phytochemistry, 17*, 1093–1098.

Søgaard, B., & von Wettstein-Knowles, P. (1987). Dissection of the cer-cqu locus. In S. Yasuda & T. Konishi (Eds.), *Barley genetics V* (pp. 161–167). Okayama: Sanyo Press Co.

Steffenson, B. J., Jin, Y., & Griffey, C. A. (1993). Pathotypes of *Puccinia hordei* with virulence for the barley leaf rust resistance gene *Rph7* in the United States. *Plant Disease, 77*, 867–869.

Steffenson, B. J., Hayes, P. M., & Kleinhofs, A. (1996). Genetics of seedling and adult plant resistance to net blotch (*Pyrenophora teres* f. *teres*) and spot blotch (*Cochliobolus sativus*) in barley. *Theoretical and Applied Genetics, 92*, 552–558.

Suprunova, T., Krugman, T., Fahima, T., Chen, I., Shams, I., Korol, A., & Nevo, E. (2004). Differential expression of dehydrin genes in wild barley, *Hordeum spontaneum*, associated with resistance to water deficit. *Plant, Cell & Environment, 27*, 1297–1308.

Tan, B. H. (1978). Verifying the genetic relationships between three leaf rust resistance genes in barley. *Euphytica, 27*, 317–323.

Teulat, B., Merah, O., Sirault, X., Borries, C., Waugh, R., & This, D. (2002). QTLs for grain carbon-isotope discrimination in field-grown barley. *Theoretical and Applied Genetics, 106*, 118–126.

Teulat, B., Zoumarou-Wallis, N., Rotter, B., Ben Salem, M., Bahri, H., & This, D. (2003). QTL for relative water content in field-grown barley and their stability across Mediterranean environments. *Theoretical and Applied Genetics, 108*, 181–188.

Teulat-Merah, B., Rotter, B., Francois, S., Borries, C., Souyris, I., & This, D. (2000). Stable QTL for plant water status, osmotic adjustment and co-location with QTLs for yield components in a Mediterranean barley progeny. *Barley Genetics, 8*, 246–248.

Thacker, S. P., Ramamurthy, V., & Kothari, R. M. (1992). Characterisation of barley *β*-amylase for application in maltose production. *Starch, 44*, 339–341.

Toubia-Rahme, H., Johnston, P. A., Pickering, R. A., & Steffenson, B. J. (2003). Inheritance and chromosomal location of *Septoria passerinii* resistance introgressed from *Hordeum bulbosum* into *Hordeum vulgare*. *Plant Breeding, 122*, 405–409.

Tuleen, N. A., & McDaniel, M. E. (1971). Location of genes *Pa* and *Pa5*. *Barley Newsletter, 15*, 106–107.

van Zee, K., Chen, F. Q., Hayes, P. M., Close, T. J., & Chen, T. H. H. (1995). Cold-specific induction of a dehydrin gene family member in barley. *Plant Physiology, 108*, 1233–1239.

von Korff, M., Wang, H., Leon, J., & Pillen, K. (2005). AB-QTL analysis in spring barley. I. Detection of resistance genes against powdery mildew, leaf rust and scald introgressed from wild barley. *Theoretical and Applied Genetics, 111*, 583–590.

Wallwork, H. (2000a). *Cereal stem and crown diseases*. Kingston: Grains Research and Development Corporation.

Wallwork, H. (2000b). *Cereal leaf and stem diseases*. Kingston: Grains Research and Development Corporation.

Walther, U., & Lehmann, C. O. (1980). Resistenzeigenschaften im Gerstenund Weizensortiment Gatersleben. 24. Prüfung von Sommerund Wintergersten auf ihr Verhalten gegenüber Zwergrost (Puccinia hordei Otth). *Kulturpflanze, 28*, 227–238.

Williams, K. J. (2003). The molecular genetics of disease resistance in barley. *Australian Journal of Agricultural Research, 54*, 1065–1079.

Young, N. D. (1996). QTL mapping and quantitative disease resistance in plants. *Annual Review of Phytopathology, 34*, 479–501.

Yun, S. J., Gyenis, L., Hayes, P. M., Matus, I., Smith, K. P., Steffenson, B. J., & Muehlbauer, G. J. (2005). Quantitative trait loci for multiple disease resistance in wild barley. *Crop Science, 45*, 2563–2572.

Zhang, Q., Webster, R. K., & Allard, R. W. (1987). Geographical distribution and associations between resistance to four races of *Rhynchosporium secalis*. *Phytopathology, 77*, 352–357.

Zhang, W. S., Li, X., & Liu, J. B. (2007). Genetic variation of *Bmy1* alleles in barley (*Hordeum vulgare* L.) investigated by CAPS analysis. *Theoretical and Applied Genetics, 114*, 1039–1050.

Zhou, F. S., Kurth, J. C., Wei, F., Elliott, C., Vale, G., Yahiaoui, N., Keller, B., Somerville, R., Wise, R., & Schulze-Lefert, P. (2001). Cell-autonomous expression of barley *Mla1* confers race- specific resistance to the powdery mildew fungus via a *Rar1*-independent signalling pathway. *The Plant Cell, 13*, 337–350.

Chapter 7
The Distribution of the *Hordoindoline* Genes and Identification of *Puroindoline b-2 Variant* Gene Homologs in the Genus *Hordeum*

Yohei Terasawa, Kanenori Takata, and Tatsuya M. Ikeda*

Abstract Barley *hordoindoline* genes (*Hina* and *Hinb*) are homologous to wheat *puroindoline* genes (*Pina* and *Pinb*). These genes are involved in grain hardness, which is an important quality characteristic for barley processing. However, the interspecific variation in the *Hin* genes in the genus *Hordeum* has not been studied in detail. We examined the variation in *Hin* genes and used it to infer the phylogenetic relationships between the genes found in two *H. vulgare* species and 10 wild relatives. Our results suggest that the *Hinb* gene duplicated during the early stages of speciation in *Hordeum*. We also identified novel variant forms of *Hina* and *Hinb* genes in the wild *Hordeum* species and preliminarily named them *Hinc*. The phylogenetic tree of *Gsp-1*, *Hin*, and *Pin* genes shown demonstrates that *Hinc* and *Pinb-2v* genes formed one cluster. Therefore, we considered that *Hinc* and *Pinb-2v* genes shared a common ancestral gene and were homologous to each other. We also studied the evolutional process of *Gsp-1*, *Hin*, and *Pin* genes. Our results suggested that *Gsp-1* genes might be the most closely related to an ancestral gene on *Ha* locus.

Keywords *Hordoindoline* • *Puroindoline* • *Hordeum* • *Phylogeny*

7.1 Introduction

In common wheat (*T. aestivum*), the *puroindoline-a* (*Pina*) and *puroindoline-b* (*Pinb*) genes are located at the *hardness* (*Ha*) locus on the short arm of chromosome 5D (Mattern et al. 1973; Law et al. 1978). Grain hardness is mainly controlled by

*Presenting author, Yohei Terasawa

Y. Terasawa • K. Takata • T.M. Ikeda (✉)
Western Region Agricultural Research Center, National Agricultural Research Organization (NARO), Fukuyama, Hiroshima, Japan
e-mail: terasaway@affrc.go.jp; tmikeda@affrc.go.jp

G. Zhang et al. (eds.), *Advance in Barley Sciences: Proceedings of 11th International Barley Genetics Symposium*, DOI 10.1007/978-94-007-4682-4_7,

these genes in common wheat (Giroux and Morris 1998; Krishnamurthy and Giroux 2001). The puroindoline proteins (PINs) encoded by the *puroindoline* (*Pin*) genes belong to a group of cysteine-rich basic proteins, but PINs are unique among plant proteins because they contain a tryptophan-rich hydrophobic domain (Blochet et al. 1993). This domain is proposed to play two important roles, to control grain hardness, which effects flour particle size (Giroux and Morris 1998; Morris 2002), and to produce antimicrobial activity (Phillips et al. 2005). Recently, Massa and Morris (2006) studied the phylogenetic relationships of the *Pina*, *Pinb*, and *grain softness protein-1* (*Gsp-1*) genes in *Triticum* and *Aegilops*.

Grain hardness in barley is also one of the most important quality characters affecting the end-use properties. There was a highly significant correlation of the grain hardness with the malting quality (Nagamine et al. 2009) and rumen dry-matter digestibility, which is one of the most important traits in feeding of beef cattle (Turuspekov et al. 2008).

The presence of *Pin* orthologs in *H. vulgare* ssp. *vulgare* (cultivated barley) has been demonstrated by molecular techniques. An orthologous *Ha* locus has been found on the short arm of the barley **5H** chromosome by Southern hybridization using wheat *puroindoline* cDNA (Rouvés et al. 1996). PCR analysis of cultivated barley using primer pairs designed based on *Pin* genes identified the *hordoindoline-a* (*Hina*) and *hordoindoline-b* (*Hinb*) genes, which appear to be orthologs of *Pina* and *Pinb*, respectively, because these hordoindoline proteins (HINs) also belong to a group of cysteine-rich basic proteins and contain a tryptophan-rich hydrophobic domain (Gautier et al. 2000). Furthermore, the *Hinb* gene in cultivated barley consists of two genes (*Hinb-1* and *Hinb-2*), which was demonstrated by Southern blot and PCR analyses (Darlington et al. 2001). A large number of *Hina* and *Hinb* allelic variations were reported among cultivated barley and wild barley (*H. vulgare* ssp. *spontaneum*) (Caldwell et al. 2006). In addition, direct evidence of the effects of *Hin* genes on grain hardness has been reported by Takahashi et al. (2010), who confirmed that the grain of the *Hinb-2* null mutant was significantly harder than that of the wild type. Yanaka et al. (2011) used barley chromosome addition lines and studied the effect of chromosome **5H** on grain characteristics in the wheat genetic background. They suggested the barley *Hin* genes located on chromosome **5H** were involved in reducing grain hardness in the wheat genetic background.

Wilkinson et al. (2008) recently found three new *Pinb*-like sequences in *T. aestivum*. Chen et al. (2010a, b) also found another *Pinb*-like sequence and assigned the chromosomal locations of four genes, *Pinb-D2v1*, *Pinb-B2v2*, *Pinb-B2v3*, and *Pinb-A2v4*, to 7**DL**, 7**BL**, 7**B**, and 7**A**, respectively. However, *Pinb*-like sequences had been not identified in the genus *Hordeum*.

There has been no study of the interspecific variation in the *Hin* genes and the *Puroindoline b-2 variant* (*Pinb-2v*) gene homolog in the genus *Hordeum*. We studied the *Hin* genes and identified *Pinb-2v* gene homolog in the genus *Hordeum*. We also discussed the evolutional process of *Gsp-1*, *Hin*, and *Pin* genes.

7.2 Materials and Methods

7.2.1 *Plant Materials*

Twelve species of *Hordeum* were used in this study (Table 7.1).[1]

7.2.2 *PCR Amplification and DNA Sequence Analyses*

Hinb-1-specific primers and *Hinb-2*-specific primers were designed based on the DNA sequences of the *Hinb-1* and *Hinb-2* genes of cultivated barley. Other primers were designed based on the nucleotide sequences identified in this study. *Hina*-specific primers were designed based on the DNA sequences of the *Hina* genes of two *H. vulgare* species (cultivated barley and *H. vulgare spontaneum*) and the *Pina* gene sequences of wheat and *Aegilops* species. Two *Hinc*-specific primers were designed based on the DNA sequences of the *Pinb-2v*, *Hina*, and *Hinb* genes (Chen et al. 2010a, 2010b; Terasawa et al. 2012). PCR was performed and sequenced directly. This primer information was referred to Terasawa et al. (2012 and data not shown).

7.2.3 *Phylogenetic Analysis*

Multiple sequence alignments of the new variant genes, related gene nucleotide sequences, and their deduced amino acid sequences used ClustalW (Thompson et al. 1994). Phylogenetic trees were constructed from the nucleotide sequences with the neighbor-joining method (Saitou and Nei 1987) using DNASIS Pro Ver.3.0 (Hitachi Software Engineering). Bootstrap analyses were performed with 1,000 samplings.

7.3 Results

7.3.1 *Analyses of Hordoindoline-a*

PCRs were carried out in all samples using *Hina*-specific primers. DNA fragments (about 560 bp) were amplified using all samples with *Hina*-specific primers. We confirmed that these DNA fragments contained *Hina* genes by sequencing them directly (data not shown).

[1] These lines were kindly provided by Dr. S. Taketa, Okayama University, Japan.

Table 7.1 *Hordeum* species used in this study and GenBank accession numbers of the *Hin* genes

		Sequence accession no.			
Species	Genome	*Hina*	*Hinb*		*Hinc*
H. bogdanii	**H**	AB605713	AB605724 (*Hinb-A*)	AB605721 (*Hinb-B*)	AB693971
H. brachyantherum ssp. californicum	**H**	AB605712	AB605725 (*Hinb-A*)	AB605722 (*Hinb-B*)	AB693972
H. roshevitzii	**H**	AB605714	AB605726 (*Hinb-A*)	AB605723 (*Hinb-B*)	AB693969
H. chilense	**H**	AB446468	AB446468		AB693973
H. comosum	**H**	AB605710	AB605727		AB693970
H. patagonicum ssp. *patagonicum*	**H**	AB605711	AB605728		AB693975
H. pusillum	**H**	AB605709	AB605729		AB693976
H. marinum ssp. *marinum*	**Xa**	AB605715	AB605719		AB693974
H. murinum ssp. *glaucum*	**Xu**	AB605716	AB605720		n/a
H. bulbosum	**I**	AB605717	AB605718		n/a
H. vulgare ssp. *spontaneum*	**I**	AB611025	AB611028 (*Hinb-1*)	AB611031 (*Hinb-2*)	n/a
H. vulgare ssp. *vulgare*, cv. Betzes	**I**	AB611024	AB611027 (*Hinb-1*)	AB611030 (*Hinb-2*)	n/a

We performed a BLAST search (NCBI) on the 12 *Hina* genes identified in this study and confirmed that 10 *Hina* were novel genes, and these sequences were registered in DDBJ (Table 7.1). The eight *Hina* gene sequences (*H. bogdanii, H. brachyantherum, H. chilense, H. comosum, H. marinum, H. pusillum, H. patagonicum,* and *H. roshevitzii*) were 447 bp in length, whereas the four *Hina* genes (*H. bulbosum, H. murinum,* cultivated barley, and *H. vulgare spontaneum*) were 450 bp long. Among the *Hina* nucleic acid sequences, 77 (17%) sites displayed variation (data not shown).

7.3.2 Analysis of Hordoindoline-b

Hinb genes amplified by PCRs using *Hinb-1*-specific primers to amplify *Hinb* were not successful in any sample except for the two *H. vulgare* species. Therefore, PCRs were carried out in all samples using *Hinb-2*-specific primers. DNA fragments were amplified from six species (*H. bulbosum, H. marinum, H. murinum, H. roshevitzii,* and two *H. vulgare* species). However, PCRs were not successful for the remaining four species (*H. chilense, H. comosum, H. patagonicum,* and *H. pusillum*). PCRs were performed with other *Hinb-2*-specific primers in all samples. DNA fragments containing *Hinb* genes were amplified from seven species (*H. bogdanii, H. brachyantherum, H. chilense, H. comosum H. patagonicum, H. pusillum,* and *H. roshevitzii*).

We performed a BLAST search (NCBI) on the all *Hinb* genes identified in this study and confirmed that 13 were novel genes, and these sequences were registered in DDBJ (Table 7.1). Comparison of the *Hinb* sequences (444-bp open reading frames: ORF) revealed the presence of two *Hinb* genes in three species (*H. bogdanii, H. brachyantherum,* and *H. roshevitzii*). Since the sequence similarity between the two *Hinb* genes in the three species was lower than that between *Hinb-1* and *Hinb-2* genes in the two *H. vulgare* species, we preliminarily designated one of the three *Hinb* genes as *Hinb-A* and other *Hinb* genes as *Hinb-B* described in Terasawa et al. (2012).

7.3.3 Identification of Pinb-2v Gene Homologs in Hordeum

To identify the presence of *Pinb-2v* gene homologs in *Hordeum*, we designed two primers based on *Pinb-2v, Hina,* and *Hinb* sequences (Chen et al. 2010a; Terasawa et al. 2012) (data not shown). With these, we successfully amplified PCR products (approximately 800 bp) in eight species, including seven **I** genome species (*H. bogdanii, H. brachyantherum, H. chilense, H. comosum, H. patagonicum, H. pusillum,* and *H. roshevitzii*) and one **Xa** genome species (*H. marinum*). We determined their nucleotide sequences and confirmed that they had not been previously reported using a BLAST search (NCBI). We preliminarily designated these novel genes *Hinc* because their nucleotide sequence identity to *Hinb* sequences was less than 75% (data not shown) and these sequences were registered in DDBJ (Table 7.1).

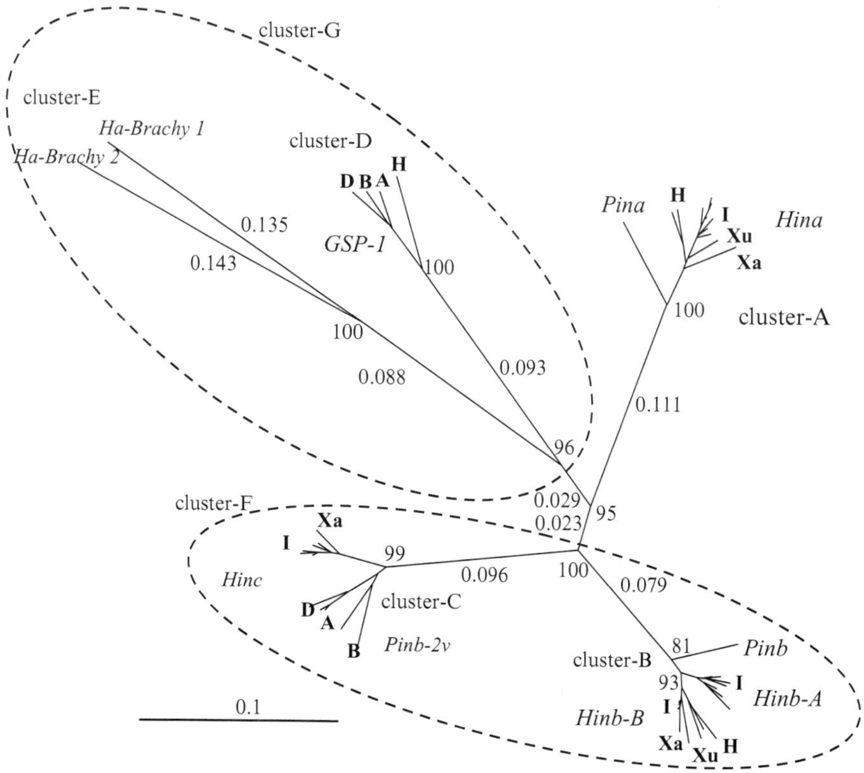

Fig. 7.1 Phylogenetic tree of seven genes based on 47 nucleotide sequences [Four *Gsp-1* (**H**: DQ269919b; *H. vulgare*, **A**: JF794544, **B:** EF109372, **D**: AY255771; *T. aestivum*), 11 *Hina*, 15 *Hinb*, eight *Hinc*, one *Pina*, one *Pinb*, five *Pinb-2v*, and two *Ha-Brachy*] from the genera *Brachypodium, Hordeum,* and *Triticum* (All sequences of *Pinb-2v* sequences were referred to Chen et al. (2010a, 2011). All sequences of *Hina*, *Hinb*, *Pina*, and *Pinb* were referred to our previous papers (Terasawa et al. 2012), and two *Ha-Brachy* were referred to sequence of *B. sylvaticum* BAC clone 37D5 (FJ234838). The *Hordeum* and *Triticum* genomes (**A**, **B**, **D**, **H**, **I,** and **Xa**) are indicated. The numbers beside the branch indicate the branch length, and the numbers beside the branch point indicate the bootstrap values (bootstrap reps = 10,000). Scale bar indicates the ratio of nucleotide substitutions. The details of phylogenetic tree of *Hina* and *Hinb* were referred to Terasawa et al. (2012))

7.3.4 Phylogenetic Analysis of Hin Genes

A phylogenetic tree for eight genes sequences (*Gsp-1*, *Ha-Brachy* and *Hina*, *Hinb*, *Hinc*, *Pina*, *Pinb*, *Pinb-2v*) was constructed using the neighbor-joining method, as shown in Fig. 7.1.

We included two *Ha-Brachy* genes of *Brachypodium sylvaticum* because these amino acid sequences showed high sequence similarity to those of *Gsp-1*, *Hins*, and *Pins* genes and were considered as homologs of these genes (Charles et al. 2009).

Table 7.2 The genetic distances among five clusters in Fig. 7.1

	Cluster-A	Cluster-B	Cluster-C	Cluster-D
Cluster-B	0.213			
Cluster-C	0.230	0.175		
Cluster-D	0.233	0.224	0.241	
Cluster-E[a]	0.367	0.358	0.375	0.320

[a]The cluster-E genetic distance is the average of the genetic distance of *Ha-Brachy-1* and *Ha-Brachy-2* genes from the diverging point

This tree was divided into small five clusters and big two clusters. One small cluster included all *Hina* and *Pina* genes, and another small cluster included *Hinb-A, Hinb-B*, and *Pinb* genes; these were defined as cluster-A and cluster-B, respectively. The *Hinc* and *Pinb-2v* genes were included in the same small cluster, defined as cluster-C. One small cluster included *Gsp-1* genes of *T. aestivum* and *H. vulgare*, and another small cluster included *Ha-Brachy* genes; these were defined as cluster-D and cluster-E, respectively. The cluster-B and cluster-C appear to form one big cluster-F (dashed circle in Fig. 7.1). Two distinct cluster-D and cluster-E diverged from common root. A bootstrap value of 96 was observed at its branching point. The cluster-D and cluster-E appear to form one big cluster-G (dashed circle in Fig. 7.1).

The genetic distances among four (A, B, C, D, and E) clusters were shown in Table 7.2. The genetic distance between cluster-B and cluster-C was the shortest among these clusters.

The phylogenic tree of *Hinc* is shown in Fig. 7.2. The tree of eight *Hinc* genes was formed divergent clusters corresponding to the two-genome groups. The *Hinc* gene of *H. marinum* was separated from the cluster of **I** genome species.

7.4 Discussion

Previous reports revealed their distribution within *H. vulgare* species (Caldwell et al. 2006). However, the interspecific variation in the *Hin* genes among the genus *Hordeum* has not been studied in detail. We determined the *Hin* genes and the distribution of the *Hin* genes in two *H. vulgare* species and 10 wild relatives.

7.4.1 The Distribution of Hin Genes

We identified 10 novel *Hina* genes and 13 novel *Hinb* genes and revealed that three species (*H. bogdanii, H. brachyantherum*, and *H. roshevitzii*) possessed two *Hinb* genes (*Hinb-A* and *Hinb-B*). We could also identify new genes: *Hinc* in eight **I** and **Xa** genome species. Wilkinson et al. (2008) and Chen et al. (2010a) reported that

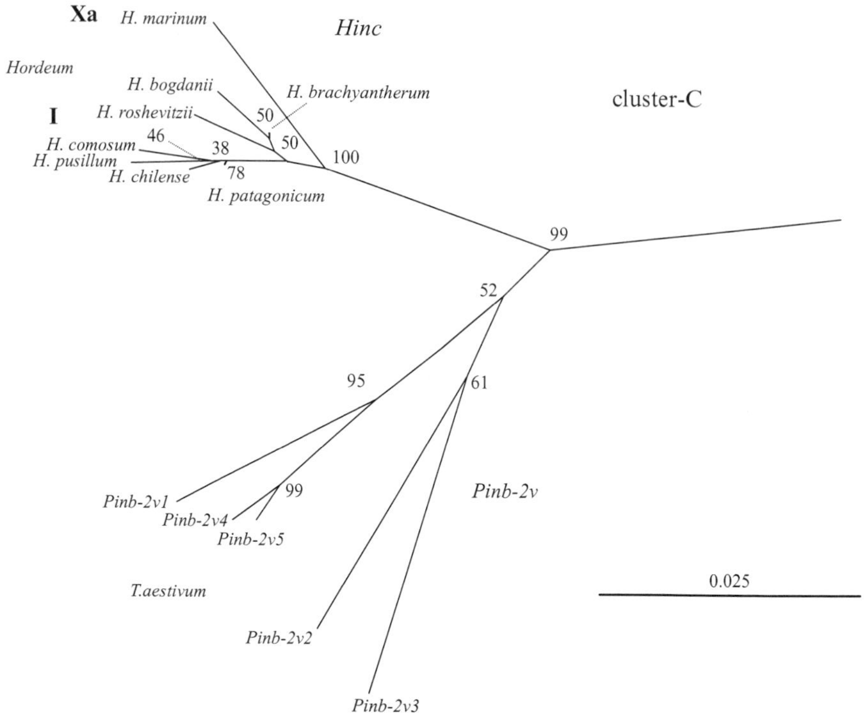

Fig. 7.2 Phylogenetic tree of *Hinc* and *Pinb-2v* genes. The description of the tree and the labels are the same as in Fig. 7.1

Pinb-2v genes were located on the homologous group 7 chromosome in *T. aestivum*. To confirm the chromosomal locations of *Hinc* gene, PCR analyses for the *H. chilense* chromosome addition lines were performed using *Hinc* gene-specific primers. It showed that *Hinc* gene in *H. chilense* was located on chromosome 7**I** (data not shown). *Hinc* gene in *H. chilense* and *Pinb-2v* in *T. aestivum* were located on the same homologous group 7 chromosome. Thus, *Hinc* gene in the other seven species may also be located on chromosomes 7**I** and 7**Xa**.

7.4.2 Evolution of Hinb Genes

Our previous study proposed the evolutional process of *Hinb* genes (Terasawa et al. 2012). We considered that that *Hinb* gene duplication occurred during the early stages of the speciation of the genus *Hordeum* and that the ancestral species of the genus *Hordeum* probably had two *Hinb* genes. *Hinb-1* and *Hinb-2* genes in *H. vulgare* seem to have been generated by a duplication of the *Hinb* gene after the split of the lineages leading to *H. vulgare* and *H. bulbosum*.

7.4.3 Evolution of Hinc Genes

The neighbor-joining phylogenetic tree shown in Fig. 7.1 demonstrates that *Hinc* and *Pinb-2v* genes formed cluster-C as that using UPGMA methods (data not shown). Therefore, we considered that *Hinc* and *Pinb-2v* genes shared a common ancestral gene and were homologous to each other. The genetic distances among *Hinc* gene were shorter than those among *Pinb-2v* genes because these *Pinb-2v* genes were located on the different genomes (Chen et al. 2010a). This phylogenic tree for *Hinc* gene was compared with those for the ITS sequences (Blattner 2004) and *Hordeum* thioredoxin-like (HTL) gene sequences (Kakeda et al. 2009). The trees for HTL and ITS showed two clearly divergent clusters that correspond to three genomes, **I** and **Xa**. The *Hinc* tree also branched according to a different genome. These results suggest that the genetic relationship among *Hinc* genes corresponds with the evolutionary processes for the two *Hordeum* species' genomes.

Based on the phylogenetic tree, we constructed the genes on *Ha* locus within the genera *Hordeum* and *Triticum* that were divided into four clusters (cluster-A, -B, -C, and -D, Fig. 7.1). This indicated that the evolutional divergences among these four cluster genes were occurred before the split of the lineages that led to the genera *Hordeum* and *Triticum*. Cluster-B and cluster-C appeared to form a larger cluster-F (Fig. 7.1), and the genetic distance between cluster-B and cluster-C was the shortest in Table 7.2. This result suggests that the cluster-B and cluster-C genes may have diverged from a common ancestral gene. However, the cluster-B and cluster-C genes were located on different homologous group chromosomes. Cluster-C genes (*Hinc* and *Pinb-2v*) were located on homologous group 7 chromosomes. In contrast, all of the other *Gsp-1*, *Hin*, and *Pin* genes were located on homologous group 5 chromosomes. Therefore, we considered that the cluster-C ancestral gene might have been generated from the cluster-B and cluster-C common ancestral gene by a gene duplication following translocation to the homologous group 7 chromosome.

7.4.4 Evolution of Genes on Ha Locus

Charles et al. (2009) showed that independent duplications and deletions of genes on *Ha* locus have occurred within grasses. Their comparative genome analysis shows that a short genomic sequence similar to *Gsp-1* gene is present in an orthologous rice locus (Caldwell et al. 2004; Chantret et al. 2004, 2005). In recent study, two homologs of hardness genes were found on *Ha* orthologous region in *B. sylvaticum*, which is distantly related to barley and wheat (Charles et al. 2009). They found two *Ha-Brachy* genes located on *Ha* orthologous region in *B. sylvaticum*. Our results revealed that the genetic distance between *Gsp-1* and *Ha-Brachy* genes was relatively close (Table 7.2 and Fig. 7.1). Cluster-D and cluster-E appeared to form a larger cluster-G and diverged from common lineage to cluster-A, which was supported by high bootstrap values (Fig. 7.1). These results suggested that *Gsp-1* genes might be the most closely related to the ancestral gene on *Ha* locus. In consideration of these findings, we propose the following evolutionary hypothesis.

First, cluster-D ancestral gene, which is related to *Gsp-1*, was located on *Ha* region in homologous group 5 chromosome. A common ancestral gene of cluster-A, -B, and -C genes may have been generated by duplication from the ancestral cluster-D gene. Subsequently, the common ancestral gene for cluster-A, -B, and -C duplicated in tandem to generate a cluster-A ancestral gene and a cluster-B and -C common ancestral gene. Finally, the cluster-C ancestral gene may have been generated from the cluster-B and cluster-C common ancestral gene by a gene duplication following translocation to the homologous group 7 chromosome. In *Hordeum*, *Hinb* gene duplication occurred during the early stages of the speciation of the genus *Hordeum*. The *Hinb-1* and *Hinb-2* genes in *H. vulgare* may have been generated by a duplication of the *Hinb* gene after the speciation of *H. vulgare*.

7.5 Conclusions and Future Work

We identified ten novel *Hina*, 13 novel *Hinb*, and eight *Hinc* genes that were homologous to *Pinb-2v* genes. These works suggested that *Gsp-1* genes might be the most closely related to an ancestral gene on *Ha* locus. We proposed the evolutional process of genes on *Ha* locus within the genera *Hordeum* and *Triticum*. Further studies are necessary to clarify the function of *Hin* genes.

Acknowledgments This work was supported by the Bio-oriented Technology Research Advancement Institution. We express our sincere gratitude to Dr. Shin Taketa for providing material support.

References

Blattner, F. R. (2004). Phylogenetic analysis of *Hordeum* (Poaceae) as inferred by nuclear rDNA ITS sequences. *Molecular Phylogenetics and Evolution, 33*, 289–299.

Blochet, J. E., Chevalier, C., Forest, E., Pebay, P. E., Gautier, M. F., Joudrier, J., Pézolet, M., & Marion, D. (1993). Complete amino acid sequence of puroindoline, a new basic and cystine-rich protein with a unique trytophan-rich domain, isolated from wheat endosperm by Triton X-114 phase partitioning. *FEBS Letters, 329*, 336–340.

Caldwell, K. S., Langridge, P., & Powell, W. (2004). Comparative sequence analysis of the region harboring the hardness locus in barley and its colinear region in rice. *Plant Physiology, 136*, 3177–3190.

Caldwell, K. S., Russell, J., Langridge, P., & Powell, W. (2006). Extreme population-dependent linkage disequilibrium detected in an inbreeding plant species, *Hordeum vulgare*. *Genetics, 172*, 557–567.

Chantret, N., Cenci, A., Sabot, F., Anderson, O. D., & Dubcovsky, J. (2004). Sequencing of the *Triticum monococcum* hardness locus reveals good microcolinearity with rice. *Molecular Genetics and Genomics, 271*, 377–386.

Chantret, N., Salse, J., Sabot, F., Rahman, S., Bellec, A., Laubin, B., Dubois, I., Dossat, C., Sourdille, P., Joudrier, P., Gautier, M. F., Cattolico, L., Beckert, M., Aubourg, S., Weissenbach, J., Caboche, M., Bernard, M., Leroy, P., & Chalhoub, B. (2005). Molecular basis of evolutionary events that shaped the hardness locus in diploid and polyploid wheat species (*Triticum* and *Aegilops*). *The Plant Cell, 17*, 1033–1045.

Charles, M., Tang, H., Belcram, H., Paterson, A., & Gornicki, P. (2009). Sixty million years in evolution of soft grain trait in grasses: emergence of the softness locus in the common ancestor of Pooideae and Ehrhartoideae, after their divergence from Panicoideae. *Molecular Biology and Evolution, 26*, 1651–1661.

Chen, F., Beecher, B., & Morris, C. F. (2010a). Physical mapping and a new variant of *Puroindoline b-2* genes in wheat. *Theoretical and Applied Genetics, 120*, 745–751.

Chen, F., Zhang, F. Y., Cheng, X. Y., Morris, C. F., Xu, H. X., Dong, Z. D., Zhan, K. H., He, Z. H., Xia, X. C., & Cui, D. Q. (2010b). Association of *Puroindoline b-B2* variants with grain traits, yield components and flag leaf size in bread wheat (*Triticum aestivum* L.) varieties of the Yellow and Huai Valleys of China. *Journal of Cereal Science, 52*, 247–253.

Chen, F., Xu, H. X., Zhang, F. Y., Xia, X. C., He, Z. H., Wang, D. W., Dong, Z. D., Zhan, K. H., Cheng, X. Y., & Cui, D. Q. (2011). Physical mapping of *puroindoline b-2* genes and molecular characterization of a novel variant in durum wheat (Triticum turgidum L.). *Molecular Breeding, 28*, 153–161.

Darlington, H. F., Rouster, J., Hoffmann, L., Halford, N. G., Shewry, P. R., & Simpson, D. J. (2001). Identification and molecular characterisations of hordoindolines from barley grain. *Plant Molecular Biology, 47*, 785–794.

Gautier, M. F., Cosson, P., Guirao, A., Alary, R., & Joudrier, P. (2000). Puroindoline genes are highly conserved in diploid ancestor wheats and related species but absent in tetraploid *Triticum* species. *Plant Science, 153*, 81–91.

Giroux, M. J., & Morris, C. F. (1998). Wheat grain hardness results from highly conserved mutations in the friabilin components puroindoline a and b. *Proceedings of the National Academy of Sciences of the United States of America, 95*, 6262–6266.

Kakeda, K., Taketa, S., & Komatsuda, T. (2009). Molecular phylogeny of the genus *Hordeum* using thioredoxin-like gene sequences. *Breeding Science, 59*, 595–601.

Krishnamurthy, K., & Giroux, M. (2001). Expression of wheat puroindoline genes in transgenic rice confers grain softness. *Nature Biotechnology, 19*, 162–166.

Law, C. N., Young, C. F., Brown, J. W. S., Snape, J. W., & Worland, A. J. (1978). The study of grain protein control in wheat using whole chromosome substitution lines. In *Seed protein improvement by nuclear techniques* (pp. 483–502). Vienna: IAEA.

Massa, A. N., Morris, C. F. (2006). Molecular evolution of puroindoline-a, puroindoline-b and grain softness protein-1 genes in the tribe triticeae. *Journal of Molecular Evolution, 63*, 526–536.

Mattern, P. J., Morris, R., Schmidt, J. W., & Johnson, V. A. (1973). Location of genes for kernel properties in wheat variety 'Cheyenne' using chromosome substitution lines. In S. R. Sears & L. M. S. Sears (Eds.), *Proceedings of the 4th International Wheat Genetics Symposium*, Columbus, MO (pp. 703–707). Agricultural Experiment Station, College of Agriculture, University of Missouri.

Morris, C. F. (2002). Puroindolines: The molecular genetic basis of wheat hardness. *Plant Molecular Biology, 48*, 633–647.

Nagamine, T., Sekiwa, T., Yamaguchi, E., Oozeki, M., & Kato, T. (2009). Relationship between quality parameters and SKCS hardness index in malting barley. *Journal of the Institute of Brewing, 109*, 129–134.

Phillips, R. L., Palombo, R. L., Panozzo, J. F., & Bhave, M. (2005). Puroindolines, Pin alleles, hordoindolines and grain softness proteins are sources of bactericidal and fungicidal peptides. *Journal of Cereal Science, 53*, 112–117.

Rouvés, S., Boef, C., Zwickert-Menteur, S., Gautier, M. F., JoudrierP, Bernard M., & Jestin, L. (1996). Locating supplementary RFLP markers on barley chromosome 7 and synteny with homeologous wheat group 5. *Plant Breed, 115*, 511–513.

Saitou, N., & Nei, M. (1987). The neighbor-joining method: A new method for reconstructing phylogenetic trees. *Molecular Biology and Evolution, 4*, 406–425.

Takahashi, A., Ikeda, T. M., Takayama, T., & Yanagisawa, T. (2010). A barley hordoindoline mutation resulted in an increase in grain hardness. *Theoretical and Applied Genetics, 120*, 519–526.

Terasawa, Y., Rahman, S. M., Takata, K., & Ikeda, T. M. (2012). Distribution of *Hordoindoline* genes in the genus *Hordeum*. *Theoretical and Applied Genetics, 124*, 143–151.

Thompson, J. D., Higgins, D. G., & Gibson, T. J. (1994). ClustalW: improving the sensitivity of progressive multiple sequence alignment through sequence weighting, position-specific gap penalties and weight matrix choice. *Nucleic Acids Research, 22*, 4673–4680.

Turuspekov, Y., Beecher, B., Darlington, Y., Bowman, J., Blake, T. K., & Giroux, M. J. (2008). Hardness locus sequence variation and endosperm texture in spring barley. *Crop Science, 48*, 1007–1019.

Wilkinson, M., Wan, Y., Tosi, P., Leverington, M., Snape, J., Mitchel, R. A. C., & Shewry, P. R. (2008). Identification and genetic mapping of variant forms of puroindoline b expressed in developing wheat grain. *Journal of Cereal Science, 48*, 722–728.

Yanaka, M., Takata, K., Terasawa, Y., & Ikeda, T. M. (2011). Chromosome 5H of *Hordeum* species involved in reduction in grain hardness in wheat genetic background. *Theoretical and Applied Genetics, 123*, 1013–1018.

Chapter 8
Exploiting and Utilizing the Novel Annual Wild Barleys Germplasms on the Qing-Tibetan Plateau

Dongfa Sun, Tingwen Xu, Guoping Zhang, Zhao Ling, Daokun Sun, and Ding Yi

Abstract Novel germplasm is becoming increasingly important for developing crop varieties with high yield potential, better quality, and abiotic or biotic resistance. Wild relatives of crops have been received increasingly more attention as gene donors, for which they have much more genetic diversity than *cultivars* and for which earlier difficulties of sexual isolation in gene transfer using conventional methods can now be overcome by the development of new biological techniques. Annual wild barleys, *H. spontaneum*, *H. agriocrithon*, and *H. paradoxon*, which mainly exist in the Near East Fertile Crescent and Qing-Tibetan Plateau, share the same genome as cultivated barley does and are completely interfertile. By using cytology, plant physiology, biochemistry, genetics, traditional genetic breeding techniques, and modern biotechnology, we have systematically evaluated important agronomic, biological, and quality traits for more than 200 annual wild barley accessions collected from Qing-Tibetan Plateau since 1978. We have successfully identified and created a batch of germplasms with early maturity, male sterility, salt tolerance, aluminum resistance, high protein, and specific malt quality, respectively. Some of them were successfully used in barley breeding program, such as Chuanluo NO. 1 (early maturity and dwarf), 88BCMS (nucleo-cytoplasmic male sterile line),

This research was supported by China Agriculture Research System

D. Sun (✉) • Z. Ling
College of Plant Science and Technology, Huazhong Agricultural University, Wuhan, Hubei 430070, China
e-mail: sundongfa1@mail.hzau.edu.cn

T. Xu
Sichuan Agricultural University, Ya'an, China

G. Zhang
Zhejiang University, Hangzhou, China

D. Sun • D. Yi
Wuhan University, Wuhan, China

G. Zhang et al. (eds.), *Advance in Barley Sciences: Proceedings of 11th International Barley Genetics Symposium*, DOI 10.1007/978-94-007-4682-4_8,

and 5199 (super early heading and cold resistance). Others being used include 69 accessions with preharvest spouting resistance, 20 accessions with aluminum resistance, seven accessions with salt tolerance, and two accessions with drought tolerance. We also examined the relationship between the annual wild barleys of Qing-Tibetan Plateau and other barley germplasm resources and the genetic improvement potential in malt quality, nutritional quality, and stress resistance. The genetic characters of some important traits including male sterility, super early heading, and dwarfing trait were characterized. Annual wild barley germplasm in Qing-Tibetan Plateau has been successfully utilized in modern barley breeding in China. We successively released 13 commercial barley varieties that have integrated the genes from annual wild barley germplasm in Qing-Tibetan Plateau during 1992–2011, indicating the great potential of the annual wild barleys in Qing-Tibetan Plateau in the genetic improvement of barley. These commercial varieties have some excellent traits, such as early maturity, dwarfism multiresistance, high-yield potential, and high quality, which can be used to exploit the prepared rows of cotton or corn as well as winter fallow field in South China. We initiated and implemented the complete idea of utilizing the prepared row of cotton and corn as well as winter fallow field using early-mature and dwarf barley varieties. By well planning, it can solve the problem of field competition of crop for barley production in the Yangtze valley, greatly promote the development and utilization of winter fallow field, and increase the efficient use of land resources in cotton and corn throughout the year.

Keywords Barley • Annual wild barleys • Germplasm • Genetic improvement • Variety • Utilization • Qing-Tibetan Plateau • Planting mode

8.1 Introduction

Gene resources, germplasm, and genetic resources are similar glossaries in crop breeding and genetics, which include cultivated plants, wild plants, and any other life forms used in crop breeding. Crop breeders utilize selected gene resources for developing new variations or improving agronomic traits like disease resistance. Genetic theory and breeding practices have indicated that exploration and utilization of germplasm resources, especially elite germplasm resources, is the key of achieving breeding objectives (Sun and Gong 2009). Abiotic stress tolerance and quality improvement in barley are mainly determined by genetic factors of breeding materials that breeders have (Ellis et al. 2002). In the long-term domestication process of cultivated barley, especially through modern breeding and intensive cultivation, genetic variation significantly reduced. This caused many genes to be lost, especially some rare alleles (Saghai Maroof et al. 1990), and modern cultivars are more sensitive to adverse environmental factors, diseases, and pests. Using SSR technology, Ellis et al. (2002) compared wild barley's genetic diversity with cultivated barley's and found that alleles in cultivated barley only represent 40 % of that in wild barley.

The monotonous genetic background of cultivated barley has become a bottleneck of crop breeding. Rich genetic diversity of wild barley can provide valuable gene source to break the bottleneck.

Wild relatives of cultivated barley have perennial and annual wild barley. Lots of researches about perennial wild barley like H. *bulbosum* and H. *chilense* have been reported. However, because of reproductive isolation and genetic burden, there is no successful example of new cultivar which have integrated beneficial trait from those species. Annual wild barleys, such as *H. spontaneum*, *H. agriocrithon*, and *H. paradoxon*, share the same genome as cultivated barley does and can be directly used in breeding program. The hybrids of *H. vulgare* × *H. spontaneum* (or *H. agriocrithon* or *H. paradoxon*) show normal chromosome pairing and segregation in meiosis, and their progeny is fully fertile. However, the spikes of the three wild barleys are fragile at maturity and the grains shatter easily, while the spikes are tough and the grains persist on them in domesticated barleys. *H. spontaneum* and cultivated two-rowed *distichum* varieties also show many morphological similarities. The main differences between them are in their modes of seed dispersal. The ears of *H. spontaneum* are brittle and at maturity disarticulate into individual arrow-like triplets. These are highly specialized devices which ensure the survival of the plant in nature. Under cultivation, this specialization was replaced by nonbrittle mutants, which were immediately selected by human for reaping, threshing, and sowing (Zohary and Hopf 1988).

Annual wild barley is mainly distributed in the Near East Fertile Crescent and Qing-Tibetan Plateau of China. Annual wild barleys *H. spontaneum*, which mainly exists in the Near East Fertile Crescent, and *H. agriocrithon* or *H. paradoxon* are widely distributed over Qing-Tibetan Plateau. Current researches on annual wild barley are mainly confined to the Middle East type. Results show that the genetic variation of annual wild barley in Middle East is very rich, especially in disease resistance and abiotic stresses including drought, cold, salt, etc. Some wild germplasm materials have successfully been used in breeding program; a number of high-quality, disease-resistant, salt-tolerant, and drought-resistant varieties have been bred (Thomas et al. 1998; Ellis et al. 2002).

Qing-Tibetan Plateau is one of the centers of origin of barley. It has tremendously rich barley germplasm resources, especially its unique annual wild barley being different from that of Middle East, which is a precious wild resource of barley world. Due to long-term growth in a variety of severe environment, suitable genetic control systems formed (Nevo 1992; Nevo et al. 1997), which showed a strong environmental stress tolerance and unique grain chemical composition. As the special geographical and social constraints, few foreign researchers' works have included large samples of the annual wild barley of Qing-Tibetan Plateau (Kaneko et al. 2000; Konishi 2001). However, Chinese scientists have systematically conducted researches on Qing-Tibetan Plateau annual wild barley germplasm resources for more than 40 years. The researches were mainly focused on the following five aspects: (1) collecting and cataloging the wild barley germplasm of Qing-Tibetan Plateau, (2) exploration and creation of the novel germplasm from annual wild barleys in Qing-Tibetan Plateau, (3) genetically characterization of target traits of the annual wild barley in Qing-Tibetan Plateau, (4) the utilization of the novel germplasm from annual wild barleys in Qing-Tibetan Plateau in the genetic improvement of barley, and (5) the application of new commercial varieties in barley production of China.

8.2 Collecting and Cataloging the Wild Barley Germplasm of Qing-Tibetan Plateau

Chinese scientists had exploration to collect wild barley from 1981 to 1985. Xu (1987) classified the collected annual wild barley into one species, two subspecies, and 32 varieties. A total of 200 accessions including degraded two-rowed wild barley, naked two-rowed wild barley, and six-rowed wild barley which are the rare varieties of world (Xu 1987) are currently stored in the laboratory of *Triticeae* crops in Huazhong Agricultural University, Chinese Academy of Agricultural Sciences, Tibetan Academy of Agricultural and Animal Husbandry Sciences, and other institutions.

8.3 The Exploration and Creation of the Novel Germplasm from Annual Wild Barleys in Qing-Tibetan Plateau

By using cytological, physiological, biochemistry, genetic, traditional breeding, and molecular technologies, we systematically characterized the main agronomical and quality traits of 200 annual wild barley accessions in Qinghai-Tibetan Plateau. A number of novel germplasm were identified, screened, and created from Qing-Tibetan Plateau annual wild barley, such as germplasm with early maturity, male sterility, salt tolerance, acid resistance, aluminum resistance, high protein, and specific malt quality. Some of them havc been successfully applied to barley breeding in China.

8.3.1 Chuanluo No. 1

Chuanluo No. 1, a widely used backbone parent, is developed from Qing-Tibetan Plateau annual wild barley. In 1982, Professor Xu from Sichuan Agricultural University successfully bred Chuanluo No. 1, a six-rowed naked barley variety by composite-crossing pedigree method. The combination is (78 Changli × Mimai 114) × (six-rowed highland barley × two-rowed wild barley). The 78 Changli and Mimai 114 are main cultivars of Ganzi Tibetan State of Sichuan Province in 1970s. Six-rowed highland barley is a landrace. Two-rowed wild barley is a barley accession collected from Shanna of Tibet by Professor Xu. Chuanluo No. 1 possesses outstanding early maturity and dwarfing and has widely been planted in Sichuan Province for many years. Since there was no barley variety approval system in Sichuan province at that time, Chuanluo No. 1 was not officially released. A lot of new barley varieties were bred using Chuanluo No. 1 as parents, such as Chuannongda No. 1, Huadamai No. 2, Huadamai No. 5, Huadamai No. 6, and Huadamai No. 7 that were all officially released.

8.3.2 88 CMS

88 CMS, another widely used backbone parent and a first nuclear-cytoplasmic male sterile line in China, was developed by using Qing-Tibetan Plateau annual wild barley. The 88 CMS and its maintainer line 88B were both selected from the offspring of cross of Qing-Tibet Plateau annual wild barley accession 236 and Composite III from Washington State University. The 88BCMS is two-rowed hulled barley with 60 cm height and 140 days of heading. Its spikelet number is about 30. The heading part can package neck. When it's flowing, hulls can half-open. Stigma is exposed less. The sterile plant rate of sterile lines was 100 % and the self-fruitful rate of sterile plants is less than 1 %. The 88BCMS was officially authorized as the first barley nucleo-cytoplasmic male sterile line in China. By using of 88BCMS, a lot of commercial barley varieties were bred and officially released, such as Chuannongda No. 2, Huadamai No. 1, Huadamai No. 4, Huadamai No. 5, Huadamai No. 6, Huadamai No. 8, Edamai 12, and Chudamai 610.

8.3.3 Line 5199

Line 5199, an especially early maturity and cold-tolerant material, was derived from early cultivated two-rowed hulled barley, Pilengbo, and an annual six-rowed wild barley accession in Qing-Tibetan Plateau. The heading date was 20 days earlier than local contrast variety and 10 days earlier than its parents. Using 5199 as parent, a barley variety with early maturity Huadamai No. 7 was bred.

8.3.4 A Number of Elite Germplasms Were Identified and Screened From Qing-Tibetan Plateau Annual Wild Barley

Using the methods of Ma et al. (2000), we screened more than 200 accessions of annual wild barley in Qing-Tibetan Plateau and 50 cultivated barley varieties and identified four salt-tolerant varieties. Their salt tolerance was significantly greater than the strongest salt tolerance cultivated barley CM72. In addition, we also identified 69 wild barley accessions with high preharvest sprouting resistance, two wild barley lines with high drought resistance, one dwarfing material, and more than ten accessions with high quality and specific malting quality.

8.4 The Genetic Characterization of the Annual Wild Barley in Qing-Tibetan Plateau for Target Traits

Using cytological and molecular methods, we have characterized genetic factors controlling male sterility, super early heading, and dwarfing trait of annual wild barley in Qing-Tibetan Plateau. Grain β-glucan, protein component content, and β-amylase activity in these accessions were also studied. We established a few of cultivar/annual wild barley DH population by using microspore culture technique. We have successfully identified SSR markers or QTL loci that linked with stress resistance and grain quality characters. We established the core germplasm of Qing-Tibetan Plateau annual wild barley based on morphological, physiological, biochemical, and molecular characters. We also took advantage of Ecotilling technology to identify superior alleles from this germplasm. These studies provide a strong theoretical foundation for breeder to efficiently use Qing-Tibetan Plateau annual wild barleys in barley breeding. Our researches have resulted in 50 publications. The results about the annual wild barley in Qing-Tibetan Plateau are outlined as follows:

8.4.1 Genetic Characteristics of Barley CMS Line 88BCMS (Sun et al. 1995)

With conventional genetic analysis, we analyzed fertility segregation in F2 populations of 88BCMS's testcross combination. The results showed that male sterility of 88BCMS was controlled by polygene. And it was obviously correlated with plant height, spike length, length of neck-panicle node, the number of effective spikes per plant, grain number per plant, grain weight per plant, and 1,000 seeds weight. By using conventional paraffin-embedded sections and photomicrographic technology, we observed flower structure, anther structure and pollen development of isonuclear alloplasmic male sterile line and their maintainer line, and photoperiod-thermosensitive male sterile line. The meiosis of pollen mother cells of 88BCMS was normal. Recession occurs in binucleus stage or monokaryotic telophase. The microspore disintegrated slowly and not thoroughly. The anther sac was rare empty. Abortion types of pollen were spherical abortion type and stained abortion type. We compared heterogeneity of esterase, peroxidase, and cytochrome oxidase in isonuclear alloplasmic male sterile line and their maintainer line and their F1. The results indicated that the differences in isozymes between sterile line's and maintainer line's anther are side effects of male sterility. There is no obvious difference between sterile and fertile cytoplasm. A new hypothesis of "variant gene hypothesis about male sterility in plant" was suggested to explain sterility, which means that male sterility in plant is the result of imbalance of some physiological and biochemical reactions during floral organ development. These were due to variant genes in cytoplasm or nucleus. This hypothesis can explain the male sterility phenomenon that previous theory can't explain.

8.4.2 *Inheritance of the Super Early Heading of Line 5199 (Sun et al. 1998)*

Line 5199 is a six-rowed super early maturity line whose heading time is about 20 days earlier than other early maturity varieties in barley. The studies on the heading dates of the six generations (F1, F2, B1, B2, P1, P2) from four crosses derived from 5199 showed that the days and effectively accumulated temperature from emerging to heading are controlled by at least three loci major genes which have different effects. Late earing time is partly dominant to early earing time. The additive effects of genes play a major role in the forming of the two characters. The interaction between these genes also contributed to them. The super early heading of 5199 is controlled by at least three recessive genes with significant recessive gene epistatic effects. There is no negative correlation between super early earing genes in 5199 and other traits.

8.4.3 *Molecular Diversity and Association Mapping of Quantitative Traits in Tibetan Wild and Worldwide Originated Barley (Hordeum vulgare L.) Germplasm (Sun et al. 2011)*

Molecular diversity of 40 accessions of Tibetan wild barley (TB), 10 Syrian (SY), 72 North American (NA), 36 European (EU), 9 South American, and 8 Australian varieties were characterized using multiple microsatellite loci. The 42 SSR primers amplified 123 SSR loci across the 175 barley accessions tested here. The average gene diversity for the whole sample was 0.3387, whereas the mean value for the each population was as follows: TB = 0.3286, SY = 0.2474, EU = 0.299, AU = 0.2867, NA = 0.3138, and SA = 0.2536. Cluster analysis based on Nei's original genetic distance showed that the EU and NA barley populations were grouped together. The TB population was well separated from the other five barley populations. Associations between microsatellite markers and 14 quantitative traits (leaf area, stem diameter, grains per plant, filled grains per plant, grain weight per plant, plant height, spikelets on main spike, grains on main spike, grain weight on main spike, length of main spike, density of main spike, length of the 1st internode, length of spike neck, and awn length) were also investigated. Significant associations were found for 64 microsatellite markers. The number of marker associated with each trait ranged from one (stem diameter, filled grains per plant, grain weight per plant, and awn length) to nine (plant height and grain weight on main spike). The percentage of the total variation explained by each marker was from 4.59% (*HVM2-2* associated with plant height) to 17.48% (*Bmac90-1* associated with density of main spike).

8.4.4 Comparative Analysis of Genetic Diversity Between Tibetan Wild and Chinese Landrace Barley

Fifty-two SSR markers were used to evaluate the genetic diversity of 33 Qinghai-Tibetan wild barley accessions, 56 landraces collected from other parts of China, and one Israeli wild barley accession. At the 52 SSR loci, 206 alleles were detected for the 90 accessions, among which 111 were common alleles. The number of alleles per locus ranged from 1 to 9, with an average of 4.0. Polymorphism information content (PIC) value ranged from 0 to 0.856 among all the markers, with an average of 0.547. The PIC value of Qinghai-Tibetan wild barley varied from 0 to 0.813 with an average of 0.543, while in landraces, the markers showed a scale of 0 to 0.790 with an average of 0.490. The SSR markers could clearly differentiate the Qinghai-Tibetan wild barley from the landraces. Twenty-five unique alleles were observed and the frequency of unique alleles in Qinghai-Tibetan wild barley was about 2.1 times higher than that in the landraces on average. Five of the seven chromosomes had more unique alleles in the Qinghai-Tibetan wild barley. Chromosome 2H of landraces showed higher unique alleles than that of Qinghai-Tibetan wild barley.

8.4.5 Allelic Diversity of a Beer Haze Active Protein Gene in Cultivated and Tibetan Wild Barley and Development of Allelic Specific Markers (Ye et al. 2011)

The objectives of this study were to determine (1) the allelic diversity of the gene controlling BTI-CMe in cultivated and Tibetan wild barley and (2) allele-specific (AS) markers for screening SE protein type. A survey of 172 Tibetan annual wild barley accessions and 71 cultivated barley genotypes was conducted, and 104 wild accessions and 35 cultivated genotypes were identified as SE + ve and 68 wild accessions and 36 cultivated genotypes as SE-ve. The allelic diversity of the gene controlling BTI-CMe was investigated. It was found that there were significant differences between the SE + ve and SE-ve types in single-nucleotide polymorphisms at 234 (SNP234), SNP313, and SNP385. Furthermore, two sets of AS markers were developed to screen SE protein type based on SNP313. Mapping analysis showed that the gene controlling the MW ~ 14 kDa was located on the short arm of chromosome 3H, at the position of marker BPB-0527 (33.302 cM) in the Franklin/Yerong DH population.

8.4.6 Genetic Variants of HvGlb1 in Tibetan Annual Wild Barley and Cultivated Barley and Their Correlation with Malt Quality (Jin et al. 2011)

Improvement of malt quality is the most important objective in malt barley breeding. The variation of malt quality characters among barley genotypes and the difference in genetic variants of HvGlb1, encoding b-glucanase isoenzyme I, between Tibetan annual wild barley and cultivated barley were investigated. The correlation between the gene variants and malt quality showed that there was a large difference in the four malt quality parameters, i.e., Kolbach index, diastatic power (DP), viscosity, and malt extract, among the analyzed barley cultivars. Kolbach index was negatively and positively correlated with viscosity and malt extract, respectively, while malt extract was negatively correlated with viscosity. Malt b-glucan content was a major determinant of malt quality and was significantly correlated with Kolbach index (0.633), malt extract (0.333), and viscosity (0.672). On the other hand, malt b-glucan content was mainly controlled by malt b-glucanase activity. The correlation analysis showed that the HvGlb1 gene was correlated with malt b-glucan content and three of four main malt quality parameters, except DP. In addition, we also found that the HvGlb1 of Tibetan barley had wider diversity in haplotype than that of the cultivated barley, supporting the hypothesis that Tibet is one of the original centers of cultivated barley.

8.4.7 *Evaluation of Salinity Tolerance and Analysis of Allelic Function of HvHKT1 and HvHKT2 in Tibetan Wild Barley (Qiu et al. 2011)*

Tibetan wild barley is rich in genetic diversity with potential allelic variation useful for salinity-tolerant improvement of the crop. The objectives of this study were to evaluate salinity tolerance and analysis of the allelic function of HvHKT1 and HvHKT2 in Tibetan wild barley. Salinity tolerance of 189 Tibetan wild barley accessions evaluated in terms of reduced dry biomass under salinity stress. In addition, Na? and K? concentrations of 48 representative accessions differing in salinity tolerance were determined. Furthermore, the allelic and functional diversity of HvHKT1 and HvHKT2 was determined by association analysis as well as gene expression assay. There was a wide variation among wild barley genotypes in salt tolerance, with some accessions being higher in tolerance than cultivated barley CM 72, and salinity tolerance was significantly associated with K?/Na? ratio. Association analysis revealed that HvHKT1 and HvHKT2 mainly control Na? and K? transporting under salinity stress, respectively, which was validated by further analysis of gene expression. These results indicated that Tibetan wild barley offers elite alleles of HvHKT1 and HvHKT2 conferring salinity tolerance.

8.4.8 *Sequence Difference of β-Glucanase Gene in Qinghai-Tibet Plateau Annual Wild Barley*

SNP analysis of 79 wild and 92 cultivated barley's β-glucanase gene revealed different mutant sites in these materials. The mutant sites were divided into nine types. β-glucanase gene in wild barley was so conservative that the mutants only appeared in two types.

Furthermore, isoelectric point and molecular weight of these nine β-glucanase mutations were measured. The results indicated that there were three kinds of isoelectric point. Although individual bases in the conserved domain of some mutations were changed, the enzyme's molecular weight and isoelectric point were not significantly impacted.

8.4.9 Chromosome NORs Solver Staining in Wild and Cultivated Barley (Ding and Song 1991, 1992; Ding et al. 1996)

Chromosome NORs were studied for two wild barley subspecies, ssp. *spontaneum*, ssp. *agriocrithon*, and a cultivar. The results show that there are no significant differences in NOR activity between Israel and China wild barley. The frequency of silver staining to 4 NORs range from 51 to 79.6%. According to their frequency, all samples were classified into three types. No interspecific differences in size and number of NORs were found, but intraspecific differences exist. The duplication and heterozygosity of NOR have been observed in cultivated and wild barley. *The relation between G-, C-, N-banding and silver-stained band of barley chromosome is significantly important to chromosome structure analysis, the chromosome banding application in researches of cytogenetic theory, and barley breeding experiment.*

8.4.10 Genetic Diversity of Hordein in Wild Relatives of Barley from Tibet (Yin et al. 2003)

We analyzed genetic diversity in the storage protein hordein encoded at Hor-1, Hor-2, and Hor-3 loci in seeds from 211 accessions of wild barley, *Hordeum vulgare* ssp. *agriocrithon and H. vulgare* ssp. *spontaneum.* A total of 32, 27, and 13 different phenotypes were found for Hor-1, Hor-2, and Hor-3, respectively. Tibetan samples possess the highest diverse level of hordein phenotypes when compared to samples from Israel and Jordan. This high degree of polymorphism supports the hypothesis that Tibet is one of the original centers of *H. vulgare* L.

8.5 The Utilization of the Novel Germplasm from Annual Wild Barleys in Qing-Tibetan Plateau in the Genetic Improvement of Barley

We pay great attention to explore specific germplasm resources and innovative breeding method for reaching the breeding targets which are the early maturity, high yield, high quality, dwarf stem, and good resistance. At the same time, we have formed the barley pyramiding hybrid, quality traits identified in early generation, and breeding technology system of pyramiding multiple genes by molecular markers. We created a lot of the compound hybrid offspring by using specific germplasm resources from the Tibetan plateau annual wild barley hybridized with local cultivated varieties or introduced varieties. The compound hybrid offspring were selected by compound cross pedigree method which was considered being much more suitable for polygene separating and restructuring. Since 1992, we have successfully bred and released 13 new commercial barley varieties with early maturity, dwarf stem, high yield, high quality, good resistance. The pedigree, characters, released time and region, novel germplasm utilized, and yield are showed in Table 8.1.

8.6 The Application of New Commercial Varieties in Barley Production of China (Sun 2011)

The lack of farmland for barley is one of the key problems of barley production in South China. The prepared rows of cotton or corn as well as winter fallow field are useful spaces for growing barley in South China. These commercial barley varieties with early maturity and dwarf can be used to exploit these spaces. We initiated and implemented the complete idea of utilizing the prepared row of cotton and corn as well as winter fallow field using early maturity and dwarf barley varieties we bred. Also, the planting mode, implementation effect, the cultivation techniques of planting barley in the prepared row of cotton and corn, and the cultivation techniques of barley plus no-tillage in winter fallow field are elaborated. In this way, it solved the space problem of developing barley production in the Yangtze valley and promoted the development and utilization of winter fallow field and efficient use of land resources in cotton and corn throughout the year. 1,500–2,250 kg per ha of yield is increased by exploiting prepared row, while 3,000–4,500 kg per ha of yield increased by exploiting winter fallow field. Besides, it also helps to improve the agroecological environment of paddy field and dry land. Based on cooperation with beer brewing companies, feed processing companies, and big national farms, these new barley varieties and the new barley planting mode were widely extended. In order to improve the application efficiency of these bred varieties, we have carried out the study for special barley cultivation, production and quality evaluation technology system, and promotion mode. At the same time, we have studied standardized cultivation technology system and barley planting mode in the field of cotton and corn based on these new varieties with high yield potential, good resistance, early maturity, and dwarf stem. All these researches play a great role in increasing the potential of barley yield in South China and income of farmers, extending the utilization of the planting mode of winter fallow field and the prepared row of cotton and corn. We will carry out demonstration and promotion to keep well standardization system of cultivation techniques of special barley new varieties and the planting mode of winter fallow field and the prepared row of cotton and corn in different areas with cooperation of special barley user enterprise, provinces agricultural technique extension station, agricultural technology department, and big national farms.

Table 8.1 The pedigree, characters, released time and region, novel germplasm utilized, and yield *of 13* released varieties

	Chuannongda No. 1	Chuannongda No. 2	Huadamai No. 1	Huadamai No. 2	Huadamai No. 3	Huadamai No. 4	Huadamai No. 5
Released time/region	1993/Sichuan	1995/Sichuan	2001/Hubei	2001/Hubei	2003/Hubei	2004/hHu-bei,2006/Anhui	2007/Anhui
Row type/hull or naked	2/hull	2/hull	2/naked	2/hull	6/hull	6/hull	2/hull
Combinations	Nature cross offspring of Chuanluo No. 1	88BCMS× {(Pilengbo× Lina)× RisΦ1508}	84-123 x Chuan-nongda No. 2	Nature cross offspring of Chuanluc No. 1	85 V24 x Chuan-nongda No. 2	{85 V24× Chuan-nongda No. 2} ×Monker	(Zhenongda3× Chuannongda No. 2)× (Chuannongda No. 1×W168)
Novel germplasms utilized	Chuanluo No. 1	88BCMS	88BCMS	Chuanluo No. 1	88BCMS	88BCMS	88BCMS, Chuanluo No. 1
区试亩产/比当地当时皮大麦对照增产%	272 kg/12 %	307.7 kg/ 11.1 %	346.3 kg/ –9.5 %	373.0 kg/ –2.6 %	464.3 kg/ 5.7 %	398.2 kg/ 16.3 %	429.4 kg/ 3.57 %

	Huadamai No. 6	Huadamai No. 7	Huadamai No. 8	Huadamai No. 9	Chudamai 610	Edamai 12
Released time/region	2008/HUbei	2008/Hubei	2009/Hubei	2011/Hubei	2010/Hubei	2008/Hubei
Row type/hull or naked	2/hull	2/hull	2/hull	2/hull	6/hull	2/naked
Combinations	(Chuan-nongda No. 2× Pilengbo) ×(Misato golden× 5199)	(Chuannongda No. 2× Plengbo) (Misato golde× 5199)	(Chuannongda No. 2号× Ganmue-rtiao) (Misato golde× Zhenon-gdano. 3)	(Misato golde× Huaai11) (Chuan-nongda No. 1× W168)	Huadamai No. 3× Edamai 7	Epi No. 2× Chuan-nongda No. 1
Novel germplasms utilized	Chuanluo No. 15199	88BCMS, 5199	88BCMS	Chuanluo No. 1	88BCMS	88BCMS
区试亩产/比当地当时皮大麦对照增产%	388.45 kg/ 3.7 %,	383.7 kg/ 8.8 %	390.35 kg/ 8.01 %	389.5 kg/ 4.01 %	400.6 kg/ 13.2 %	356.31 kg/ 1.01 %

Days to mature	170	163	176	174	175	175	182	182	173	175	184	184	175
Days early to local wheat standard variety	15	20	18	20	15	15	15	15	20	18	13	13	18
Plant height (cm)	80	80	80	80	80–85	85	90	90	90	91	80	86	90
Purpose and reaching national standard	Both/ reaching	Feed/ exceeding	Feed/ exceeding	Malt/ exceeding	Feed/ reaching	Feed/ reaching	Malt/exceeding	Malt/ exceeding	Malt/ exceeding	Feed/ exceeding	Feed/ exceeding	Feed/ reaching	Feed/ exceed-ing
Resistance to disease	Yellow rust, fusarium	Yellow rust	All rusts and powdery mildew	All rusts and powdery mildew	All rusts and powdery mildew	All rusts and powdery mildew	All rusts and powdery mildew	All rusts and powdery mildew	All rusts and powdery mildew	All rusts and powdery mildew	Powdery mildew, fusarium	All rusts	All rusts and powdery mildew

By cooperation with these units, we have established a few core demonstration areas of barley production which can guarantee the quality and yield of barley. At present, this technology system utilizing the prepared row of cotton and corn as well as winter fallow field has widely applied.

References

Ding, Y., & Song, Y. C. (1991). Comparisons between G-,C-,N-banded and silver stained chromosomes in barley(Hordeum vulgare L.). *Chinese Science Bulletin, 36*(19), 1641–1645.

Ding, Y., & Song, Y. C. (1992). Studies of chromosome NORs silver staining in wild and cultivated barley. In *Proceedings of the Second Sino-Japanese Symposium on Plant Chromosomes: Plant Chromosome Research 1992* (pp. 147–152). Beijing: International Academic Publishers, 1994

Ding, Y., Son, Y. C., & Yu, X. J. (1996). Analysis of G-band fluctuation and correspondence of G-banding patterns in barley. *Chinese Journal of Genetics, 23*(4), 295–302.

Ellis, R. P., Forster, B. P., Gordon, D. C., Handley, L. L., Keith, R., Lawrence, P., Meyer, R. C., Powell, W., Robinson, D., Scrimgeour, C. M., Young, G. R., & Thomas, W. T. B. (2002). Phenotype/genotype associations of yield and salt tolerance in a barley mapping population segregating for two dwar.ng genes. *Journal of Experimental Botany, 53*, 1163–1176.

Jin, X. L., Cai, S. G., Qiu, B. Y., Qiu, L., Wu, D. Z., Ye, L. Z., & Zhang, G. P. (2011). Genetic variants of HvLDI in Tibetan annual wild and cultivated barleys and their association with LD activity and malt quality. *Journal of Cereal Science, 53*, 59–64.

Kaneko, T., Kihara, M., Ito, K., & Takeda, K. (2000). Molecular and chemical analysis of beta-amylase-less mutant barley in Tibet. *Plant Breeding, 119*, 383–387.

Konishi, T. (2001). Genetic diversity in *Hordeum agriocrithon E. Aberg*, six-rowed barley with brittle rachis, from Tibet. *Genetic Resource and Crop Evolution, 48*, 27–34.

Ma, Z., Steffenson, B. J., Prom, L. K., & Lapitan, N. L. (2000). Mapping of quantitative trait loci for *Fusarium* head blight resistance in barley. *Phytopathology, 90*, 1079–1088.

Massa, A. N., & Morris, C. F. (2006). Molecular evolution of puroindoline-a, puroindoline-b and grain softness protein-1 genes in the tribe triticeae. *Journal of Molecular Evolution, 63*, 526–536.

Nevo, E. (1992). Origin, evolution, population genetics and resources for breeding of wild barley, *Hordeum spontaneum* in the Fertile Crescent. In P. R. Shewry (Ed.), *Barley genetics, biochemistry, molecular biology and biotechnology* (pp. 19–43). CAB International: Wallingford.

Nevo, E., Apelbaum-Elkaher, I., Garty, J., & Beiles, A. (1997). Natural selection causes microscale allozyme diversity in wild barley and lichen at 'Evolution Canyon' Mt Carmel Iseael. *Heredity, 78*, 373–382.

Qiu, L., Wu, D. Z., Ali, S., Cai, S. G., Dai, F., Jin, X. L., Wu, F. B., & Zhang, G. P. (2011). Evaluation of salinity tolerance and analysis of allelic function of HvHKT1 and HvHKT2 in Tibetan annual wild barley. *Theoretical and Applied Genetics, 122*, 695–703.

Saghai Maroof, M. A., Allard, R. W., & Zhang, Q. (1990). Genetic diversity and ecogeographical differentiation among ribosomal alleles in wild and cultivated barley. *Proceedings of the National Academy of Sciences, USA, 87*, 8486–8490.

Sun, D. F. (2011). The planting mode and cultivation techniques of barley in utilizing winter fallow field and the prepared row of cotton and corn. *Journal of Zhejiang Agricultural Science, 1*, 75–76.

Sun, D. F., & Gong, X. (2009). Barley germplasm and utilization. In G. Zhang & C. Li (Eds.), *Genetics and improvement of barley malt quality*. Berlin/Heidelberg: Springer. ISBN 978-3-642-01278-5:18–62.

Sun, D. F., Xu, T. W., Jiang, H. R., & Zhao, L. (1995). Selection and genetic analysis of nucleo-cytoplasmic male sterile line 88BCms in barley. *Scientia Agricultra Sinica, 28*, 37–42.

Sun, D. F., et al. (1998). Inheritance of special early heading of a super early maturing line 5199 in barley. *Journal of Huazhong Agricultural University, 17*(5), 418–424.

Sun, D. F., Ren, W. B., Sun, G. L., & Peng, J. H. (2011). Molecular diversity and association mapping of quantitative traits in Tibetan wild and worldwide originated barley (Hordeum vulgare L.) germplasm. *Euphytica, 178*, 31–43.

Thomas, W. T. B., Barid, E., Fuller, J., Lawrence, P., Young, G. R., Russel, J. R., Ramsay, L., Waugh, R., & Powell, W. (1998). Identification of a QTL decreasing yield in barley linked to Mlo powder mildew resistance. *Molecular Breeding, 4*, 381–393.

Xu, T. (1987). Origin and phylogeny of cultivated barley in China. In *Barley Genetics V* (pp. 91–99). Okayama: Academic Societies of Japan.

Ye, L. Z., Dai, F., Qiu, L., Sun, D. F., & Zhang, G. P. (2011). The allelic diversity of the haze active protein gene in cultivated and Tibetan wild barley and the development of the allelic specific markers. *Journal of Agriculture and Food Chemistry, 49*, 7218–7223.

Yin, Y. Q., Ma, D. Q., & Ding, Y. (2003). Analysis of genetic diversity of hordein in wild close relatives of barley from Tibet. *Theoretical and Applied Genetics, 107*, 837–842.

Zohary, D., & Hopf, M. (1988). *Domestication of plants in the Old World*. Oxford: Clarendon.

Chapter 9
Agronomic and Quality Attributes of Worldwide Primitive Barley Subspecies

Abderrazek Jilal, Stefania Grando, Robert James Henry, Nicole Rice, and Salvatore Ceccarelli

Abstract Old barley germplasm from the primary gene pool (landraces and wild relative) provides a broad representation of natural variation not only in agronomically important traits but also in nutraceuticals. Five hundred and twenty barley landraces including 36 wild barley relatives belonging to 33 countries were subject to agronomic and quality screening. The mean values for the four implemented environments (two sites by 2 years) revealed that the subspecies *H. spontaneum* was a great source of important traits (spike length, plant height, protein content and β-glucan content) comparing to *H. vulgare* (the top in grain yield and heading date) and *H. distichon* (first in TKW and particle size index). The blue aleurone colour was dominated for the most studied accessions. The ANOVA between subspecies associated with canonical variate analysis and hierarchical clustering confirms the finding to be used in barley breeding through incorporation of the candidate gene into the commercial varieties.

Keywords Barley • Germplasm • *H. spontaneum* • Agronomic and quality attributes

---

A. Jilal (✉) • S. Grando • S. Ceccarelli
Grain Foods CRC, International Centre for Agricultural Research in the Dry Areas (ICARDA), P.O. Box 5466, Aleppo, Syria
e-mail: abderrazek_2001@yahoo.fr

R.J. Henry
Grain Foods CRC, Centre for Plant Conservation Genetics (CPCG), Southern Cross University, P.O. Box 157, Lismore, NSW 2480, Australia

Queensland Alliance for Agriculture and Food Innovation, The University of Queensland , Brisbane, QLD 4072, Australia

N. Rice
Grain Foods CRC, Centre for Plant Conservation Genetics (CPCG), Southern Cross University, P.O. Box 157, Lismore, NSW 2480, Australia

G. Zhang et al. (eds.), *Advance in Barley Sciences: Proceedings of 11th International Barley Genetics Symposium*, DOI 10.1007/978-94-007-4682-4_9,

9.1 Introduction

Natural selection accompanied by conscious selection of desired genotypes by farmers during centuries of cultivation has resulted in landraces that are genetically variable for qualitative and quantitative characters, have good adaptation to specific environmental conditions and give dependable yields (Harlan 1992; Ceccarelli 1996). This advantage of barley landraces may be explained by their buffering ability, wherein better-adapted individuals of the population compensate for the losses of less-adapted lines under harsh conditions. These landraces formed the foundation material in breeding programmes, which started some 150 years ago with the development of new technology and methods that have increased the genetic diversity even further and turned barley into the universal, highly diverse crop it is today (von Bothmer et al. 2003). Wild barley, *Hordeum vulgare* L. ssp. *spontaneum (C. Koch)*, is the progenitor of cultivated barley (Harlan and Zohary 1966). It populates diverse habitats ranging from high-rainfall regions to deserts (Volis et al. 2002). *Hordeum vulgare* ssp. *spontaneum C. Koch* is a valuable source of new genes for breeding. These include grain protein content (Jaradat 1991), earliness (Korff et al. 2004; Li et al. 2005), biomass (Lu et al. 1999) and plant height under drought (Baum et al. 2003). The objective of this study was to disclose the agronomic and quality richness of the primitive barley subspecies.

9.2 Materials and Methods

9.2.1 Experimental Materials

Out of the 576 barley landraces from 43 countries worldwide conducted at tow ICARDA's experimental stations (Tel Hadya and Breda) during the cropping seasons 2005–2006 and 2006–2007, 520 of them from 33 countries were selected for this detailed study by excluding countries with low number of samples (Fig. 9.1). These accessions comprised three subspecies *H. vulgare ssp. vulgare convar. vulgare, H. vulgare ssp. vulgare convar. distichon* and *H.* ssp. *spontaneum* referred hereafter as *H. vulgare, H. distichon* and *H. spontaneum,* respectively. The trial was carried out at Tel Hadya (36°01'N; 37°20'E, elevation 284 m asl, referred as TH) and Breda (35°56'N; 37°10'E, elevation 354 m asl, referred as BR) featuring a long-term average rainfall of 340 (30 seasons) and 275 mm (25 seasons), respectively. The experimental design was a spatial row and column design with four repeated checks. Plots were 1.8 m wide (6 rows 30 cm apart) and 2.5 m long.

9.2.2 Field Observations

Five agronomic parameters were evaluated. Heading date (HD) was determined as the number of days from emergence to the day when the awns appeared in 50% of the plot. Plant height (PH: height from the soil surface to the top of the spike,

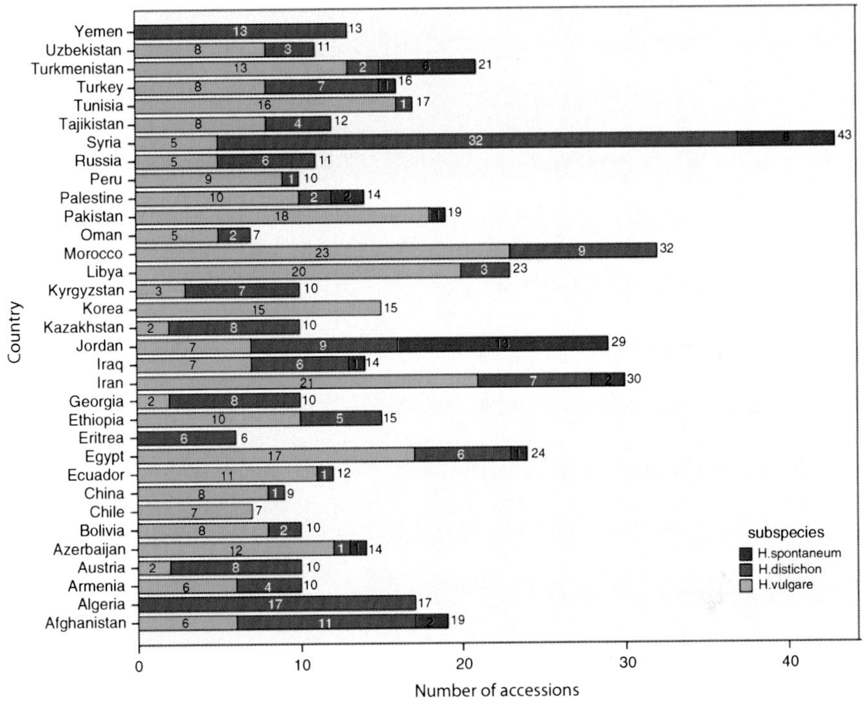

Fig. 9.1 Number of investigated barley accessions per subspecies within country origin

excluding the awns) and spike length (SL: length of the main spike of the tallest culm) were measured at maturity based on the average of three measures each. Grain (GY) and biomass (BY) yields were obtained from the four middle rows of the plot (1.5 m^2). Once the grains have been harvested, thousand kernel weight (TKW) was measured via grain counter.

9.3 Quality Evaluations

Aleurone colours (Al) scored on pearled barley as white, green, blue, brown or black kernels are depicted in Fig. 9.2. Grain hardness was determined via particle size index (PSI) as detailed by Williams et al. (1988), and protein (P) and β-glucan (BG) content were evaluated using near-infrared spectroscopy (Williams and Norris 2001).

9.4 Statistical Analysis

The adjusted BLUP data, best linear unbiased predictors, based on the repeated checks was produced using SPUR (spatial analysis for unreplicated design) in GenStat software (VSN International Ltd 2003). Based on the mean values of the

Fig. 9.2 The five aleurone colours of the investigated barley accessions

four environments, descriptive statistics (Range, Mean, CV%, Mode and Rank) per subspecies associated with the analysis of variance between subspecies were performed for the agronomic attributes (HD, PH, SL, GY, BY and TKW) and the quality attributes (Al, PSI, P and BG). The only nominal variable (Al) was subject to chi-square test over subspecies. Canonical variate analysis (CVA) was performed for the adjusted data using subspecies as the defined groups of accessions, and the clustering were carried out using Ward's method and presented via DARwin software (version 5.0.130) (Perrier et al. 2003).

9.5 Results

Descriptive statistics associated with the analysis of variance between subspecies for the agronomic and quality attributes were presented in Table 9.1. Based on the highly significance of ANOVA and the mean values of each parameter, the comparison of the three subspecies *H. vulgare, H. distichon* and *H. spontaneum* featured high grain yields and long-cycle crop for the six-row cultivated barleys, high thousand kernel weight and high particle size index for the two cultivated barleys and high spike length, earliness (low heading date), high plant height, high protein content and high β-glucan for the wild species. Aleurone colour, highly significant between subspecies through chi-square, dominated blue colour in *H. vulgare* (52.8%) and *H. spontaneum* (44.4%) accessions, while white colour reigned the *H. distichon* (48.2%). The biomass yield was found similar between the three subspecies ($p > 0.05$).

Canonical discriminant analysis using subspecies as classifying variable explained 100% of the variation with the two first functions (CVA-1 with 75.20% and CV-2 with 24.80%) as depicted in Fig. 9.3. The three subspecies were clearly distinguishable by the convex hull with an overlapping between the cultivated barleys. The canonical variate analysis biplot revealed that the subspecies *H. spontaneum* (red colour) featured the high values of the parameters (P, BG, PH and SL) with also low PSI index and low GY confirming the results in Table 9.1. Same outputs were supported for the two cultivated subspecies.

Cluster analysis of the investigated data using Ward's method (Fig. 9.4) revealed the distinct group (circles) of the wild species (*H. spontaneum*), coloured in red, in addition to some accessions grouping the cultivated barleys (two-row barley in blue colour and six-row barley in green colour.)

Table 9.1 Descriptive statistics associated with *F*-ANOVA between subspecies of the investigated agronomic and quality attributes

	Item	Barley subspecies			
		H. vulgare	*H. distichon*	*H. spontaneum*	
	N	314	170	36	F-ANOVA between subspecies
Agronomic attributes	*Grain yield (g/m²)*				
	Range	207–800.89	218.35–791.95	100.46–525.99	20.65***
	Mean	**387.09**	380.88	249.2	
	CV (%)	0.337	0.294	0.391	
	Thousand kernel weight (g)				
	Range	20.47–47.92	24.57–46.01	25.77–42.2	28.54***
	Mean	33.3	**35.77**	32.42	
	CV (%)	0.113	0.101	0.105	
	Spike length (cm)				
	Range	5.64–9.56	6.22–8.79	6.48–9.34	110.19***
	Mean	7.07	7.71	**8.08**	
	CV (%)	0.078	0.064	0.087	
	Biomass yield (g/m²)				
	Range	730.86–1736.99	732.89–1644.44	734.48–1401.09	2.18 ns
	Mean	**1041.75**	993.66	1014.26	
	CV (%)	0.245	0.235	0.157	
	Heading date (days)				
	Range	87.58–112.58	88.69–108.79	88.81–102	23.97***
	Mean	**97.54**	95.49	92.67	
	CV (%)	0.05	0.048	0.031	
	Plant height (cm)				
	Range	48.45–70.81	50.07–74.84	56.89–69.72	13.65***
	Mean	60.13	60.24	**63.95**	
	CV (%)	0.068	0.073	0.061	

(continued)

Table 9.1 (continued)

	Item	Barley subspecies			
		H. vulgare	*H. distichon*	*H. spontaneum*	
	N	314	170	36	F-ANOVA between subspecies
Quality attributes	*Particle size index (%)*				
	Range	38.54–48.05	40.32–47.67	40.25–46.47	39.02***
	Mean	45.12	**45.75**	42.75	
	CV (%)	0.045	0.035	0.032	
	Protein content (%)				
	Range	11.44–16.8	11.36–15.14	13.37–18.78	141.9***
	Mean	13.5	12.98	**15.7**	
	CV (%)	0.061	0.068	0.083	
	Beta glucan content (%)				
	Range	4.88–6.01	5.06–6.05	5.44–6.18	53.92***
	Mean	5.39	5.35	**5.73**	
	CV (%)	0.039	0.034	0.034	
	Aleurone colour (code)[a]				
	Range	1–5	1–5	1, 2, 3, 5	*Chi-square*
	Mode	**3** (52.8%)	**1** (48.2%)	**3** (44.4%)	109.88***
	Rank	3,1, 2, 5, 4	1, 3, 5, 4, 2	3, 2, 1, 5	

[a]Aleurone colour: 1 (*white*); 2 (*green*); 3 (*blue*); 4 (*brown*); 5 (*black*)
*** the standard of data accuracy
Bold means Maximum data

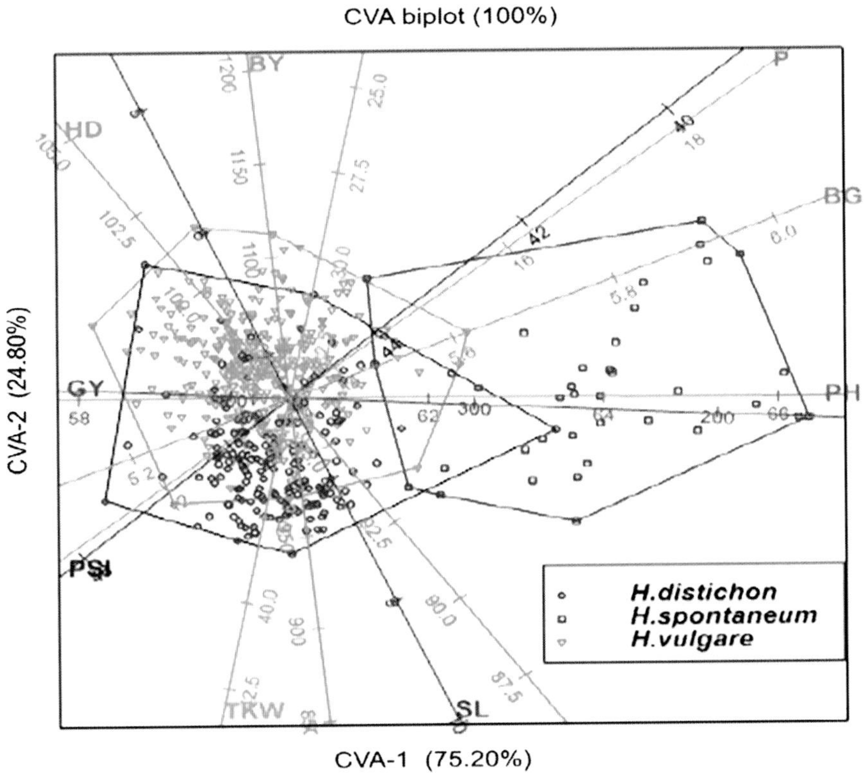

Fig. 9.3 Canonical variate analysis biplot of the 520 barley accessions over the ten agronomic and quality parameters

9.6 Discussions

The wild barley *H. spontaneum*, a source of protein content as reported by Friedman and Atsmon (1988), Nevo et al. (1985) and Corke and Atsmon (1990), confirms the outputs disclosed through assessing the 520 barley accessions likewise for the spike length and the plant height cited by Pillen et al. (2003) and Al-Saghir et al. (2009). The high β-glucan in wild barley depicted in this study was reported by Leon et al. (2000) who observe that some *H. spontaneum* lines had glucan contents twofold higher than those of the cultivars. These findings reveal the great source of the wild barley relatives "*H. spontaneum*" in terms of protein, β-glucan, spike length and plant height which can be used in barley breeding via the incorporation of the candidate gene of interest through crossing and high throughput genotyping from the wild barley to the cultivated one. β-glucan content in two-row barleys resulted lower than in six-row barleys as reported by Narasimhalu et al. (1994). High TKW in *H. distichon* comparing to *H. vulgare* was supported by Newman and Newman (2008).

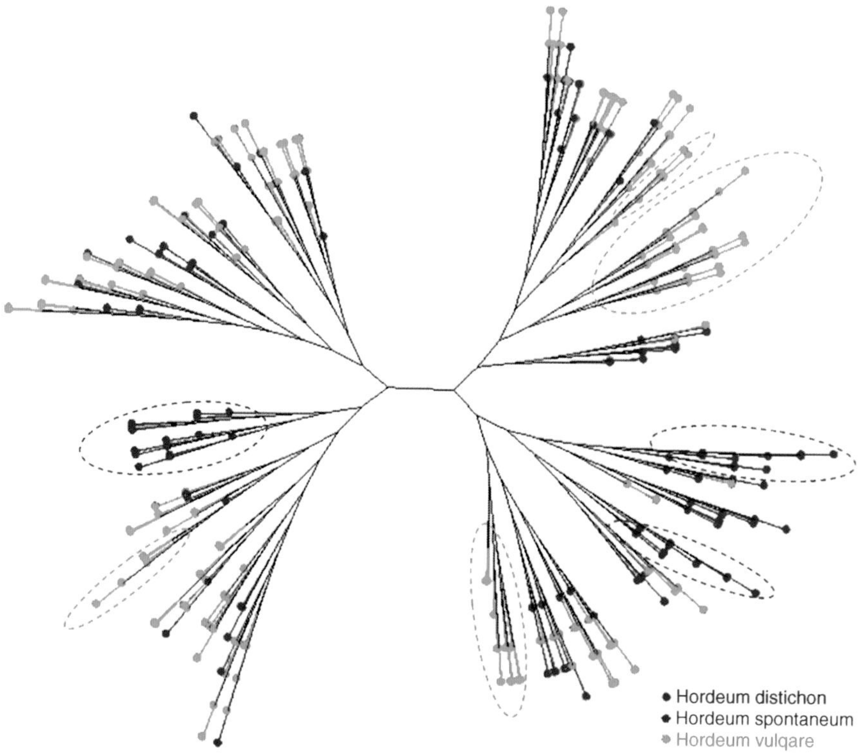

Fig. 9.4 Dendrogram of the cluster analysis of the ten parameters investigated for the worldwide barley subspecies

The high grain yield and the long-cycle crop of the six-row barleys overweighting the other subspecies in addition to high TKW and high PSI index (soft kernel) in two-row barleys could be exploited by using these landraces with elite trait phenotyping to be used in barley breeding for candidate genes or through advanced backcrosses.

References

Al-Saghir, M. G., Malkawi, H. I., & El-Oqlah, A. (2009). Morphological diversity in Hordeum spontaneum C. Koch of Northern Jordan (Ajloun Area). *Middle East Journal of Scientific Research, 4*(1), 24–27.

Baum, M., Grando, S., Backes, G., Jahoor, A., Sabbagh, A., & Ceccarelli, S. (2003). QTL for agronomic traits in the Mediterranean environment identified in recombinant inbred lines of the cross 'Arta' 9H. *spontaneum* 41–1. *Theoretical and Applied Genetics, 107*, 1215–1225.

Ceccarelli, S. (1996). Adaptation to low/high input cultivation. *Euphytica, 92*, 203–214.

Corke, H., & Atsmon, D. (1990). Endosperm protein accumulation in wild and cultivated barley and their cross grown in spike culture. *Euphytica, 48*, 225–231.

Friedman, M., & Atsmon, D. (1988). Comparison of grain composition and nutritional quality in wild barley (*Hordeum spontaneum*) and in a standard cultivar. *Journal of Agricultural and Food Chemistry, 36*, 1167–1172.

Harlan, J. R. (1992). *Crops and man*. Madison: American Society of Agronomy.

Harlan, J. R., & Zohary, D. (1966). Distribution of wild wheats and barley. *Science, 153*, 1074–1080.

Jaradat, A. A. (1991). Grain protein variability among populations of wild barley (*Hordeum spontaneum C. Koch*) from Jordan. *Theoretical and Applied Genetics, 83*, 164–168.

Korff, V. M., Wang, H., Leon, J., & Pillen, K. (2004) Detection of QTL for agronomic traits in an advanced backcross population with introgression from wild barley (Hordeum vulgare ssp. spontaneum). In: *Genetic variation for plant breeding, Proceeding of the 17th EUCARPIA General Congress, 8–11 September, 2004, Tulln* (pp. 207–211). Vienna: BOKU – University of Natural Resources and Life Sciences.

Leon, J., Silz, S., & Harloff, H. J. (2000). β-D-glucan content during grain filling in spring Barley and its wild progenitor H. vulgare ssp. Spontaneum. *Journal of Agronomy and Crop Science, 185*, 1–8.

Li, J. Z., Huang, X. O., Heinrichs, F., & Ganal, M. W. (2005). Analysis of QTLs for yield, yield components, and malting quality in a BC3-DH population of spring barley. *Theoretical and Applied Genetics, 110*, 356–363.

Lu, Z., Neumann, P. M., Tamar, K., & Nevo E (1999) Physiological characterization of drought tolerance in wild barley (*Hordeum spontaneum*) from the Judean Desert. http://wheat.pw.usda.gov/ggpages/bgn/29/a29-09.html

Narasimhalu, P., Kong, D., Choo, T. M., Ferguson, T., Therrien, M. C., Ho, K. M., May, K. W., & Jiu, P. (1994). Effects of environment and cultivar on total mixed-linkage,6-glucan content in Eastern and Western Canadian barleys (Hordeum Vulgare L.). *Canadian Journal of Plant Science, 75*, 371–376.

Nevo, E., Atsmon, D., & Beiles, A. (1985). Protein resources in wild barley, *Hordeum spontaneum*, in Israel: Predictive method by ecology and allozyme markers. *Plant Systematics and Evolution, 150*, 205–222.

Newman, R. K., & Newman, C. W. (2008). *Barley for food and health, science, technology and products*. Hoboken: Wiley. ISBN 978-0-470-10249-7.

Perrier, X., Flori, A., & Bonnot, F. (2003). Data analysis methods. In P. Hamon, M. Seguin, X. Perrier, & J. C. Glaszmann (Eds.), *Genetic diversity of cultivated tropical plants* (pp. 43–76). Enfield: Science Publishers.

Pillen, K., Zacharias, A., & Leon, J. (2003). Advanced backcross QTL analysis in barley (*Hordeum vulgare* L.). *Theoretical and Applied Genetics, 107*, 340–352.

Volis, S., Mendlinger, S., & Ward, D. (2002). Adaptive traits of wild barley of Mediterranean and desert origin. *Oecologia, 133*, 131–138.

Von Bothmer, R., Sato, K., Knüpffer, H., & van Hintum, T. (2003). Diversity in barley – *Hordeum vulgare. Developments in Plant Genetics and Breeding, 7*, 280.

VSN International Ltd. (2003). *GenStat Release 7.1*. Rothamsted: Lawes Agricultural Trust. ISBN 1-904375-11-1.

Williams, P., & Norris, K. (2001). *Near infrared technology in the agricultural and food industries*. St. Paul: American Association of Cereal Chemists.

Williams, P. C., Jaby El-Haramein. F., Nakkoul, H., & Rihawi, S. (1988). *Crop quality evaluation, methods and guidelines* (Technical Manual No. 14). Aleppo: ICARDA.

Chapter 10
Differences between Steely and Mealy Barley Samples Associated with Endosperm Modification

Barbara Ferrari, Marina Baronchelli, Antonio M. Stanca, Luigi Cattivelli, and Alberto Gianinetti

Abstract Structurally, different areas may occur in the endosperm of the barley grain, and they can be visually classified as either mealy or steely. Barleys with a high proportion of grains that are mostly steely often show uneven physical–chemical modification of the endosperm during malting. To study the relationship between steeliness and endosperm modification, two samples of barley cv. Scarlett with contrasting malting quality were analysed. The proportions of steely grains were 77 and 46% in the two samples, which were then defined as steely sample and mealy sample, respectively. The steely sample showed slower modification during malting (in terms of β-glucan degradation, friability increase, and Calcofluor staining), lower hot water extract (HWE) and acrospire growth, and higher extract viscosity. Endosperm permeation to large molecules (tested with the fluorescein isothiocyanate–dextran conjugate, FITC-D) closely followed cell wall modification in the steely sample, but this was not so in the mealy sample. Higher steeliness was associated with higher levels of C hordeins in the grain of barley cv. Scarlett. It is proposed that such hordeins can increase the permeability to large molecules (FITC-D) but slow modification. Like steeliness and the level of C hordeins, permeability to FITC-D appears to be

B. Ferrari
CRA – Fodder and Diary Productions Research Centre,
Viale Piacenza, 29, 26900 Lodi, Italy

M. Baronchelli • L. Cattivelli • A. Gianinetti
CRA Genomics Research Centre, Via San Protaso 302,
29017 Fiorenzuola d'Arda (PC), Italy

A.M. Stanca (✉)
CRA Genomics Research Centre, Via San Protaso 302,
29017 Fiorenzuola d'Arda (PC), Italy

Department of Agricultural and Food Sciences, University of Modena
and Reggio Emilia, Reggio Emilia, Italy
e-mail: michele@stanca.it

G. Zhang et al. (eds.), *Advance in Barley Sciences: Proceedings of 11th International Barley Genetics Symposium*, DOI 10.1007/978-94-007-4682-4_10,

more linked to environmental rather than genetic effects. Although a more general association of C hordeins with steeliness of malting barley still has to be ascertained, the negative role of C hordeins in malting quality has been confirmed.

Keywords Barley • Malt

10.1 Background

In the germinating barley grain, breakdown of endosperm cell walls and partial digestion of the protein matrix render starch granules accessible to amylolytic enzymes. These two processes make the starch susceptible to degradation during mashing (Lewis and Young 2002). Degradation of cell walls (mainly constituted by β-glucans) represents the essential step of endosperm modification and occurs at the boundary between two compartments: the modified cell wall spatial network and the unmodified endosperm (Gianinetti 2009). Some predictions on the β-glucan profile in the modifying grain can be derived from theoretical considerations (Gianinetti 2009), but actual analysis of β-glucan levels in the modified and unmodified endosperm would be useful to check details of β-glucan breakdown. To partition the grain into two fractions that are closely associated with modification compartments, the friability test can be used. The friabilimeter separates a friable portion of the modified grain from a non-friable residue (Briggs 1998). Friability is linked to the extent of cell wall modification (Anderson 1996; Wentz et al. 2004). Thus, the two fractions produced by this device should provide good indications on the breakdown of β-glucans in the germinating barley grain.

Physical modification of the barley endosperm, although closely related to β-glucan breakdown (Aastrup and Erdal 1980), does not coincide with this single hydrolytic step (Fretzdorff et al. 1982; de Sá and Palmer 2004). Following β-glucan degradation, the change in the chemical properties of the grain is greatly intensified by the digestion of the protein matrix (mainly hordeins) that envelops starch granules (Lewis and Young 2002; Slack et al. 1979). It is well known that the higher the level of protein in the barley grain, the lower is the extract yield of the malt (Bishop 1930). In addition to this negative relationship with starch level, storage proteins have been suggested to play a significant part in determining the rate at which the whole during malting (Palmer 1989). Hordeins (barley prolamines) are water-insoluble but alcohol-soluble proteins and represent the main storage protein fraction in barley grains. They account for 35–55% of the total protein in the mature grain and may be classified into different groups, based on their electrophoretic mobility (Shewry 1993). The B and C groups account for 70–80% and 10–20% of total hordeins, respectively, whereas the D and γ hordeins are minor components (Shewry and Darlington 2002).

Although the levels of hydrolytic enzymes have an important role in malting quality, the structural properties of the endosperm may also be a crucial factor (Palmer 1989; Palmer and Harvey 1977; Brennan et al. 1997; Holopainen et al. 2005; Gianinetti et al. 2007). Some lots of malting barleys have a hard, vitreous endosperm and are difficult to malt; the term ‘steely’ is used to stress these features.

On the contrary, 'mealy' barleys modify quickly (Palmer 1989; Chandra et al. 1999). The endosperm of steely grains tends to have a more compact protein matrix than that of mealy grains, and the starch granules are smaller with less free spaces (Palmer 1989). Barleys with a high proportion of steely grains are often associated with a higher protein and β-glucan content, as well as uneven modification of the endosperm during malting (Chandra et al. 1999).

Differences in the speed of cell wall modification can be due to a difference in the initial β-glucan content, in the level of β-glucanase developed during germination, or in the apparent β-glucan degradability (Gianinetti et al. 2007). The latter can be quantified but not yet explained (Gianinetti 2009). However, accessibility of β-glucanase to its substrate has been suggested to play a pivotal role in determining cell wall modification, and matrix proteins could affect this interaction (Palmer 1989; Brennan et al. 1997; Holopainen et al. 2005). Total protein content does not correlate with the apparent β-glucan degradability (Gianinetti et al. 2007).

10.2 Differences Between Steely and Mealy Barley Endosperm

To obtain clues on the possible effect of hordeins on cell wall modification, we studied the constitutive differences between two samples of barley cv. Scarlett, one steely and the other mealy, and we observed the changes that β-glucans and hordeins undergo during malting (Ferrari et al. 2010).

The proportions of steely grains were 77 and 46% in the two samples (Ferrari et al. 2010), which were then defined as steely sample and mealy sample, respectively. The steely sample showed slower modification during malting (in terms of β-glucan degradation, friability increase, and Calcofluor staining), lower hot water extract (HWE) and acrospire growth, and higher extract viscosity (Ferrari et al. 2010). Endosperm permeation to large molecules (tested with fluorescein isothiocyanate–dextran conjugate, FITC-D; Gianinetti 2009) closely followed cell wall modification in the steely sample, but this was not so in the mealy sample (Ferrari et al. 2010).

Our results (Ferrari et al. 2010) suggest that the region of the endosperm that has undergone cell wall modification can be conveniently represented by the friable flour obtained from the friabilimeter test, whereas the non-friable residue gives a biased representation of the unmodified endosperm, because it includes some portions of the modified endosperm. Notwithstanding this, some interesting observations can be made about endosperm modification. In accordance with what was expected by theoretical considerations (Gianinetti 2009), low but significant amounts of β-glucans were still present in the modified region of the endosperm (friable flour) of the germinating barley grains (Ferrari et al. 2010). In addition, the β-glucan levels of the friable flours were very similar in the two samples. In fact, β-glucans are expected to be quickly degraded by an excess of enzyme in the modified endosperm (Gianinetti 2009), so that the different β-glucan contents in the unmodified endosperms of the two samples are levelled off in the modified compartments.

The Scarlett sample with a higher proportion of steely grains had higher protein and β-glucan contents as well (Ferrari et al. 2010), as is frequently found (Chandra et al. 1999). During malting, friability increases in close association with β-glucan breakdown (Anderson 1996; Wentz et al. 2004): once the cell wall framework is broken down, the endosperm becomes a loosely packed agglomerate of naked storage cells and easily crumbles. However, when barley samples from different genotypes and locations are considered together, friability of the finished malt shows a negative correlation with protein content (Gianinetti et al. 2005). Since proteins should not play a direct role in holding together the naked storage cells, this result confirms that total protein content can affect the speed of cell wall modification (Palmer 1989) and, in turn, the friability that a sample acquires within the prefixed time of malting. In fact, for a given genotype, higher protein levels appear to cause reduced breakdown of β-glucans (Darlington and Palmer 1996).

Well-made malts often only contain about half the quantity of hordein originally present in the unmalted barley (Briggs 1998). This was observed in our experiment too (Ferrari et al. 2010). The steely Scarlett sample had a whole-grain hordein content about 0.5% greater than that of the mealy sample, and this higher level was maintained throughout modification (Ferrari et al. 2010). However, in the friable flours of the two Scarlett samples, the hordein contents were relatively similar, analogously to what was observed for β-glucans, and the hordein content of the non-friable residue was even slightly higher in the mealy than in the steely sample. The higher whole-grain hordein content of the steely sample was mostly due to the reduced progress of modification, so that in this sample the non-friable residue contributed more to the overall grain content than occurred in the case of the mealy sample, analogously to what already observed for β-glucans. Remarkably, higher levels of C hordeins were found in the steely sample (Ferrari et al. 2010). Since these hordeins are those that are more slowly modified (Celus et al. 2006), they appear to be responsible for the steadily greater hordein content in the steely sample. This points to a possible role of the higher level of C hordeins in the slower modification of the steely sample.

In a previous study (Gianinetti 2009), permeation with FITC–dextran was performed to show that endosperm permeabilisation to enzymes is closely associated with β-glucan degradation. That study employed the Scarlett steely sample that was further characterised in the work of Ferrari et al. (2010). However, when the same technique was applied to the mealy sample, a different, unexpected, result was observed: after three days of malting, the dye was no longer able to reach the front of cell wall modification, even if cell walls appeared to be fully degraded in the modified area (Ferrari et al. 2010). Indeed, hordeins can form a gel (Smith and Lister 1983), and differences in hordeins were observed between these samples. These differences might then be linked to different endosperm permeability to large molecules, but since the protein matrix of the endosperm is enclosed by the cell walls, breakdown of wall β-glucans has to take place prior to hordein digestion. The latter is thus dependent on the former (Lewis and Young 2002), and a reversal effect is not easy to explain. In addition, reduced permeability of the modified endosperm was observed in the mealy sample, which had a faster cell wall modification than

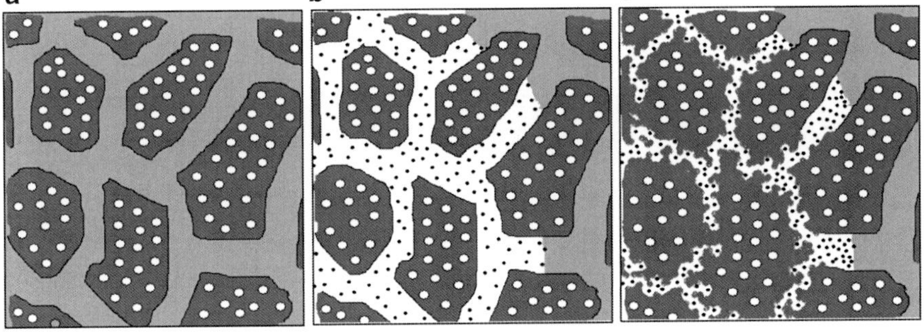

Fig. 10.1 Schematic illustration of the mechanism proposed to explain why permeation to large molecules is hindered in the mealy sample, which has faster modification. (**a**) Undegraded barley endosperm (cell walls are *light grey*, the protein matrix is *dark grey*, *white spots* are starch granules). (**b**) The barley endosperm is partially degraded (*light grey* cell walls turn *white*) by β-glucanase (*black spots*) and other enzymes: cell wall degradation is represented as a two-compartment process; in one compartment, cell walls are solubilised so that this space is fully accessible to large molecules; in the other compartment, cell walls are undegraded and enzymes cannot penetrate, and cell walls are dismantled at the boundary between the two compartments (Gianinetti 2009). (**c**) A low level of C hordeins would favour in vivo protein gelation and swelling, so that the space freely accessible to the enzymes is restricted; therefore, β-glucanase concentration is higher and modification speed is increased

the steely sample. Two questions therefore arise: Why was permeation to large molecules hindered in the mealy sample, which is that with less hordeins? and why should this have a positive effect, if any, on β-glucan degradability? We can only propose hypotheses to account for these findings. Firstly, the mealy sample had less C hordeins, and these proteins contain no cysteine residues and cannot form disulphide bridges (Shewry and Darlington 2002). These hordeins might then hinder disulphide cross-linking gelation of sulphur-rich hordeins. The latter can, in fact, form a gel (Smith and Lister 1983; Skerritt and Janes 1992) within which C hordeins are trapped (Celus et al. 2006). High levels of C hordeins could then prevent the formation of a hordein gel. Secondly, the hypothesised formation of a gel in the mealy sample would correspond to a smaller volume of endosperm being accessible to β-glucanase and to a greater enzyme concentration in the free volume (Fig. 10.1). In other words, as protein would gelatinise and swell in the modified endosperm, the enzyme molecules would be pushed against the modification front where cell walls could then be more rapidly dismantled (Ferrari et al. 2010).

The endosperm region that, although already free of cell wall β-glucans, appears inaccessible to FITC-D is visible after the third day of malting. However, for the mealy sample, the strongest improvement in malting quality parameters occurred prior to this time point (Ferrari et al. 2010). This is particularly evident for friability, which is directly linked to cell wall modification. Hence, the notably extension of this inaccessible region in the quickly modifiable sample seems to reveal a difference vs. the steely sample that has already explicated most of its action before it becomes visible with the perfusion test with FITC-D. Indeed, this region becomes impermeable to large molecules from the third day on, when most β-glucanase has already been produced (Gianinetti et al. 2007), so that the presence of this barrier does not have a negative effect on the initial diffusion of the enzyme molecules towards the modification front. Instead, it can be supposed that before this region becomes completely occluded, protein swelling can only partially restrict the volume freed from cell walls, excluding enzyme molecules from part of the space that would be otherwise available to them, but not yet precluding their diffusion. A high B to C hordein ratio would increase in vivo gelation and modification speed and should therefore be desirable for malting quality, in addition to a low total protein content (Ferrari et al. 2010). Given that the B to C hordein ratio generally decreases with increasing nitrogen content (Griffiths 1987; Molina-Cano et al. 2001; Qi et al. 2006), it seems clear that, at high nitrogen contents, two negative effects occur: a slower modification because of the higher proportion of C hordeins and a reduction in extractable sugars (since more protein means less starch). Differences in the experimental conditions may lead one or the other effects to prevail and could thus explain the contrasting findings reported in literature on the relation between B hordeins and malting quality.

Higher steeliness was associated with higher levels of C hordeins in the grain of barley cv. Scarlett (Ferrari et al. 2010). It was proposed that such hordeins can increase the permeability to large molecules (FITC-D) but slow modification (Ferrari et al. 2010). Like steeliness and the level of C hordeins, permeability to FITC-D appears to be more linked to environmental rather than genetic effects (Ferrari et al. 2010). Although a more general association of C hordeins with steeliness of malting barley still has to be ascertained, the negative role of C hordeins in malting quality was confirmed (Ferrari et al. 2010).

References

Aastrup, S., & Erdal, K. (1980). Quantitative determination of endosperm modification and its relationship to the content of 1,3:1,4-β-glucans during malting of barley. *Carlsberg Research Communications, 45*, 369–379.

Anderson, I. W. (1996). The Cambridge Prize Lecture 1995: Indirect and direct contributions of barley malt to brewing. *Journal of the Institute of Brewing, 102*, 409–413.

Bishop, L. R. (1930). The nitrogen content and quality of barley. *Journal of the Institute of Brewing, 36*, 352–369.

Brennan, C. S., Amor, M. A., Harris, N., Smith, D., Cantrell, I., Griggs, D., & Shewry, P. R. (1997). Cultivar differences in modification patterns of protein and carbohydrate reserves during malting of barley. *Journal of Cereal Science, 26*, 83–93.

Briggs, D. E. (1998). *Malts and malting*. London: Blackie Academic & Professional.

Celus, I., Brijs, K., & Delcour, J. A. (2006). The effects of malting and mashing on barley protein extractability. *Journal of Cereal Science, 44*, 203–211.

Chandra, G. S., Proudlove, M. O., & Baxter, E. D. (1999). The structure of barley endosperm – An important determinant of malt modification. *Journal of the Science of Food and Agriculture, 79*, 37–46.

Darlington, H. F., & Palmer, G. H. (1996). Homogeneity of the friable flour of malting barley. *Journal of the Institute of Brewing, 102*, 179–182.

de Sá, R. M., & Palmer, G. H. (2004). Assessment of enzymatic endosperm modification of malting barley using individual grain analyses. *Journal of the Institute of Brewing, 110*, 43–50.

Ferrari, B., Baronchelli, M., Stanca, A. M., & Gianinetti, A. (2010). Constitutive differences between steely and mealy barley samples associated with endosperm modification. *Journal of the Science of Food and Agriculture, 90*, 2105–2113.

Fretzdorff, B., Pomeranz, Y., & Bechtel, D. B. (1982). Malt modification assessed by histochemistry, light microscopy, and transmission and scanning electron microscopy. *Journal of Food Science, 47*, 786–791.

Gianinetti, A. (2009). A theoretical framework for β-glucan degradation during barley malting. *Theory in Biosciences, 128*, 97–108.

Gianinetti, A., Toffoli, F., Cavallero, A., Delogu, G., & Stanca, A. M. (2005). Improving discrimination for malting quality in barley breeding programmes. *Field Crops Research, 94*, 189–200.

Gianinetti, A., Ferrari, B., Frigeri, P., & Stanca, A. M. (2007). *In vivo* modeling of β-glucan degradation in contrasting barley (*Hordeum vulgare* L.) genotypes. *Journal of Agricultural and Food Chemistry, 55*, 3158–3166.

Griffiths, D. W. (1987). The ratio of B to C hordeins in barley as estimated by high performance liquid chromatography. *Journal of the Science of Food and Agriculture, 38*, 229–235.

Holopainen, U. R. M., Wilhelmson, A., Salmenkallio-Marttila, M., Peltonen-Sainio, P., Rajala, A., Reinikainen, P., Kotaviita, E., Simolin, H., & Home, S. (2005). Endosperm structure affects the malting quality of barley (*Hordeum vulgare* L.). *Journal of Agricultural and Food Chemistry, 53*, 7279–7287.

Lewis, M. J., & Young, T. W. (2002). *Brewing* (2nd ed., pp. 191–204). New York: Kluwer Academic Plenum Publishers.

Molina-Cano, J. L., Polo, J. P., Romera, E., Araus, J. L., Zarco, J., & Swanston, J. S. (2001). Relationships between barley hordeins and malting quality in a mutant of cv. Triumph I. Genotype by environment interaction of hordein content. *Journal of Cereal Science, 34*, 285–294.

Palmer, G. H. (1989). Cereals in malting and brewing. In G. H. Palmer (Ed.), *Cereal science and technology* (pp. 61–242). Aberdeen: Aberdeen University Press.

Palmer, G. H., & Harvey, A. E. (1977). The influence of endosperm structure on the behaviour of barleys in the sedimentation test. *Journal of the Institute of Brewing, 83*, 295–299.

Qi, J. C., Zhang, G. P., & Zhou, M. X. (2006). Protein and hordein content in barley seeds as affected by nitrogen level and their relationship to *beta*-amylase activity. *Journal of Cereal Science, 43*, 102–107.

Shewry, P. R. (1993). Barley seed proteins. In J. MacGregor & R. Bhatty (Eds.), *Barley: Chemistry and technology* (pp. 131–197). St. Paul: American Association of Cereal Chemists.

Shewry, P. R., & Darlington, H. (2002). The proteins of the mature barley grain and their role in determining malting performance. In G. A. Slafer, J. L. Molina-Cano, R. Savin, J. L. Araus, & I. Romagosa (Eds.), *Barley science: Recent advances from molecular biology to agronomy of yield and quality* (pp. 503–521). Oxford: The Haworth Press.

Skerritt, J. H., & Janes, P. W. (1992). Disulphide-bonded 'gel protein' aggregates in barley: Quality-related differences in composition and reductive dissociation. *Journal of Cereal Science, 16*, 219–235.

Slack, P. T., Baxter, E. D., & Wainwright, T. (1979). Inhibition by hordein of starch degradation. *Journal of the Institute of Brewing, 85*, 112–114.

Smith, D. B., & Lister, P. R. (1983). Gel-forming proteins in barley grain and their relationship with malting quality. *Journal of Cereal Science, 1*, 229–239.

Wentz, M. J., Horsley, R. D., & Schwarz, P. B. (2004). Relationships among common malt quality and modification parameters. *Journal of the American Society of Brewing Chemists, 62*, 103–107.

Chapter 11
Genotypic Difference in Molecular Spectral Features of Cellulosic Compounds and Nutrient Supply in Barley: A Review

Peiqiang Yu

Abstract The objective of this chapter was to review four studies from our lab on genotypic difference in molecular spectral features of cellulosic compounds in hull, endosperm and whole seeds of barley grown in Western Canada using advanced molecular spectroscopy techniques – synchrotron-based SR-IMS and DRIFT techniques and nutrient utilization availability profiles in barley. The emphasis of this review is to focus on molecular spectral analysis methodology, results summary and implications. In study 1, the synchrotron experiments were carried out in Canadian Light Source (CLS, Saskatoon) and National Synchrotron Light Sources (NSLS) and Brookhaven National Laboratory (New York, US Department of Energy) to study genotypic effect on spectral feature in endosperm tissue within cellular and subcellular dimension. In study 2, the DRIFT experiments were carried out in Saskatchewan Structure Science Center (SSSC, University of Saskatchewan) to study genotypic effect on spectral features in hull and whole seed tissues. The molecular spectral features included ca. 1,293–1,212 cm^{-1} which were attributed mainly to cellulosic compounds in the hull, ca. 1,273–1,217 cm^{-1} which were attributed mainly to cellulosic compounds in the endosperm and ca. 1,269–1,217 cm^{-1} which were attributed mainly to cellulosic compound in the whole seeds. Spectral analyses included univariate and multivariate molecular analyses including an agglomerative hierarchical cluster and principal component spectral analyses. In study 3, in situ and in vitro experiments were carried out to study genotypic effect on degradation kinetics, degradation rate and effective degradability. In study 4, nutrient modelling was applied to study genotypic effect on predicted nutrient supply to ruminants. Our results showed that (1) the molecular spectral techniques (synchrotron SR-IMS and DFIRT) plus agglomerative hierarchical cluster and principal component spectral analyses were able to reveal cellulosic compounds spectral features and were able

P. Yu, Ph.D. (✉)
College of Agriculture and Bioresources, University of Saskatchewan,
6D10 Agriculture Building, 51 Campus Drive, Saskatoon, SK, S7N 5A8 Canada
e-mail: peiqiang.yu@usask.ca

G. Zhang et al. (eds.), *Advance in Barley Sciences: Proceedings of 11th International Barley Genetics Symposium*, DOI 10.1007/978-94-007-4682-4_11,

to identify molecular structure spectral differences at a spatial resolution among the barley genotypes; (2) genotype did have significant effect on nutrient utilization and nutrient supply to ruminants.

Keywords Synchrotron • DRIFT • Molecular structure • Endosperm • Hull • Cellulosic compounds • Genotype

11.1 Research Motivation and Research Background

Different genotypes of barley may have different degradation fermentation behaviour (Liu and Yu 2010a). The carbohydrate polymers include structural and non-structural carbohydrate. Both have different degradation kinetics in terms of degradation rate and extends (Hart et al. 2008; Yu et al. 2009). Cellulosic compounds are structural carbohydrates that exist in hull, endosperm and whole seeds. However, the structural conformation or make-up in hull, endosperm and whole seeds may be different due to component make-up.

Ultra-spatially resolved synchrotron-based SR-IMS technique and DRIFT molecular spectroscopy are two types of molecular spectral analysis techniques (Wetzel et al. 1998; Doiron et al. 2009). They are able to reveal inherent structural make-up and conformation. Nutrient profiles in barley for ruminant usually include in situ degradation kinetics, degradation rate and effective degradability, intestinal digestions and truly absorbable nutrient supply to ruminants in terms of DVE and OEB values (Hart et al. 2011).

The objective of this chapter was to review our previous studies on genotypic difference in molecular spectral features of cellulosic compounds in hull, endosperm and whole seeds of barley grown in Western Canada using advanced molecular spectroscopy techniques – synchrotron-based SR-IMS and DRIFT techniques and study genotype effects of barley on nutrient utilization and availability in ruminants. The hypothesis for these studies was that the spectral feature of cellulosic compounds in hull, endosperm and whole seeds was different among barley genotypes which results different molecular spectral profile. The genotype did have significant effect on barley nutrient availability and supply to ruminants.

11.2 Methodology

11.2.1 Genotypes of Barley Used in Previous Studies

Six barley varieties were used in following studies, including AC Metcalfe (malting purpose), CDC Dolly (feed purpose), McLeod (feed purpose), CDC Helgason (feed purpose), CDC Trey (feed purpose) and CDC Cowboy (feed and forage dual purpose), provided by B.G. Rossnagel, Crop Development Centre (Liu and Yu 2010a, b).

11.2.2 Study 1: Synchrotron Radiation SR-IMS Experiment

In study 1 (Liu and Yu 2010a), the synchrotron SR-IMS experiments were performed at U2B beamline, National Synchrotron Light Source, Brookhaven National Laboratory, US Department of Energy (NSLS-BNL, Upton, NY) with Thermo Nicolet Magna 860 Step-Scan FTIR (Thermo Fisher Scientific Inc., Waltham, MA) spectrometer equipped with a Spectra Tech Continuum IR Microscope (Spectra Tech Inc., Shelton, CT) and liquid nitrogen-cooled mercury cadmium telluride (MCT) detector. Spectral data analyses were used: OMNIC, SAS and Statistica software (Liu and Yu 2010a).

11.2.3 Study 2: The DRIFT Experiment

In study 2 (Liu and Yu 2010b), the DRIFT experiment was performed at the Saskatchewan Structure Sciences Center (SSSC), University of Saskatchewan, Canada using a Bio-Rad FTS-40 with a ceramic IR source and MCT detector (Bio-Rad Laboratories, 95 Hercules, CA, USA). Data were collected using Win-IR software (Bio-Rad Digilab, Cambridge, USA). Agglomerative hierarchical cluster analysis (AHCA), using Ward's algorithm method without prior parameterization, and principal component analysis (PCA) were performed (Yu 2005). Spectral data analyses were used: OMNIC, SAS and Statistica software. The detailed methodology was reported in Liu and Yu (2010b).

11.2.4 Study 3: In Situ and In Vitro Experiments

In study 3 (Hart et al. 2008), in situ ruminal degradation characteristics were determined using the standard departmental in situ procedure. The hourly effective degradation synchronization ratio was determined (Yu et al. 2009). The models used for estimating degradation kinetics were from McDonald et al. (1999). The detailed methodology was reported in Hart et al. (2008).

11.2.5 Study 4: Modelling Nutrient Supply

In study 4 (Hart et al. 2011), the detailed predicted nutrient supply was carried out by Hart and Yu (2011) using DVE/OEB system and NRC-2001 model. The major nutrient supply included rumen microbial protein synthesis, endogenous protein in small intestine, total metabolizable protein and degraded protein balance.

11.3 Experimental Result Summary

11.3.1 Study 1: Genotype Effect on Spectral Feature of Cellulosic Compounds in Endosperm Tissue of Barley

The genotype effect on spectral feature of cellulosic compounds in endosperm tissue of barley was reported by Liu and Yu (2010a) in detail. The brief summary is presented in Table 11.1 in terms of the absorbance peak area intensities of cellulosic compounds. There were significantly different ($P<0.05$) in cellulosic compound spectral area which varied from 0.48 (McLeod) to 0.54 (CDC Helgason).

11.3.2 Study 2: Genotype Effect on Spectral Feature of Cellulosic Compounds in Hull and Whole Seed Tissue of Barley

The genotype effect on spectral feature of cellulosic compounds in hull and whole seed tissue of barley was reported by Liu and Yu (2010b). The summary is presented in Table 11.1 in terms of the absorbance peak area intensities of cellulosic compounds. In hull tissue, there were significantly different ($P<0.05$) in cellulosic compound spectral area which varied from 1.18 (Dolly) to 1.75 (CDC Helgason). In whole seed tissue, there were significantly different ($P<0.05$) in cellulosic compound spectral area which varied from 0.84 (CDC Helgason) to 0.97 (CDC Metcalfe). From studies 1 and 2, you will see that molecular spectral features of cellulosic compound were different in terms of peak centre, region and baseline (Table 11.2) which indicate that the structural make-up or conformation of cellulosic compounds are different at different layers in barley seeds. The intensities and structural profiles may be highly related to nutrient availability.

Table 11.1 Genotypic effect of barley on cellulosic compound characteristics in the endosperm, hull and whole seed tissue of barley, revealed using synchrotron SR-IMS and DRIFT spectroscopy (Source: Liu and Yu 2010a, b)

	Barley-absorbed peak area intensity		
Studies	Range	SEM	Genotype *P* value
Study 1. Using SR-IMS			
Endosperm tissue: cellulosic compound	0.47 ~ 0.54	0.020	0.003
Study 2. Using DRIFT			
Hull tissue: cellulosic compound	1.34 ~ 1.75	0.037	<0.001
Whole barley seed: cellulosic compounds	0.84–0.97	0.015	<.0001

Table 11.2 Spectral feature of cellulosic compounds in the endosperm, hull and whole seed tissue of barley, revealed using synchrotron SR-IMS and DRIFT (Source: Liu and Yu 2010a, b)

Studies	Peak (cm^{-1})	Region (cm^{-1})	Baseline (cm^{-1})
Study 1. Using SR-IMS			
Endosperm tissue: cellulosic compound	~1,240	1,273–1,217	1,273–1,217
Study 2. Using DRIFT			
Hull tissue: cellulosic compound	~1,245	1,293–1,212	1,293–1,212
Whole barley seed: cellulosic compounds	~1,246	1,269–1,217	1,269–1,217

Table 11.3 Genotypic effect of barley on in situ degradation kinetics, degradation ratio (Fn/FCHO) and predicted nutrient supply (Source: Hart et al. 2008, 2011; Yu et al. 2009)

	Barley nutrient features		
Studies	Range	SEM	Genotype *P* value
Study 3. Degradation kinetics of dry matter			
Degradation ratio of N to energy (FN/FCHO)	16.6 ~ 19.0	0.32	<0.001
Degradation rate ($\% h^{-1}$)	14.1 ~ 17.8	1.16	<0.001
Degradation extend ($g\ kg^{-1}$ DM)	355 ~ 406	10.9	<0.001
Study 4. Modelling nutrient supply ($g\ kg^{-1}$ DM)			
Hull tissue			
Metabolizable protein	85.4 ~ 92.3	0.90	<0.001
Degraded CP balance	−52.3 ~ −58.1	1.82	<0.001

11.3.3 Study 3: Genotype Effect on Degradation Kinetics and Degradation Rate in Barley

The genotype effect on degradation kinetics (Hart et al. 2008) and degradation synchronization ratio (Yu et al. 2009) of barley was reported and summarized in Table 11.3. The results indicate that different genotypes have significantly different ($P<0.05$) degradation ratio, which ranged from 16.6 to 19.0 (Yu et al. 2009). The results also indicate that different genotypes have significantly different ($P<0.05$) degradation kinetics in terms of degradation rate ranged from 14.1 to 21.7 $\% h^{-1}$ and effective degradability ranged from 355 to 406 $g kg^{-1}$ DM (Hart et al. 2008). The results indicate that genotype affects degradation behaviours and rumen bypasses nutrient.

11.3.4 Study 4: Genotype Effect on Predicted Nutrient Supply in Barley

The genotype effect on predicted nutrient supply (Hart et al. 2011) of barley was reported and summarized in Table 11.3. The results indicate that different genotypes have significantly different ($P<0.05$) metabolizable protein, ranged from 85.4 to 92.3 $g\,kg^{-1}$ DM, and degraded protein balance, ranged from −52.3 to −58.1 $g\,kg^{-1}$ DM. The results indicate that all barley exhibits negative OEB value, which indicates shortage of N source. The MP values were different among the genotypes but not to a great extent.

11.4 Conclusion, Implication and Future Study

In conclusion, the genotype significantly affects spectral feature of cellulosic compounds in endosperm, hull and whole seed of barley. The conformation and make-up of cellulosic compounds were different between endosperm, hull and whole seeds. The implication of these studies was that different barleys have different degradation kinetics and nutrient supply. Future study is needed to quantify the relationship between cellulosic compounds spectral features and carbohydrate utilization.

Acknowledgments The author thanks Saskatchewan Agriculture Strategic Research Chair Program and Agriculture Development Fund (ADF) for the financial support.

References

Doiron, K. J., Yu, P., McKinnon, J. J., & Christensen, D. A. (2009). Heat-induced protein structures and protein subfractions in relation to protein degradation kinetics and intestinal availability in dairy cattle. *Journal of Dairy Science, 92*, 3319–3330.

Hart, K. J., Rossnegal, B. G., & Yu, P. (2008). Chemical characteristics and in situ ruminal parameters of barley for cattle: Comparison of the malting cultivar AC Metcalfe and five feed cultivars. *Canadian Journal of Animal Science, 88*, 711–719.

Hart, K. J., Rossnegal, B. G., & Yu, P. (2011). Investigate the magnitude of differences in total metabolizable protein among different genotypes of barley grown for three consecutive years. *Cereal Research Communications*. in press.

Liu, N., & Yu, P. (2010a). Characterize microchemical structure of seed endosperm within a cellular dimension among six barley varieties with distinct degradation kinetics, using ultraspatially resolved synchrotron-based infrared microspectroscopy. *Journal of Agricultural and Food Chemistry, 58*, 7801–7810.

Liu, N., & Yu, P. (2010b). Using DRIFT molecular spectroscopy with uni- and multi-variate molecular spectral techniques to detect plant protein molecular structure difference among different genotypes of barley. *Journal of Agricultural and Food Chemistry, 58*, 6264–6269.

McDonald, B. A., Zhan, J., & Burdon, J. J. (1999). Genetic structure of *Rhynchosporium secalis* in Australia. *Phytopathology, 89*, 639–645.

Wetzel, D. L., Eilert, A. J., Pietrzak, L. N., Miller, S. S., & Sweat, J. A. (1998). Ultraspatially-resolved synchrotron infrared microspectroscopy of plant tissue *in situ*. *Cellular and Molecular Biology (Noisy-le-Grand, France), 44*, 145–168.

Yu, P. (2005). Applications of hierarchical cluster analysis (CLA) and principal component analysis (PCA) in feed structure and feed molecular chemistry research, using synchrotron-based Fourier transform infrared (FTIR) microspectroscopy. *Journal of Agricultural and Food Chemistry, 53*, 7115–7127.

Yu, P., Hart, K., & Du, L. (2009). An investigation of carbohydrate and protein degradation ratios, nitrogen to energy synchronization, and hourly effective rumen digestion of barley: Effect of variety and growth year. *Journal of Animal Physiology and Animal Nutrition, 93*, 555–567.

Chapter 12
Use of Barley Flour to Lower the Glycemic Index of Food: Air Classification β-Glucan Enrichment and Postprandial Glycemic Response After Consumption of Bread Made with Barley β-Glucan-Enriched Flour Fractions

Francesca Finocchiaro, Alberto Gianinetti, Barbara Ferrari, Antonio Michele Stanca, and Luigi Cattivelli

Abstract Barley is the world's fourth most important cereal after wheat, rice and corn. This cereal can be moreover appreciated for its potential beneficial properties when utilized as "functional" food. "Functional foods" may provide health benefits after consumption of appropriate quantities in a normal diet. β-glucans are polysaccharides known for their beneficial effects against high-glycemic index (GI) food. Aim of this research was to evaluate both the optimization of air classification for the production of β-glucan-enriched flour and the effect of the ingestion of bread made with flour enriched in β-glucans on human plasma glucose concentration. We found that, including barley flour in blend with wheat flour, the overall quality of bread is slightly worse, but positive consequences on glycaemia are obtained with a normal starch barley. However, the effect was quite less marked with a *waxy* barley, despite the higher β-glucan concentration. Thus, the barley starch type can affect the GI of bread, as amylopectin induces a higher glycemic response than amylose.

Keywords Functional foods • Barley • β-glucans • Glycaemic index • Waxy starch

F. Finocchiaro • A. Gianinetti • L. Cattivelli
CRA Genomics Research Centre, Via San Protaso 302, 29017 Fiorenzuola d'Arda (PC), Italy

B. Ferrari
CRA – Fodder and Diary Productions Research Centre, Viale Piacenza, 29, 26900 Lodi, Italy

A.M. Stanca (✉)
Department of Agricultural and Food Sciences, University of Modena and Reggio Emilia, Reggio Emilia, Italy
e-mail: michele@stanca.it

G. Zhang et al. (eds.), *Advance in Barley Sciences: Proceedings of 11th International Barley Genetics Symposium*, DOI 10.1007/978-94-007-4682-4_12,

12.1 Introduction

Whole grain, low glycemic index and high dietary fibre (particularly soluble fibre) are associated with reduced risk of development of type 2 diabetes and heart disease (Liu et al. 2002; Liu et al. 2000; Nilsson et al. 2008). Foods high in soluble dietary fibre are considered low-glycemic index foods (Björck et al. 2000). The GI is a measure of the increment in blood glucose concentration that occurs after the ingestion of foods rich in carbohydrates. The GI value of a food is determined by feeding human subjects a portion of such food containing 50 g of digestible carbohydrate and then measuring the effect on their blood glucose levels over the next 2 h. For this test food, the area under the 2-h blood glucose response (area under the curve, AUC) is then measured for each person. On another occasion, the same people consume an equal-carbohydrate portion (50 g) of glucose sugar (the reference food), and their 2-h blood glucose response (glucose AUC) is also measured. A GI value for the test food is then calculated for each person by dividing the glucose AUC for the test food by the glucose AUC for the reference food (glucose). The final GI value for the test food is the average GI value for the tested subjects (Brand-Miller et al. 2009). The GI is a useful tool to determine the speed at which the food carbohydrates are digested and absorbed as glucose: foods with a high-GI score contain rapidly digested carbohydrates, which produce a rapid and large fluctuation in the level of blood glucose; in contrast, foods with a low-GI score contain slowly digested carbohydrates, which produce a slower, and relatively lower, fluctuation in the level of blood glucose (Augustin et al. 2002; Brand-Miller et al. 2009).

Cereals rich in soluble dietary fibre, such as barley and oat, reduce the rate at which glucose is released to the intestine (Anderson and Chen 1979) and to the blood (Björck et al. 2000). This is thought to be due to soluble fibre as β-glucans, which, swelling in the presence of water to form gel, delays gastric emptying rate (Benini et al. 1995).

With its high starch and low fibre contents, bread is considered high glycaemic foods, while pasta has lower glycemic index, due to the structure after the extrusion process, which leads to a protein network entrapping starch granules and delaying enzymic attacks; incorporation of β-glucans into these so popular and widely consumed foods could be furtherly positive in reducing glycaemic response, but usually cause undesirable effects on taste, colour (darker, like whole wheat – Knuckles et al. 1997), crumb texture and loaf volume for the gluten dilution (gluten network cannot develop very well and the loaf results more compact). Pasting properties of flours depend on starch characteristics (swelling potential after water absorption, gelatinization degree, re-association of amylose and amylopectin after granule disruption). Several studies evidenced how the presence of a large amount of amylose could decrease significantly the GI, thanks to the high level of resistant starch. The use of barley-soluble fibre (i.e. β-glucans)-enriched fractions evidenced the possibility to lower the GI of bread (Cavallero et al. 2002); moreover, there's a major flour hydration, because β-glucan strongly interacts with much water during dough development. So, combination of barleys with different starches and of high amounts of non-starch polysaccharides should balance the negative effects associated with gluten dilution (Izydorczyk et al. 2001).

In this study, two β-glucan-enriched flour fractions previously obtained (Ferrari et al. 2009) from a hull-less barley with normal starch type and a high-β-glucan waxy genotype were mixed with bread wheat flour. The two flour blends, as well as a 100% wheat flour control, were evaluated for rheological properties, and the post-prandial glucose response was determined. The sensory characteristics were also evaluated for acceptability by a laboratory panel.

12.2 Material and Methods

12.2.1 Barley Flours

β-glucan-enriched flours from two barley genotypes (Priora, hull-less, and CDC Alamo, hull-less and waxy) were obtained as reported by Ferrari et al. (2009). β-glucan content and amylase content of these flours are reported in Finocchiaro et al. (2011).

12.2.2 Rheological Evaluation

Brabender extensograph was utilized to record dough resistance to stretching with constant speed before breaking and to evaluate plasticity (AACC Method 54-10.01 2009). Registered parameters on dough were extensibility (mm), resistance at constant extension of 5 cm and maximum resistance till laceration.

Barley/wheat flour blends (30 and 50% of barley flour) and wheat flour were also evaluated with Brabender Farinograph using a 50-g mixer according to ICC Method 115-D (2008). Registered parameters were water absorption (%), development (min), stability (min) and degree of softening.

12.2.3 Bread Making

To test the bread-making characteristics of barley/wheat blends, baking tests were made using flour mix consisted in 30 and 50% of barley-enriched fraction and 70 and 50% of commercial wheat flour. A 100% wheat flour was used as control.

The flour blends and a wheat flour control were baked according to AACC Method 10-10.03 (2009) with minor modifications (Finocchiaro et al. 2011). Pilot plant baking tests were performed for the glycaemic index evaluation. The β-glucan-enriched flour consisted of 40% barley-enriched fractions (from cultivars

Priora and CDC Alamo) and 60% of commercial wheat flour. All ingredients were mixed together in a straight dough (Finocchiaro et al. 2011). A short rest time was preferred because during prolonged fermentation, β-glucan chains can be cleaved (Andersson et al. 2004). Hence, three breads were obtained: CDC Alamo (40%), Priora (40%) and wheat (100%) (control, obtained with 100% of commercial wheat flour). The servings of the three breads were frozen (−18°C) until glycaemic index determination.

12.2.4 Chemical Analyses

For the determination of β-glucan content, bread samples were desiccated at 35°C overnight, then grounded in Cyclotec Sample Mill (Foss Italia S.p.A., Padova, Italy) equipped with a 0.5-mm screen. Their water content was determined with a Precisa HA60 IR moisture analyser (Precisa Instruments, Diekinton, Germany) and fixed at about 9%. Total β-glucans were determined enzymatically, with a mixed-linkage β-glucan detection assay kit (Megazyme™ International Ltd., Wicklow, Ireland), according to McCleary and Codd (1991).

Bread samples were also analysed for the determination of protein, fat and starch content. Crude protein was determined with Kjeldhal method, using the nitrogen composition on dry weight base (N × 6.25), as recommended by Official Journal of the European Union (L. 179/9, 22/07/1993). Fat was determined by Soxhlet method as recommended by the Official Journal of the European Union (CE 52/2009, 27/01/2009). Starch was determined using the polarimetric method (ICC Standard Method 123/1 2008). Results are expressed on dry basis. All the analyses were done in duplicate. Acceptable maximum difference among duplicate results was 0.2% for protein, starch and fat.

12.2.5 Glycaemic Index Test

Portions containing 50 g of available carbohydrate, determined according to Holm, of the three different breads, obtained as reported above, were fed to subjects to evaluate the glycaemic index. Nine healthy, non-diabetic volunteers (five males and four females), 40.9 ± 6.6 (mean ± standard error of the mean – SEM) years old, with a body mass index of 23.7 ± 1.3 (mean ± SEM), participated in the study. All subjects gave their written informed consent, according to Helsinki declaration on human rights. Six tests (three breads and three reference tests with 50 g of glucose) were served, as breakfast, in random order to each subject in nonconsecutive days spaced an average of 3 days apart during a 3-week period. Subjects were asked not to consume food or beverages (except water) 12 h before the test. The dinner before each study day was standardized for quantity and quality of food items (low-fibre and high-GI carbohydrate sources) for all the subjects. Before every test, each subject was screened by a questionnaire for the general health status and the characteristics of the last meal consumed. Subjects consumed the meal, served with 500 ml of water, over a 12–15-min period. Finger-prick capillary blood samples were taken prior to the meal and at 15, 30, 45, 60, 90 and 120 min after meal consumption. Blood glucose was determined using a glucose analyser with a combined enzymatic-electrochemical system (YSI 2300 Stat Plus, Yellow Spring Instruments, OH, USA). The glycaemic indices were calculated from 120-min incremental postprandial blood glucose areas under the curves, obtained by plotting blood glucose concentration against time, using 50 g of pure glucose as reference food.

The results were expressed as means ± SEM and were statistically analysed according to a two-way analysis of variance (ANOVA) by using the Systat 9.0 software (SPSS, Chicago, IL, USA). The significance of differences was assessed with the Tukey's test for pairwise comparisons ($P \leq 0.05$).

12.2.6 Bread Sensorial Evaluation

During the glycaemic index test, a questionnaire was given to each subject to evaluate the acceptance of the bread tested, judging some features: aroma, bread size (shape and loaf volume), texture, bread colour, crumb colour, crumb grain, crust firmness, taste, aftertaste, rusticness – which means the ability to remember a good traditional product – and a synthetic score, which gives an overall estimation of the bread product. For each bread feature, a score from 1 (awful) to 5 (excellent) was assigned. The product feature was judged acceptable if the score was above 3. Data were analysed according to a Kruskal–Wallis analysis of variance by using the Systat 9.0 software (SPSS Inc., Chicago, IL, USA).

The physiological sensations of hunger and satiety were evaluated through a questionnaire before the consumption of the meal and after 30, 60 and 120 min. The subjects were asked to rate each sensation by drawing a cross along a line of 10 cm (0 cm = nothing, 10 cm = extremely). Data were analysed according to univariate analysis of variance by using the Systat 9.0 software (SPSS Inc., Chicago, IL, USA).

12.3 Results and Discussion

12.3.1 Flour Blend Evaluation and Baking Trials

Flour blends with increasing barley flour percentages (from 0 to 80%) were studied with the extensograph. Increasing percentages of barley flour cause a progressive disruption of gluten network in the blend and, therefore, a worsening in the rheological properties of the flour mix (Knuckles et al. 1997; Cavallero et al. 2002;

Gill et al. 2002; Jacobs et al. 2008). In fact, extensibility and strength progressively decreased with higher percentages of barley flour (Finocchiaro et al. 2011).

Water absorption increased proportionately with the increment of the barley percentage in the flour blend (Finocchiaro et al. 2011) because of the presence of β-glucans which strongly interact with water (Lazaridou and Biliaderis 2007; Holtekjølen et al. 2008). However, the higher water absorption was not associated with better rheological properties, as instead occurs for wheat, for which water absorption is linked to an increase in protein content and gluten quality (Pomeranz 1987). Nonetheless, with 20% barley flour, the resistance value was higher with respect to 100% wheat flour (0% barley, Finocchiaro et al. 2011), probably because the mild gluten dilution was not sufficient to disrupt the gluten network and the higher viscosity of the added of β-glucans caused the higher resistance of the dough. When the percentage of barley flour incorporated into the dough was further increased, resistance dropped too. These results were also confirmed by the farinograph test (Finocchiaro et al. 2011): it is evident how the replacement of wheat flour with 30% of barley-enriched fraction caused a significant decrease in the dough development, despite a higher water absorption. Farinograph stability decreased in all blends, with the exception of the blend with 50% of CDC Alamo. The high stability of the blend with 50% CDC Alamo was probably due to the very high level of β-glucans and to the low-amylose starch which gave high viscosity to the dough, as also evidenced by the value of degree of softening which dropped to zero, due to the high viscosity of the dough. These negative rheological features are also reflected in the baking test (Finocchiaro et al. 2011): the dilution of wheat gluten consequent to the addition of barley flour enriched in β-glucan caused a significant decrease in the bread volume and height, while the weight increased due to the great water-holding capacity of β-glucans. Similar negative effects were also shown by Brennan and Cleary (2007) who used Glucagel® in β-glucan enrichment of breads: this barley-isolated fibre was included at 2.5 and 5% and lowered dough quality and baking performance.

The β-glucan content of baked breads (Finocchiaro et al. 2011) slightly decreased with respect to their original flour mixes. Indeed, bread making can cause a significant loss of β-glucans during mixing and fermentation, mainly due to the action of endogenous β-glucanases (Andersson et al. 2004; Anttila et al. 2004) and, to a less extent, to thermal degradation during baking (Symons and Brennan 2004; Trogh et al. 2004).

12.3.2 *Effect on Blood Glucose and Glycaemic Index*

Breads obtained by blending 40% β-glucan-enriched barley flour fractions and 60% of a commercial wheat flour were tested for their effect on the GI. The blends had been formulated to reach a final concentration of β-glucans in bread of about 6%, which was reported to be the minimal concentration to significantly reduce the bread GI (Cavallero et al. 2002), calculated from 120-min incremental postprandial blood glucose areas under the curves (Fig. 12.1).

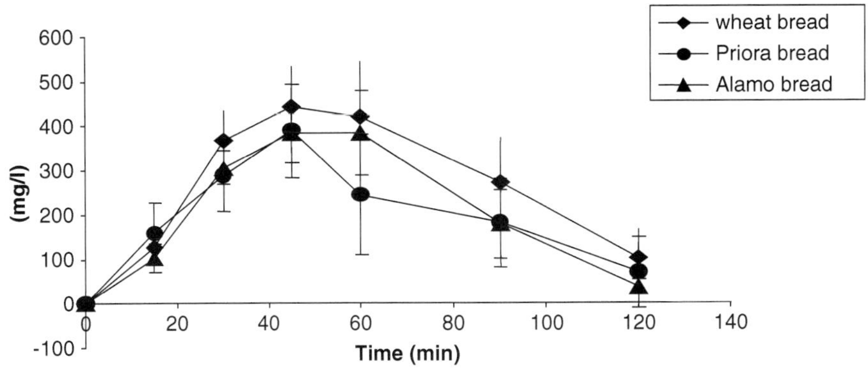

Fig. 12.1 Blood glucose increment in healthy volunteers after ingestion of a wheat bread (100% wheat flour) and the barley breads tested, obtained with a flour blends of 40% of barley flour (from cultivars Priora and CDC Alamo). Values are means, $n=9$

As shown by Finocchiaro et al. (2011), only the bread obtained with flour from cv. Priora showed a GI significantly lower than the control (100% wheat flour). Indeed, the bread obtained with the flour from cv. CDC Alamo, despite its high β-glucan content, showed a GI not significantly different from the control (Finocchiaro et al. 2011). This unexpected result could be justified by the different amylose/amylopectin ratios present in the two varieties of barley analysed; in fact, the β-glucan-enriched fraction of cv. Priora contains more amylose with respect to the analogue fraction of cv. CDC Alamo, total starch being almost equal (see Finocchiaro et al. 2011). This difference was the only macroscopic diversity in nutritional composition between the two cultivars and is, therefore, the most verisimilar responsible for their different GI scores. Indeed, the role of amylose/amylopectin ratio on gelatinisation changes (Fredriksson et al. 1998; Morell et al. 2003), and plasma glucose responses of cereals (Behall and Hallfrisch 2002; Behall et al. 2006; Alminger and Eklund-Jonsson 2008) has been already clearly demonstrated. Kabir et al. (1997) evaluated the effects of consumption of two starches characterized by different amylose/amylopectin content (high-amylose starch from mung bean and a waxy cornstarch) on glucose metabolism in adipocytes of normal and diabetic rats. These authors found that the meal with low-amylose starch (waxy cornstarch) showed a higher glycaemic index than the meal with high-amylose starch.

On the other hand, Liljeberg et al. (1996) and Östman et al. (2006) demonstrated that the waxy barley Prowashonupana significantly lowered the GI of the products (breads and porridge). However, the breads tested in those works (Liljeberg et al. 1996; Östman et al. 2006) had β-glucan contents from about 8–14%, that is, quite higher than that adopted in our experiment (about 6%). It appears, therefore, that the high β-glucan contents used in those experiments overcome any counteracting effect of the high-amylopectin starch provided by Prowashonupana flours. In contrast, the GI-increasing effect of the high-amylopectin starch provided by the waxy barley CDC Alamo was able to counteract the GI-reducing action of β-glucans in our experiment, wherein a low β-glucan level was used. Thus, even though having a β-glucan content is similar to that of the normal starch barley cv. Priora, the higher proportion of amylopectin in the waxy barley CDC Alamo significantly reduced the capability of its flour to lower the bread GI.

12.3.3 Sensorial Test

No significant differences were observed for the sensorial characteristics of the breads tested, with the exception of the aftertaste, and no significant differences were reported for the physiological sensations of hunger and satiety ($P \leq 0.05$; Finocchiaro et al. 2011).

12.4 Conclusions

These results confirm that the use of β-glucan-enriched barley flour fractions can reduce the glycaemic index of bread without substantially reducing its overall quality and palatability. The effectiveness of bread enriched with β-glucans in reducing GI is, however, influenced by the amylose/amylopectin ratio of the barley used. The development of barley genotypes with high β-glucan content associated with a starch having normal or high amylose content appears to be an important condition in the production of functional foods aimed to lowering the glycaemic response.

References

AACC International Methods 10–10.03 and 54–10.01. (2009). *AACC international approved methods of analysis* (11th ed.). St. Paul: AACC Press.

Alminger, M., & Eklund-Jonsson, C. (2008). Whole-cereal products based on a high-fibre barley or oat genotype lower post-prandial glucose and insulin responses in healthy humans. *European Journal of Nutrition, 47*, 294–300.

Anderson, J. W., & Chen, W. L. (1979). Plant fiber: Carbohydrate and lipid metabolism. *American Journal of Clinical Nutrition, 32*, 346–363.

Andersson, A. A. M., Armö, E., Grangeon, E., Fredriksson, H., Andersson, R., & Åman, P. (2004). Molecular weight and structure units of (1→3) (1→4)-β-glucans in dough and bread made from hull-less barley milling fractions. *Journal of Cereal Science, 40*, 195–204.

Anttila, H., Sontag-Strohm, T., & Salovaara, H. (2004). Viscosity of beta-glucan in oat products. *Agricultural and Food Science, 13*, 80–87.

Augustin, L. S., Franceschi, S., Jenkins, D. J. A., Kendall, C. W. C., & La Vecchia, C. (2002). Glycemic index in chronic disease: A review. *European Journal of Clinical Nutrition, 56*, 1049–1071.

Behall, K. M., & Hallfrisch, J. (2002). Plasma and insulin reduction after consumption of breads varying in amylose content. *European Journal of Clinical Nutrition, 56*, 913–920.

Behall, K. M., Scholfield, D. J., Hallfrisch, J., & Liljeberg, H. G. M. (2006). Consumption of both resistant starch and β-glucans improves postprandial plasma glucose and insulin in women. *Diabetes Care, 29*, 976–981.

Benini, L., Castellani, G., Brighenti, F., Heaton, K. W., Brentegani, M. T., Casiraghi, M. C., Sembenini, C., Pellegrini, N., Fioretta, A., Minniti, G., Porrini, M., Testolin, G., & Vantini, I. (1995). Gastric emptying of a solid meal is accelerated by the removal of dietary fibre naturally present in food. *Gut, 36*, 825–830.

Björck, I., Liljeberg, H., & Östman, E. (2000). Low glycaemic-index foods. *British Journal of Nutrition, 83*, S149–S155.

Brand-Miller, J. C., Stockman, K., Atkinson, F., Petocz, P., & Denyer, G. (2009). Glycemic index, postprandial glycemia, and the shape of the curve in healthy subjects: analysis of a database of more than 1000 foods. *American Journal of Clinical Nutrition, 89*, 97–105.

Brennan, C. S., & Cleary, L. J. (2007). Utilisation Glucagel® in the β-glucan enrichment of breads: A physicochemical and nutritional evaluation. *Food Research International, 40*, 291–296.

Cavallero, A., Empilli, S., Brighenti, F., & Stanca, A. M. (2002). High (1→3) (1→4)-β-glucan barley fractions in bread making and their effects on human glycemic response. *Journal of Cereal Science, 36*, 59–66.

Ferrari, B., Finocchiaro, F., Stanca, A. M., & Gianinetti, A. (2009). Optimization of air classification for the production of β-glucan enriched barley flours. *Journal of Cereal Science, 50*, 152–158.

Finocchiaro, F., Ferrari, B., Gianinetti, A., Spazzina, F., Pellegrini, N., Caramanico, R., Salati, C., Shirvanian, V., & Stanca, A. M. (2011). Effects of barley β-glucan-enriched flour fractions on the glycaemic index of bread. *International Journal of Food Sciences and Nutrition*. doi:10.3109/09637486.2011.593504.

Fredriksson, H., Silverio, J., Andersson, R., Eliasson, A.-C., & Åman, P. (1998). The influence of amylopectin characteristics on gelatinization and retrogradation properties of different starches. *Carbohydrate Polymers, 35*, 119–134.

Gill, S., Vasanthan, T., Ooraikul, B., & Rossnagel, B. (2002). Wheat bread quality as influenced by the substitution of waxy and regular barley flours in their native and extruded forms. *Journal of Cereal Science, 36*, 219–237.

Holtekjølen, A. K., Olsen, H. H. R., Færgestad, E. M., Uhlen, A. K., & Knutsen, S. H. (2008). Variations in water absorption capacity and baking performance of barley varieties with different polysaccharide content and composition. *LWT – Food Science and Technology, 41*, 2085–2091.

ICC-Standards. Methods 115/1, 123/1. (2008). *Standards methods of the International Association for Cereal Science and Technology, Wien*. Detmold: Verlag Moritz Schäfer.

Izydorczyk, M. S., Hussain, A., & McGregor, A. W. (2001). Effect of barley and barley components on rheological properties of wheat dough. *Journal of Cereal Science, 34*, 251–260.

Jacobs, M. S., Izydorczyk, M. S., Preston, K. R., & Dexter, J. E. (2008). Evaluation of baking procedures for incorporation of barley roller milling fractions containing high levels of dietary fibre into bread. *Journal of the Science of Food and Agriculture, 88*, 558–568.

Kabir, M., Rizkalla, S. W., Champ, M., Luo, J., Boillot, J., Bruzzo, F., & Slama, G. (1997). Dietary amylose-amylopectin starch content affects glucose and lipid metabolism in adipocytes of normal and diabetic rats. *The Journal of Nutrition, 128*, 35–432.

Knuckles, B. E., Hudson, C. A., Chiu, M. M., & Sayre, R. N. (1997). Effect of β-glucan barley fractions in high-fiber bread and pasta. *Cereal Food World, 42*, 94–99.

Lazaridou, A., & Biliaderis, C. G. (2007). Molecular aspects of cereal β-glucan functionality: physical properties, technological application and physiological effects. *Journal of Cereal Science, 46*, 101–118.

Liljeberg, H. G. M., Granfeldt, Y. E., & Björck, I. M. E. (1996). Products based on a high fiber barley genotype, but not on common barley and oats, lower postprandial glucose and insulin response in healthy humans. *The Journal of Nutrition, 126*(2), 458.

Liu, S., Manson, J. E., Stampfer, M. J., Hu, F. B., Giovannucci, E., Colditz, G. A., Hennekens, C. H., & Willett, W. C. (2000). A prospective study of whole-grain intake and risk of type 2 diabetes mellitus in US women. *American Journal of Public Health, 90*, 1409–1415.

Liu, S., Manson, J. E., Buring, J. E., Stampfer, M. J., Willett, W. C., & Ridker, P. M. (2002). Relation between a diet with a high glycemic load and plasma concentrations of high-sensitivity C-reactive protein in middle-aged women. *American Journal of Clinical Nutrition, 75*, 492–498.

Ma, Z., Steffenson, B. J., Prom, L. K., & Lapitan, N. L. (2000). Mapping of quantitative trait loci for *Fusarium* head blight resistance in barley. *Phytopathology, 90*, 1079–1088.

McCleary, B. V., & Codd, R. (1991). Measurement of (1→3), (1→4)-β-D-glucan in barley and oats: a streamlined enzymic procedure. *Journal of the Science of Food and Agriculture, 55*, 303–312.

Morell, M. K., Kosar-Hashemi, B., Cmiel, M., Samuel, M. S., Chandler, P., Rahman, S., Buleon, A., Batey, I. L., & Li, Z. (2003). Barley sex6 mutants lack starch synthase IIa activity and contain a starch with novel properties. *The Plant Journal, 34*, 173–185.

Nilsson, A. C., Östman, E. M., Granfeldt, Y., & Björck, I. M. E. (2008). Effect of cereal test breakfasts differing in glycemic index and content of indigestible carbohydrates on daylong glucose tolerance in healthy subjects. *American Journal of Clinical Nutrition, 87*, 645–654.

Östman, E., Rossi, E., Larsson, H., Brighenti, F., & Björck, I. (2006). Glucose and insulin response in healthy men to barley bread with different levels of (1→3;1→4)−β−glucans; predictions using fluidity measurements of in vitro enzyme digests. *Journal of Cereal Science, 43*, 230–235.

Pomeranz, Y. (1987). Grain quality. In Y. Pomeranz (Ed.), *Modern cereal science and technology* (pp. 72–149). New York: VCH.

Symons, L. J., & Brennan, C. S. (2004). The influence of (1→3;1→4)-β-D-glucan-rich fractions from barley on the physicochemical properties and in vitro reducing sugar release of white wheat breads. *Journal of Food Science, 69*, 463–467.

Trogh, I., Courtin, C. M., Andersson, A. A. M., Åman, P., Sørensen, J. F., & Delcour, J. A. (2004). The combined use of hull-less barley flour and xylanase as a strategy for wheat/hull-less barley flour breads with increased arabinoxylan and (1→3;1→4)- β-D-glucan levels. *Journal of Cereal Science, 40*, 257–267.

Chapter 13
Food Preparation from Hulless Barley in Tibet

Nyima Tashi, Tang Yawei, and Zeng Xingquan

Abstract Barley occupied over 65% of the total food production in Tibet Autonomous Region. It is cultivated in the valleys and in the higher land on Tibet. Currently, the total cultivated barley area was about 117,900 ha that makes up more than 69.7% of the total area of grain in Tibet. Barley was traditionally cultivated in spring, but winter barley is nowadays relatively common. All cultivated barley varieties are hulless barley, also called naked barleys. With a long history of hulless barley use, Tibetans have established unique ways of preparing food products from it which are culturally and economically adapted to the primitive mountainous region of the Tibetan Plateau. This chapter introduces the traditional ways of preparation of hulless barley into Tsangpa (roasted barley flour) which is the main product, chang brewed from hulless barley that is the major alcoholic beverage, and barley also made into cakes, soups, porridge, and snack foods. The significance of hulless barley is also remarkable in Tibetans' daily life. This chapter provides that reasons for the choice of barley as a staple food crop by Tibetans are the adaptation mechanism to survive in the harsh condition of mountain area.

Keywords Hulless barley • Tibet food preparation • Traditional ways

---

N. Tashi (✉)
Tibet Academy of Agricultural and Animal husbandry Sciences (TAAAS), Lhasa 850002, Tibet, China
e-mail: Nyima_tashi@163.com

T. Yawei • Z. Xingquan
Agricultural Research Institute, Tibet Academy of Agricultural and Animal husbandry Sciences (TAAAS), Lhasa 850002, Tibet, China

G. Zhang et al. (eds.), *Advance in Barley Sciences: Proceedings of 11th International Barley Genetics Symposium*, DOI 10.1007/978-94-007-4682-4_13,

13.1 Introduction

There is no other crop more important than hulless barley (Latin name) as the staple food crop in Tibet. In 2008, hulless barley accounted for over 65%, increased from 56% in 1980, of the total food production in Tibet. In many other areas of the world, as living standard improved, barley consumption has decreased due to increased intake of rice and wheat.

Barley has been cultivated in the valleys and in the highlands of Tibet for centuries. In 2008, the total cultivated barley area was about 117,900 ha, which is more than 69.7% of the total area of grain in Tibet. The total production was nearly 618,200 t with an average yield at 5 t/ha, which is much lower in contrast to the potential yield estimated at 6.5 t/ha. The main reason for this gap is caused by varietal difference. Most of the improved barley varieties are grown in the central part of Tibet with an average yield in this area at 6.0–7.0 t/ha. However, there is rarely any elite barley variety in the rest 44% of total barley cultivating area. Instead, local traditional varieties yielding at 3–4 t/ha have been used. High rate of seeding, sowing depth, poor land and fertilizer management, and lack of weed and pest control are among the factors that limit the yield of barley.

Barley was traditionally cultivated in spring, but winter barley is becoming relatively common nowadays. At present, spring barley represents 93% of the total cultivated barley. Seeds are sown in April and harvested in August. The major varieties are Zangqing no. 320, Zangqing no. 148, and Ximala no. 19. However, due to their special quality for food, landraces Chachu, Yangsun, and Lhazi Ziqingke are also quite popular in spite of their lower yield (3 t/ha).

Winter barley such as frost-tolerant varieties were developed by researchers in Tibetan Academy of Agricultural and Animal husbandry Sciences (TAAAS) and cultivated in the lower land where temperatures are milder. The main varieties are Guoluo, Dongqing no. 1, and Dongqing no. 8. Winter barleys are usually sown in October and are harvested in July in the following year. Longer season results in higher yield (6 t/ha) than that of spring varieties at 5 t/ha.

All Tibetan-cultivated barley varieties are hulless barley, also called naked barleys. Hulled barley varieties will not be accepted by farmers due to the difficulty for cooking and low quality for Tibetan dishes. Malting barley has been cultivated in Tibet since the Lhasa beer factory was built during 1980s. There were about 350 ha of malting barley in 1986 and 1987. However, because of the high price for malting barley and high cost for the malt processing in Tibet, at present, all the malts have been imported from other provinces near Tibet. No hulled barley was growing in Tibet at the moment.

With a long history of cultivating and using hulless barley, Tibetans have established unique ways of preparing barley food products which are culturally and economically adapted to the primitive mountainous region of the Tibetan Plateau. Hulless barley is used for food in several ways. Tsangpa, a type of roasted barley flour, is the main product. Chang brewed from hulless barley is the major alcoholic beverage. Also, many varieties of cakes, soups, porridge, and snack foods are made of barley.

Barley varieties differ according to their use. For example, purple barley is preferred for chang preparation, while barley with white or yellow grains is used for Tsangpa. The traditional landraces have special characters appreciated by consumers. For instance, cultivar Garsha is regarded as the best for preparing Tsangpa, and cultivar Lhazi Ziqingke (purple seed) is renowned for its special quality for chang.

13.2 Food Products

13.2.1 Preparation of Tsangpa

Roasted grain flour Tsangpa is commonly made from hulless barley. Hulless barley grain is carefully cleaned and washed. Barley grain is roasted with fine sand for distributing the heat evenly in order to prevent the barley kernels from burning. Fine sand is heated in a large, heavy pan at about 100–150°C. Grain is then poured into the heated fine sand and mixed for 2–3 min. The sand is sieved off and the remaining roasted barley grain is called Yue (Fig. 13.1). Yue is cleaned again and ground into Tsangpa (Fig. 13.2) using a water mill.

Nowadays, a couple of large-scale Tsangpa-processing factories were established in Tibet. Cleaning the barley seed is mechanized, as well as the process of roasting, but fine sand is still used for roasting. Modern milling techniques are suggested; however, water milling is preferred by the Tibetan consumers for maintaining the taste and flavor of tsampa. Nicely packaged tsampa is getting more popular in order to meet the demand for the urban dwellers. Thus, tsampa can be found in the supermarket in Lhasa, the capital of Tibet; Kathmandu in Nepal; and in some other cities in China.

Hulless barley is processed into many different food products such as barley meal, barley noodle, pealed barley, and barley tea. These products can be found in supermarkets in many cities in China. However, the predominant product is Tsangpa that is eaten in many different ways. It can be added to tea, skim milk, chang, and cold water, stirred up and consumed as a beverage. Many people even like to mix Tsangpa

Fig. 13.1 Roasted barley grain: *Yue*

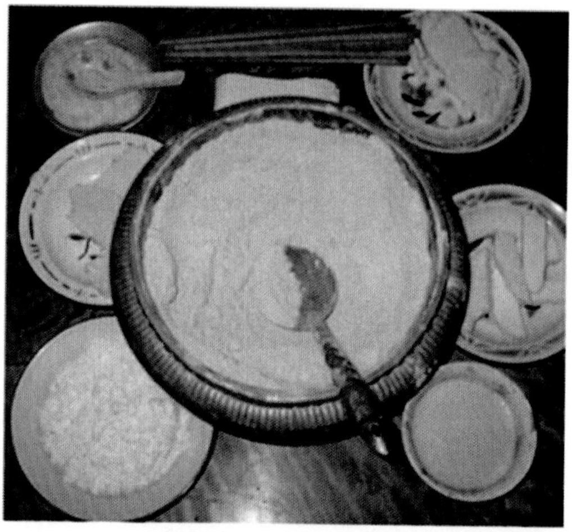

Fig. 13.2 *Tsangpa*

Fig. 13.3 *Ba*

with sugar and eat it directly. In most cases, however, Tsangpa is mixed with tea, and then kneaded into doughlike balls called ba (Fig. 13.3). Ba is cooked in various methods and used in many religious occasions and ceremonies such as weddings and harvesting festivals by tossing small amounts of cooked ba into the sky.

13.2.2 Preparation of Cakes

Chima is often made during festivals from the Tsangpa mixed with butter, dried cheese, and sugar. Chima symbolizes happiness and is prepared particularly for the New Year festival. Usually, a small amount is made for presentation during the festival. Family members and visitors share the Chima and offer best wishes to each other.

Magsan is usually prepared during the summer time in the central Tibet for picnics. It is made from Tsangpa mixed with skim milk, dried cheese, and brown sugar. The dough is kneaded well to make stiff cake, which is then top dressed with butter and brown sugar.

Tsog is cake for religious occasions, presented for worship and eaten as holy food. It is a mixture of Tsangpa, tea, dried cheese, dried grapes, and brown sugar.

13.2.3 Preparation of Porridge

Yuetub is made from coarse particles of roasted barley grain, yak or lamb meat, and vegetable. The barley grain is roasted roughly and crushed into 1/4 size of the grain. Traditionally, nettle leaves are used as the vegetable for yuetub. Most Tibetans prefer yuetub with fresh cheese and sugar. Yuetub is widely acclaimed in central Tibet as food for breakfast.

Sanchak Tukba. Sanchak means steeped and mashed barley grain, and tukba means porridge in Tibetan. To make sanchack tukba, barley grain is soaked overnight and usually mashed by hand. It can be either cooked fresh or dry after several months' storage. Sanchak is boiled in water and then mixed with lamb or yak meat, dried cheese, and green beans or peas. This porridge is usually prepared around the Tibetan New Year in February.

13.2.4 Preparation of Soup

Changuel is made from chang, Tsangpa, cooked rice, dried cheese, and sugar. The mixture is boiled for about 10 min. Changuel is normally prepared for breakfast with fried cookies in the Tibetan New Year festival in Lhasa and Shigatse Changuel, made via cooking mixtures of cheese and cooked rice, for same reason.

Tsangtub is cooked from tsang and water. Beef or yak meat, lamb, peas, dried cheese, and various vegetables are added to enhance both the nutrition and taste. Tsangtub is often served for dinner and occasionally for lunch.

13.2.5 Preparation of Snack Food

Yue, roasted or popped barley, is the most popular snack food among Tibetans for daily consumption. Some Tibetans like to roast barley grain with sugar and a little butter for a better taste. However, it has not been commercially produced and little is sold in the market.

Drubdrub is made from immature barley spikes. People consume the immature seed in small quantities after removing awns and glumes.

13.2.6 Preparation of Beverage

Chang, known as Qingke Jiu in Chinese, is the major alcoholic beverage in Tibet. The preparation of chang begins with cleaning and washing of raw barley kernels. Purple barley grain is used for desirable flavor and color. Greater uniformity of kernels is always considered to ensure uniform processing. Currently, many brewers mix malting barley with hulless barley in a 1:2 ratio, a specially prepared barely grain called changdru. Changdru is then boiled with water for 2–3 h. This boiled barley grain named poub is very tasty when mixed with sugar and butter but is often consumed in small amount. Poub is cooled and yeast powder is added to allow fermentation for 3–5 days. There is no free liquid remaining at this point. This fermented barley grain is known as lenmar and is also directly consumed in a small quantity or sometimes fried in oil with added sugar as a delicacy. Lenmar and water is usually mixed in an earthenware pot and steeped for 6–10 h; it is then filtered to produce chang. Since water is added three to four times during the process, the alcoholic content of the chang depends on the amount of water added. Alcohol content of the first run of filtered chang is usually about 7%, which is reduced to about 5% by the second time. Chang that people normally drink has about 5% or less alcoholic content. Chang tastes like white wine but a little bit sweet when it is just filtered.

Despite many local Tibetan still home brewing their chang, with the increasing demand for chang, the process of chang is now completely industrialized by some factories. At present, there are brands of chang such as Jintian, Shi-Dse, Amachangma, etc. These are now found in supermarkets in many cities of China.

Chang is employed in many cultural and religious occasions. Changuel, made from chang and mixture tsampa and others, is mostly prepared for breakfast during Tibetan New Year festival. Also, chang is used as offerings to the Buddha when Tibetan Buddhists pray in monasteries. It is also presented to the Gods in holy manner during Tibetan New Year festival in special container together with holy water and holy tea before any family member can consume. Although beer, Chinese spirits, wines, and even whiskey have overtaken chang in Tibetan alcoholic beverage market, occasions using chang mentioned above are uniquely conserved by the Tibetans for cultural and religious purposes. Moreover, locally made chang has its own unique market for the Tibetans and tourists.

13.2.7 Preparation of Sanchang

Sanchang is a slightly alcoholic beverage brewed from Tsangpa. Tsangpa is mixed with warm and clean water and powdered yeast is added. The mixture is fermented for about 3 days. Half-fermented Tsangpa taste sweet and is known as sanchang. It is usually cut into square pieces and dried. Many people like chewing it as a snack food, while some steep it in water and then drink it. Sanchang is widely acclaimed in the western Tibetan region where there is limited fuel for boiling barley grain for chang and the boiling pot of water is at about only 80°C.

13.3 Significance of Hulless Barley in Tibetan Life

Tibetan traditional medicine believes that roasted barley flour (Tsangpa) has most jue (safety and nutrition) and is regarded as a mengarbu (white medicine). Tsangpa and chang are occasionally made from wheat and a mixture of wheat and barley, but Tibetans believe it is not a healthy food. Instead, Tsangpa and chang made from pure barley grain are good to offer to others. Nevertheless, there is no big difference in their flavor, and actually wheat can be grown in most of the major agricultural land.

Studies on barley have confirmed that barley has medicinal properties and value. In particular, it has the potential to contribute soluble and insoluble dietary fibers such as beta-glucan to the diet and to reduce the influence of cholesterol absorption so as to prevent diseases such as colon cancers and heart diseases (Mcintosh et al. 1992). Due to the high altitude and cool temperature, not many vegetables and fruit trees are suitable to grow in Tibet. Therefore, sources of vegetable fibers are limited in most regions. Dairy products, especially butter, and meat are widely consumed in Tibet, which will normally result in high cholesterol content in blood, high blood pressure, and heart disease in the modern society. However, large quantities of hulless barley are consumed, approximately about 155 kg per person each year, which is the only significant source of dietary fiber in Tibetan diet. The beta-glucan content in major Tibetan cultivars is not only significantly higher than that of Harrington, CDC Richard, and CDC Buck in Canada but also significantly higher than that of wheat and rice (Tashi 1993). This may be one of the main reasons for which heart diseases and colon cancers occur at a very low rate in Tibet than expected.

The significance of hulless barley is also remarkable in Tibetans' daily life. Atmospheric pressure on the Tibetan Plateau is very low due to the high altitude. Therefore, food cannot be quickly cooked. Fuel wood is quite limited in most areas where barren land and pastoral land occupy a major proportion of the land. Tsangpa is a ready to eat food. It is also very convenient for storing and handling. Moreover, for example, the unique ways of processing barley for food such as using sand during the Tsangpa preparation both to distribute the heat evenly and to prevent the barley kernels from burning. Also, using sand for roasting barley can preserve heat for significantly saving of fuel wood, which will otherwise cause drastic loss of habitat in shrub land and deforestation. Food in Tibet is generally simple and natural but nutritious. A typical Tibetan meal usually consists of cooked yak or lamb meat, butter tea (cooked tea blended with a little butter), chang and Tsangpa. It provides all the nutrients the human requires. The choice of barley as a staple food crop by Tibetans is an adaptation mechanism to survive in the harsh condition of mountainous area.

References

Mcintosh, G., Jorgensen, L., Royle, P., & Kerry, A. (1992, September 7–10). *A role of barley foods in human health and nutrition*. Barley for Food and Malt: ICC/SCF International Symposium, the Swedish University of Agricultural Science, Uppsala, Sweden, p. 152.

Tashi, N. (1993). *Training report on barley germplasm appraisal and breeding*. Submitted to the Department of Crop Sciences, College of Agriculture, University of Saskatchewan, Saskatoon, Canada.

Chapter 14
Screening Hull-less Barley Mutants for Potential Use in Grain Whisky Distilling

John Stuart Swanston and Jill Elaine Middlefell-Williams

Abstract Most Scotch whisky is marketed as blends between malt and grain whiskies, the latter comprising between 50 and 80% of the blend. These are produced using high-diastase barley malt as a source of enzymes to break down starch from a cereal-based adjunct, prior to fermentation and distillation. As the husk dilutes expression of endosperm components, hull-less types have potential for higher enzyme activity. Beta-amylase, a major part of diastatic activity, can be measured in un-malted grain, providing a rapid screening test. Lines from a mutant population in the hull-less variety, Penthouse, were grown, in replicated trial, over two seasons. Analysis of variance showed highly significant effects of genotype and season, plus significant genotype × season interaction for beta-amylase activity, but not grain nitrogen. Six lines, showing a range of beta-amylase levels, were malted, using four different steep regimes, followed by 4 or 5 days germination. Highest levels of diastatic power occurred with the longest malting regime, while one line showed relatively high diastatic power, despite moderate levels of grain beta-amylase. Phenotypic testing will be extended to a wider population of the mutants, as it appears likely that lines with enhanced enzyme activity can be detected.

Keywords Beta-amylase • Diastatic activity • Malt • Nitrogen content • Starch breakdown • Whisky

---

Presenting author, John Stuart Swanston

J.S. Swanston (✉) • J.E. Middlefell-Williams
The James Hutton Institute, Invergowrie, Dundee DD2 5DA, Scotland, UK
e-mail: stuart.swanston@hutton.ac.uk

G. Zhang et al. (eds.), *Advance in Barley Sciences: Proceedings of 11th International Barley Genetics Symposium*, DOI 10.1007/978-94-007-4682-4_14,

14.1 Introduction

Two types of whisky are distilled in Scotland, malt and grain (Bathgate 1989), the latter legally defined as being distilled from a mash which consists only of a cooked cereal subsequently saccharified by the actions of enzymes from added malted barley. Blended whiskies, which comprise mixtures of grain and malt and in which grain whisky is usually the much larger component, currently provide the largest volume in global sales of Scotch whisky (Bringhurst et al. 2010). As the malted barley typically comprises less than 15% of the cereal in the mash (Bathgate 1989) and is the only source of starch-degrading enzymes, high levels of diastatic power are required to degrade the starch from the large quantity of un-malted adjunct (Briggs 1998).

Traditionally, barleys with high diastase activity were imported from either North America or Scandinavia (Bathgate 1989). Alternative ways of increasing enzyme activity are to use thinner grain, with lower starch content, or to add gibberellic acid during malting (Schwarz and Li 2011), but the latter is not permitted by the regulations for Scotch whisky production. Generally, therefore, grain is obtained with higher protein levels than would normally be used for malting (Bathgate 1989), as a positive association between protein content and diastatic power has been long established (Hayter and Riggs 1978). Additionally, malts are lightly kilned to preserve enzyme activity.

Agu et al. (2009) investigated the use of hull-less barley to produce malts with higher diastatic activity. As the husk contributes no enzymes, it effectively dilutes enzyme activity, and hull-less barley has also been shown to malt more quickly (Bhatty 1996; Agu et al. 2009) with potential reduction in both energy and water use.

The absence of husk particles to form a filter bed is not problematic as the whole mash may be pumped, after cooling, to the fermentation vessel (Bathgate 1989). Diastatic power comprises the activity of several enzymes, but beta-amylase, which is considered to form the largest and most important component (Arends et al. 1995), is not synthesised during malting but during grain development. Measurement of beta-amylase activity in the mature grain can, therefore, be used as a surrogate for diastatic power (Bendelow 1981), e.g. in breeding programmes, where rapid assessment of large numbers of progeny is required.

Hayter and Allison (1976) described the use of mutation breeding in an attempt to produce enhanced diastatic activity in barley suited to Scottish growing conditions. More recently, Swanston and Molina-Cano (2001) observed higher beta-amylase activity in mutants derived from Triumph, compared to those of the parent variety. Screening hull-less barley lines that had been subjected to mutagenesis, for enhanced beta-amylase activity, could therefore be effective in identifying potentially useful types for grain distilling. Swanston and Middlefell-Williams (2010) showed, in initial analyses of lines derived from the hull-less variety Penthouse, that some lines appeared to show improved malting quality. Further analysis concentrated on the content of beta-glucan in the cell walls and its degradation during malting (Swanston et al. 2011), to consider potential utility in malt distilling. However, it is also possible that the population may contain lines with higher levels of starch-degrading enzymes. The work described here, therefore, comprised screening advanced lines which had been included in replicated trial, in more than one season.

14.2 Materials and Methods

14.2.1 Trials Over Two Seasons

A population of mutant lines, in the hull-less variety Penthouse was derived by treatment with ethyl-methane sulphinate (EMS), with subsequent selection of M3 rows exhibiting phenotypic differences from the parental variety (Swanston et al. 2011). An initial trial of 40 entries, in two replicates (including Penthouse and the malting variety, Optic), was grown at the former Scottish Crop Research Institute (SCRI), Dundee, in 2008, with plot sizes and treatments as described by Swanston et al. (2011). These lines were included in a further trial in 2010, with similar replications, plot sizes and treatments. Following harvest, grain was cleaned and nitrogen content determined by near-infrared transmission, with an Infratec 1241 grain analyser (Foss UK, Warrington). Grain samples were then finely milled and beta-amylase activity was determined as described by Swanston and Molina-Cano (2001).

14.2.2 Malting and Diastatic Power Analysis

From the initial population, six lines were chosen for further malting studies, as they represented a range in beta-glucan contents (Swanston and Middlefell-Williams, submitted). As these lines also differed considerably in beta-amylase contents, the malt samples were also used for estimates of diastatic power (i.e. the combined activity of the starch-degrading enzymes in the malt). Four different steeping regimes were used, all at 16°C, steep A being the standard procedure and steep B the single 8-h immersion, which were both used by Agu et al. (2009). Steep C comprised two 8-h immersions, separated by a 16-h air rest, while steep D was a single immersion of 16 h (Bryce et al. 2010). From all 4 steeping regimes, samples were allowed to germinate for 4 and 5 days, prior to kilning. Diastatic power was determined by a scaled-down and modified version of the Institute of Brewing recommended method (IOB 1982), with 0.5-g grist extracted in 10 ml of water and reducing sugar content determined by use of a dinitrosalicylic acid (DNS) reagent (Swanston and Molina-Cano 2001).

14.2.3 Free and Thermostable Beta-Amylase

In addition to total beta-amylase, the proportions of free and thermostable enzyme activity were determined on the grain samples of the six lines used for malting. Free beta-amylase was determined in the same manner as total activity except that the flour samples were extracted in water rather than 1% papain (Swanston and Molina-Cano 2001). Thermostability was determined by placing a portion of the extract for total beta-amylase in a water bath at 60 °C, for 10 min, prior to estimation of enzyme activity (Eglinton et al. 1998), and calculating the percentage of enzyme activity that remained intact.

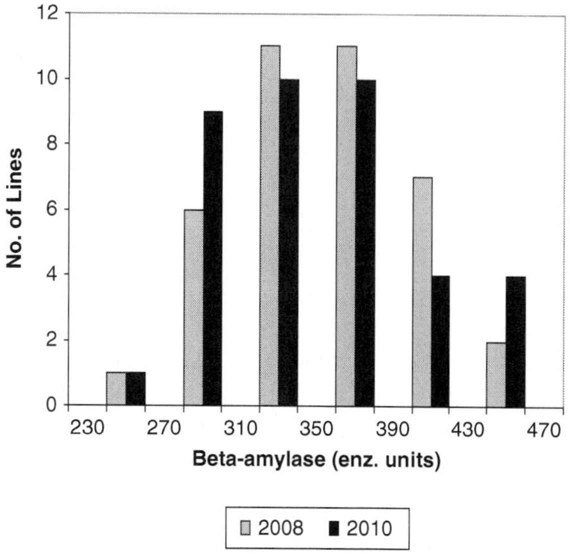

Fig. 14.1 The range of values for beta-amylase activity, in a population of Penthouse mutants, over two seasons

Table 14.1 Analysis of variance for grain nitrogen and beta-amylase activity in a population of Penthouse mutants

		MS	
Source of variation	df	Grain nitrogen	Beta-amylase activity
Rep	1	0.348	28.5
Genotype	37	0.482	51.3***
Season	1	0.031	591.6***
Genotype × season	37	0.281	28.5*
Residual	75	0.964	16.7
Total	151		

*Significant at the 5% level; ***significant at the 0.1% level

14.3 Results

14.3.1 Grain Nitrogen and Beta-Amylase Results

The Penthouse mutant lines showed a normal distribution for beta-amylase activity (Fig. 14.1), covering a fairly broad range of values, with analysis of variance showing a highly significant effect of genotype (Table 14.1). While the range was generally similar in both years, the overall mean was slightly higher in 2008, leading to a significant effect of season, while there was also a slight, but significant, genotype × season interaction, reflecting some changes in ranking order between the lines. By contrast, although there also appeared to be a range of values for grain nitrogen, analysis of variance showed no significant effects or interactions (Table 14.1). This was probably due to both trials comprising only two replications

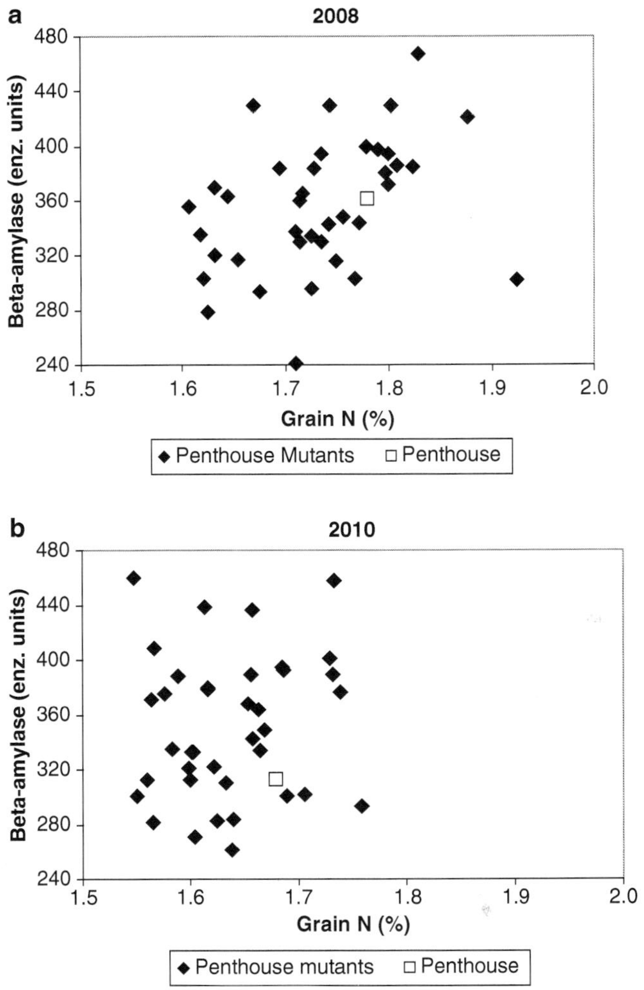

Fig. 14.2 Beta-amylase activity plotted against grain nitrogen, for a population of mutants in the variety Penthouse in 2008 (**a**) and 2010 (**b**)

and replicate differences between grain nitrogen levels being observed. This most likely resulted from yield dilution effects as, for some lines, there were considerable replicate differences in grain yield.

When beta-amylase activities were plotted against grain nitrogen levels (Fig. 14.2a and b), there was no significant correlation between the two characters in either year, indicating that lines with a range of beta-amylase levels could be obtained at any given level of nitrogen. The parent variety, Penthouse, had grain nitrogen levels, in both seasons, which were slightly higher than the mean values of the mutant lines. For beta-amylase activity, Penthouse was similar to the mean value

Table 14.2 Analysis of variance for malt diastatic power in six Penthouse mutant lines

Source of variation	df	MS
Rep	1	0.0032
Genotype	5	0.0125***
Steep	3	0.1828***
Germination time	1	0.0805***
Genotype × steep	15	0.0012***
Genotype × germination	5	0.0013***
Steep × germination	3	0.0032***
Genotype × steep × germination	15	0.0009***
Residual	47	0.0002
Total	95	

*** Significant at the 0.1% level

of mutant lines in 2008 (Fig. 14.2a) but slightly below the mean value in 2010 (Fig. 14.2b). This initial group of lines appeared, therefore, to contain some with enhanced beta-amylase activity compared to the parental variety. There was no correlation between beta-amylase activity and plot yield (data not shown), suggesting that higher enzyme levels did not impose a penalty on grain yield.

14.3.2 Analysis of Malt Samples

Analysis of variance for diastatic power in the malt samples (Table 14.2) showed highly significant effects of genotype, steep regime and germination time, with highest levels observed after 5-day germination, following the standard steep. For all the steep regimes, higher beta-amylase activity was observed after 5 compared to 4-day germination, but proportionally, there was a greater increase for the single 8-h immersion, leading to a significant germination time × steep regime interaction. Additionally, there were significant interactions between genotype and both steep regime and germination time, which was likely to reflect differences in the rate of release of bound beta-amylase and the synthesis of the other starch-degrading enzymes. When grain beta-amylase levels were plotted against malt diastatic power after 5-day germination, following the standard steep regime, there was a positive relationship, but no significant correlation (Fig. 14.3). This was not surprising, given the small sample size and the fact that the complete range of beta-amylase values (Fig. 14.2) was not covered. However, with the exception of one sample, which gave high diastatic power, despite a relatively moderate grain beta-amylase activity, the grain test was a reasonable indicator as to whether malt diastatic activity could be classed as high, medium or low. Further testing will, however, be required with a much larger population, to verify the utility of the grain beta-amylase test as a rapid screening procedure for diastatic power, in this population of hull-less barley lines.

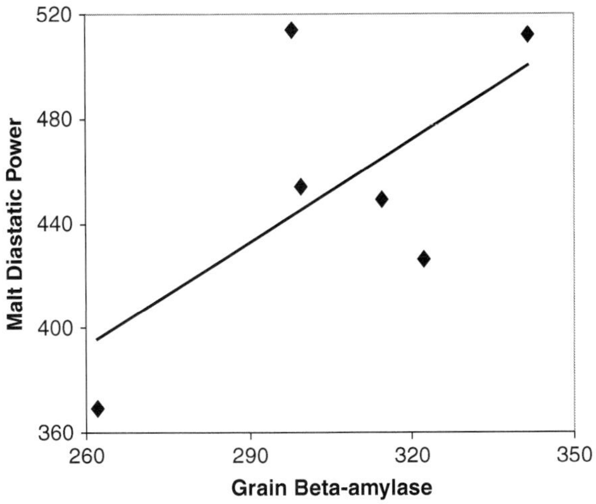

Fig. 14.3 Diastatic power after malting, plotted against grain beta-amylase activity for six lines from the Penthouse mutant population (Values for both characters are given in enzyme units, calculated from samples, of known activity, and used to create a standard curve)

Table 14.3 Proportions of free beta-amylase and thermostable beta-amylase, classed as high, medium or low, for six mutant lines derived from Penthouse and three varieties

Genotype	Proportion of free beta-amylase	Thermostable proportion
11	High	Medium
13	High	Medium
20	High	Medium
21	High	Medium
24	High	Medium
33	High	Medium
Optic	Low	Medium
Penthouse	High	Medium
Golden Promise	High	Low

14.3.3 Free and Thermostable Beta-Amylase

Allison and Swanston (1974) noted that varieties with the Sd2 electrophoretic phenotype were characterised by having a high proportion, approximately two-thirds, of their total beta-amylase activity in a free (water soluble) form, while the Sd1 phenotype had about one-third. Here, Penthouse and the six mutant lines tested all had high levels of free beta-amylase (Table 14.3). To test the thermostability, an additional variety, Golden Promise, known from previous results (Swanston, unpublished data) to be of the Sd2L type, as classified by Eglinton et al. (1998), was included. Results for Penthouse and all 6 mutant lines, however, showed the thermostable proportion to be moderate, similar to Steptoe (Kaneko et al. 2001) and Morrell (Eglinton et al. 1998).

14.4 Discussion

Mutants induced in the malting cultivar, Triumph, were screened in germination tests and selected either for speed of germination (Molina-Cano et al. 1989) or for capacity to germinate in the presence of ABA (Molina-Cano et al. 1999). While this screening was intended to select genotypes with less dormancy, two mutants were subsequently shown to have higher levels of beta-amylase than their parental variety (Swanston and Molina-Cano 2001). Similarly, Hayter and Allison (1976) used germination in the presence of ABA to screen barley mutants and obtained some with enhanced starch-degrading activity, although this was generally only expressed at high grain nitrogen levels (Allison et al. 1979). The mutant lines observed, here, had not been subjected to any screening for germination properties, but lines with higher beta-amylase than the parent variety were detected. Future work, with the wider population, will therefore contain some initial screening, with the intention of detecting lines with even higher levels of beta-amylase activity.

The absence of any significant association, here, between beta-amylase and grain nitrogen was unexpected, given the general relationship between the two factors, but the range of nitrogen levels was fairly restricted, particularly in 2010, while outliers could have resulted from mutation at loci influencing either factor. Swanston (1980) associated differences in beta-amylase activity at given grain nitrogen levels with electrophoretic patterns, subsequently shown to result from allelic variation in amino-acid sequences at the *Bamy*1 locus on chromosome 4H (Li et al. 2002). Lines showing widely different beta-amylase activity from that of the parent variety will, therefore, be investigated for genotypic differences at relevant loci. The activity and thermostability of β-amylase are of major importance to the fermentability of the hot water extract (Evans et al. 2005), and the mutant lines tested showed a potentially useful combination of high free beta-amylase combined with moderate thermostability. Kaneko et al. (2001) demonstrated Steptoe, which shares this combination, to have an allele for higher thermostability than Morex at the *Bamy1* locus on chromosome 4H, but this may relate to a slightly higher thermostability in free compared to bound beta-amylase (Swanston and Molina-Cano 2001).

Additional enzymes contribute to starch degradation during mashing. These are synthesised during germination, so, while the faster malting of hull-less types (Bhatty 1996; Agu et al. 2009) may be advantageous, poor or uneven germination, resulting from damage to embryos during grain handling, is a potential problem. Malting and enzyme assays applied to the lines in this experiment suggested that germination and enzyme development were not, in general, adversely affected by the malting process, although the hull-less types did not show any advantages over Optic, associated with the shorter steeping regimes or germination times. However, if hull-less types are identified, ultimately, as suitable for commercial grain distilling, careful harvesting will be essential. This should be achievable since it represents a relatively small, niche market for malted barley in Scotland, so grain may be obtained through contracts with individual growers.

Barley is generally malted prior to use in brewing or distilling, whether it represents the total cereal component of the mash or, as in grain distilling, a source of enzymes to degrade a starch-based adjunct. However, alcoholic beverages can also be produced from any source of starch and extraneous enzymes. The advantages of this are partly from a significantly reduced use of resources such as energy and water but also from less time and cost. Aastrup (2010) described a beer made entirely from un-malted barley. The enzymes normally produced during malting were supplied from a commercially developed product, although starch degradation was still, in part, dependent on the beta-amylase in the barley grain. The end product was also of acceptable flavour, so it is possible that a significant future proportion of mass-produced beer will be made from un-malted raw materials. Potential advantages both in extract and higher beta-amylase activity would make hull-less barley desirable for this use, as mash filters can overcome any filtration problems resulting from lack of husk particles. In conclusion, therefore, it is likely that hull-less barley will attract increasing interest, in future, as a raw material for distilled beverage production.

Acknowledgements The James Hutton Institute is supported by the Scottish Government Rural and Environment Science and Analysis Services (RESAS) Division. The authors also thank the Scottish Society for Crop Research (SSCR) for the additional funding.

References

Aastrup, S. (2010). Beer from 100% barley. *Scandinavian Brewers' Review, 67*, 28–33.

Agu, R. C., Bringhurst, T. A., Brosnan, J. M., & Pearson, S. (2009). Potential of hull-less barley malt for use in malt and grain whisky production. *The Journal of the Institute of Brewing, 115*, 128–133.

Allison, M. J., & Swanston, J. S. (1974). Relationships between β-amylase polymorphisms in developing, mature and germinating grains of barley. *The Journal of the Institute of Brewing, 80*, 285–291.

Allison, M. J., Ellis, R. P., Hayter, A. M., & Swanston, J. S. (1979). Breeding for malting quality at the Scottish Plant Breeding Station. *Scottish Plant Breeding Annual Report, 58*, 92–139.

Arends, A. M., Fox, G. P., Henry, R. J., Marschke, R. J., & Symons, M. H. (1995). Genetic and environmental variation in the diastatic power of Australian barley. *Journal of Cereal Science, 21*, 63–70.

Bathgate, G. N. (1989). Cereals in Scotch whisky production. In G. H. Palmer (Ed.), *Cereal science and technology* (pp. 243–278). Aberdeen: Aberdeen University Press.

Bendelow, V. M. (1981). Selection for quality in malting barley breeding. In *Proceedings of the Fourth International Barley Genetics Symposium* (pp. 181–185). Edinburgh: Edinburgh University Press.

Bhatty, R. S. (1996). Production of food malt from hull-less barley. *Cereal Chemistry, 73*, 75–80.

Briggs, D. E. (1998). *Malts and malting*. New York: Blackie Academic and Professional.

Bringhurst, T. A., Glasgow, E., Agu, R. C., Brosnan, J. M., & Thomas, W. T. B. (2010). Varietal and site interactions in the growing of malting barley for brewing and distilling. In *Proceedings of the Tenth International Barley Genetics Symposium* (pp. 416–427). Aleppo: International Center for Agricultural Research in the Dry Areas.

Bryce, J. H., Goodfellow, V., Agu, R. C., Brosnan, J. M., Bringhurst, T. A., & Jack, F. R. (2010). Effect of different steeping conditions on endosperm modification and quality of distilling malt. *Journal of the Institute of Brewing, 116*, 125–133.

Eglinton, J. K., Langridge, P., & Evans, D. E. (1998). Thermostability variation in alleles of barley *beta*-amylase. *Journal of Cereal Science, 28*, 301–309.

Evans, D. E., Collins, H., Eglinton, J., & Wilhelmson, A. (2005). Assessing the impact of the level of diastatic power enzymes and their thermostability on the hydrolysis of starch during wort production to predict wort fermentability. *Journal of the American Society of Brewing Chemists, 63*, 185–198.

Hayter, A. M., & Allison, M. J. (1976). Breeding for high diastatic power. In H. Gaul (Ed.), *Proceedings of the Third International Barley Genetics Symposium* (pp. 612–619). Karl Thiemig: International Center for Agricultural Research in the Dry Areas.

Hayter, A. M., & Riggs, T. J. (1978). The inheritance of diastatic power and alpha-amylase contents in spring barley. *Theoretical and Applied Genetics, 52*, 251–256.

IOB. (1982). *Recommended methods of analysis*. London: Institute of Brewing. Revision of 1977 Edition.

Kaneko, T., Zhang, W., Takahashi, H., Ito, K., & Takeda, K. (2001). QTL mapping for enzyme activity and thermostability of β-amylase in barley (*Hordeum vulgare* L.). *Breeding Science, 51*, 99–105.

Li, C. D., Langridge, P., Zhang, X. Q., Eckstein, P. E., Rossnagel, B. G., Lance, R. C. M., Lefol, E. B., Lu, M. Y., Harvey, B. L., & Scoles, G. J. (2002). Mapping of barley (*Hordeum vulgare* L.) *beta*-amylase alleles in which an amino acid substitution determines *beta*-amylase isozyme type and the level of free *beta*-amylase. *Journal of Cereal Science, 35*, 39–50.

Molina-Cano, J.-L., Roca de Togores, F., Royo, C., & Perez, A. (1989). Fast germinating low β-glucan mutants induced in barley with improved malting quality and yield. *Theoretical and Applied Genetics, 78*, 748–754.

Molina-Cano, J.-L., Sopena, A., Swanston, J. S., Casas, A. M., Moralejo, M. A., Ubieto, A., Lara, I., Perez-Vendrell, A. M., & Romagosa, I. (1999). A mutant induced in the malting barley cv Triumph with reduced dormancy and ABA response. *Theoretical and Applied Genetics, 98*, 347–355.

Schwarz, P., & Li, Y. (2011). Malting and brewing uses of barley. In S. E. Ullrich (Ed.), *Barley: Production improvement and uses* (pp. 478–521). Chichester: Wiley-Blackwell.

Swanston, J. S. (1980). The use of electrophoresis in testing for high diastatic power in barley. *The Journal of the Institute of Brewing, 86*, 81–83.

Swanston, J. S., & Middlefell-Williams, J. E. The influence of steep regime and germination period on the malting properties of some hull-less barley lines. *The Journal of The Institute of Brewing*. submitted.

Swanston, J. S., & Middlefell-Williams, J. E. (2010, April 6–8). *Hulless barley mutants may improve alcohol yields and reduce energy use in malt whisky distilling*. Abstract Proceedings of the EUCARPIA Cereals Meeting, Cambridge, UK.

Swanston, J. S., & Molina-Cano, J.-L. (2001). *Beta*-amylase activity and thermostability in two mutants derived from the malting barley cv. Triumph. *Journal of Cereal Science, 33*, 155–161.

Swanston, J. S., Middlefell-Williams, J. E., Forster, B. P., & Thomas, W. T. B. (2011). Effects of grain and malt β-glucan on distilling quality in a population of hull-less barley. *The Journal of the Institute of Brewing, 117*, 389–393.

Chapter 15
Natural Variation in Grain Iron and Zinc Concentrations of Wild Barley, *Hordeum spontaneum*, Populations from Israel

Jun Yan, Fang Wang, Rongzhi Yang, Tangfu Xiao, Tzion Fahima, Yehoshua Saranga, Abraham Korol, Eviatar Nevo, and Jianping Cheng

Abstract Wild barley (*Hordeum spontaneum*), the progenitor of cultivated barley, is an important genetic resource for cereal improvement. Iron (Fe) and zinc (Zn) are essential minerals for human good health. In the current study, the grain Fe and Zn concentrations (GFeC and GZnC) of 92 *H. spontaneum* genotypes collected from nine populations in Israel, and ten barley cultivars (*Hordeum vulgare*) from five provinces in China were investigated. Remarkable variations in GFeC and GZnC were found between and within wild barley populations, ranging from 10.8 to 329.1 and 66.3 to 493.9 mg kg^{-1} among the 92 wild genotypes with an average of 74.3 and 173.9 mg kg^{-1}, respectively. The mean value of GFeC and GZnC in each population varied from 48 to 146 and 96 to 291 mg kg^{-1}, respectively. Significant correlations

J. Yan (✉)
Faculty of Biotechnology Industry, Chengdu University, Chengdu, China

Institute of Triticeae Crops, Guizhou University, Guiyang, China

Institute of Evolution and the Department of Evolutionary and Environmental Biology, University of Haifa, Haifa, Israel
e-mail: yanjun6622@gmail.com

F. Wang • R. Yang • J. Cheng
Institute of Triticeae Crops, Guizhou University, Guiyang, China
e-mail: agr.jpcheng@gzu.edu.cn

T. Xiao
State Key Laboratory of Environmental Geochemistry, Institute of Geochemistry, Chinese Academy of Sciences, Guiyang, China

T. Fahima • A. Korol • E. Nevo
Institute of Evolution and the Department of Evolutionary and Environmental Biology, University of Haifa, Haifa, Israel

Y. Saranga
The Robert H. Smith Institute of Plant Sciences and Genetics in Agriculture, The Hebrew University of Jerusalem, Rehovot, Israel

G. Zhang et al. (eds.), *Advance in Barley Sciences: Proceedings of 11th International Barley Genetics Symposium*, DOI 10.1007/978-94-007-4682-4_15,

were found among four ecogeographical factors out of the 14 studied, including both GFeC and GZnC. Wild barley exhibited higher values and greater diversity of GFeC and GZnC than its cultivated counterparts. The higher Fe and Zn grain concentrations found in *H. spontaneum* suggest that wild barley germplasm confers higher abilities for mineral uptake and accumulation, which can be used for genetic studies of barley nutritional value and for further improvement of domesticated cereals.

Keywords Grain iron concentration • Grain zinc concentration • *Hordeum spontaneum* • Israel • Wild barley

15.1 Introduction

The essential trace minerals, such as iron (Fe) and zinc (Zn), are of fundamental importance to human health. Micronutrient malnutrition, also known as "hidden hunger," is a specific type of malnutrition caused by a lack of minerals and vitamins in the diet, such as zinc, iron, and vitamin A (http://www.harvestplus.org). Estimates suggest that some 815 million households worldwide suffer from "hidden hunger" (Underwood 2003). During the past 40 years, it became apparent that deficiency of Fe and Zn in humans is quite prevalent and may affect more than two billion subjects in the developing world (Welch and Graham 2004; Uauy et al. 2006; Prasad 2008). Fe and Zn deficiencies rank fifth and sixth among the ten most important risk causes of illness and disease in low-income countries, respectively (WHO 2002). Recently, Fe and Zn deficiencies together with vitamin A deficiency have been identified among the top priority global problems facing the world (http://www.copenhagenconsensus.com) and represent a major cause of child death in the world (Black et al. 2008). Fe and Zn enter the food chain mainly through plants. In many countries, soils are often low in available micronutrients; hence, the food systems are deficient in Fe and Zn (Cakmak et al. 2000, 2004). As a main source of calorie intake, cereal-based foods are extensively consumed in the developing world. It makes up 29–30% of the world's total cereal production and is humans' most important source of vegetarian protein and micronutrient. However, cereal crops such as wheat, barley, rye, and oats are inherently very poor both in concentration and bioavailability of Fe and Zn in seeds, particularly when grown on these mineral-deficient soils (Cakmak et al. 2000, 2004). Breeding programs directed toward increased yield have narrowed the genetic basis of modern crop plants. FAO estimates 75% of crop diversity was lost between 1900 and 2000 (FAO 2010). Developing genetically micronutrient-enriched cereals and improving their bioavailability (biofortification) using genetics and genomics tools are considered as promising and cost-effective approaches for diminishing malnutrition (Uauy et al. 2006). Existence of large genetic variation for micronutrients in grain is essential for a successful breeding program aiming at development of micronutrient-rich new plant genotypes.

Wild barley, *Hordeum spontaneum* C. Koch, is the progenitor of cultivated barley, *Hordeum vulgare* L. ssp. *vulgare* (Nevo 1992). It is a wild winter annual barley widely distributed over the eastern Mediterranean rim and western Asian countries (Nevo 1992). The center of diversity for *H. spontaneum* and the primary site of domestication are considered to be the Fertile Crescent of the Near East (Zohary 1969; Lev-Yadun et al. 2000). In Israel, *H. spontaneum* is abundant, occupying an extraordinarily large diversity of habitats ranging from the mesic Mediterranean to the xeric southern steppes (Nevo et al. 1979, 1984). As the wild progenitor of cultivated *H. vulgare* L., *H. spontaneum* exhibits distinctive differences from its progeny. Apart from its ecological features, cultivated barley differs from its wild progenitor primarily in the gene-controlling brittleness of the rachis. Wild barley exhibits a huge amount of polymorphisms in morphological traits (Nevo et al. 1979); in agronomic traits such as earliness, biomass and yield (Nevo et al. 1984), kernel weight, size and color (Chen et al. 2004b), and seed dormancy (Gutterman and Nevo 1994; Zhang et al. 2005; Yan et al. 2008); in growth characteristics (Van Rijn et al. 2000); and in resistance to diseases (Moseman et al. 1983), drought (Chen et al. 2002, 2004a, 2010, 2011; Zhang et al. 2005; Nevo and Chen 2010), and salt (Yan et al. 2008; Nevo and Chen 2010). Also, *H. spontaneum* is polymorphic at the DNA and protein levels such as storage proteins (Nevo et al. 1983), isozymes (Nevo et al. 1979, 1986), ribosomal DNA variation (Saghai Maroof et al. 1984), chloroplast DNA variation (Clegg et al. 1984), restriction fragment length polymorphism (RFLP) (Peterson et al. 1994), random amplified polymorphic DNA (RAPD) (Owuor et al. 1999), amplified fragment length polymorphism (AFLP) (Pakniyat et al. 1997; Turpeinen et al. 2003), and simple sequence repeat (SSR) (Forster et al. 1997; Baek et al. 2003). Most importantly, wild barley populations in Israel contain large amounts of unexplored characteristics including grain mineral contents. Moreover, barley has a self-fertile, diploid ($2n=2x=14$) genetic system and can therefore serve as a model species for genetic and physiological studies in Triticeae species (Brantestam 2005). Thus, the genetic resources of the progenitor display striking morphological and physiological performances, which are of great economic importance for improving cultivated barley, even closely related cereal like wheat or other crops (Nevo et al. 1984; Nevo 1996, 1992). The objectives of the present study were to analyze the variation in grain Fe and Zn concentrations in Israeli *H. spontaneum* populations and to choose a set of donor parents for the breeding of Fe- and Zn-enriched barley cultivars.

15.2 Materials and Methods

15.2.1 Plant Materials

Ninety-two genotypes from nine Israeli populations of the wild barley *H. spontaneum* and ten barley cultivars from five provinces in China were used in the present study. The detail was in Yan et al. (2011).

Table 15.1 Operating parameters of the ICP-MS and ICP-OES for Zn and Fe determination

ICP-MS		ICP-OES	
Parameters	Setting	Parameters	Settings
Power	1.4 kW	RF power	1.0 kW
Plasma gas flow	15.0 L/min	Plasma gas flow	15 L/min
Auxiliary gas flow	1.0 L/min	Auxiliary gas flow	1.5 L/min
Carrier gas flow	1.12 L/min	Nebulizer argon flow	0.75 L/min
Sampling rate	0.4 mL/min	Sample uptake rate	1 mL/min
Sampling depth	7 mm	Viewing height	12 mm
Orifice of sampling cone	1 mm	Integration time	5 s
Orifice of skimmer cone	0.4 mm	Fe emission line	238.2 nm
Data acquisition mode	Quantitative analysis		
Integration time	0.3 s/isotope		
Cerium oxide/cerium	<0.5%		
Doubly charge	<2%		

15.2.2 Determination of Fe and Zn in Barley Seeds

Barley sample preparation for chemical analyses was same as Yan et al. (2011). Zn in the digested samples of barley seeds was determined using a Finnigan MAT Element inductively coupled plasma mass spectrometry (ICP-MS) (Thermo Scientific, USA) according to the protocol similar to Górecka et al. (2006), while Fe was determined by an inductively coupled plasma optical emission spectroscopy (ICP-OES) (Varian Inc., USA) according to the protocol similar to Górecka et al. (2006) and Souza et al. (2007). All of the specifications met the installation requirements including sensitivity, background, oxide, doubly charge, stability, etc. The operating parameters are shown in Table 15.1 for ICP-MS and ICP-OES.

Zn and Fe concentrations were expressed as mg kg^{-1} dry weight. Quality control of the analytical processes was assured throughout by the use of a certified shrub leaves reference (GBW-07602 with Zn = 20.6 ± 2.2 mg kg^{-1} and Fe = 1,020 ± 40 mg kg^{-1}), a blank, and duplicate samples (three duplication per ten samples). The measured values of GBW-07602 were consistent with the reference values. The blanks were below the detection limits. The results of the duplicate samples were relatively good with the coefficient of variation (CV) below 15%.

15.2.3 Statistical Analysis

JMP 6.0 (SAS Institute) software was used to perform ANOVA. Tukey-Kramer's honestly significant difference (HSD) test was used to compare means of all pairs (significance level, 5%). Spearman's rho correlation was used to analyze the multivariate correlations.

15.3 Results

15.3.1 *Variation in H. spontaneum GFeC and GZnC*

Large variations were observed on *H. spontaneum* GFeC and GZnC among the 92 genotypes and the selected nine populations, respectively (Table 15.2; Figs. 15.1 and 15.2). Significant differences on the GFeC and GZnC ($p = <0.0001$) were found among the nine *H. spontaneum* populations.

GFeC of the 92 genotypes ranged from 10.8 to 329.1 mg kg^{-1}, with an average of 74.3 mg kg^{-1}. The genotype 20_2 originated from the Sede Boqer population was the highest one, while the 32_16 originated from the Ein-Zukim population was the lowest one. The average GFeC value of each population was in the range of 48 mg kg^{-1} (Mt. Hermon population) to 146 mg kg^{-1}(Sede Boqer population). In the Sede Boqer population, the GFeC value varied between 62.9 and 329.1 mg kg^{-1}, while in the Mt. Hermon population, the value changed from 21.8 to 84.3 mg kg^{-1}. The variation coefficient (CV) of each population is in the range of 28% (the Maalot population) to 57% (the Sede Boqer and the Ein-Zukim population).

GZnC among the 92 genotypes varied from 66.3 to 493.9 mg kg^{-1}, with an average of 173.9 mg kg^{-1}. The highest value was found in the genotype 32_2 originated from the Ein-Zukim, while the lowest one was found in the 24_50 originated from Akhziv population. Among the 9 populations, the GZnC of the Mt. Hermon population was the highest with an average of 291 mg kg^{-1} (range of 96.1 to 412.5 mg kg^{-1}), while in the Maalot population was the lowest with an average of 96 mg kg^{-1} (range of 80 to 122.8 mg kg^{-1}). The CV of each population changes from 12% for the Maalot to 71% for the Sede Boqer population.

15.3.2 *GFeC and GZnC of Cultivated Barley from China as Compared with H. spontaneum*

GFeCs of the ten barley cultivars ranged from 36.5 to 140.9 mg kg^{-1} (Table 15.2). Yunpi 6 from Yunnan Province had the highest value, while Ganpi 3 from Gansu province had the lowest value. In comparison, the highest GFeC found in the wild barley genotype 32-2 (329.1 mg kg^{-1}) was 2.3 times of the cultivar Yunpi 6, and in total five *H. spontaneum* genotypes (three from the Sede Boqer and two from Caesarea, respectively) exhibited statistically significant higher GFeC than Yunpi 6 (Fig. 15.1). The CV of the ten cultivars is 43%, which is lower than the value (66%) obtained for wild barley genotypes.

GZnCs of the ten barley cultivars ranged from 48 to 98.7 mg kg^{-1}. Supi 4 from Jiangsu province had the highest value, while Ganpi 3 from Gansu province had the lowest value. The highest GZnC in the wild barley genotype 20-2 (493.9 mg kg^{-1}) was five times greater than in the cultivar Supi 4. Sixty-eight *H. spontaneum* genotypes

Table 15.2 Summary of grain Fe and Zn concentrations of the nine *H. spontaneum* populations and ten *H. vulgare* cultivars

ID	Populations/ variety	Samples	GFeC (Mean ± SD) (mg kg^{-1})	GFeC Range (mg kg^{-1})	GFeC CV (%)	GZnC (Mean ± SD) (mg kg^{-1})	GZnC Range (mg kg^{-1})	GZnC CV (%)
1	Mt. Hermon	9	48.0 ± 22.0 (d[a])	21.8–84.3	46	291.1 ± 127.3 (a)	96.1–412.5	44
10	Maalot	10	50.8 ± 14.3 (d)	36.3–79.1	28	96.2 ± 11.5 (c)	80.0–122.8	12
20	Sede Boqer	10	146.2 ± 82.8 (a)	62.9–329.1	57	172.0 ± 121.5 (abc)	85.1–374.0	71
22	Mehola	8	73.4 ± 21.7 (bcd)	43.8–107.1	30	103.2 ± 13.8 (c)	79.7–124.2	13
24	Akhziv	8	77.0 ± 34.9 (bcd)	42.9–142.6	45	102.7 ± 18.0 (c)	66.3–127.5	17
25	Atlit	8	72.7 ± 33.7 (bcd)	42.1–146.7	46	127.4 ± 16.7 (bc)	103.5–158.9	13
26	Caesarea	10	118.1 ± 41.3 (abc)	65.3–214.3	35	113.2 ± 19.0 (c)	85.9–150.7	17
32	Ein-Zukim	22	49.9 ± 28.5 (d)	10.8–114.2	57	245.7 ± 141.3 (ab)	93.7–493.9	58
37	Evolution-Canyon	7	53.0 ± 19.7 (d)	16.2–77.0	37	213.2 ± 94.6 (abc)	164.1–426.9	44
	All wild barley	92	74.3 ± 49.3	10.8–329.1	66	173.9 ± 114.3	66.3–493.9	66
Zj1	Yunpi 2	1 × 3	104.3 ± 4.2 (abcd)			57.8 ± 1.2 (bc)		
Zj2	Yunpi 6	1 × 3	140.9 ± 5.1(ab)			66.4 ± 1.6 (bc)		
Zj3	Zhexiu 12	1 × 3	97.3 ± 2.1 (abcd)			64.8 ± 2.1 (bc)		
Zj4	E32380	1 × 3	41.7 ± 6.1 (bcd)			61.6 ± 1.4 (bc)		
Zj5	Yangnongpi 5	1 × 3	84.1 ± 4.4 (abcd)			65.5 ± 4.1 (bc)		
Zj6	Baoshan8640-1	1 × 3	44.7 ± 4.4 (cd)			55.5 ± 3.5 (c)		
Zj7	Yancheng 01094	1 × 3	100.0 ± 6.0 (abcd)			64.9 ± 3.6 (bc)		
Zj8	Zhexiu33	1 × 3	108.8 ± 8.0 (abcd)			63.1 ± 2.6 (bc)		
Zj9	Ganpi 3	1 × 3	36.5 ± 5.1 (cd)			48.0 ± 2.0 (c)		
Zj10	Supi 4	1 × 3	48.2 ± 7.4 (bcd)			98.7 ± 4.1 (abc)		
	All cultivars	10 × 3	80.6 ± 34.7	36.4–140.9	43	64.6 ± 13.0	48.0–98.7	20

[a]Different letters in bracket indicate significant differences ($p < 0.05$ by Tukey–Kramer HSD test)

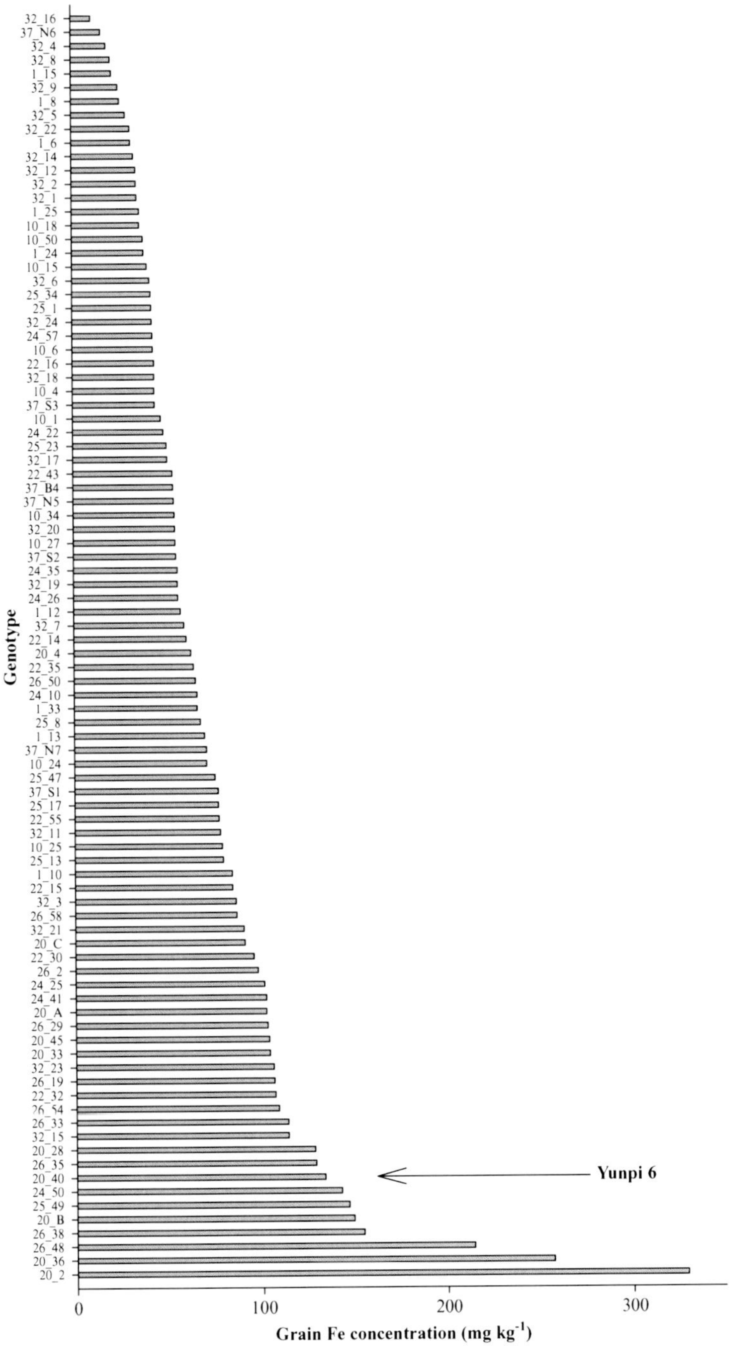

Fig. 15.1 GFeC of the 92 *H. spontaneum* genotypes

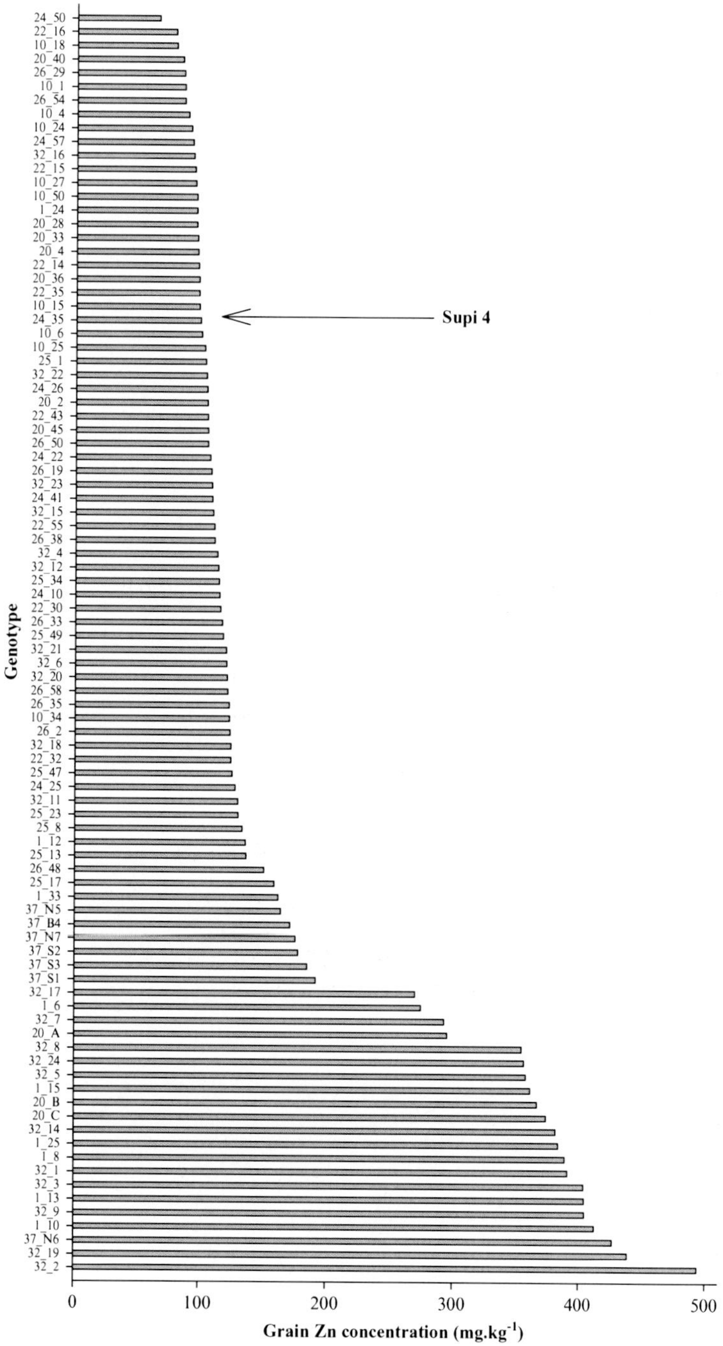

Fig. 15.2 GZnC of the 92 *H. spontaneum* genotypes

Nonparametric: Spearman's ρ

Variable	by variable	Spearman ρ	Prob>|ρ|
GFeC	Ln	-0.4150	<.0001
GFeC	Td	-0.3767	0.0002
GFeC	Ev	-0.3001	0.0037
GFeC	Huan	0.2732	0.0084
GZnC	Tm	0.2106	0.0439
GZnC	Tj	0.2075	0.0471
GZnC	Td	0.2386	0.0220
GZnC	Tdd	-0.2669	0.0101

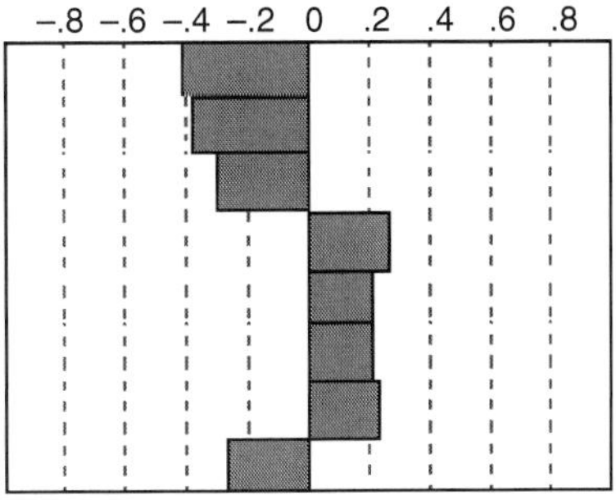

Fig. 15.3 Spearman's rho significant ($p<0.05$) correlations between the grain Fe and Zn concentrations and ecogeographical data (Yan et al. 2011) of the original sites of nine *H. spontaneum* populations in Israel

(74% of the 92 genotypes) exhibited statistically significant higher GZnC than Supi 4 (Fig. 15.2). The CV of the ten cultivars is 20%, which is lower than the value obtained for wild barley (66%).

15.3.3 Association of Grain Mineral Concentrations of *H. spontaneum* with Ecogeographic Factors

Twenty-six potential correlations were tested between the grain Fe and Zn concentrations and 13 ecogeographical factors, of which eight correlations were found significant (Fig. 15.3). GFeC was negatively correlated with longitude (Ln), mean seasonal temperature difference (Td), and mean annual evaporation (Ev) and positively correlated with mean annual humidity (Huan). Positive correlations were observed between GZnC and mean annual temperature (Tm), mean January temperature (Tj), as well as mean seasonal temperature difference (Td), while negative correlations were found between GZnC and mean daily temperature difference (Tdd). Moreover, nine *H. spontaneum* populations were derived from five soil types (Fig. 15.4), and one-way ANOVA indicated significant effect of soil type on the GFeC ($p=0.05$) and GZnC ($p=< 0.0001$). The GFeC of the *H. spontaneum* originated from sandy loam was the highest, while the one originated from terra rossa was the lowest. Tukey-Kramer's honest significant difference (HSD) test showed a significant difference in GFeC between populations originating in sandy loam and terra rossa (Fig. 15.4). The GFeC in populations originating in the other three soil types were at the middle level and did not show significant difference from sandy loam or terra rossa soil types. The populations originated from terra rossa exhibited the highest GZnC, followed by loess, while the rendzina and alluvium populations were the lowest.

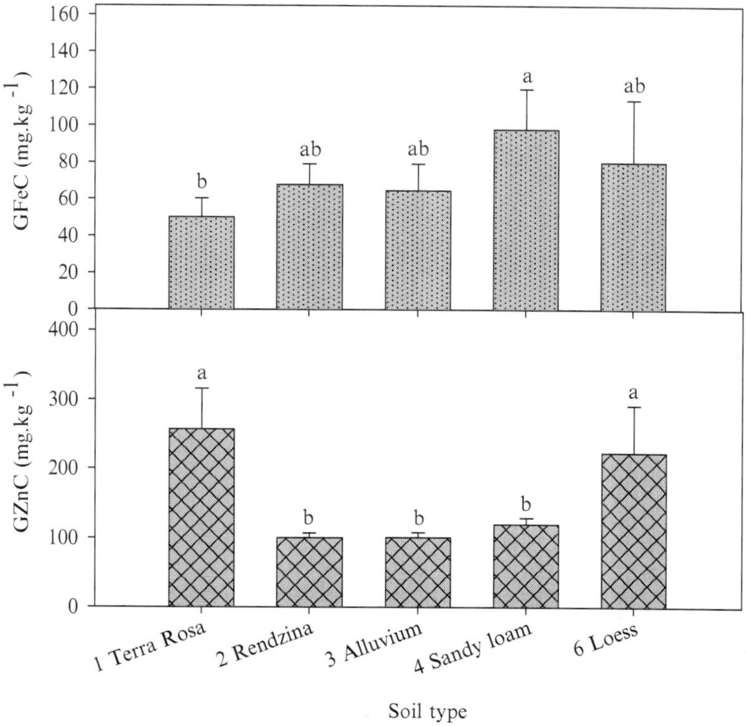

Fig. 15.4 Comparison of grain Fe and Zn concentrations of the *H. spontaneum* originated from five soil types (*Bars* represent standard deviation. Different letters above *bars* indicate significant differences ($p < 0.05$ by Tukey–Kramer HSD test))

Tukey-Kramer's HSD test shows that populations from terra rossa and loess had higher level of GZnC, while sandy loam, rendzina, and alluvium showed lower levels.

15.4 Discussion

15.4.1 Genetic Diversity of Micronutrient in H. spontaneum

The 92 wild barley genotypes examined in the current study were collected from nine sites across a relatively small geographical area from the mesic north mountainous and Mediterranean regions to the xeric southern steppes in Israel. Nevertheless, these sites are quite diverse in their ecogeographical conditions (Nevo et al. 1979, 1984). A wide range and very high grain Fe and Zn concentrations were found in 92 genotypes of *H. spontaneum* from nine populations. GFeCs and GZnCs of 7 and 70 *H. spontaneum* genotypes were higher than that of the highest cultivars out of ten Chinese cultivars, respectively. The genotypes of *H. spontaneum* with the highest

GFeCs and GZnCs had 2.3 and 5 times higher mineral concentrations than those of the highest cultivars, respectively (Table 15.2, Figs. 15.1, 15.2). Furthermore, in the current set of experiments, the grains were harvested from plants that had been grown in the same location, at the same time, under the same conditions. Therefore, the variation in these mineral concentrations revealed genetic differences resulting from the adaptation of *H. spontaneum* to different eco-environments. Existence of large genetic variation for micronutrients in grain is essential for a successful breeding program aiming at development of micronutrient-rich new plant genotypes. Our results demonstrate the huge potential of wild barley for improvement of GFeCs and GZnCs. These high-GFeC and high-GZnC *H. spontaneum* genotypes can be utilized as a set of donor parents for biofortification of barley cultivars, and as optimized model plants for the improvement of cereal micronutrient concentration via gene introgression, toward the development of micronutrient-enriched cereals in the future.

15.4.2 Associations Between Micronutrient of H. spontaneum Populations and the Ecogeographical Conditions of Their Collection Sites

The environment is a key contributor to the range of morphological variation found in various organisms (van Valen 1965). Associations between the performance of wild barley populations and the ecogeographical conditions of their collection sites can shed light on the major driving forces contributing to the ecological fitness of these populations and, more specifically, to the development of their mineral uptake and accumulation mechanisms. In the current study, Spearman correlation analysis showed that of 14 ecogeographic factors (Yan et al. 2011), four factors correlated with both GFeC and GZnC. The positive correlations between GZnC and four temperature factors (Tm, Tj, Td, and Tdd) were observed, while the correlations between GFeC and Td and Ev were negative (Fig. 15.3). These results implied that the ecogeographical impacts were not homogenous for the different minerals during a long evolutionary history.

Soil type at the *H. spontaneum* collection site was also an important factor affecting its ability to take up minerals and to accumulate micronutrients in its grains. Several previous reports drew particular attention to the wide range of mineral concentrations in different types of foods and revealed that the variations are due to differences in these minerals' availability in the soil on which an animal is raised or a plant is grown (Welch 1999; Welch and Graham 2004; Reilly 2006). In the present study, nine *H. spontaneum* populations were derived from five soil types. The GFeC of the *H. spontaneum* originated from sandy loam soil was the highest, and the second highest was one from loess soil, but the difference between the two was not significant. The lowest GFeC was of the *H. spontaneum* population originated from terra rossa soil, and it was significantly different from all other soil types. For GZnC of *H. spontaneum*, both those originating from loess soil and sandy loam were the highest and had significant difference from the other three kinds of soil types (Fig. 15.4). Our results suggest that loess soil has the lowest mineral availability, and therefore the wild barley growing there had to develop strong mineral uptake and accumulation abilities. In a similar study of wild emmer, Chatzav et al. (2010) found negative correlations between soil-extractable Zn and grain Zn concentrations, suggesting that populations originating from Zn-deficient soils have evolved a better capacity to accumulate the deficient mineral in their grains, which may be of critical importance for seedling germination and establishment. Bonfil and Kafkafi (2000) also suggested that when a certain nutrient is naturally deficient in the soil, genotypes which store a higher concentration of that nutrient in the seed will have an advantage under such conditions. The amount of minerals in the seed depends on a plethora of processes, including absorption from the soil, uptake by the roots, translocation and redistribution within the plant tissues, and remobilization to the seed (Grusak and Cakmak 2005; Cakmak et al. 2010).

Cultivated barley contains, on average, 40% *H. spontaneum* alleles (Ellis et al. 2000). Because *H. spontaneum* and cultivated barley are interfertile, *H. spontaneum* can be used to increase the genetic diversity of cultivated barley by crosses (Nevo and Chen 2010). In the past, breeders were reluctant to use exotic germplasm in their breeding programs due to linkage drag of negative traits. A negative relationship between cereal grain yield and grain protein concentration (GPC) has been observed. This phenomenon has been called "growth dilution" and exists also for nutrients other than nitrogen (Pleijel et al. 1999). However, modern genomic technologies have led to the development of marker-assisted selection (MAS) approaches that enable efficient transfer of only small chromosome segments carrying the target gene and therefore avoiding linkage drag of negative traits. The *Triticum dicoccoides* and *H. spontaneum* genotypes of high GFeC and GZnC can be utilized as a set of donor parents for the improvement of cereal nutrients concentrations via gene introgression. Future cloning of the genes controlling these traits will allow also the use of transgenic approaches for manipulation of these traits and improvement of wheat and barley crops. A perfect example of this approach is provided by the cloning of the high grain protein gene, *Gpc-B1*, associated with increased grain zinc and iron content, derived from wild emmer wheat (Uauy et al. 2006). It is now being applied in wheat breeding (Uauy et al. 2006). Also, a homologous barley gene, designated *HvNAM-1*, was shown to be responsible for the grain protein QTL located on barley chromosome 6H (Distelfeld et al. 2008). The identification of quantitative trait loci (QTL) may assist these breeding programs, and as more information on the biochemistry of mineral accumulation becomes available, genetic modification can be applied (Grusak 2002). Therefore, the results presented in the current and previous studies demonstrate the potential of wild cereal populations as a rich source for grain mineral content genes for crop improvement.

Acknowledgments This study was funded by Training Programme Foundation for the Youth Talents in Science and Technology by Guizhou province in China (2007–2011). The authors are greatly indebted to Lingling Zhang, Zhouyun Pan, Xiuwen Wei, Mingliang Hu, Junbo Gou,

Changmin Xu, Yunchang Wang, and Bin Wan at Guizhou University in China, for their excellent help in fieldwork. The authors also wish to express their thanks to State Key Laboratory of Environmental Geochemistry, Institute of Geochemistry, Chinese Academy of Sciences, and Hangzhou National Barley Improvement Center, China. E. N. thanks Ancell-Teicher Research Foundation for Genetics and Molecular Evolution for the financial support.

References

Baek, H. J., Beharav, A., & Nevo, E. (2003). Ecological-genomic diversity of microsatellites in wild barley, *Hordeum spontaneum*, populations in Jordan. *Theoretical and Applied Genetics, 106*, 397–410.

Black, R. E., Lindsay, H. A., Bhutta, Z. A., Caulfield, L. E., De Onnis, M., Ezzati, M., Mathers, C., & Rivera, J. (2008). Maternal and child undernutrition: Global and regional exposures and health consequences. *Lancet, 371*, 243–260.

Bonfil, D. J., & Kafkafi, U. (2000). Wild wheat adaptation in different soil ecosystems as expressed in the mineral concentration of the seeds. *Euphytica, 114*, 123–134.

Brantestam, A. K. (2005). *A century of breeding – Is genetic erosion a reality? Temporal diversity changes in Nordic and Baltic barley*. Ph.D. thesis submitted to Swedish University of Agricultural Sciences, Alnarp.

Cakmak, I., Ozkan, H., Braun, H. J., Welch, R. M., & Romheld, V. (2000). Zinc and iron concentrations in seeds of wild, primitive, and modern wheats. *Food and Nutrition Bulletin, 21*, 401–403.

Cakmak, I., Torun, A., Millet, E., Feldman, M., Fahima, T., Korol, A., Nevo, E., Braun, H. J., & Ozkan, H. (2004). *Triticum dicoccoides*: An important genetic resource for increasing zinc and iron concentration in modern cultivated wheat. *Soil Science & Plant Nutrition, 50*, 1047–1054.

Cakmak, I., Pfeiffer, W. H., & McClafferty, B. (2010). Biofortification of durum wheat with zinc and iron. *Cereal Chemistry, 87*, 10–20.

Chatzav, M., Peleg, Z., Ozturk, L., Yazici, A., Fahima, T., Cakmak, I., & Saranga, Y. (2010). Genetic diversity for grain nutrients in wild emmer wheat: potential for wheat improvement. *Annals of Botany, 105*, 1211–1220.

Chen, G., Krugman, T., Fahima, T., Korol, A. B., & Nevo, E. (2002). Comparative study of morphological and physiological traits related to drought resistance between xeric and mesic *Hordeum spontaneum* lines in Israel. *Barley Genetics Newsletter, 32*, 22–33.

Chen, G., Suprunova, T., Krugman, T., Fahima, T., & Nevo, E. (2004a). Ecogeographic and genetic determinants of kernel weight and colour of wild barley (*Hordeum spontaneum*) populations in Israel. *Seed Science Research, 14*, 137–146.

Chen, G., Krugman, T., Fahima, T., Zhang, F., & Nevo, E. (2004b). Differential patterns of germination and desiccation tolerance of mesic and xeric wild barley (*Hordeum spontaneum*) in Israel. *Journal of Arid Environments, 56*, 95–105.

Chen, G., Krugman, T., Fahima, T., Chen, K., Hu, Y., Röder, M., Nevo, E., & Korol, A. (2010). Chromosomal regions controlling seedling drought resistance in Israeli wild barley *Hordeum spontaneum* C. Koch. *Genetic Resources and Crop Evolution, 57*, 85–99.

Chen, G., Komatsuda, T., Ma, J. F., Nawrath, C., Pourkheirandish, M., Tagiri, A., Hu, Y. G., Sameri, M., Li, X., Zhao, X., Liu, Y., Li, C., Ma, X., Wang, A., Nair, S., Wang, N., Miyao, A., Sakuma, S., Yamaji, N., Zheng, X., & Nevo, E. (2011). An ATP-binding cassette subfamily G full transporter is essential for the retention of leaf water in both wild barley and rice. *Proceedings of the National Academy of Sciences of the United States of America, 108*(30), 12354–12359.

Clegg, M. T., Brown, A. H. D., & Whitfeld, P. R. (1984). Chloroplast DNA diversity in wild and cultivated barley: implications for genetic conservation. *Genetical Research, 43*, 339–343.

Distelfeld, A., Korol, A. B., Dubcovsky, J., Uauy, C., Blake, T., & Fahima, T. (2008). Colinearity between the barley grain protein content (GPC) QTL on chromosome arm 6HS and the wheat *Gpc-B1* region. *Molecular Breeding, 22*, 25–38.

Ellis, R., Foster, B., Handley, L., Handley, L. L., Gordon, D. C., Russell, J. R., & Powell, W. (2000). Wild barley: a source of genes for crop improvement in the 21st century? *Journal of Experimental Botany, 51*, 9–17.

FAO, Food and Agriculture Organization of the United Nations. (2010). http://faostat.fao.org/faostat. Retrieved December 20, 2010.

Forster, B. P., Russel, J. R., Ellis, R. P., Handley, L. L., Hackett, C. A., Nevo, E., Waugh, R., Gordon, D. C., Keith, R., & Powell, W. (1997). Locating genotypes and genes for abiotic stress tolerance in barley: A strategy using maps, markers and the wild species. *New Phytologist, 137*, 141–147.

Górecka, H., Chojnacka, K., & Górecki, H. (2006). The application of ICP-MS and ICP-OES in determination of micronutrients in wood ashes used as soil conditioners. *Talanta, 70*, 950–956.

Grusak, M. A. (2002). Enhancing mineral content in plant food products. *Journal of the American College of Nutrition, 21*, 178S–183S.

Grusak, M. A., & Cakmak, I. (2005). Methods to improve the crop-delivery of minerals to humans and livestock. In M. R. Broadley & P. J. White (Eds.), *Plant nutritional genomics* (pp. 265–286). Oxford: Blackwell.

Gutterman, Y., & Nevo, E. (1994). Temperatures and ecological-genetic differentiation affecting the germination of *Hordeum spontaneum* caryopses harvested from three populations: The Negev Desert and opposing slopes on Mediterranean Mount Carmel. *Israel Journal of Plant Sciences, 42*, 183–195.

Lev-Yadun, S., Gopher, A., & Abbo, S. (2000). The cradle of agriculture. *Science, 288*, 1602–1603.

Moseman, J. G., Nevo, E., & Zohary, D. (1983). Resistance of *Hordeum spontaneum* collected in Israel to infection with *Erysiphe graminis hordei*. *Crop Science, 23*, 1115–1119.

Nevo, E. (1992). Origin, evolution, population genetics and resources for breeding of wild barley, *Hordeum spontaneum*, in the Fertile Crescent. In P. Shewry (Ed.), *Barley: Genetics, molecular biology, and biotechnology* (pp. 19–43). Wallingford: CAB International.

Nevo, E. (1996). 'Evolution Canyon', Nahal Oren, Mount Carmel, Israel: Predictions and tests of evolutionary theory across phylogeny. *Israel Journal of Zoology, 42*, 77.

Nevo, E., & Chen, G. X. (2010). Drought and salt tolerances in wild relatives for wheat and barley improvement. *Plant, Cell & Environment, 33*, 670–685.

Nevo, E., Brown, A. H. D., & Zohary, D. (1979). Genetic diversity in the wild progenitor of barley in Israel. *Experientia, 35*, 1027–1029.

Nevo, E., Beiles, A., & Storch, N. (1983). Microgeographic differentiation in Hordein polymorphisms of wild barley. *Theoretical and Applied Genetics, 64*, 123–132.

Nevo, E., Beiles, A., Gutterman, Y., Storch, N., & Kaplan, D. (1984). Genetic resources of wild cereals in Israel and the vicinity: II Phenotypic variation within and between populations of wild barley, *Hordeum spontaneum*. *Euphytica, 33*, 737–756.

Nevo, E., Beiles, A., Kaplan, D., Golenberg, E. M., Olsvig-Whittaker, L., & Naveh, Z. (1986). Natural selection of allozyme polymorphisms: A microsite test revealing ecological genetic differentiation in wild barley. *Evolution, 40*, 13–20.

Owuor, E. D., Fahima, T., Beharav, A., Korol, A., & Nevo, E. (1999). RAPD divergence caused by microsite natural selection. *Genetics, 105*, 177–192.

Pakniyat, H., Powell, W., Baird, E., Handley, L. L., Robinson, D., Sorimgeour, C. M., Nevo, E., Hackett, C. A., Caligari, P. D. S., & Forster, B. P. (1997). AFLP variation in wild barley (*Hordeum spontaneum* C. Koch) with reference to salt tolerance and associated ecogeography. *Genome, 40*, 332–341.

Peterson, L., Ostergard, H., & Giese, H. (1994). Genetic diversity among and cultivated barley as revealed by RFLP. *Theoretical and Applied Genetics, 89*, 676–681.

Pleijel, H., Mortensen, L., Fuhrer, J., Ojanpera, K., & Danielsson, H. (1999). Grain protein accumulation in relation to grain yield of spring wheat (*Triticum aestivum* L.) grown in open-top chambers with different concentrations of ozone, carbon dioxide and water availability. *Agriculture, Ecosystems & Environment, 72*, 265–270.

Prasad, A. S. (2008). Zinc in human health: effect of zinc on immune cells. *Molecular Medicine, 14*, 353–357.

Reilly, C. (2006). *Selenium in food and health* (2nd ed., p. 198). Springer: New York.

Saghai Maroof, M. A., Soliman, K. M., Jorgensen, R. A., & Allard, R. W. (1984). Ribosomal DNA spacer-length polymorphisms in barley: Mendelian inheritance, chromosomal location, and population dynamics. *Proceedings of the National Academy of Sciences of the United States of America, 81*, 8014–8018.

Souza, R. M., Saraceno, A. L., Duyck, C., Silveira, C. L., & Aucélio, R. Q. (2007). Determination of Fe, Ni and V in asphaltene by ICP OES after extraction into aqueous solutions using sonication or vortex agitation. *Microchemical Journal, 87*, 99–103.

Turpeinen, T., Vanhala, T., Nevo, E., & Nissila, E. (2003). AFLP genetic polymorphism in wild barley (*Hordeum spontaneum*) populations in Israel. *Theoretical and Applied Genetics, 106*, 1333–1339.

Uauy, C., Distelfeld, A., Fahima, T., Blechl, A., & Dubcovsky, J. (2006). A NAC gene regulating senescence improves grain protein, zinc, and iron content in wheat. *Science, 314*, 1299–1301.

Underwood, B. A. (2003). Scientific research: essential, but is it enough to combat world food insecurities? *Journal of Nutrition, 133*, 1434S–1437S.

Van Rijn, C. P. E., Heersche, I., Van Berkel, Y. E. M., Nevo, E., Lambers, H., & Poorter, H. (2000). Growth characteristics in *Hordeum spontaneum* populations from different habitats. *New Phytologist, 146*, 471–481.

van Valen, L. (1965). Morphological variation and width of ecological niche. *The American Naturalist, 99*, 377–390.

Welch, R. M. (1999). Importance of seed mineral nutrient reserves in crop growth and development. In Z. Rengel (Ed.), *Mineral nutrition of crops: Fundamental mechanisms and implications* (pp. 205–226). New York: Food Products Press.

Welch, R. M., & Graham, R. D. (2004). Breeding for micronutrients in staple food crops from a human nutrition perspective. *Journal of Experimental Botany, 55*, 353–364.

WHO. The World Health Report. (2002). *Reducing risks, promoting healthy life* (pp. 1–168). Geneva: World Health Organization.

Yan, J., Chen, G. X., Cheng, J. P., Nevo, E., & Gutterman, Y. (2008). Phenological and phenotypic differences and correlations among genotypes of *Hordeum spontaneum* originating from different locations in Israel. *Genetic Resources and Crop Evolution, 55*(7), 995–1005.

Yan, J., Wang, F., Qin, H. B., Chen, G. X., Nevo, E., Fahima, T., & Cheng, J. P. (2011). Natural variation in grain selenium concentration of wild barley, *Hordeum spontaneum* derived from Israel. *Biological Trace Element Research, 142*, 773–786.

Zhang, F., Chen, G., Huang, Q., Orion, O., Krugman, T., Fahima, T., Korol, A. B., Nevo, E., & Gutterman, Y. (2005). Genetic basis of barley caryopsis dormancy and seedling desiccation tolerance at the germination stage. *Theoretical and Applied Genetics, 110*, 445–453.

Zohary, D. (1969). The progenitors of wheat and barley in relation to domestication and agricultural dispersal in the old world. In P. J. Ucko & G. W. Dimbelby (Eds.), *Domestication and exploitation of plants and animals* (pp. 47–66). London: Gerald Duckworth.

Chapter 16
Genes Controlling Low Phytic Acid in Plants: Identifying Targets for Barley Breeding

Hongxia Ye, Chengdao Li, Matthew Bellgard, Reg Lance, and Dianxing Wu

Abstract Phytic acid (myo-inositol 1, 2, 3, 4, 5, 6-hexakisphosphate) is the most abundant form of phosphorus in plant seeds. It is indigestible by both humans and nonruminant livestock and can contribute to human mineral deficiencies. The degradation of phytic acid in animal diets is necessary to overcome both environmental and nutritional issues. The development of plant cultivars with low phytic acid content is therefore an important priority. More than 25 low-phytic acid mutants have been developed in rice, maize, soybean, barley, wheat, and bean, from which

H. Ye
State Key Lab of Rice Biology and Key Lab of the Ministry of Agriculture for Nuclear-Agricultural Sciences, Zhejiang University, Hangzhou 310029, People's Republic of China

Department of Agriculture and Food, Government of Western Australia, 3 Baron-Hay Court, South Perth, Perth, WA 615, Australia

C. Li (✉)
Department of Agriculture and Food, Government of Western Australia, 3 Baron-Hay Court, South Perth, Perth, WA 615, Australia

The State Agricultural Biotechnology Centre, Murdoch University, Murdoch 6150, Australia
e-mail: Chengdao.li@agric.wa.gov.au

M. Bellgard
Centre for Comparative Genomics, Murdoch University, Perth, WA, Australia

R. Lance
Department of Agriculture and Food, Government of Western Australia, 3 Baron-Hay Court, South Perth, Perth, WA 615, Australia

D. Wu
State Key Lab of Rice Biology and Key Lab of the Ministry of Agriculture for Nuclear-Agricultural Sciences, Zhejiang University, Hangzhou 310029, People's Republic of China
e-mail: dxwu@zju.edu.cn

G. Zhang et al. (eds.), *Advance in Barley Sciences: Proceedings of 11th International Barley Genetics Symposium*, DOI 10.1007/978-94-007-4682-4_16,

11 genes, belonging to six gene families, have been isolated and sequenced from maize, soybean, rice, and *Arabidopsis*. Forty-one members of the six gene families were identified in the rice genome sequence. A survey of genes coding for enzymes involved in the synthesis of phytic acid identified candidate genes for the six barley mutants with low phytic acid through comparison with syntenic regions in sequenced genomes.

Keywords Low phytic acid • Comparative mapping • Orthology • Barley • Rice • Synteny • Candidate gene

16.1 Introduction

Phytic acid (PA), myo-inositol 1, 2, 3, 4, 5, 6-hexakisphosphate or Ins P6, is the main storage form of phosphorus (P) in plant seeds. It usually exists in the form of a mixed salt (phytate or Phytin) which accumulates in seed protein bodies either dispersed throughout the bodies or in dense inclusions called globoids. PA binds several important mineral cations such as calcium, magnesium, potassium, iron, and zinc. It typically represents 65–85% of the phosphorus in seeds and accounts for 0.5–5% of the seed dry weight (Lott et al. 2000; Raboy 1997). Depending on the plant species, phytate globoids are localized predominantly in the aleurone layer (rice, wheat, and barley) or in the embryo (maize). PA content can vary substantially from plant to plant, as well as from node to node within a plant. The range for PA is 0.86–1.06% in cereal grains, 0.55–1.70% in legume, and high variability in oil seeds from below 1–4.71% (Lott et al. 2000). In cereal grains, there are many factors that affect the PA content. Genotype and environment significantly affect the PA content. It is reported that the PA content in 100 barley genotypes ranged from 0.385–0.985%, and the environment had the larger effect than genotypes on the PA content (Dai et al. 2007).

There are several possible roles for PA. It serves as a major storage form for myo-inositol, phosphorus, and mineral cations for use during seedling growth and also controls inorganic phosphate (Pi) levels in both developing seeds and seedlings in plants (Strother 1980). In addition, PA and its pyrophosphate-containing derivative have also been implicated in mRNA export (York et al. 1999), DNA repair (Hanakahi and West 2002), cell signaling (Sasakawa et al. 1995), endocytosis, and cell vesicular trafficking and ATP regeneration (Safrany et al. 1999; Saiardi et al. 2002).

In terms of human health and nutrition, PA has both negative and positive outcomes. PA as an antioxidant and anticarcinogen is suggested to have some beneficial effects on human health. However, the negative impact of PA should be of more concern. PA is considered to be an antinutritional substance in animal feed and human diets, since it has the potential to form very stable complexes with minerals and proteins, which are poorly utilized to monogastric animals due to the absence or insufficient amount of the phytate-degrading enzymes in their digestive system. As a result, it contributes to human mineral deficiency, especially in populations throughout the developing world that rely on grains and legumes as staple foods. In addition, it also contributes to the eutrophication of lakes and rivers, since large amounts of phytate-P are excreted to the environment with the animal waste.

In recent years, several alternative methods have been used to deal with the dietary and environmental problems associated with PA. PA can, for example, be enzymatically hydrolyzed by phytases to a series of lower myo-inositol phosphates and phosphate. During certain food processes including soaking, seed germination, and fermentation, the activity of the enzyme phytase has shown to enhance the degradation of PA (Hotz and Gibson 2007). Microbial phytases are widely used as a feed additive to increase phosphorus availability of plant-based feed to animals and reduce phosphorus excretion in manure (Brinch-Pedersen et al. 2002). Although these approaches have improved the phosphorus bioavailability and reduced the phosphorus excretion in the areas of intensive livestock, they all incur additional cost and labor. Breeding low-phytic acid (lpa) mutants through chemical (EMS, NaN_3) and physical (^{60}Co-γ) mutagenesis provided one of the more cost-effective, simple, and sustainable approaches to develop new cultivars with low phytic acid levels. Mutant lines with the lpa phenotype have already been isolated in several crops. Transgenic crops that decrease phytic acid in seeds have also been generated (Kuwano et al. 2009; Kuwano et al. 2006). Low-phytic acid grain in feed not only improves phosphate bioavailability but also reduces the amount of phosphorus supplementation required in animal feeds which has the net effect of reducing phosphorus pollution to the environment. Grain which possesses this quality would also increase in the availability and uptake of iron and zinc, thereby significantly improving human nutrition (Mendoza et al. 1998). Feeding trials with low-phytic acid cereal grains have been shown to greatly increase the availability of phosphorus and minerals for fish and poultry, associated with improvements in animal growth and/or reductions in the excretion of phosphorus (Li et al. 2001a, b; Overturf et al. 2003).

In this chapter, we briefly review the metabolic pathways for synthesis of phytic acid, the genetic basis of low phytic acid, and the genes controlling low phytic acid. The genes from maize, soybean, rice, and *Arabidopsis* and their homologous genes were further mapped on to the rice genome. The candidate genes for the 6 barley *lpa* mutants were predicted by comparative analyses with the rice genome. All of these might be used to identify candidate genes controlling low phytic acid in barley (*Hordeum vulgare* L.) and bread wheat (*Triticum aestivum* L.).

16.2 Isolation of Low-Phytic Acid Mutants in Plant

Over the past decade, induced low-phytic acid (lpa) mutants have been isolated in various crop species. The first *lpa* mutant crop was reported in maize and was followed by isolation of mutants in barley, rice, soybean, wheat, and bean. The low-phytic acid mutants have lowered phytate levels and increased free inorganic phosphate compared to the wild type (new mutants with low phytic acid could potentially be first identified by screening seeds with high inorganic phosphorus levels). Three types of low-phytic acid mutations have been identified in maize: The first type of *lpa* mutants, designated as *lpa1*, is characterized by a decrease in phytic acid content and a corresponding increase in inorganic phosphorus (Pi). The *lpa2* mutant has reduced phytic acid content in seeds, accompanied by an accumulation of lower inositol phosphates (Raboy et al. 2001). In *lpa3* mutant, phytic acid content was also reduced in seeds, and lower inositol phosphates were detected, but myo-inositol content was increased (Shi et al. 2005). These three types of *lpa* mutation have also been identified in barley and rice (Dorsch et al. 2003; Larson et al. 1998, 2000; Rasmussen and Hatzack 1998; Roslinsky et al. 2007).

16.3 Metabolic Pathways for Synthesis of Phytic Acid

Mutations that block the synthesis or accumulation of phytic acid during seed development are often referred to as low-phytic acid (*lpa*) mutations (Raboy 2007). Reduced phytic acid accumulation in a given mutant could be due to blocks in the synthesis or supply of the two substrates myo-inositol (Ins) and inorganic phosphorus (Pi), or blocks in their conversion to phytate, or blocks in various transport or regulatory functions important to this process. Thus, identifying stages that might be inhibited in the key metabolic pathways for synthesis of phytic acid is a starting point for decreasing phytic acid content in grain. The metabolic pathways for synthesis of phytic acid, which are necessary for the study of low-phytic acid mutants, are summarized in Fig. 16.1. In reverse, new low-phytic acid mutants also provide an ideal tool to refine our understanding of the synthetic pathways of phytic acid in developing grains.

In most eukaryotic cells, the basic structural pathways for the synthesis of phytic acid are similar (Dorsch et al. 2003). However, particular steps or their relative activity or physiological importance may differ among species, and large genetic differences exist between closely related genomes. To date, several genes involved in phytic acid biosynthesis have been cloned and characterized (Hitz et al. 2002; Josefsen et al. 2007; Stevenson-Paulik et al. 2005; Suzuki et al. 2007). Based on the characterization of cloned genes combined with the knowledge from mammalian and yeast systems, several pathways have been suggested for phytic acid synthesis in plants. However, the biosynthetic pathway leading to phytic acid is still poorly understood, particularly with respect to the sequential phosphorylation steps and the regulative mechanisms. Recent studies on low-phytic acid mutants have provided some evidence for synthesis of phytic acid pathways (Kim et al. 2008a, b; Shi et al. 2003, 2005; Stevenson-Paulik et al. 2005).

Myo-inositol and inorganic phosphorus are the two main substrates in the synthesis of phytic acid, and the pathways contributing to phytic acid consist of two parts: the early pathway representing Ins and/or Ins P1 synthesis and the late pathway representing Ins phosphate metabolism that converts Ins to InsP6. Ins is synthesized from glucose via a two-step process. First, glucose 6-phosphate (G6P) is converted to D-Ins(3)P1(Ins with a single P ester at the "3" position) catalyzed by

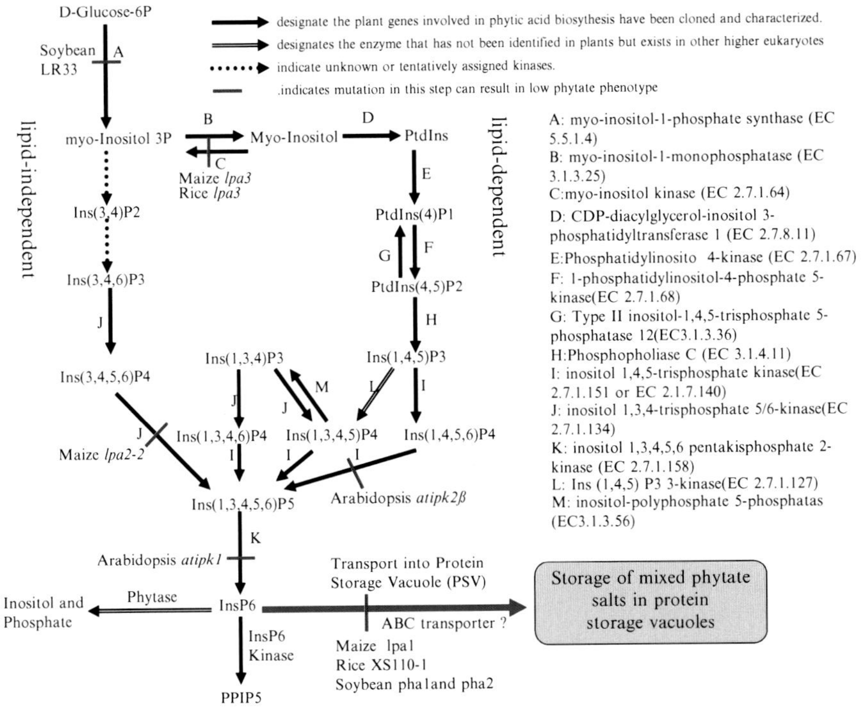

Fig. 16.1 Biochemical pathways of phytic acid in plant metabolism

myo-inositol-3-phosphate [Ins(3)P] synthase (MIPS; EC5.5.1.4). The myo-inositol-3-phosphate synthase gene has been identified in several plant species including barley (Larson et al. 2000). Studies of the soybean "LR33" mutation and "low-phytic acid" transgenic rice antisense repression of the 1D-myo-inositol-3-phosphate synthase gene demonstrated that seed-specific blocks in Ins(3)P1 synthase activity clearly blocks seed phytic acid accumulation. Conversion of Ins(3)P1 to free Ins is catalyzed by myo-inositol monophosphatase (IMP, EC3.1.3.25). There are several families of sequences encoding potential Ins monophosphatases in most plant genomes (Torabinejad and Gillaspy 2006).Recently, Fu (Fu et al. 2008) cloned the barley IMP-1 gene. But there is no genetic evidence to date that seed-specific Ins monophosphatase activity is critical to phytic acid synthesis, since no mutations that block Ins monophosphatase activity have been shown to block seed phytic acid accumulation.

The latter steps of conversion, Ins/InsP1 to InsP3, are consistent with sequential ATP-dependent phosphorylations of soluble Ins phosphates mediated by several kinases and phospholipase C-dependent conversion of phosphatidylinositol phosphate intermediates to InsP3 (Loewus and Murthy 2000; Raboy 2003, 2009). Two pathways that begin with Ins might then lead to Ins P6. It is considered to proceed via an inositol lipid-dependent (PLC-dependent) pathway or lipid-independent

(PLC-independent) pathway, or perhaps via a combination of both pathways (Stevenson-Paulik et al. 2005). The "lipid-dependent" pathway begins with the conversion of Ins and phosphatidic acid to phosphatidylinositol (PtdIns). Two subsequent phosphorylations yield PtdIns(4,5)P2, which is then broken down to diacylglycerol and Ins(1,4,5)P3 via the action of a specific phospholipase C. In the lipid-independent pathway, Ins is phosphorylated to Ins(3)P1, or perhaps to other Ins monophosphates, via myo-inositol kinase activity (MIK, EC2.7.1.64). Isolation of maize *lpa3* mutations and rice "N15-186" mutations and cloning the mutant gene "MIK" encodes Ins kinase provided genetic evidence for its existence and its role in seed phytic acid synthesis (Kim et al. 2008a, b; Shi et al. 2005). But the following steps of conversion InsP1 to InsP2 or InsP3 have been still unknown.

Finally, Ins(1,4,5)P3 or other Ins P3 produced via the lipid-dependent pathway or lipid-independent pathway are then further phosphorylated to Ins P6 by the action of three Ins polyphosphate kinases: Ins(1,4,5)P3 3-/6-kinase(IPK2, EC2.7.1.151 or EC2.1.7.140), Ins(1,3,4)P3 5-/6-kinase(ITP5/6K, 2.7.1.134), and Ins polyphosphate 2-kinase (IPK1, 2.7.1.158). Previous studies of AtIPK1 and AtIPK2β mutants in *Arabidopsis* confirmed that inositol polyphosphate kinases, AtIPK1 and AtIPK2β, are important to seed Ins P6 synthesis (Stevenson-Paulik et al. 2005). Furthermore, the analyses of the maize *lpa2* mutation and cloning the mutant gene "ZmIPK" which encodes "ITP5/6K" confirmed its importance to seed phytic acid synthesis, indicating that all three enzymes work in concert to yield Ins P6 (Shi et al. 2003).

Phytic acid is believed to be synthesized primarily in the cytoplasm and for storage as mixed salts in membrane-bound storage microbodies referred to as protein storage vacuoles. Using map-based cloning, genes "MRP4," "MRP5," and "MRP4/5" encoding an ABC transporter as the site of mutations that confer low-phytic acid phenotypes in maize, rice, and soybean were identified, respectively (Gillman et al. 2009; Shi et al. 2007; Xu et al. 2009). The exact function of the ABC transporter ("MRP4" and "MRP5") in seed phytic acid synthesis, transport, or storage has not been determined, but one possibility is that it functions in transport of phytic acid into the PSV for storage.

16.4 Genes Isolated from Plants for Controlling Low Phytic Acid

Studies have shown that all the low-phytic acid mutations are controlled by recessive genes. Genetic analyses in the maize, barley, and rice *lpa* phenotypes indicated that they resulted from single locus mutations; soybean *lpa* phenotypes could be controlled by recessive genes at one locus or two independent loci exhibiting duplicate dominant epistasis (Oltmans et al. 2003); and the wheat *lpa* phenotypes were suggested to have two or more genes (Guttieri et al. 2003). About 20 *lpa* mutant genes have been mapped to different positions of various chromosomes, and 11 genes controlling low phytic acid in plant have been cloned and characterized recently.

To date, three *lpa* genes in maize have already been cloned. The maize *lpa1* gene encodes a multidrug resistance protein (MRP) ATP-binding cassette (ABC) transporter and appears to be involved in the transport or accumulation of phytic acid in the seed rather than its biosynthesis (Shi et al. 2007). This gene was shown to be a homologue of rice MRP5. The maize *lpa2* gene encodes a myo-inositol phosphate kinase (ZmIpk) belonging to the Ins (1,3,4)P3, 5/6-kinase gene family (Shi et al. 2003). The gene responsible for the maize *lpa3* mutants encodes a myo-inositol kinase (MIK) (Shi et al. 2005) and, like maize *lpa2*, is directly involved in the inositol polyphosphate biosynthetic pathway. This gene was shown to be a homologue of rice MIK.

In rice, four genes resulting in low phytic acid were recently identified via positional cloning. The gene (OsLpa1) represented by *lpa1* mutation encodes a sequence with no known homology to any previously describe eukaryotic gene, and the biochemical function of the protein it encodes and its role in seed phytic acid synthesis is not yet known (Kim et al. 2008a, b; Zhao et al. 2008). The *lpa* N15-186 mutant gene encoding myo-inositol kinase (OsMIK) is orthologue of maize *lpa3* (Kim et al. 2008a, b). Os-*lpa*-MH86-1 and Os-*lpa*-Z9B mutant gene encodes a sulfate transporter, but its role in seed phytic acid synthesis is not yet known (Zhao, unpublished). The nonlethal mutant Os-*lpa*-xs110-2 (HIPj2) and the lethal mutant Os-*lpa*-xs110-3 (HIPj3) are orthologues of the maize *lpa1* gene, which encodes a multidrug resistance-associated protein (MRP) ATP-binding cassette (ABC) transporter (Xu et al. 2009).

In soybean, "LR33" mutant resulted from a single-base substitution in the MIPS1 gene and Gm-lpa-TW-1 lpa mutation resulted from a 2-bp deletion in the MIPS1 gene lead to a decreased capacity for the synthesis of myo-inositol hexaphosphate (phytic acid) and a concomitant increase in inorganic phosphate (Hitz et al. 2002; Yuan et al. 2007). The "CX1834" mutant which is controlled by two independent loci has two recessive mutations in the soybean homologs of the maize lpa1 gene (MRP4). One nonsense mutation on LG N/chr3 Glyma03 g32500 and another novel missense mutation on LG L/chr19 Glyma19 g35230 were identified, respectively. The results provide clear evidence that two recessive MRP4/MRP5 mutant loci are required for the soybean low-phytic acid phenotype (Gillman et al. 2009).

In *Arabidopsis*, mutants carrying T-DNA insertions that disrupted either ATPK1 or ATPK2β confirmed that these two enzymes are important to seed Ins P6 synthesis (Stevenson-Paulik et al. 2005).

16.5 Comparative Analyses of Cloned Genes

Comparative genomics utilizes evolutionary conservation of individual genes, or conservation of gene order in the chromosomes of related species. Eleven genes controlling low phytic acid in different plant species have already been cloned. They are ZmIPK, MRP4, and MIK in maize; OsMIK, MRP5, OsLPA1, and OsSultr3 in rice; MIPS and MRP4/5 in soybean; and AtIPK1 and AtIPK2β in *Arabidopsis*.

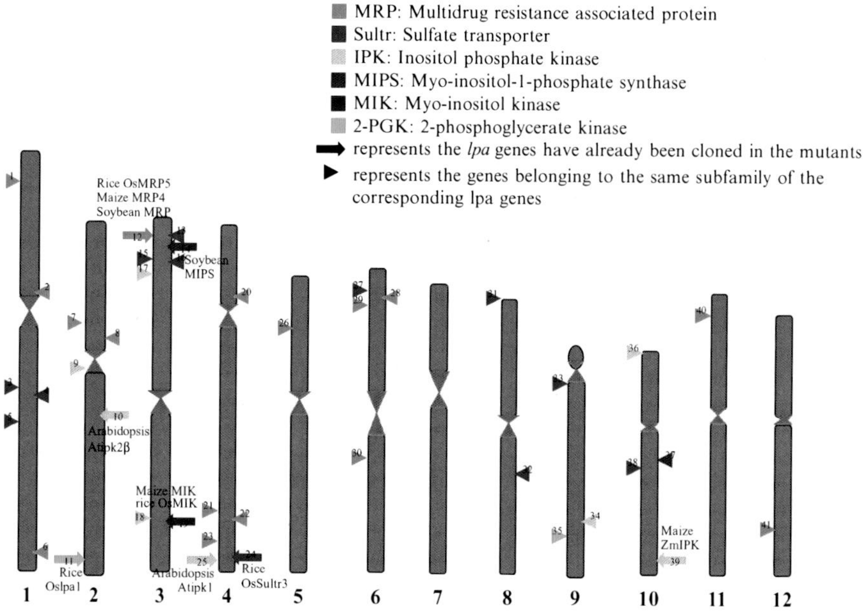

Fig. 16.2 Chromosomal locations of the cloned genes and their homologous genes for low phytic acid from different crops. The position of genes resulted in low phytic acid is indicated according to the Nipponbare genome sequence information (TIGR Rice Database, Release V5)

These genes belong to six different gene families. Forty-one members of the six gene families were identified in the rice genome sequence. The projection of the orthologous/homologous genes for low phytic acid from various crops onto the rice chromosomes is provided in Fig. 16.2. These genes were distributed to 11 chromosomes except chromosome 7 in rice. The rice chromosome 3 is clearly the hot spot for low-phytic acid genes, on which six genes belonged to three gene families have been isolated from rice, maize, and soybean.

The MIK in maize is homologous to osMIK in rice on chromosome 3; MRP4 in maize and MRP4/5 in soybean are homologous to MRP5 in rice located on chromosome 3 and also have homology with another gene in rice chromosome 5; there are 17 genes (OsMRP1–OsMRP17) belonging to the multidrug resistance-associated protein (MRP) subfamily of ABC transporters in rice (Klein et al. 2006); and AtIPK1 and AtIPK2β in *Arabidopsis* had the orthologous gene in rice chromosome 4 (OsIPK1) and 2 (OsIPK2), respectively. The maize ZmIPK gene has an orthologous gene in rice chromosome 10(OsITP5/6K-5), and the other five rice cDNA clones, OsITP5/6K-1, OsITP5/6K-2, OsITP5/6K-3, OsITP5/6K-4, and OsITP5/6K-6 from rice EST and genomic databases were identified as the same family of ITP5/6K (Suzuki et al. 2007); MIPS in soybean have orthologous genes in rice chromosome 3 (RINO1) and 10 (RINO2); OsLPA1 in rice was located in chromosome 2 (2PGK-1) and also has an orthologous gene in rice chromosome 9 (2PGK-2); OsSultr3 in rice was located in chromosome 4 (OsSultr3; 3) and also has a homologous gene in rice chromosome 5 (OsSultr3; 4). In addition, 14 genes belonging to sulfate transporter gene family from rice that encode putative sulfate transporters have been identified (Buchner et al. 2004).

16.6 Identification of Putative Barley Low-Phytic Acid Genes via Comparative Genomics Analysis with Rice

In barley, six nonallelic loci controlling low phytic acid have been identified by genetic analysis, but no candidate gene has been identified. Syntenic regions of the low phytic acid were identified through BLASTN analysis of molecular markers associated with the low-phytic acid loci (Fig. 16.3).

The barley *lpa1-1* mutation was localized to chromosome 2H by linkage to a dominant aMSU21 STS-PCR marker and ABC153 SCAR marker (approximately 16.8 cM) (Larson et al. 1998). BLASTN analysis showed that nine markers at the 2H terminal region controlling low phytic acid were aligned to the rice genome sequences along the terminal region of rice chromosome 4 (Fig. 16.3). The candidate gene for *lpa1-1* may be either the sulfate transporter (OsSultr3; 3) or Inositol 1.4.5-trisphosphate 3-/6-kinase (OsIPK1) based on the barley and rice comparison (Figs. 16.2 and 16.3).

The *lpa2-1* is located within a recombination interval of approximately 30 cM between two AFLP markers (KgE40M48 and KgE37M59) that were subsequently mapped to barley chromosome 7H by integration with the same NABGMP population (Larson et al. 1998). Comparative mapping showed the relevant region of barley was collinear with a region on rice chromosome 6 (Fig. 16.3). The multidrug resistance-associated protein gene (OsMRP11 and OsMRP12) may be the candidate gene for *lpa2-1* (Figs. 16.2 and 16.3).

The *lpa3-1* and "M955" are situated between ISSR marker LP75 and Un8-700R (approximately 24.4 cM) that were mapped to barley chromosome 1H (Roslinsky et al. 2007). Comparative mapping showed the relevant region of barley was collinear with a region on rice chromosome 5 (Fig. 16.3). There was no candidate gene identified in the rice syntenic region.

The mutation "M678" was mapped between EBmac701 and Bmag714B on chromosome 4H (Oliver et al. 2009). Comparative mapping showed the relevant region of barley was collinear with a region on rice chromosome 3 (Fig. 16.3). The candidate gene may be the multidrug resistance-associated protein (OsMRP5), sulfate transporter (OsSultr3; 2), or the myo-inositol 1-phosphate synthase (RINO1) gene (Figs 16.2 and 16.3).

"M640" linked to *lpa2-1* were localized to chromosome 7H between Bmag120 and AWB22 (Oliver et al. 2009). Comparative mapping showed the relevant region of barley was collinear with a region on rice chromosome 6 (Fig. 16.3). A multidrug resistance-associated protein (OsMRP11) gene was mapped in the similar region (Figs. 16.2 and 16.3).

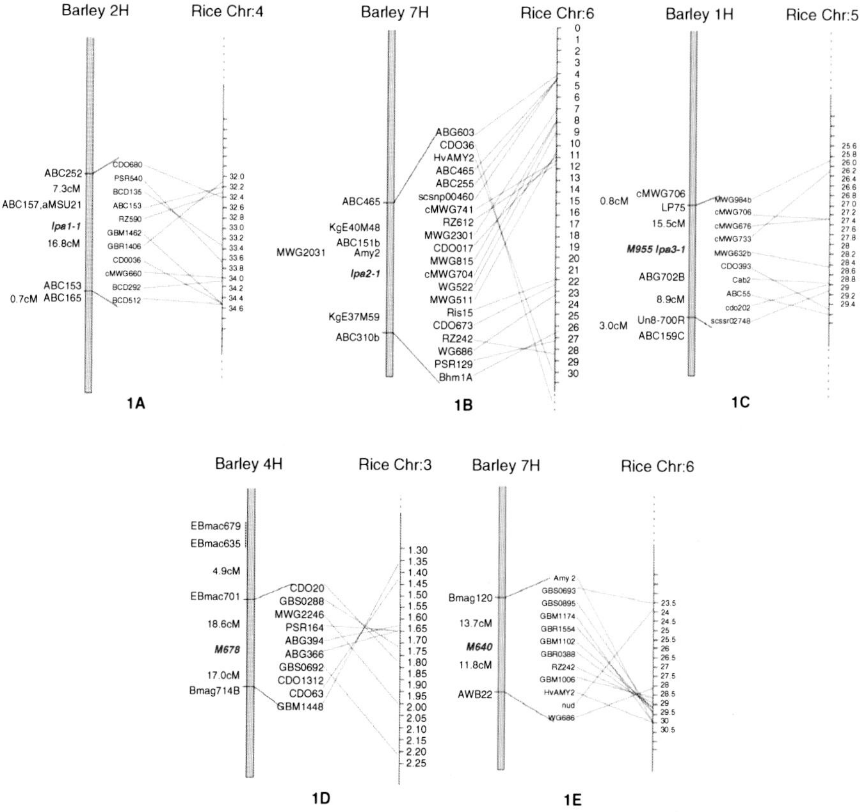

Fig. 16.3 Alignment of barley molecular markers linked to barley low phytic acid with rice chromosomes based on BLAST analysis. (**a**) The region around the barley *lpa1-1* gene was syntenic to a segment of rice chromosome 4. (**b**) The region around the barley *lpa2-1* gene was syntenic to a segment of rice chromosome 6. (**c**) The region around the barley *lpa3-1* gene and "M955" was syntenic to a segment of rice chromosome 5. (**d**) The region around the barley "M678" was syntenic to a segment of rice chromosome 3. (**e**) The region around the barley "M640" was syntenic to a segment of rice chromosome 6

16.7 Conclusions: Candidate Genes Provide the Basis for the Design of High-Throughput Diagnostics

The primary objective of this study was identification of putative barley low-phytic acid genes for the six barley *lpa* mutants via comparative genomics analysis with rice candidate genes. For example, comparative analysis revealed that barley *lpa1-1* genes are found at syntenic chromosomal locations with the rice mutant gene encodes a sulfate transporter. Furthermore, a 211-bp sulfate transporter genomic DNA amplification product, obtained using the primer ST1F

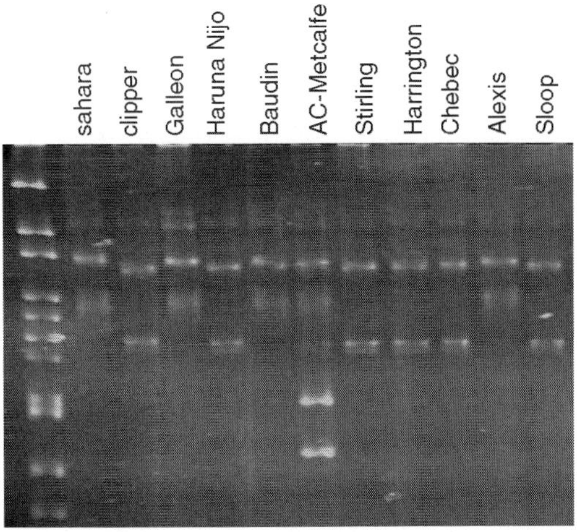

Fig. 16.4 Barley doubled-haploid mapping population parents amplified using the ST1F/ST1R primer of sulfate transporter gene

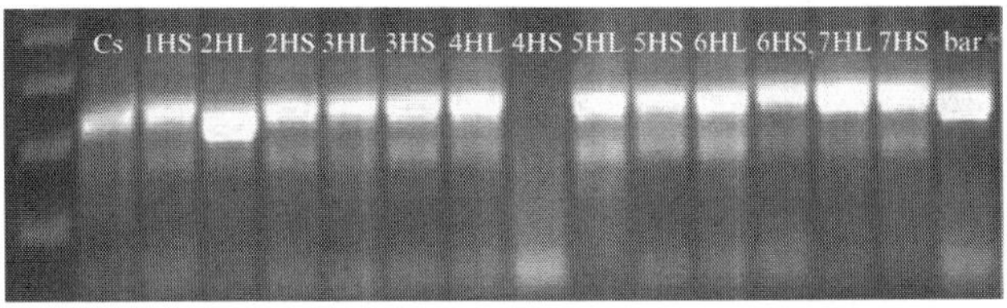

Fig. 16.5 DNA fragment amplified from wheat-barley chromosome addition lines amplified using the ST1F/ST1R primer of sulfate transporter gene

(TCTTGCCACAATAATTCCGC) and ST1R (TCCTTCACCTTCCCCTTCAT), showed SSCP polymorphic that distinguish in the doubled-haploid populations of Clipper/Sahara, Galleon/Haruna Nijo, Baudin/AC Metcalfe, and Alexis/Sloop (Fig. 16.4). By analysis of the wheat-barley chromosome addition lines, we did localize this barley sulfate transporter genomic DNA amplification product to chromosome 2HL (Fig. 16.5). And we have determined that this SSCP polymorphism is between bcd266 and cdo665 on chromosome 2H in the 150 doubled-haploid line of Clipper/Sahara mapping population. Thus, we found this barley sulfate transporter gene does correspond with the barley *lpa1-1* locus that was mapped to chromosome 2H, and the mapping results provide more support of our assertion that barley *lpa1-1* gene is orthologous to rice OsSultr3;3. For other candidate genes, more experiments are required to test in barley using wheat-barley additional lines and mapping populations.

Acknowledgment This project is supported by the Australian Grain Research and Development Corporation and DAFWA Australia-China Fund. Valuable suggestion from Professor Rudi Appels is appreciated.

References

Brinch-Pedersen, H., Sørensen, L. D., & Holm, P. B. (2002). Engineering crop plant getting a handle on phosphate. *Trends in Plant Science, 7*, 118–125.

Buchner, P., Takahashi, H., & Hawkesford, M. J. (2004). Plant sulphate transporters: Co-ordination of uptake, intracellular and long-distance transport. *Journal of Experimental Botany, 55*, 1765–1773. doi:10.1093/Jxb/Erh206.

Dai, F., Wang, J. M., Zhang, S. H., Xu, Z. Z., & Zhang, G. P. (2007). Genotypic and environmental variation in phytic acid content and its relation to protein content and malt quality in barley. *Food Chemistry, 105*, 606–611. doi:10.1016/j.foodchem.2007.04.019.

Dorsch, J. A., Cook, A., Young, K. A., Anderson, J. M., Bauman, A. T., Volkmann, C. J., Murthy, P. P. N., & Raboy, V. (2003). Seed phosphorus and inositol phosphate phenotype of barley low phytic acid genotypes. *Phytochemistry, 62*, 691–706.

Fu, J. M., Peterson, K., Guttieri, M., Souza, E., & Raboy, V. (2008). Barley (*Hordeum vulgare* L.) inositol monophosphatase: gene structure and enzyme characteristics. *Plant Molecular Biology, 67*, 629–642.

Gillman, J. D., Pantalone, V. R., & Bilyeu, K. (2009). The low phytic acid phenotype in soybean line CX1834 is due to mutations in two homologs of the maize low phytic acid gene. *The Plant Genome, 2*, 179–190.

Guttieri, M., Bowen, D., Dorsch, J. A., Raboy, V., & Souza, E. (2003). Identification and characterization of a low phytic acid wheat. *Crop Science, 44*, 418–424.

Hanàkahi, L. A., & West, S. C. (2002). Specific interaction of IP6 with human Ku70/80, the DNA-binding subunit of DNA-PK. *EMBO Journal, 21*, 2038–2044.

Hitz, W. D., Carlson, T. J., Kerr, P. S., & Sebastian, S. A. (2002). Biochemical and molecular characterization of a mutation that confers a decreased raffinosaccharide and phytic acid phenotype on soybean seeds. *Plant Physiology, 128*, 650–660. doi:10.1104/Pp. 010585.

Hotz, C., & Gibson, R. S. (2007). Traditional food-processing and preparation practices to enhance the bioavailability of micronutrients in plant-based diets. *Journal of Nutrition, 137*, 1097–1100.

Josefsen, L., Bohn, L., Sorensen, M. B., & Rasmussen, S. K. (2007). Characterization of a multifunctional inositol phosphate kinase from rice and barley belonging to the ATP-grasp superfamily. *Gene, 397*, 114–125.

Kim, S. I., Andaya, C. B., Goyal, S. S., & Tai, T. H. (2008a). The rice OsLpa1 gene encodes a novel protein involved in phytic acid metabolism. *Theoretical and Applied Genetics, 117*, 769–779.

Kim, S. I., Andaya, C. B., Newman, J. W., Goyal, S. S., & Tai, T. H. (2008b). Isolation and characterization of a low phytic acid rice mutant reveals a mutation in the rice orthologue of maize MIK. *Theoretical and Applied Genetics, 117*, 1291–1301.

Klein, M., Burla, B., & Martinoia, E. (2006). The multidrug resistance-associated protein (MRP/ABCC) subfamily of ATP-binding cassette transporters in plants. *FEBS Letters, 580*, 1112–1122. doi:10.1016/j.febslet.2005.11.056.

Kuwano, M., Ohyama, A., Tanaka, Y., Mimura, T., Takaiwa, F., & Yoshida, K. T. (2006). Molecular breeding for transgenic rice with low-phytic-acid phenotype through manipulating myo-inositol 3-phosphate synthase gene. *Molecular Breeding, 18*, 263–272.

Kuwano, M., Mimura, T., Takaiwa, F., & Yoshida, K. T. (2009). Generation of stable 'low phytic acid' transgenic rice through antisense repression of the 1d-myo-inositol 3-phosphate synthase gene (RINO1) using the 18-kDa oleosin promoter. *Plant Biotechnology Journal, 7*, 96–105.

Larson, S. R., Young, K. A., Cook, A., Blake, T. K., & Raboy, V. (1998). Linkage mapping of two mutations that reduce phytic acid content of barley grain. *Theoretical and Applied Genetics, 97*, 141–146.

Larson, S. R., Rutger, J. N., Young, K. A., & Raboy, V. (2000). Isolation and genetic mapping of a non-lethal rice (*Oryza sativa* L.) low phytic acid 1 mutation. *Crop Science, 40*, 1397–1405.

Li, Y. C., Ledoux, D. R., & Veum, T. L. (2001a). Low phytic acid barley improves performance, bone mineralization, and phosphorus retention in turkey poults. *Journal of Applied Poultry Research, 10*, 178–185.

Li, Y. C., Ledoux, D. R., Veum, T. L., Raboy, V., Zyla, K., & Wikiera, A. (2001b). Bioavailability of phosphorus in low phytic acid barley. *Journal of Applied Poultry Research, 10*, 86–91.

Loewus, F. A., & Murthy, P. P. N. (2000). Myo-inositol metabolism in plants. *Plant Science, 150*, 1–19.

Lott, J. N. A., Ockenden, I., Raboy, V., & Batten, G. D. (2000). Phytic acid and phosphorus in crop seeds and fruits: a global estimate. *Seed Science Research, 10*, 11–33.

Mendoza, C., Viteri, F. E., Lonnerdal, B., Young, K. A., Raboy, V., & Brown, K. H. (1998). Effect of genetically modified, low-phytic acid maize on absorption of iron from tortillas. *American Journal of Clinical Nutrition, 68*, 1123–1127.

Oliver, R. E., Yang, C., Hu, G., Raboy, V., & Zhang, M. (2009). Identification of PCR-based DNA markers flanking three low phytic acid mutant loci in barley. *Journal of Plant Breeding and Crop Science, 1*, 087–093.

Oltmans, S. E., Fehr, W. R., Welke, G. A., & Cianzio, S. R. (2003). Inheritance of low-phytate phosphorus in soybean. *Crop Science, 44*, 433–435.

Overturf, K., Raboy, V., Cheng, Z. J., & Hardy, R. W. (2003). Mineral availability from barley low phytic acid grains in rainbow trout (Oncorhynchus mykiss) diets. *Aquaculture Nutrition, 9*, 239–246.

Raboy, A. (1997). Accumulation and storage of phosphate and minerals. In B. A. Larkins & I. K. Vasil (Eds.), *Cellular and molecular biology of plant seed development* (pp. 441–477). Dordrecht: Kluwer Academic.

Raboy, V. (2003). Myo-Inositol-1,2,3,4,5,6-hexakisphosphate. *Phytochemistry, 64*, 1033–1043.

Raboy, V. (2007). The ABCs of low-phytate crops. *Nature Biotechnology, 25*, 874–875.

Raboy, V. (2009). Approaches and challenges to engineering seed phytate and total phosphorus. *Plant Science, 177*, 281–296.

Raboy, V., Young, K. A., Dorsch, J. A., & Cook, A. (2001). Genetics and breeding of seed phosphorus and phytic acid. *Journal of Plant Physiology, 158*, 489–497.

Rasmussen, S. K., & Hatzack, F. (1998). Identification of two low-phytate barley (*Hordeum vulgare* L.) grain mutants by TLC and genetic analysis. *Hereditas, 129*, 107–112.

Roslinsky, V., Eckstein, P. E., Raboy, V., Rossnagel, B. G., & Scoles, G. J. (2007). Molecular marker development and linkage analysis in three low phytic acid barley (*Hordeum vulgare*) mutant lines. *Molecular Breeding, 20*, 323–330.

Safrany, S. T., Caffrey, J. J., Yang, X. N., & Shears, S. B. (1999). Diphosphoinositol polyphosphates: The final frontier for inositide research? *Biological Chemistry, 380*, 945–951.

Saiardi, A., Sciambi, C., McCaffery, J. M., Wendland, B., & Snyder, S. H. (2002). Inositol pyrophosphates regulate endocytic trafficking. *Proceedings of the National Academy of Sciences of the United States of America, 99*, 14206–14211.

Sasakawa, N., Sharif, M., & Hanley, M. R. (1995). Metabolism and biological-activities of inositol pentakisphosphate and inositol hexakisphosphate. *Biochemical Pharmacology, 50*, 137–146.

Shi, J. R., Wang, H. Y., Wu, Y. S., Hazebroek, J., Meeley, R. B., & Ertl, D. S. (2003). The maize low-phytic acid mutant 1pa2 is caused by mutation in an inositol phosphate kinase gene. *Plant Physiology, 131*, 507–515.

Shi, J. R., Wang, H. Y., Hazebroek, J., Ertl, D. S., & Harp, T. (2005). The maize low-phytic acid 3 encodes a myo-inositol kinase that plays a role in phytic acid biosynthesis in developing seeds. *The Plant Journal, 42*, 708–719.

Shi, J. R., Wang, H. Y., Schellin, K., Li, B. L., Faller, M., Stoop, J. M., Meeley, R. B., Ertl, D. S., Ranch, J. P., & Glassman, K. (2007). Embryo-specific silencing of a transporter reduces phytic acid content of maize and soybean seeds. *Nature Biotechnology, 25*, 930–937.

Stevenson-Paulik, J., Bastidas, R. J., Chiou, S. T., Frye, R. A., & York, J. D. (2005). Generation of phytate-free seeds in Arabidopsis through disruption of inositol polyphosphate kinases. *Proceedings of the National Academy of Sciences of the United States of America, 102*, 12612–12617.

Strother, S. (1980). Homeostasis in germinating-seeds. *Annals of Botany, 45*, 217–218.

Suzuki, M., Tanaka, K., Kuwano, M., & Yoshida, K. T. (2007). Expression pattern of inositol phosphate-related enzymes in rice (*Oryza sartiva* L.): Implications for the phytic acid biosynthetic pathway. *Gene, 405*, 55–64.

Torabinejad, J., & Gillaspy, G. (2006). Functional genomics of inositol metabolism. In B. B. Biswas & A. L. Majumder (Eds.), *Biology of inositols and phosphoinositides*. New York: Springer.

Xu, X. H., Zhao, H. J., Liu, Q. L., Frank, T., Engel, K. H., An, G. H., & Shu, Q. Y. (2009). Mutations of the multi-drug resistance-associated protein ABC transporter gene 5 result in reduction of phytic acid in rice seeds. *Theoretical and Applied Genetics, 119*, 75–83.

York, J. D., Odom, A. R., Murphy, R., Ives, E. B., & Wente, S. R. (1999). A phospholipase C-dependent inositol polyphosphate kinase pathway required for efficient messenger RNA export. *Science, 285*, 96–100.

Yuan, F. J., Zhao, H. J., Ren, X. L., Zhu, S. L., Fu, X. J., & Shu, Q. Y. (2007). Generation and characterization of two novel low phytate mutations in soybean (Glycine max L. Merr.). *Theoretical and Applied Genetics, 115*, 945–957.

Zhao, H. J., Liu, Q. L., Ren, X. L., Wu, D. X., & Shu, Q. Y. (2008). Gene identification and allele-specific marker development for two allelic low phytic acid mutations in rice (*Oryza sativa* L.). *Molecular Breeding, 22*, 603–612.

Chapter 17
Correlation Analysis of Functional Components of Barley Grain

Tao Yang, Ya-Wen Zeng, Xiao-Ying Pu, Juan Du, and Shu-Ming Yang

Abstract Barley is a medicinal and edible crop that is not only the most stress tolerant and comprehensively utilized cereal, but is also a functional food crop used in strategies for chronic disease prevention, such as reduction of obesity and Type II diabetes. Using spectrophotometry methods, four functional components in 830 accessions of Yunnan improved landraces barley grains have been detected: resistant starch, total flavonoids, total alkaloids, and γ-aminobutyric acid. The average content of resistant starch was 3.57 ± 2.72 mg/100 g. The average content of total flavonoids was 133.09 ± 27.65 mg/100 g. The average content of total alkaloids was 21.26 ± 9.18 mg/100 g. The average content of γ-aminobutyric acid was 5.70 ± 2.82 mg/100 g. Sixty accessions of high functional components of high superior landraces were detected, including 14 accessions with high content of total flavonoids (content >195 mg/100 g), 17 accessions with high content of resistant starch (content >12%), 11 accessions of high content of total alkaloids (content >45.19 mg/100 g), 18 accessions of high content of γ-aminobutyric acid (content >13.03 mg/100 g). These superior barley functional landraces offer excellent landraces for barley breeding. By correlation analysis of four functional components of barley grain, total flavonoids and γ-aminobutyric acid have a significant negative correlation ($r = -0.154^{**}$, $n = 830$), total flavonoids and resistant starch have a significant negative correlation ($r = -0.097^{**}$, $n = 830$), total alkaloids and γ-aminobutyric acid have a significant negative correlation ($r = -0.096^{**}$, $n = 830$), and γ-aminobutyric acid and resistant starch have a significant negative correlation($r = -0.122^{**}$, $n = 830$). By a principal component analysis of functional components, the first principal component is γ-aminobutyric acid and the second is resistant starch, followed be flavonoids. Content of γ-aminobutyric acid changes the most, followed by resistant starch and flavonoids.

T. Yang • Y.-W. Zeng (✉) • X.-Y. Pu • J. Du • S.-M. Yang
Biotechnology and Genetic Germplasm Institute, Yunnan Academy of Agricultural Sciences/ Agricultural Biotechnology Key Laboratory of Yunnan Province, Kunming 650205, China
e-mail: zengyw1967@126.com

G. Zhang et al. (eds.), *Advance in Barley Sciences: Proceedings of 11th International Barley Genetics Symposium*, DOI 10.1007/978-94-007-4682-4_17,

Keywords Barley • Resistant starch • Total flavonoids • Total alkaloids • γ-Aminobutyric acid

Barley is the most widely used cereal crop, with medicinal and nutritional benefits, and also plays an important role in the beer brewing industry, animal husbandry, and industrial restructuring. Studies have shown that barley contains flavonoids, polyphenols, ergot compounds, and other active ingredients with beneficial health effects (Yang et al. 2007) and also contains resistant starch, alkaloids, and γ-aminobutyric acid.

Resistant starch is a low-calorie functional food base and a form of dietary fiber that is not absorbed by the small intestine. Resistant starch undergoes intestinal microbial fermentation arising from short-chain fatty acids such as butyric acid, an acid that can prevent the growth and reproduction of cancer cells. Studies have shown that resistant starch can promote the growth of beneficial intestinal bacteria, including bifidobacteria, and can increase stool volume and promote intestinal peristalsis for prevention of constipation, hemorrhoids, colon cancer, and other diseases. Human intake of foods high in resistant starch can lower insulin reaction and delay the postprandial rise in blood sugar, thus assisting in diabetes control. Resistant starch also increases lipid excretion, the partial exclusion of lipids in food, to reduce calorie intake. Resistant starch itself is almost calorie-free and is effectively used as a low-calorie food additive for weight control. In short, resistant starch can be effective in improving glucose and lipid metabolism, promoting weight loss, increasing insulin sensitivity, and preventing and treating metabolic syndrome (Lian and Li 2008; Li 2008; Zhao et al. 2002).

Flavonoids include catechin, myricetin, quercetin, and kaempferol (Kim et al. 2007). They have been found to have antibacterial, antiviral, and anti-tumor properties. They also inhibit increases in cholesterol, lipids, and glucose; prevent cardiovascular disease; boost immunity; improve brain function; and strengthen liver function, scavenging free radicals and tissue malondialdehyde (MDA) and reducing free radical damage to liver tissue (Chen 2003; Tang et al. 2006; Tan et al. 2005; Yu et al. 1988).

Plant alkaloids, also known as alkali, are alkaline nitrogen-containing organic compounds. Their ring structure is insoluble in water and they can form salts with acids. Most are bitter, colorless crystals; a small number are liquids. There are numerous alkaloids and many have toxic or physiological effects on other organisms (Qin and Jiang 2006; Liang et al. 2007).

γ-Aminobutyric acid (GABA) is a non-protein amino acid. Glutamic acid is the enzyme glutamic acid de-completion of a product, widely present in the action, the plant kingdom, is the mammalian brain and spinal cord in with the inhibitory neurotransmitter-mediated inhibition of 40% or more neurotransmitters. GABA is also present in bacteria, fungi, ferns, and some higher plants (He et al. 2007; Zhao et al. 2009). It acts on the spinal cord vasomotor center, and have anti-dilation of blood vessels postsynaptic GABA receptors and the role of the sympathetic nerve endings have presynaptic inhibition of GABA receptors. By inhibiting secretion of the anti-diuretic hormone vasopressin, it effectively promotes vasodilation and lowers blood pressure (He et al. 2007).

In this study of improved strains of barley grain, component testing of the four functional ingredients was performed. Correlation analysis will assist in choosing between functional components, in seed selection, and in testing of high germplasm and high functional components, contributing to the development of better barley and broadening the barley industrial system.

17.1 Materials and Methods

17.1.1 Test Materials

830 accessions grains of improved Yunnan barley lines were tested.

17.1.2 Reagents and Instruments

γ-Aminobutyric acid, glucose reagent colorimetric box, rutin, and berberine supplied by Sigma-Aldrich Company. Other reagents were of analytical grade and supplied by Tianjin Chemical Reagent Factory.

Beckman DU UV – visible spectrophotometer, HY-2-type reciprocating oscillator, Beckman J2-21 high-speed centrifuge, FA1104 electronic balance (accuracy ±1 mg) supplied by Scientific Instrument Co., Ltd., Shanghai, China; Finland Decker pipette gun; SHZ-88 water bath oscillator; all chemical reagents were of analytical grade purity.

17.1.3 Test Methods

17.1.3.1 Preparation of Test Materials

Barley (200–400 g) was machine-powdered for testing.

17.1.3.2 Determination of Resistant Starch of Barley Grain

Resistant starch was determined according the methods of Goñi (Goñi et al. 1996). A 0.5 g sample of barley powder was placed in 10 mL centrifuge tube, to which was added 2 mL (pH5.8) phosphate buffer solution and 0.6 mL thermostable α-amylase. The sample was placed in a boiling water bath for 30 min, cooled to room temperature, and the supernatant centrifuged (5,000 r/min, 20 min). Next, 4 ml distilled water was added, the precipitate washed, and the supernatant centrifuged (this step was repeated at least once); 1.2 mL 2 mol/L KOH solution was added to the precipitate at room temperature and oscillated for 30 min. To promote the precipitation of dissolved in 2 mol/L HCL pH adjusted to neutral by adding 1 mL acetate buffer (pH4.4) and 0.2 mL glucoamylase. The sample was then placed in a 60°C water bath for 45 min and then centrifuged (5,000 r/min, 20 min). Supernatants were collected, and then 10 ml distilled water was added to the precipitant (this step is repeated twice), which was centrifuged (5,000 r/min, 20 min), and the supernatants again collected. The combined supernatants were collected in a fixed volume in a 100 mL flask. Glucose test kit with samples of glucose, 0.9 is multiplied by the amount of resistant starch.

17.1.3.3 Extraction of Flavonoids and Method for Determination of Barley Grain

The method of Zhuang et al. (1992) was used to determine the aluminum nitrate complex of flavonoids. Barley samples weighing 500 mg ± 1 mg were placed in 10 ml centrifuge tubes with lids; to each tube 5 ml 50% ethanol was add, and the tubes oscillated (200 times/min) for 2 h. The supernatant was 1.5 mL and was placed in another centrifuge tube and centrifuged (5,000 r/min) for 3 min. 1 mL supernatant was obtained after centrifugation in the 10 mL centrifuge tube, to which was added 0.4 mL 5% $NaNO_2$, which was mixed after standing 6 min and then 0.4 mL 10% of $AL(NO_3)_3$ was added. After mixing and standing 6 min, 4 mL 5% NaOH was added, mixing after standing 15 min. The sample was measured at a wavelength of 500 nm absorbance. Rutin standard sample concentration (by 0, 200, 400, 600, 800, 1,000 mg/100 mL) and its absorbance showed a good linear relationship between: $y = 242.9x + 57.096$, the correlation coefficient $R^2 = 0.9994$; the standard line drawing experiment was repeated three times. Therefore, the determination of total flavonoids of barley is feasible, the testing accuracy high, and the results stable.

17.1.3.4 Alkaloid Extract and Determination Method of Barley Grain

Standard solutions of 1, 2, 3, 4, and 5 mL were drawn and placed in 20 mL test tubes to which water was added up to 5 mL. Bromocresol green (2 mL) and 10 mL phosphate buffer solution of chloroform was added and shaken for 1 min, then poured into a separatory funnel. After standing 1 h, for the sub-chloroform layer, the measured absorbance at this wavelength value to alkaloid content of the abscissa, the absorbance value of the vertical axis, the standard curve to be linear relationship: $y = 1.3002x + 0.0309$, correlation coefficient: $R^2 = 0.9972$.

Five grams of barley powder were weighed and extracted by adding 95% ethanol and diluted hydrochloric acid to adjust pH = 3. After shock extraction for 1 h, the sample was filtered, and 5 mL filtrate was obtained. The filtrate was adjusted to neutral with concentrated ammonia solution (pH = 7.0). Added to the filtrate were 2 mL bromocresol green, 2 mL phosphate buffer, and 10 mL chloroform solution. After 1 min shaking, the sample was poured into a separatory funnel; after standing 1 h, points sub-chloroform layer, measured at 425 nm wavelength absorbance values, with chloroform method was blank. The alkaloid content was calculated using the standard curve.

17.1.3.5 Extraction and Determination of γ-Aminobutyric Acid of Barley Grain

GABA content was determined by the method of Inatomi and Slaughter (1971), in which GABA with phenol under the action of sodium hypochlorite forms a blue-green compound that can be identified by the assay content of data produced by GABA Sigma standards for the conversion from the standard line. The samples labeled powder 0.5 g were weighed with 5 mL distilled water into volume, and then underwent vibration extraction (2 h), filtration, and centrifugation. Filtrate (1 mL) was added to 0.6 mL pH 9.0 borate buffer, 2 mL 5% solution of redistilled phenol, and 1 mL of 7% available chlorine of sodium hypochlorite and then underwent full oscillation, a boiling water bath (10 min) followed immediately by 20 min on ice and oscillation until a blue-green compound appeared, then 2 mL 60% ethanol was added at 645 nm than the color, the measured absorbance (A), the standard curve by GABA (take 5 0, 5, 10, 15, 20 mg/100 g) calculated concentrations of standard samples the standard curve equation $y = 444.99x + 2.1032$, x is the concentration of GABA, y for the 645 nm visible light absorbance, $R^2 = 0.995$ derived GABA content.

17.1.3.6 Data Processing

Data processing and statistical analysis were performed with Microsoft Excel and SPSS 16.0 software

17.2 Results and Analysis

17.2.1 Distribution of Functional Components Content of Barley Grain

Resistant starch, flavonoids, alkaloids, and GABA were tested in the 830 improved strains of Yunnan landraces barley grain. Fig. 17.1 shows that the average resistant starch content is 3.57 ± 2.72 mg/100 g, total flavonoids 133.09 ± 27.65 mg/100 g, alkaloids 21.26 ± 9.18 mg/100 g, and GABA 5.70 ± 2.82 mg/100 g. Sixty high grain,

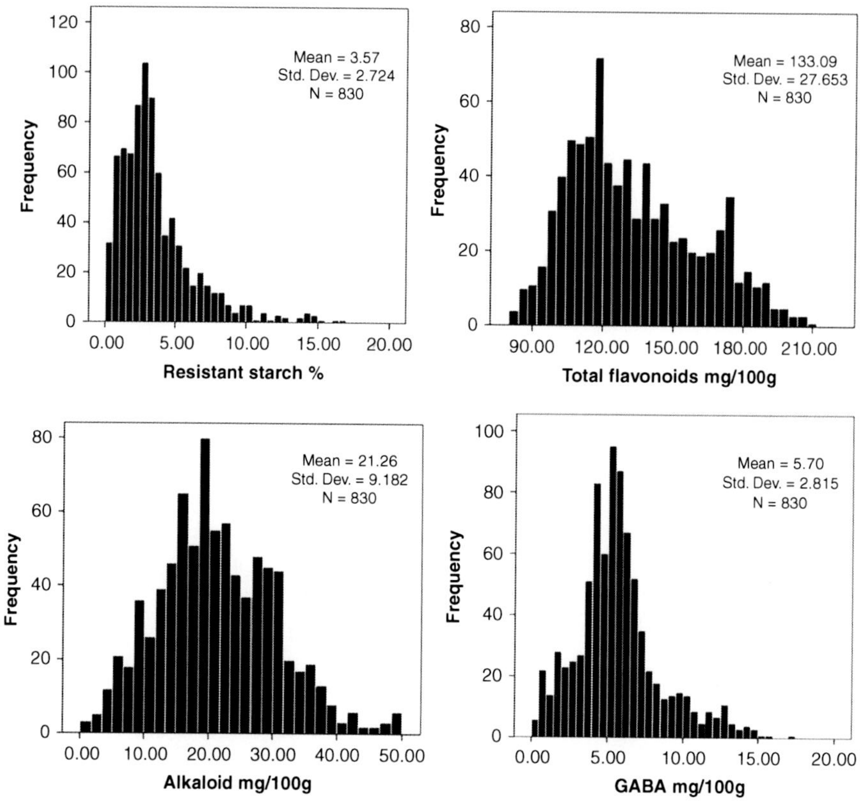

Fig. 17.1 Distribution of four functional components of barley grain

high germplasm functional components were selected and detected, including resistant starch grains >12% of 17 accessions (12.01–16.89%), total flavonoid content >195 mg/100 g of 14 accessions(195.37–208.82 mg/100 g), alkaloid content >45 mg/100 g of 11 accessions (45.19–49.98 mg/100 g), and GABA content >13 mg/100 g of 18 accessions (13.03–17.32 mg/100 g).

17.2.2 Correlation Analysis of Four Functional Components of Barley Grain

Flavonoids (Table 17.1) and γ-aminobutyric acid and resistant starch were significantly negatively correlated, with correlation coefficients of –0.154** and –0.097**, respectively. Alkaloids and γ-aminobutyric acid were significantly negatively correlated, with a correlation coefficient of –0.096**. γ-Aminobutyric acid and resistant starch were significantly negatively correlated, with a correlation coefficient of –0.122**. The greatest correlation coefficient is for flavonoids and γ-aminobutyric acid,

Table 17.1 Relevancy of four functional components of barley grain ($N=830$)

Items	Flavonoids	Alkaloid	GABA	Resistant starch
Flavonoids	1	0.016	−0.154**	−0.097**
Alkaloid	0.016	1	−0.096**	0.012
GABA	−0.154**	−0.096**	1	−0.122**
Resistant starch	−0.097**	0.012	−0.122**	1

**Correlation is significant at the 0.01 level (two-tailed)

Table 17.2 Analysis of main functional components of barley grain

	Common factors		
	1	2	Component source
Characteristic	1.20	1.10	
Contribution rate	29.91	27.37	
Accumulative percentage	29.91	57.28	
Eigenvector	0.26	0.81	Resistant starch
	0.53	−0.65	Flavonoids
	0.45	0.10	Alkaloid
	−0.80	−0.11	GABA

followed by γ-aminobutyric acid and resistant starch; the lowest correlation coefficient is flavonoids and resistant starch, and alkaloids and γ-aminobutyric acid.

17.2.3 Principal Component Analysis of Functional Components of Barley Grain

To further clarify the function of the four functional components of barley, the size of the contribution of components was analyzed. For the principal component analysis of the four functional components of barley grain, each factor and the cumulative percentage of eigenvalues and eigenvectors are listed in Table 17.2. Table 17.2 shows only two principal components, with the cumulative contribution rate of 57.28%. The first principal component factor is 1.20, with a contribution rate of 29.91%, mainly corresponding to the flavonoids, alkaloids, large feature vectors of scores, respectively 0.53, 0.45, indicating that 2-like compared to the greatest contribution, the relationship between the composition and function more. γ-Aminobutyric acid has a large negative vector, indicating that γ-aminobutyric acid content of the functional components are also more active ingredients. The second principal component factor eigenvalue of 1.10, the contribution rate of 27.37%, corresponding to the eigenvectors of resistant starch score greater, up to 0.81, and functional ingredients content.

17.3 Discussion and Conclusions

Resistant starch, flavonoid, alkaloid, and γ-aminobutyric acid composition of 830 accessions grains of Yunnan barley improved lines were tested, and 60 grain with high lines of functional components were selected; thus, superior barley functional lines for barley breeding were detected.

By correlation analysis of four functional components of barley grain, flavonoids and γ-aminobutyric acid were found to have a significant negative correlation, flavonoids and resistant starch were found to have a significant negative correlation, alkaloids and γ-aminobutyric acid were found to have a significant negative correlation, and γ-aminobutyric acid and resistant starch were found to have a significant negative correlation, indicating that in barley grain, when content of one functional component is high, the content of the others is low.

By principal component analysis of the functional components of barley grain, the first principal component factor eigenvalue is 1.20 and the second principal component factor eigenvalue is 1.10, greater than 1, indicating that in barley grain, when content of one functional component is high, the content of the others is low. The first principal components are mainly flavonoids and alkaloids; the γ-aminobutyric acid negative vector is large compared with the results of the analysis related to validation, indicating that flavonoids, alkaloids, and γ-aminobutyric acid content were negatively correlated with the absolute value of which maximum of γ-aminobutyric acid, shows, γ- aminobutyric acid contents in the content of the four kinds of functional ingredients most affected.

Secondly, the second principal component mainly to resistant starch and flavonoids in γ-aminobutyric acid content changes based on the flavonoid content of resistant starch and change in composition of four functional levels greater impact.

Acknowledgments This research was supported by China Agriculture Research System (CARS-05). We are grateful for many valuable suggestions from Professor Benxun Li; Miss Sha Li assisted with some determinations and analyses for functional components in barley grains.

References

Chen, Q. Q. (2003). Functionality and its market development of catechins. *Global Food Industry, 10*, 28–29.

Goñi, I., García-Diaz, L., Mañas, E., et al. (1996). Analysis of resistant starch: A method for food products. *Food Chemistry, 56*, 445–449.

He, X. P., Zhang, M., Li, J. F., et al. (2007). The physiological function of γ-aminobutyric acid and the general research about γ-aminobutyric acid. *Journal of Guangxi University (Natural Sciences Edition), 32*, 464–466.

Inatomi, K., & Slaughter, J. C. (1971). The role of glutamate decarbozylase and γ-aminobutyric acid in germinating barley. *Journal of Experimental Botany, 22*, 561–571.

Kim, M. J., Hyun, J. N., Kim, J. A., et al. (2007). Relationship between phenolic compounds, anthocyanins content and antioxidant activity in colored barley germplasm. *Journal of Agricultural and Food Chemistry, 55*(12), 4802–4809.

Li, M. (2008). Advances on the physiologicalfunction of resistantstarch. *Journal of Hygiene Research, 37*(5), 640–643.

Lian, X. J., & Li, J. Y. (2008). Research advance on resistant starch physiological function. *Cereals Oils, 7*, 3–5.

Liang, D., Li, Z. W., Li, M., et al. (2007). Research and application of natural alkaloids. *Journal of Anhui Agricultural Sciences, 35*(35), 11340–11342.

Qin, Z. L., & Jiang, Y. R. (2006). Supercritical fluid extraction in progress extraction of alkaloids. *Chemical Industry Times, 20*(7), 60–63.

Tan, X. R., Zhang, W. M., & Gong, Z. L. (2005). Research on quercetin functional properties. *Cereals Oils, 11*, 46–48.

Tang, L., Zhang, L. J., & Wang, M. Q. (2006). Research advances on myricetin in Bayberry. *Chinese Traditional Patent Medicine, 28*(1), 121–122.

Yang, T., Zeng, Y. W., Xiao, F. H., et al. (2007). Research progress on medicinal barley. *Journal of Triticeae Crops, 27*(6), 1154–1158.

Yu, P. Z., Yao, L. Y., & Wang, L. P. (1988). HPLC method for the determination dysosma injection Zhongshan Chennai phenol content. *Chinese Traditional Patent Medicine, 20*(6), 19–20.

Zhao, K., Zhang, S. W., & Fang, G. Z. (2002). Study on physiological functions of resistant starch and its application in food industry. *Journal of Harbin University Communications (Natural Sciences Edition), 18*(6), 661–663.

Zhao, D. W., Pu, X. Y., Zeng, Y. W., et al. (2009). Determination of the γ-aminobutyric acid in barley grain. *Journal of Triticeae Crops, 29*(1), 69–72.

Zhuang, X. P., Yu, X. Y., Yang, G. S., et al. (1992). Such as *Ginkgo biloba* flavonoids and extraction method. *Chinese Traditional Herbal Drugs, 23*(3), 122–124.

Chapter 18
Genome-Wide Association Mapping Identifies Disease-Resistance QTLs in Barley Germplasm from Latin America

Lucía Gutiérrez, Natalia Berberian, Flavio Capettini, Esteban Falcioni, Darío Fros, Silvia Germán, Patrick M. Hayes, Julio Huerta-Espino, Sibyl Herrera, Silvia Pereyra, Carlos Pérez, Sergio Sandoval-Islas, Ravi Singh, and Ariel Castro

Abstract Diseases are the main problem for barley in Latin America. Spot blotch (caused by *Cochliobolus sativus*), stripe rust (caused by *Puccinia striiformis* f. sp. *hordei*), and leaf rust (caused by *Puccinia hordei*) are three of the most important diseases that attack the crop in the region. Chemical control of those diseases is both economically and environmentally inappropriate, making the development of durable resistant varieties a priority for breeding programs. However, the availability of new resistance sources is a limiting factor. The objective of this work was to detect genomic regions associated to durable resistance to spot blotch, stripe rust, and leaf rust in Latin American germplasm. Associations between disease severities measured in several environments across the Americas and 1,536 SNPs (belonging to the barley OPA1) in a population of 360 genotypes were used to identify genomic regions associated with disease. Several models for association mapping with mixed

L. Gutiérrez • N. Berberian
Departamento de Biometría, Estadística y Computación, Facultad de Agronomía, UDELAR, Montevideo, Uruguay

F. Capettini ICARDA, Aleppo, Syria

E. Falcioni • J. Huerta-Espino • S. Herrera • C. Pérez • R. Singh • A. Castro (✉)
Departamento de Producción Vegetal, Facultad de Agronomía, UDELAR, Montevideo, Uruguay
e-mail: vontruch@fagro.edu.uy

D. Fros
Instituto Nacional de Investigación Agropecuaria, Montevideo, Uruguay

S. Germán • S. Pereyra
Department of Crop and Soil Science, Oregon State University, Corvallis, OR 97331-3002, USA

P.M. Hayes
CIMMYT, El Batan, Mexico

S. Sandoval-Islas
COLPOS, Chapingo, Mexico

G. Zhang et al. (eds.), *Advance in Barley Sciences: Proceedings of 11th International Barley Genetics Symposium*, DOI 10.1007/978-94-007-4682-4_18,

models were compared. These models considered either the structure of the population (Q) through PCA analysis, the identity by descent through coancestry information (K), or both. Results show significant marker-trait associations for spot blotch and leaf and stripe rust. Associations are environment specific.

Keywords Association mapping • QTL • Model comparison • Mixed models

18.1 Introduction

Barley (*Hordeum vulgare*, L.) is the fourth most important cereal crop in the world in terms of total world production (FAOSTAT 2008), and malting quality is one of its most relevant traits. The main use of this crop in Latin America is for malt production (German 2004). One of the main limitations for achieving high yields and malt of good quality is the presence of fungal diseases (German 2007).

Fungal diseases affect barley production both directly (i.e., affecting grain weight and germinability) and indirectly (i.e., foliar diseases that affect photosynthesis decreasing yield; Nutter et al. 1985). Spot blotch (caused by *Cochliobolus sativus*), stripe rust (caused by *Puccinia striiformis* f. sp. *hordei*), and leaf rust (caused by *Puccinia hordei*) are three of the most important diseases that attack the crop in Latin America (Pereyra 1996). Chemical control of those diseases is both economically and environmentally inappropriate, making the development of durable resistant varieties a priority for breeding programs. However, the availability of new resistance sources is a limiting factor.

Genome-wide association mapping (GWA) can be used to detect genomic regions associated to relevant traits (Jannink et al. 2001) and provide targets for MAS in relevant germplasm (Mather et al. 1997). The advantages of GWA include simultaneous assessment of broad diversity, higher resolution for fine mapping, effective use of historical data, and immediate applicability to cultivar development because the genetic background in which QTLs are estimated is directly relevant for plant breeding (Kraakman et al. 2004). Additionally, GWA has been proposed as a promising tool for selfing-species crop improvement and specifically barley breeding (Kraakman et al. 2004, 2006; Hayes and Szücs 2006; Stracke et al. 2009; Waugh et al. 2009; Roy et al. 2010; Bradbury et al. 2011; von Zitzewitz et al. 2011). The objective of this work was to detect genomic regions associated to durable resistance to spot blotch, stripe rust, and leaf rust in Latin American germplasm.

18.2 Materials and Methods

A total of 360 genotypes from ICARDA and national breeding programs were used. Each genotype was evaluated for disease resistance in several environments. Severity of spot blotch infection (in percentage of covered area from 0 to 100%) was evaluated in three environments: EEMAC (northern Uruguay) in 2009 and 2010 and INIA-LE (west Uruguay) in an off-season nursery in 2009–2010. Severity of leaf rust infection (in a scale from 1 to 10) was measured in four environments: EEMAC 2009 and 2010 and INIA-LE 2009 and 2010. Stripe rust was evaluated in two environments: Mexico 2010 and Peru 2011.

All accessions were genotyped for 1,536 SNPs using one Illumina GoldenGate oligonucleotide pool assays (OPAs) developed for the barley CAP. The development and application of the barley OPAs (BOPA1) is described in detail in Close et al. (2009) and Szücs et al. (2009). In this research, we used the consensus SNP map developed by Close et al. (2009), available by downloading the 1.77 version of the barley HarvEST database (2011).

Several models for association mapping with mixed models were compared to use the most appropriate model in each case. Modifications of the general mixed model equation were used:

$$\mathbf{Y} = \mathbf{X}\beta + \mathbf{Q}\upsilon + \underline{\mathbf{Zu}} + \mathbf{e} \tag{18.1}$$

where **Y** is the phenotypic vector, **X** is the molecular marker matrix, β is the unknown vector of allele effects to be estimated, **Q** is the population structure represented by the scores of the relevant axis of a principal components analysis, υ is the vector of population effects (parameters), **Z** is a matrix that relates each measurement to the individual from which it was obtained (an identity matrix in our case), **u** is the vector of random background polygenic effects, and **e** is the residual errors. Random effects are underlined. We compared the following mainstream models: (1) naïve; a simple test of association (Kruskal–Wallis) with no correction for population structure ($\mathbf{Y}=\mathbf{X}\beta+\mathbf{e}$); (2) fixed, a fixed-effects model using population structure as fixed covariate ($\mathbf{Y}=\mathbf{X}\beta+\mathbf{Q}\upsilon+\mathbf{e}$); (3) kinship, a mixed model including the coancestry matrix among genotypes as a random effect ($\mathbf{Y}=\mathbf{X}\beta+\mathrm{Zu}+\mathbf{e}$ following Parisseaux and Bernardo 2004); (4) Price, a mixed-effects model including population structure but as a random effect ($\mathbf{Y}=\mathbf{X}\beta+\mathrm{Qu}+\mathbf{e}$ following Price et al. 2006); and (5) Yu, a mixed-effects model including both population structure and coancestry among genotypes ($\mathbf{Y}=\mathbf{X}\beta+\mathbf{Q}\upsilon+\mathrm{Zu}+\mathbf{e}$ following Yu et al. 2009).

18.3 Results and Discussion

There is not a single best model for all the traits. A mixed model with population structure either as random (Price) or fixed (Yu) effect was the best model for spot blotch (Fig. 18.1a) in all environments. However, a mixed model without population structure and with coancestry information (kinship) was the best model for leaf rust (Fig. 18.1b) in all environments. The Price and Yu models performed relatively well for leaf rust.

We detected several QTL for all the diseases studied (Table 18.1). Eleven significant marker-trait associations were found for leaf rust. Associations were found on

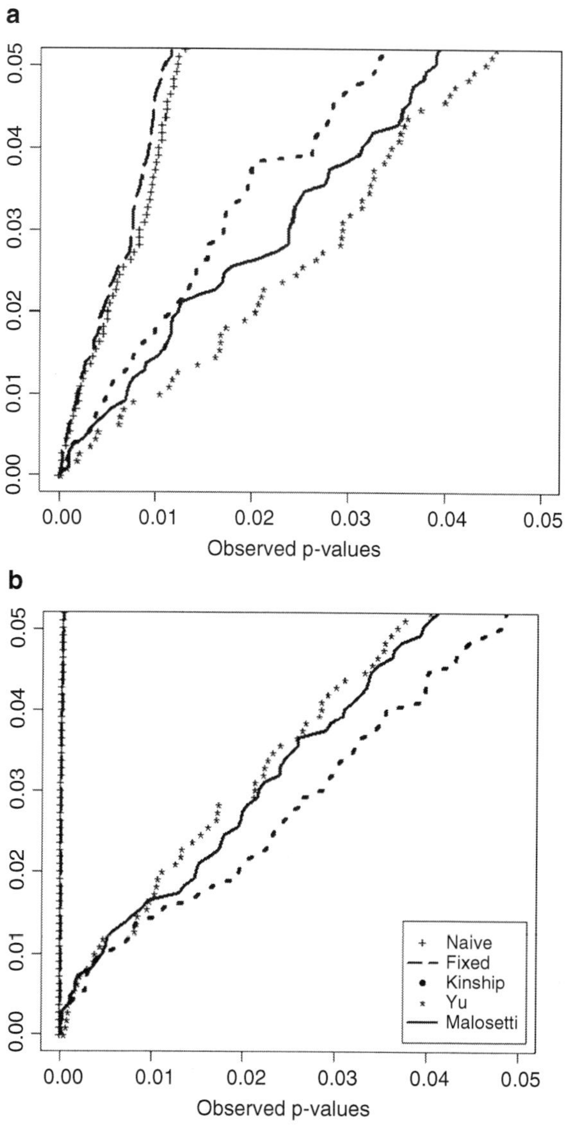

Fig. 18.1 Cumulative distribution functions (cdf) of *p* values in genome-wide scans for spot blotch (**a**) and leaf rust (**b**) diseases in the EEMAC 2009 environment

chromosomes 2H, 3H, 5H, 6H, and 7H. Nine significant marker-trait associations were found for spot blotch. Associations were found on chromosomes 1H, 3H, 4H, 5H, 6H, and 7H. Only one significant marker-trait association was detected for stripe rust.

Although we expect that the underlying distribution of disease scoring be normal, because of the way these diseases are measured (i.e., in a scale), not all of the traits follow a normal distribution. Additionally, some markers had heterogeneous variances.

Table 18.1 Markers with a significant marker-trait association in the barley Latin American germplasm

Trait	Environment	Ch	Pos	Marker
Leaf rust	EELE 2009	2	8,6	11_21377
Leaf rust	EELE 2009	2	63,5	11_21399
Leaf rust	EELE 2009	2	156,7	11_10085
Leaf rust	EEMAC 2009	3	59,9	11_21511
Leaf rust	EEMAC 2009	3	91,9	11_10253
Leaf rust	EELE 2010	5	87,4	11_20645
Leaf rust	EEMAC 2009	5	100,3	11_11473
Leaf rust	EEMAC 2009	5	100,3	11_20097
Leaf rust	EEMAC 2009	5	159,8	11_21024
Leaf rust	EELE 2010	6	60,2	11_10189
Leaf rust	EELE 2010	6	60,2	11_20058
Leaf rust	EELE 2010	6	60,2	11_21310
Leaf rust	EEMAC 2010	6	71,9	11_11459
Leaf rust	EEMAC 2010	7	0,0	11_10971
Spot blotch	EELE 09-10	1	45,5	11_10275
Spot blotch	EELE 09-10	1	49,7	11_10470
Spot blotch	EELE 09-10	1	50,0	11_11478
Spot blotch	EEMAC 2009	3	43,2	11_21533
Spot blotch	EEMAC 2009	4	23,1	11_11136
Spot blotch	EEMAC 2010	4	92,4	11_20732
Spot blotch	EEMAC 2010	4	119,1	11_10610
Spot blotch	EELE 09-10	5	80,6	11_20236
Spot blotch	EELE 09-10	5	151,4	11_10080
Spot blotch	EELE 09-10	6	65,0	11_10124
Spot blotch	EELE 09-10	7	21,1	11_10025
Stripe rust	PERU	7	141,0	11_10867

The most common transformations of the original variable were used to fit final models when the untransformed data did not perform properly. Further study of these traits with more complex models would be beneficial.

18.4 Conclusions

We were able to find significant marker-trait associations for leaf rust, spot blotch, and stripe rust. Although the best correction for population structure was dependent on the trait studied, significant associations were consistent across models. These preliminary results show candidate QTL that should be further studied and compared to previous reports. Evaluations in more environments are being conducted to improve the estimation of QTL. Association mapping is a promising tool to detect candidate QTL to include in marker-assisted selection programs.

References

Bradbury, P., Parker, T., Hamblin, M. T., & Jannink, J. L. (2011). Assessment of power and false discovery rate in genome-wide association studies using the barley CAP germplasm. *Crop Science, 51*, 52–59.

Close, T. J., Bhat, P. R., Lonardi, S., Wu, Y., Rostoks, N., Ramsay, L., Druka, A., Stein, N., Svensson, J. T., Wanamaker, S., Bozdag, S., Roose, M. L., Moscou, M. J., Chao, S., Varshney, R., Szucs, P., Sato, K., Hayes, P. M., Matthews, D. E., Kleinhofs, A., Muehlbauer, G. J., DeYoung, J., Marshall, D. F., Madishetty, K., Fenton, R. D., Condamine, P., Graner, A., & Waugh, R. (2009). Development and implementation of high-throughput SNP genotyping in barley. *BMC Genomics, 10*, 582.

FAOSTAT. (2008). Available at http://faostat.fao.org/site/339/default.aspx. Accessed 5 May 2011.

German, S. (2004). Available at http://www.cazv.cz/service.asp?act=print&val=32831. *Breeding malting barley under stress conditions in South America* (pp: 140–144). Colonia, Uruguay.

German, S. (2007). Roya de la hoja en cultivos de invierno: Epidemiologia de la enfermedad y comportamiento varietal. Jornada de cultivos de invierno, Young, Serie actividades de difusion 484.

HarvEST. (2011). Available at http://harvest.ucr.edu/HBarley178.exe. Accessed 5 May 2011.

Hayes, P., & Szücs, P. (2006). Disequilibrium and association in barley: Thinking outside the glass. *Proceedings of the National Academy of Sciences of the United States of America, 103*(49), 18385–18386.

Jannink, J. L., Bink, M. C. A. M., & Jansen, R. C. (2001). Using complex plant pedigrees to map valuable genes. *Trends in Plant Science, 6*, 337–342.

Kraakman, A. T. W., Niks, N. E., van den Berg, P. M. M. M., Stam, P., & van Eeuwijk, F. A. (2004). Linkage disequilibrium mapping of yield and yield stability in modern spring barley cultivars. *Genetics, 168*, 435–446.

Kraakman, A. T. W., Martinez, F., Mussiraliev, B., van Eeuwijk, F. A., & Niks, R. E. (2006). Linkage disequilibrium mapping of morphological, resistance, and other agronomically relevant traits in modern spring barley cultivars. *Molecular Breeding, 17*, 41–58.

Mather, D. E., Tinker, N. A., Laberge, D. E., Edney, M., Jones, B. L., Rossnagel, B. G., Legge, W. G., Briggs, K. G., Irvine, R. B., Falk, D. E., & Kasha, K. J. (1997). Regions of the genome that affect grain and malt quality in a North American two row barley cross. *Crop Science, 37*, 544–554.

Nutter, E. W., Pederson, V. D., & Foster, A. E. (1985). Effect of inoculations with Cochliobolus sativus at species growth stages on grain yield and quality of malting barley. *Crop Science, 25*, 933–938.

Parisseaux, B., & Bernardo, R. (2004). Insilico mapping of quantitative trait loci in maize. *Theoretical and Applied Genetics, 109*, 508–514.

Pereyra, S. (1996). Manejo de enfermedades en cereales de invierno y pasturas, Montevideo, INIA. *Serie Tecnica, 74*, 110–111.

Price, A. L., Patterson, N. J., Plenge, R. M., Weinblatt, M. E., Shadick, N. A., & Reich, D. (2006). Principal components analysis corrects for stratification in genome-wide association studies. *Nature Genetics, 38*, 904–909.

Roy, J. K., Smith, K. P., Muehlbauer, G. J., Chao, S., Close, T. J., & Steffenson, B. J. (2010). Association mapping of spot blotch resistance in wild barley. *Molecular Breeding, 26*, 243–256.

Stracke, S., Haseneyer, G., Veyrieras, J.-B., Geiger, H. H., Sauer, S., Graner, A., & Piepho, H.-P. (2009). Association mapping reveals gene action and interactions in the determination of flowering time in barley. *Theoretical and Applied Genetics, 118*, 259–273.

Szűcs, P., Blake, V. C., Bhat, P. R., Close, T. J., Cuesta-Marcos, A., Muehlbauer, G. J., Ramsay, L. V., Waugh, R., & Hayes, P. M. (2009). An integrated resource for barley linkage map and malting quality QTL alignment. *Plant Genome, 2*, 134–140.

von Zitzewitz, J., Cuesta-Marcos, A., Condon, F., Castro, A. J., Chao, S., Corey, A., Filichkin, T., Fisk, S. P., Gutierrez, L., Haggard, K., Karsai, I., Muehlbauer, G. J., Smith, K. P., Veisz, O., & Hayes, P. M. (2011). The genetics of winter hardiness in barley: Perspectives from genome-wide association mapping. *Plant Genome, 4*, 76–91.

Waugh, R., Jannink, J.-L., Muehlbauer, G. J., & Ramsay, L. (2009). The emergence of whole genome association scans in barley. *Current Opinion in Plant Biology, 12*, 218–222.

Yu, P., Hart, K., & Du, L. (2009). An investigation of carbohydrate and protein degradation ratios, nitrogen to energy synchronization, and hourly effective rumen digestion of barley: Effect of variety and growth year. *Journal of Animal Physiology and Animal Nutrition, 93*, 555–567.

Chapter 19
The CC-NB-LRR-type *Rdg2a* Resistance Gene Evolved Through Recombination and Confers Immunity to the Seed-Borne Barley Leaf Stripe Pathogen in the Absence of Hypersensitive Cell Death

Chiara Biselli, Davide Bulgarelli, Nicholas C. Collins, Paul Schulze-Lefert, Antonio Michele Stanca, Luigi Cattivelli, and Giampiero Valè

Abstract Leaf stripe disease on barley is caused by the seed-transmitted hemi-biotrophic fungus *Pyrenophora graminea*. Race-specific resistance to leaf stripe is controlled by two known *Rdg* (*r*esistance to *D*rechslera *g*raminea) genes: the *H. spontaneum*-derived *Rdg1a*, mapped to chromosome 2HL and *Rdg2a*, identified in *H. vulgare*, mapped on chromosome 7HS. Both resistance genes have been extensively used in classical breeding. The positional cloning and molecular characterization of the *Rdg2a* locus is described here. BAC and cosmid libraries, respectively, derived from barley cvs. *Morex* (susceptible to leaf stripe) and *Thibaut* (the donor of the *Rdg2a* allele) were used for physical mapping of *Rdg2a*. At the *Rdg2a* locus, three sequence-related coiled-coil, nucleotide-binding site and leucine-rich repeat (CC-NB-LRR) encoding genes were identified. Sequence comparisons suggested that paralogs of this resistance locus evolved through recent gene duplication and were subjected to frequent sequence exchange. Transformation of the leaf stripe susceptible cv. *Golden Promise* with two *Rdg2a* candidates identified a member of the CC-NB-LRR gene family that conferred resistance against the *Dg2* leaf stripe isolate, towards

---

*Presenting author, Antonio Michele Stanca

C. Biselli • A.M. Stanca • G. Valè (✉)
Genomic Research Center, CRA-GPG,
Via S. Protaso 302, 29017, Fiorenzuola d'Arda, (PC), Italy
e-mail: michele@stanca.it; giampiero.vale@ entecra.it

L. Cattivelli
CRA Genomics Research Centre, Via San Protaso 302,
29017 Fiorenzuola d'Arda (PC), Italy

D. Bulgarelli • P. Schulze-Lefert
Department of Plant Microbe Interactions, Max Planck Institute für Züchtungsforschung,
Carl-von-Linné-Weg 10, 50829 Köln, Germany

N.C. Collins
Australian Centre for Plant Functional Genomics, School of Agriculture Food and Wine,
University of Adelaide, Glen Osmond, SA 5064, Australia

G. Zhang et al. (eds.), *Advance in Barley Sciences: Proceedings of 11th International Barley Genetics Symposium*, DOI 10.1007/978-94-007-4682-4_19,

which the *Rdg2a* gene is effective. Histological analysis demonstrated that *Rdg2a*-mediated leaf stripe resistance prevents pathogen colonisation in the embryos without any detectable hypersensitive cell death response, indicating an unusual resistance mechanism for a CC-NB-LRR protein.

Keywords Leaf stripe • Resistance genes • Barley • Hypersensitive response • Embryo

19.1 Introduction

Leaf stripe disease on barley (*H. vulgare*), caused by the seed-transmitted hemibiotrophic fungus *Pyrenophora graminea* (anamorph *Drechslera graminea*), leads severe yield reductions at high infection rates, especially in organic farming systems (Delogu et al. 1995; Mueller et al. 2003). The fungal mycelia survive in the pericarp, in the hull and the seed coat, but not in the embryo (Platenkamp 1976). During seed germination, the hyphae begin to grow intercellularly within the coleorhizae and then into the embryo structures, the roots and scutellar node, to establish infection in the seedling. During this first colonisation phase, the pathogen behaves as a biotroph and degrades host-cell walls using hydrolytic enzymes without causing cellular necrosis (Hammouda 1988; Haegi et al. 2008; Platenkamp 1976). Once infection spreads into the young leaves, growth switches to a necrotrophic phase with the production of a host-specific glycosyl toxin (Haegi and Porta-Puglia 1995) that causes longitudinal dark brown necrotic stripes between the leaf veins, as well as spike sterility. Spores produced on the infected leaves of susceptible plants spread to infect nearby plant spikes.

To date, only two *P. graminea* race-specific resistance genes are known: *Rdg1a* (*r*esistance to *D*rechslera *g*raminea) and *Rdg2a*. These genes are responsible of hyphal degeneration in the basal part of the coleorhiza and prevent stripe symptoms from appearing on leaves of young or old plants (Bulgarelli et al. 2004; Haegi et al. 2008; Platenkamp 1976). *H. spontaneum*-derived *Rdg1a* has been mapped to the long arm of chromosome 2H (Biselli et al. 2010) while *Rdg2a*, identified in *H. vulgare*, has been mapped on the short arm of chromosome 7HS (Tacconi et al. 2001). Both resistance genes have been extensively used in classical breeding.

Histological characterization of the *Rdg2a*-dependent resistance response (Haegi et al. 2008) showed the termination of *P. graminea* growth at the scutellar node and basal region of provascular tissue of barley embryos. The immune response was associated with cell wall reinforcement through accumulation of phenolic compounds and enhanced transcription of genes involved in *r*eactive *o*xygen *s*pecies (ROS) production and detoxification/protection, but no localised *p*rogrammed *c*ell *d*eath (PCD), which is typically seen in race-specific immune responses (Zhou et al. 2001), was apparent.

In this study, the cloning of *Rdg2a*, the molecular characterization of the *Rdg2* locus and the possible *Rdg2a* resistance functions are described.

19.2 Materials and Methods

19.2.1 Plant and Fungal Materials

93 F_2 recombinants for the 3.47 cM *Rdg2a* marker interval ABG704-ScOPQ9 were used for the genetic mapping (Bulgarelli et al. 2004). NIL3876 contains the *Rdg2a* gene from Thibaut backcrossed into the Mirco genetic background (Tacconi et al. 2001). The susceptible variety Golden Promise was used for transformation tests. The leaf stripe isolates Dg2, the most virulent in a collection of monoconidial isolates (Gatti et al. 1992), and Dg5 were used. Fungi were grown on PDA (Liofilchem, Italy), in Petri dishes at 20°C for 10 days in the dark. Seeds were infected using the "sandwich" technique (Pecchioni et al. 1996).

19.2.2 Transgenic Barley Lines

Nbs1-Rdg2a and *Nbs2-Rdg2a*, with their native promoter and terminator regions, were PCR-amplified from cosmid 95-9-3 (*Nbs1-Rdg2a*) and cosmid 17-1-1 (*Nbs2-Rdg2a*) using Phusion HF Taq DNA polymerase. The 6-kb amplicons were subcloned in pDONR201 (Invitrogen) and transferred to pWBVec8 (Invitrogen). Immature embryos of cv. Golden Promise were agro-infiltrated. Transgenes were detected by PCR with a gene specific primer pair that amplified fragments of different dimension in Thibaut and Golden Promise.

19.2.3 TUNEL Assay

Sections of inoculated (14, 22 and 26 dai (days after infection)) and control embryos of NIL3876 were used for histological analyses. The TUNEL (terminal deoxynucleotidyl *t*ransferase-mediated d*U*TP *n*ick *e*nd *l*abelling) assay was performed following manufacturer's instructions (Roche Diagnostics, Germany). Three independent replicate experiments were performed, and for each experiment, six embryos were observed per time point of inoculation. A negative control was provided by omitting terminal deoxynucleotidyl transferase and a positive control was provided by treating samples with DNaseI. Samples were observed with an Olympus BX51 microscope with excitation at 451–490 nm and emission at 491–540 nm for fluorescein, or excitation at 335–380 nm and emission at >420 nm for autofluorescence.

19.2.4 Additional Materials and Methods

Additional information on genetic materials and molecular biology procedures are available on Bulgarelli et al. (2010).

19.3 Results

19.3.1 Genetic and Physical Map of the Rdg2a Locus

Leaf stripe isolate Dg2, which is recognised by *Rdg2a* (Tacconi et al. 2001), is virulent on cv. Morex, indicating that this cv. does not contain a functional *Rdg2a* allele. However, due to the availability of a Morex BAC library (Yu et al. 2000), we took advantage of this resource for marker development. Screening of the library with a probe derived from the CAPS marker MWG851 (0.07 cM from *Rdg2a*; Bulgarelli et al. 2004) allowed the identification of BAC clones 146G20, 244G14 and 608H20 that were subjected to end and low-pass (0.3-fold) shotgun sequencing, and nine additional molecular markers were identified (Fig. 19.1a). Two of these (146.60-1-2 and 146.9-5-6) showed complete linkage with *Rdg2a* (Bulgarelli et al. 2010). To clone the region containing the *Rdg2a* resistance gene, a genomic cosmid library of the *Rdg2a*-containing cv. Thibaut was constructed and screened using markers 146.9-5-6 and 608.32-3-4, leading to the identification of clones 95-3-3 and 17-1-1. Analysis of these two clones with other PCR markers from the region indicated that the clones spanned the *Rdg2a* interval bounded by the closest flanking genetic markers (Fig. 19.1c). The two cosmids, which overlapped by 5.9 kb, were sequenced, providing a contiguous sequence of 72,630 bp. In BLASTX analyses, the region was shown to contain three genes similar to plant *R* genes encoding NB-LRR proteins (GenBank accession number HM124452). The three NB-LRR-encoding genes were designated *Nbs1-Rdg2a*, *Nbs2-Rdg2a* and *Nbs3-Rdg2a* with their relative locations shown in Fig. 19.1c.

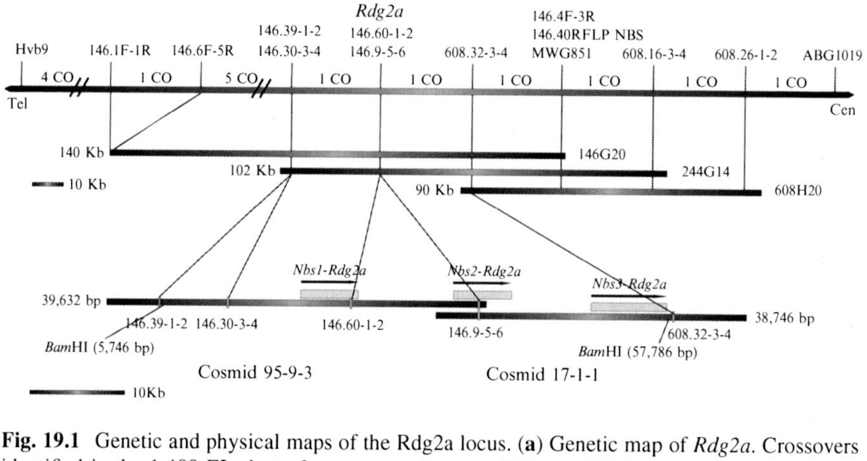

Fig. 19.1 Genetic and physical maps of the Rdg2a locus. (**a**) Genetic map of *Rdg2a*. Crossovers identified in the 1,400 F2 plants from a cross between Thibaut (*Rdg2a*) and Mirco (*rdg2a*) are shown at the *top* (CO). Orientation is indicated by Tel (telomere) and Cen (centromere). (**b**) Contig of Morex BAC clones. (**c**) Thibaut cosmid contig and genes at the *Rdg2a* locus. Transcription directions of the genes are indicated by *arrows*

19.3.2 Structure of Rdg2a Candidate Genes

All the three *Rdg2a* candidates were found to be transcribed in resistant embryos and the transcript structures were determined by random amplification of cDNA ends (RACE) and RT-PCR (Bulgarelli et al. 2010). *Nbs1-Rdg2a* and *Nbs2-Rdg2a* encodes predicted full-length NB-LRR protein products of 1,232 and 1,158 amino acids, respectively. The *Nbs3-Rdg2a* transcript contained a repeat structure, comprising similarity to a full-length NB-LRR protein followed by similarity to part of a NB domain and a full LRR domain. However, several observations led us to conclude that *Nbs3-Rdg2a* encodes only predicted truncated proteins and was excluded from being an *Rdg2a* candidate (Bulgarelli et al. 2010).

Apart from the major structural differences, the ORFs of the three genes were 87–90% identical to one another and 81–86% identical and 91–93% similar at protein level. Comparisons of the 5′ untranscribed regions showed that *Nbs2-Rdg2a* and *Nbs3-Rdg2a* were 93% identical in the 1,040 bp-long region upstream the transcription start site, apart from a 347-bp insertion in *Nbs2-Rdg2a*, located at 145 bp from the ATG. These findings suggest that the *Rdg2a* locus arose by gene duplication.

Semi-quantitative RT-PCR was performed using primer combinations specific for the *Nbs1-Rdg2a* and *Nbs2-Rdg2a* genes in either cv. Mirco or NIL3876-*Rdg2a* (Fig. 19.2a). In the susceptible cv. Mirco, neither gene showed detectable expression in embryos or leaves. In NIL3876-*Rdg2a*, expression of both genes was observed in infected and uninoculated control embryos and in leaves of pathogen-free plants. Quantitative RT-PCRs, performed comparing the transcription levels of the genes in control and inoculated embryos of NIL3876-*Rdg2a* at five time points (7, 14, 18, 22 and 28 dai (*d*ay *a*fter *i*nfection)), indicate that *Nbs2-Rdg2a* expression was significantly increased by inoculation at 7, 14 and 18 dai ($P<0.05$) and was unresponsive by 22 dai; while *Nbs1-Rdg2a* expression was not appreciably altered by leaf stripe inoculation (Fig. 19.2b).

19.3.3 Identification of Rdg2a

Genomic clones of the two *Rdg2a* candidates containing their native 5′ and 3′ regulatory sequences were used to transform the leaf stripe susceptible barley cv. Golden Promise (Bulgarelli et al. 2010). Within T_1 families, resistance to the same isolate co-segregated with the *Nbs1-Rdg2a* transgene and its expression (Fig. 19.3). These lines were susceptible to leaf stripe isolate Dg5, which is not recognised by *Rdg2a*. T_1 lines containing the *Nbs2-Rdg2a* transgene were fully susceptible to both the leaf stripe isolates, although RT-PCRs confirmed the transgene was expressed (Bulgarelli et al. 2010). As the *Nbs1-Rdg2a* gene could confer the same resistance specificity as *Rdg2a* in transgenic plants, we concluded that *Nbs1-Rdg2a* is *Rdg2a*.

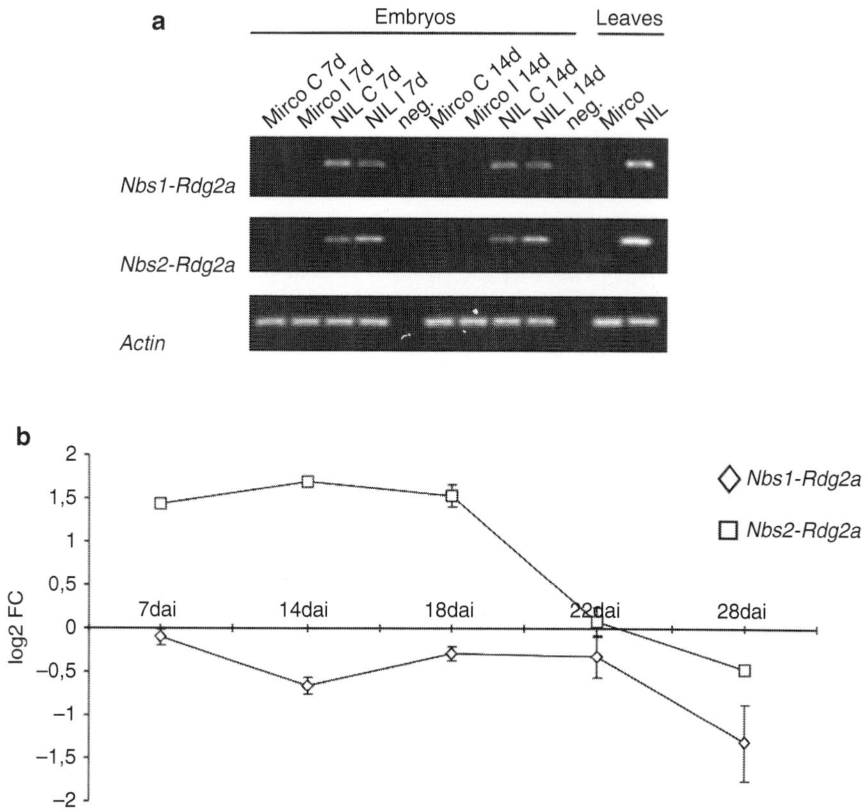

Fig. 19.2 Analysis of the *Rdg2a*-candidate gene transcript regulation. (**a**) semi-quantitative RT-PCR analysis of the *Rdg2a*-candidate gene expression using gene-specific primers. (**b**) Quantitative RT-PCR at 7, 14, 18, 22 and 28 dai for the two *Rdg2a* candidates in embryos of NIL3876. Values are expressed as $\log_2$ fold changes of transcript levels in the inoculated samples with respect to the transcript levels in un-inoculated barley embryos. Error bars represent SD across all RT-PCR replicates

19.3.4 The RDG2A Protein

The predicted *Rdg2a* product of 1,232 amino acids has an estimated molecular weight of 139.73 kDa. It contains all the conserved NB domain motifs of NB-LRR proteins defined by Meyer and co-workers (1999, 2003).

The RDG2A and NB2-RDG2A proteins are 75.3% identical. Differences include a deletion of three consecutive LRRs in NB2-RDG2A. Similarity is higher in the CC region than in the NB or LRR domains (92.6 versus 73–74%), and the proportion of non-conservative amino acid substitutions is lower in the NB domain (75/104 = 72%) than in the LRR domain (57/71 = 80%). Similarly, the ratio of non-synonymous (*Ka*) to synonymous (*Ks*) nucleotide substitutions between *Rdg2a*, *Nbs2-Rdg2a*

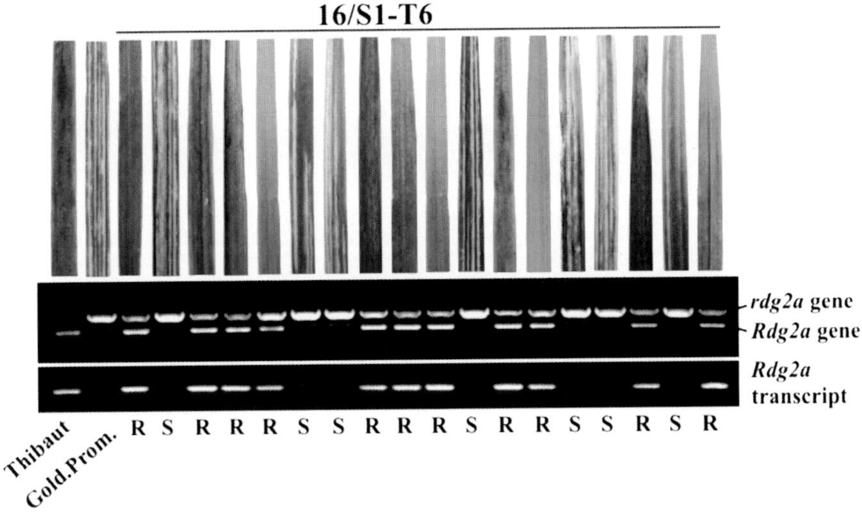

Fig. 19.3 Analysis of T1 family 16/S1-T6 segregating for the *Nbs1-Rdg2a* transgene. T_1 seeds were inoculated with *P. graminea* isolate Dg2 and plants analysed for disease symptoms in leaves (*upper panel*), an STS marker for *Rdg2a* (*middle panel*; upper band represents the *rdg2a* susceptibility allele from cv. Golden Promise while the lower band represents the *Rdg2a* transgene or endogenous gene) and *Rdg2a* transgene or endogenous gene expression by RT-PCR (*lower panel*). Resistance (R) or susceptibility (S) status of the plants is indicated underneath

and *Nbs3-Rdg2a* (longest ORF) is 0.99, 2.13 and 2.63 for the CC, NB and LRR regions, respectively. These comparisons indicate that *Rdg2a* and its paralogues have been subjected to the highest level of diversifying selection in the LRR-coding region, consistent with the LRR domain being an important determinant of resistance specificity (DeYoung and Innes 2006).

19.3.5 Rdg2a Resistance Does Not Involve Hypersensitive Cell Death

Rdg2a-mediated resistance involves cell wall-associated host cell autofluorescence in tissues containing hyphae (Haegi et al. 2008). Since whole-cell autofluorescence is regarded as an indicator of *h*ypersensitive *r*esponse (HR) in race-specific resistance to powdery mildew (Görg et al. 1993; Hückelhoven et al. 1999), we used terminal deoxynucleotidyl *t*ransferase-mediated d*U*TP *n*ick *e*nd *l*abelling (TUNEL) to verify whether serial sections of NIL3876-*Rdg2a* barley embryos inoculated with leaf stripe showed HR-reacting tissues (Fig. 19.4). In non-inoculated embryos, no autofluorescence was observed under UV light (Bulgarelli et al. 2010). In inoculated embryos, UV-autofluorescent tissues were observed at the scutellar node and provascular tissue at 14, 22 and 26 dai (Fig. 19.4g, h and i). TUNEL analysis

14 dai

22 dai

26 dai

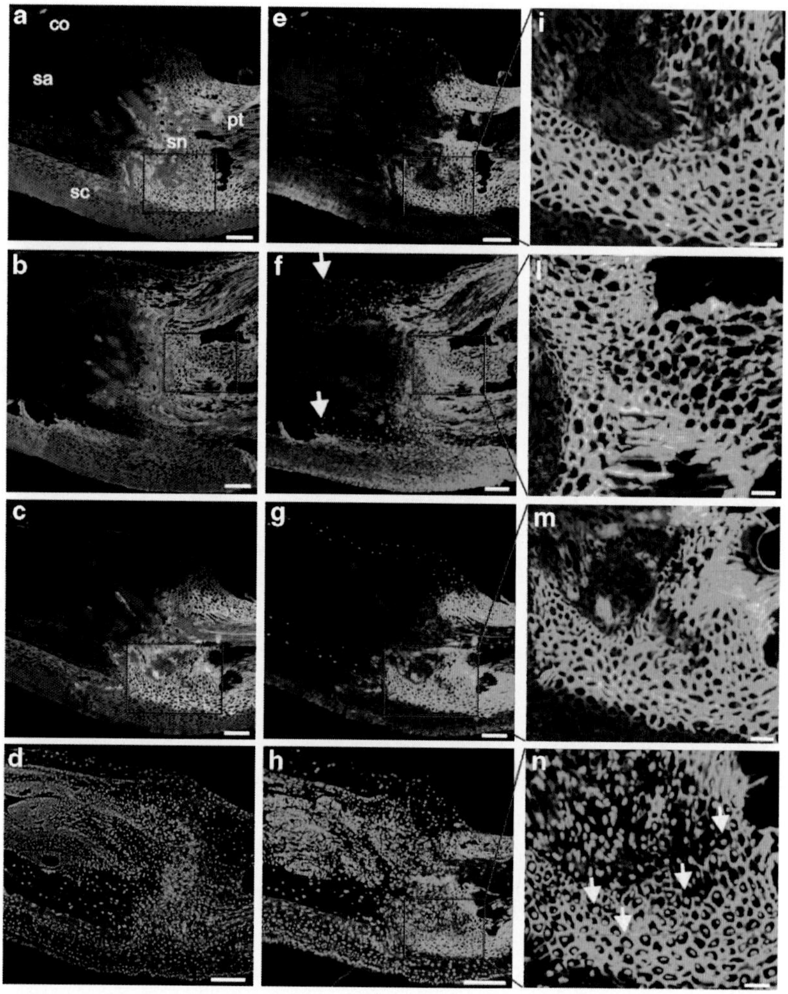

Fig. 19.4 TUNEL assay. (**a**)–(**c**) sections of embryos inoculated with leaf stripe isolate Dg2 and observed under UV excitation. (**e**)–(**g**) sections in (**a**)–(**c**) subjected to TUNEL analysis; the bright fluorescence at the level of scutellar node and provascular tissue is due to cell wall autofluorescence. (**i**)–(**m**) magnified views of the boxed regions in (**a**)–(**c**) and (**e**)–(**g**). (**d**) and (**h**), respectively, sections of control and inoculated embryos at 26 dai, treated with DNaseI and subjected to TUNEL analysis. (**n**) A magnified view of the region boxed in (**h**). *White arrows* in (**f**) and (**n**) indicate TUNEL positive nuclei. Scale bars represent 200 μM (**a**–**h**) and 50 μM (**i**–**n**) (*co* coleoptile, *pt* provascular tissue, *sa* shoot apex, *sn* scutellar node)

revealed basically no TUNEL positive nuclei in the inoculated samples. Moreover, inoculation had no detectable effects on the frequency of the TUNEL signals when non-inoculated and inoculated samples were compared (Fig. 19.4m–o; Bulgarelli et al. 2010).

19.4 Discussion

Rdg2a resides in a gene cluster, as does many other resistance genes. This organisation can promote unequal recombination, which results in sequence exchange between paralogs and generation of recombinant genes with new resistance specificities, as well as expansion/contraction of gene copy number (Leister 2004). At the *Rdg2a* locus, paralogs appear to be the result of relatively recent gene duplication as indicated by the strong DNA sequence identity between the three genes. The unusual structure of *Nbs3-Rdg2a*, in which sequences encoding part of the NB and the LRR regions are duplicated, together with the deletion of the region containing three complete LRR units in NB2-RDG2A relative to RDG2A, provide further examples of variation at *Rdg2* locus generated by recombination. Comparing the sequences of the putative RDG2A and NB2-RDG2A proteins, we found that the *Rdg2a* LRR-encoding domain is subjected to diversifying selection (Bulgarelli et al. 2010). This finding is consistent with a model in which the *R* gene co-evolves with the pathogen effector(s) gene, due to direct interaction of the two gene products.

Strikingly, neither *Nbs1-rdg2a* nor *Nbs2-rdg2a* was transcribed in the susceptible cv. Mirco. Given the fitness cost of expressing some *R* genes (Tian et al. 2003), unnecessary *R* genes may become rapidly inactivated (Michelmore and Meyers 1998). Rearrangements in the promoter region caused by insertion/deletion of transposable elements may explain the lack of expression of the Mirco genes (Bulgarelli et al. 2010).

While the *Rdg2a* resistance allele from cv. Thibaut is used in breeding and still provides useful field resistance against leaf stripe disease, it is not effective against all isolates (Gatti et al. 1992). Therefore, identification of further alleles with different resistance specificity should have value, by broadening the range of resistance genes available to breeders and thus delaying the spread of virulent isolates. The cloning of *Rdg2a* should facilitate this task, by enabling sequencing and expression analysis of homologues from both wild and cultivated barley.

While HR is a common component of resistance gene-mediated defence, there are a few known cases of NB-LRR genes conferring resistance without PCD, at least based on the failure to observe macroscopically visible host cell death (Bendahmane et al. 1999; Bieri et al. 2004; Gassmann 2005; Freialdenhoven et al. 1994). In the current study, it was observed that inoculation did not increase the frequency or distribution of TUNEL signals. Autofluorescence at the junction of the scutellum and scutellar node regions was observed in the response to leaf stripe, but was essentially confined to the cell walls and only occasionally observed throughout a whole cell (Bulgarelli et al. 2010; Haegi et al. 2008). HR deprives obligate biotrophic pathogens of living host cells required for successful colonisation, but may be favourable to the hemi-biotrophic leaf stripe pathogen, which obtains nutrients at latter stages of colonisation by means of hydrolytic degradation of host cell walls. *Rdg2a* resistance terminates *P. graminea* mycelium growth at the scutellar node and basal regions of provascular tissue of barley embryos and is associated with the accumulation of phenolic compounds in the cell walls of the invaded host tissues. These phenolic compounds are the likely source of the cell wall localised autofluorescence. We therefore propose that inducible secretory immune responses, leading to physical and chemical barriers to infection in the cell walls and intercellular spaces of the barley embryo tissues, represent mechanisms by which the CC-NB-LRR-encoding *Rdg2a* gene mediates resistance to leaf stripe.

19.5 Conclusion (and Future Work)

In this study, we were able to clone and characterise the first seed-borne disease resistance gene: the *Rdg2a* gene that confers resistance against the hemi-biotrophic fungus *Pyrenophora graminea* isolate Dg2. To date, only Thibaut *Rdg2a* and Vada *Rdg1a* are used in breeding programmes, but they are not effective towards all isolates (Biselli et al. 2010; Bulgarelli et al. 2010; Gatti et al. 1992). The complete elucidation of the whole barley-*P. graminea* interactions could sustain the crop improvement by taking in consideration the factors that act in this pathosystem in order to obtain new practical applications to contrast other isolates. The new rapid and cheap methods of sequencing based on the next-generation sequencing represent a powerful tool to extensively analyse the genomes of crop species. In this view, the *Rdg2a* sequence could be the starting point for allele mining experiments on wild and cultivated barley varieties to better understand the mechanisms at the base of the evolution of the *Rdg2a* locus and, mainly, to identify further alleles with different specificity to *P. graminea* isolates.

Acknowledgements We thank Sabine Haigis (Max Planck Institute für Züchtungsforschung) and Donata Pagani (CRA-GPG) for excellent technical assistance.

References

Bendahmane, A., Kanyuka, K., & Baulcombe, D. C. (1999). The Rx gene from potato controls separate virus resistance and cell death responses. *The Plant Cell, 11*, 781–791.

Bieri, S., Mauch, S., Shen, Q. H., Peart, J., Devoto, A., Casais, C., Ceron, F., Schulze, S., Steinbiß, H. H., Shirasu, K., & Schulze-Lefert, P. (2004). RAR1 positively controls steady state levels of barley MLA resistance proteins and enables sufficient MLA6 accumulation for effective resistance. *The Plant Cell, 16*, 3480–3495.

Biselli, C., Urso, S., Bernardo, L., Tondelli, A., Tacconi, G., Martino, V., Grando, S., & Valè, G. (2010). Identification and mapping of the leaf stripe resistance gene *Rdg1a* in Hordeum spontaneum. *Theoretical and Applied Genetics, 120*, 1207–1218.

Bulgarelli, D., Collins, N. C., Tacconi, G., Dall'Aglio, E., Brueggeman, R., Kleinhofs, A., Stanca, A. M., & Valè, G. (2004). High-resolution genetic mapping of the leaf stripe resistance gene Rdg2a in barley. *Theoretical and Applied Genetics, 108*, 1401–1408.

Bulgarelli, D., Biselli, C., Collin, N. C., Consonni, G., Stanca, A. M., Schulze-Lefert, P., & Valè, G. (2010). The CC-NB-LRR-Type Rdg2a Resistance gene confers immunity to the seed-borne barley leaf stripe pathogen in the absence of hypersensitive cell death. *PLoS One, 5*, e12599.

Delogu, G., Porta-Puglia, A., Stanca, A. M., & Vannacci, G. (1995). Interaction between barley and Pyrenophora graminea: An overview of research in Italy. *Rachis, 14*, 29–34.

DeYoung, B. J., & Innes, R. W. (2006). Plants NBS-LRR proteins in pathogen sensing and host defense. *Nature Immunology, 7*, 1243–1249.

Freialdenhoven, A., Scherag, B., Hollricher, K., Collinge, D. B., Thordal-Christensen, H., & Schulze-Lefert, P. (1994). Nar-1 and Nar-2, two loci required for Mla12-specified race-specific resistance to powdery mildew in barley. *The Plant Cell, 6*, 983–994.

Gassmann, W. (2005). Natural variation in the Arabidopsis response to the avirulence gene hopPsyA uncouples the hypersensitive response from disease resistance. *Molecular Plant-Microbe Interactions, 18*, 1054–1060.

Gatti, A., Rizza, F., Delogu, G., Terzi, V., Porta-Puglia, A., & Vannacci, G. (1992). Physiological and biochemical variability in a population of Drechslera graminea. *Journal of Genetics and Breeding, 46*, 179–186.

Görg, R., Hollricher, K., & Schulze-Lefert, P. (1993). Functional analysis and RFLP-mediated mapping of the Mlg resistance locus in barley. *The Plant Journal, 3*, 857–866.

Haegi, A., & Porta-Puglia, A. (1995). Purification and partial characterization of a toxic compound produced by Pyrenophora graminea. *Physiological and Molecular Plant Pathology, 46*, 429–444.

Haegi, A., Bonardi, V., Dall'Aglio, E., Glissant, D., Tumino, G., Collins, N. C., Bulgarelli, D., Infantino, A., Stanca, A. M., Delledonne, M., & Valè, G. (2008). Histological and molecular analysis of Rdg2a barley resistance to leaf stripe. *Molecular Plant Pathology, 9*, 463–478.

Hammouda, A. M. (1988). Variability of Drechslera graminea, the causal fungus of leaf stripe of barley. *Acta Phytopathologica Academiae Scientiarum Hungaricae, 23*, 73–80.

Hückelhoven, R., Fodor, J., Preis, C., & Kogel, K. H. (1999). Hypersensitive cell death and papilla formation in barley attacked by the powdery mildew fungus are associated with hydrogen peroxide but not with salicylic acid accumulation. *Plant Physiology, 119*, 1251–1260.

Leister, D. (2004). Tandem and segmental gene duplication and recombination in the evolution of plant disease resistance genes. *Trends in Genetics, 20*, 116–122.

Meyers, B. C., Dickerman, A. W., Michelmore, R. W., Sivaramakrishnan, S., Sobral, B. W., & Young, N. D. (1999). Plant disease resistance genes encode members of an ancient and diverse protein family within the nucleotide-binding superfamily. *The Plant Journal, 20*, 317–332.

Meyers, B. C., Kozik, A., Griego, A., Kuang, H., & Michelmore, R. W. (2003). Genome-wide analysis of NBS-LRR-encoding genes in Arabidopsis. *The Plant Cell, 15*, 809–834.

Michelmore, R. W., & Meyers, B. C. (1998). Clusters of resistance genes in plants evolve by divergent selection and a birth-and-death process. *Genome Research, 8*, 1113–1130.

Mueller, K. J., Valè, G., & Enneking, D. (2003). Selection of resistant spring barley accessions after natural infection with leaf stripe (*Pyrenophora graminea*) under organic farming conditions in Germany and by sandwich test. *Journal of Plant Pathology, 85*, 9–14.

Pecchioni, N., Faccioli, P., Toubia-Rahme, H., Valè, G., & Terzi, V. (1996). Quantitative resistance to barley leaf stripe (*Pyrenophora graminea*) is dominated by one major locus. *Theoretical and Applied Genetics, 93*, 97–101.

Platenkamp, R. (1976). Investigations on the infections pathway of Drechslera graminea in germinating barley. In Royal Veterinary and Agricultural University, Yearbook. pp. 49–64.

Tacconi, G., Cattivelli, L., Faccini, N., Pecchioni, N., Stanca, A. M., & Valè, G. (2001). Identification and mapping of a new leaf stripe resistance gene in barley (*Hordeum vulgare* L.). *Theoretical and Applied Genetics, 102*, 1286–1291.

Tian, D., Traw, M. B., Chen, J. Q., Kreitman, M., & Bergelson, J. (2003). Fitness costs of R-gene-mediated resistance in Arabidopsis thaliana. *Nature, 423*, 74–77.

Yu, Y., Tomkins, J. P., Waugh, R., Frisch, D. A., Kudrna, D., Kleinhofs, A., Brueggeman, R. S., Muehlbauer, G. J., Wise, R. P., & Wing, R. A. (2000). A bacterial artificial chromosome library for barley (*Hordeum vulgare* L.) and the identification of clones containing putative resistance genes. *Theoretical and Applied Genetics, 101*, 1093–1099.

Zhou, F., Kurth, J., Wei, F., Elliot, C., Valè, G., Yahiaoui, N., Keller, B., Somerville, S., Wise, R., & Schulze-Lefert, P. (2001). Cell-autonomous expression of barley Mla1 confers race-specific resistance to the powdery mildew fungus via a Rar1-independent signaling pathway. *The Plant Cell, 13*, 337–350.

Chapter 20
Increased Auxin Content and Altered Auxin Response in Barley Necrotic Mutant *nec1*

Anete Keisa, Ilva Nakurte, Laura Kunga, Liga Kale, and Nils Rostoks

Abstract The role of hormone crosstalk in plant immunity is lately emerging as significant topic of plant physiology. Although crosstalk between salicylic acid and auxin affects plant disease resistance, molecular mechanisms of this process have not yet been uncovered in details. Mutations disrupting cyclic nucleotide-gated ion channel 4 (CNGC4) affect SA-mediated disease resistance in barley *Hordeum vulgare* and in *A. thaliana*. Significantly, decreased stomatal apertures of barley CNGC4 mutant *nec1* and dwarfed stature of *A. thaliana* CNGC4 mutant *dnd2* suggest that nonfunctional CNGC4 might be affecting also auxin signaling. Excised coleoptile elongation, stomatal conductance, and cell size measurements assaying physiological effect of exogenous auxin treatment suggested altered auxin signaling in *nec1* mutant. Real-time qPCR analysis identified significant change in mRNA abundance of four auxin-related genes – *YUCCA1, VT2, HVP1*, and *TIR1*. Analysis of endogenous auxin content of *nec1* plants detected ca. fourfold increase in indole acetic acid (IAA) content in *nec1* leaves and roots compared to wt plants, as measured by HPLC. These results suggest that apart from SA-related disease resistance, CNGC4 functions also in auxin signaling in barley; therefore, barley *nec1* mutant could serve as model system revealing role of SA-auxin crosstalk in plant disease resistance.

Keywords Barley • *nec1* mutant • Lesion mimic mutant • Auxin • Disease resistance

---

Presenting author, Anete Keisa

A. Keisa • I. Nakurte • L. Kunga • L. Kale • N. Rostoks (✉)
Faculty of Biology, University of Latvia, Riga, Latvia
e-mail: anetekeisa@yahoo.com; nils.rostoks@lu.lv

G. Zhang et al. (eds.), *Advance in Barley Sciences: Proceedings of 11th International Barley Genetics Symposium*, DOI 10.1007/978-94-007-4682-4_20,

20.1 Introduction

Hormone crosstalk in plant immunity is lately emerging as significant topic of plant physiology (Spoel and Dong 2008; Pieterse et al. 2009; Grant and Jones 2009; Verhage et al. 2010). Although well-known antagonistic effect of salicylic acid (SA) and jasmonate/ethylene signaling is certainly a central backbone of hormonal control of plant immunity (Glazebrook 2005), several other phytohormones feed into this central signaling pathway. Auxin has previously been mainly associated with plant growth and development, but recently, it has become apparent that it also plays a significant role in plant immunity (Spaepen and Vanderleyden 2011). For example, resistance of *A. thaliana* to virulent bacteria *Pseudomonas syringae* pv. *tomato* DC3000 (Pst DC3000) requires suppression of auxin signaling, and exogenously applied auxins render *A. thaliana* more susceptible to infection (Navarro et al. 2006). AvrRpt2 alters host's auxin signaling to ensure virulence on *A. thaliana* lacking corresponding R-gene (Chen et al. 2007). Limited evidence suggests that crosstalk between auxin signaling and immunity is operating also in monocots. For example, constitutive expression of auxin-conjugating enzyme GH3-8 in rice promotes resistance against *Xanthomonas oryzae* pv. *oryzae* (Ding et al. 2008) and over-expression of OsWRKY31 enhances resistance to *Magnaporthe grisea* and altered auxin responsiveness of rice. In addition, SGT1, a protein required for auxin signaling and disease resistance in *A. thaliana* (Azevedo et al. 2006), also participates in auxin signaling and disease resistance of rice (Wang et al. 2008). SGT1 is also required for certain R-gene mediated powdery mildew resistance in barley (Azevedo et al. 2002).

Repression of auxin signaling is thought to be a common mechanism employed by SA-mediated disease resistance (Wang et al. 2007). Auxin interferes with *PR*-gene up-regulation in response to SA (Iglesias et al. 2011) and vice versa – SA represses auxin receptors and thereby stabilizes auxin signaling repressors (Wang et al. 2007). Despite the evidence supporting antagonistic effect of auxin and plant disease resistance, auxin signaling components have recently been shown to participate in systemic acquired resistance (Truman et al. 2010) indicating a complexity of mechanisms underlying SA-auxin crosstalk.

SA-auxin crosstalk is also supported by the fact that several plant mutants exhibiting altered disease resistance and overaccumulating SA display phenotype resembling distorted auxin signaling (Wang et al. 2007). Therefore, SA-overaccumulating mutants can help to advance our knowledge of role of SA-auxin crosstalk in plant immunity. Mutations disrupting cyclic nucleotide-gated ion channel 4 (CNGC4) affect SA-mediated signaling and plant resistance in barley (Keisa et al. 2011) and in *A. thaliana* (Balague et al. 2003; Jurkowski et al. 2004). Barley necrotic mutant *nec1* containing defective CNGC4 exhibits increased concentration of SA and H_2O_2 (Keisa et al. 2011). In this study, we show that *nec1* mutation also affects auxin signaling in barley supporting the role of crosstalk between SA and auxin in monocot immunity.

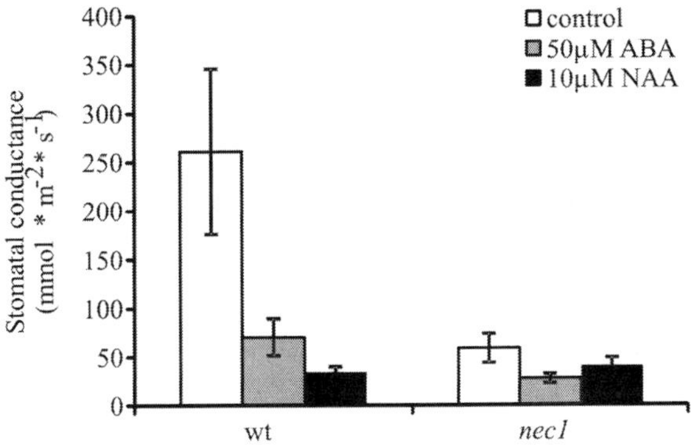

Fig. 20.1 Stomata of *nec1* and parental cv. "Parkland" respond differentially to synthetic auxin 1-naphthalene acetic acid (NAA). Plants were pretreated with 50 μM abscisic acid (ABA) and after that subjected to treatment with 10 μM NAA. Error bars represent the standard deviation of means

20.2 Materials and Methods

20.2.1 Plant Material

All experiments were conducted with barley necrotic mutant GSHO 1284 (further in text – *nec1*) carrying mutant allele *nec1.c*. *nec1.c* allele contains an insertion of a transposable element within an intron of the gene that causes alternative splicing and predicted nonfunctional protein coding sequence (Rostoks et al. 2006). GSHO 1284 is essentially isogenic to its parental variety Parkland (Keisa et al. 2011).

20.2.2 Stomatal Response to Exogenous Auxin

Stomatal responses to exogenous abscisic acid and 1-naphthalene acetic acid (NAA) were determined by measuring stomatal conductance using Li-6400 Portable Photosynthesis System (Li-Cor Inc., USA). Second leaves from 20 plants (five plants per pot) from Parkland as well as *nec1* were used for measurements. MES-KCl buffer (Bright et al. 2006) was sprayed several times on 2-week-old plants to open stomata. After 4 h, stomatal conductance was measured first time (Fig. 20.1, control). Then plants were sprayed with MES + 50 μM ABA solution and measured again. Third measurement was performed after plants were sprayed with MES + 50 μM ABA + 10 μM NAA solution (Fig. 20.1).

20.2.3 Coleoptile Elongation in Response to NAA

Coleoptile elongation was assayed as described by Kotake et al. (2000) with minor modifications. Briefly, 5-mm-long coleoptile segments (3 mm below the tip off the

Table 20.1 Genes and corresponding primers used for qRT-PCR analysis

Gene/harvest 35 unigene	*A. thaliana* homologue	Primer sequence 5′–3′
YUCCA1	*AtYUC1*	F:ATGGAGGTCTCCCTGGACCTGT
6837	AT4G32540	R:TCACCTTGTCCACGAACCAGAG
VT2	*AtTAA1*	F:GTATCCTGCCGTGACGGACTTC
11280	AT1G70560	R:ACTGCGGCCAGTAGTAGGCAAG
NIT2	*AtNIT1*	F:GTACCTGGGTAAGCACCGCAAG
17606	AT3G44310	R:GTGCTGTCCTTAACAGTGGCATC
HVP1	*H⁺PPase*	F:AGCCACAGAATTCGTGAAAGAACTG
534	AT1G15690	R:GCCCAATAAAGACATTAGGTGTCAGG
H⁺-ATP	*H⁺ATPase*	F:TCTGGCTCTTCAGCATTGTG
14211	AT4G30190	R:ACGGTAGCTGCTCTTGTCGT
TIR1	*AtTIR1*	F:TGCCAAGCTGGAGACAATGC
2793	AT3G62980	R:CGGGCTATCATCCGGAAGTG
SGT1	*AtSGT1a*	F:TCTGGTGACTCAAGGTTTACTCGTC
1570	AT4G23570	R:GTGCATTCTCCATATTTTCACCATC
ABP1	*AtABP1*	F:TCCTGTCGCTCAGAATAGCA
19159	AT4G02980	R:CCAGACAAAGGGGAACTTCA
GAPDH		F:CGTTCATCACCACCGACTAC
19159		R:CAGCCTTGTCCTTGTCAGTG

seedling) were excised from 3-day-old barley seedlings grown in dark at 22°C. Excised segments were floated in dark for 4 h in 5 μM NAA solution containing 5 mM KCl, 50 μM $CaCl_2$, and 10 mM MES, pH 6.5.

20.2.4 Gene Expression Analysis Using Real-Time qRT-PCR

For RNA extraction, segments of cotyledon leaf from two-week-old plants of necrotic mutant *nec1* and parental cv. RNA extraction, cDNA synthesis, and quantitative real-time PCR were performed as described in Keisa et al. (2011). Primers used for qRT-PCR analysis are listed in Table 20.1. Relative quantification was performed using $2^{-\Delta\Delta Ct}$ method as described by Livak and Schmittgen (2001). Transcript levels of the studied genes were normalized to *HvGAPDH* transcript value in the same sample.

20.2.5 IAA Detection and Quantification Using HPLC

For leaf tissue samples, seedlings were grown in soil at 22°C under long-day (16 h day, 8 h night), medium-light (ca. 150 μmol m^{-2} s^{-1}) conditions. For root samples, plants were hydroponically grown in Knop's salt solution at 22°C under long-day (16 h day, 8 h night) conditions. Leaf and root samples were taken from 14-day-old plants.

Extraction and purification for auxin was performed as described by (Dobrev and Kaminek 2002) with minor modifications (Nakurte et al. unpublished). Briefly, the plant material was ground in liquid nitrogen and extracted with 100% methanol. Samples were pre-concentrated by SPE using AccuBOND II ODS-C18 200 mg 3 ml SPE (Agilent). Chromatographic analysis was performed on a modular HPLC system, Agilent 1100 series (Agilent Technologies, Germany). HPLC separations were achieved by using a reverse-phase Zorbax Eclipse XDB-C8 (Agilent Technologies, Germany) column 4.6 × 150 mm, 5 μm. Mobile phase composed of methanol and 1% acetic acid (60:40 $v\,v^{-1}$) in isocratic mode at a flow rate of 1 ml min^{-1}. The detection was monitored at 282 nm (Ex) 360 nm (Em) (IAA, IPA). The developed method was validated in terms of accuracy, precision, linearity, limit of detection, limit of quantification, and robustness. Standards of indole-3-acetic acid (>99%) and indole-3-pyruvic acid (99%) were purchased from Sigma-Aldrich (St. Louis, USA).

20.3 Results

20.3.1 nec1 Phenotype Suggests Changes in Auxin Signaling

We performed several physiological auxin sensitivity tests to assay *nec1* auxin response. Since auxin is one of the regulators of stomatal conductance, we measured stomatal conductance of *nec1* and parental line Parkland. *nec1* mutant showed significantly reduced stomatal conductance at otherwise permissive conditions. This prompted us to examine *nec1* stomatal response to exogenously applied synthetic auxin 1-naphthalene acetic acid (NAA). Since auxin is known to inhibit abscisic acid (ABA)-induced stomatal closure (Tanaka et al. 2006), we examined NAA effect on ABA-treated barley leaves. Further reduction of stomatal conductance in *nec1* plants was observed suggesting that *nec1* plants respond to ABA. In contrast to ABA response, *nec1* showed significantly different response to NAA compared to wt plants, that is, NAA reversed the ABA-induced stomatal closure in *nec1* plants, but closed them even further in cv. Parkland (Fig. 20.1).

We also examined three physiological responses traditionally serving as indicators for altered auxin response – root gravitropic response, cell size, and coleoptile elongation. Endogenous auxin regulates cell expansion (Perrot-Rechenmann 2010). We used leaf impression method (Khazaie et al. 2011) to determine epidermal cell size. Epidermal cells from middle part of adaxial surface were measured. Unlike *dnd2* in *A. thaliana*, *nec1* does not exhibit dwarfed stature. However, average perimeter of epidermal cells in *nec1* was somewhat shifted toward lower values (Fig. 20.2a). Root tip curving in response to gravistimulation at 90° to the vertical was not significantly altered in *nec1* (Fig. 20.2b). Coleoptile elongation is one of the tests routinely used for characterizing auxin sensitivity in plants. Exogenously applied auxin triggers excised coleoptile elongation in barley in dose-dependent manner (Kotake et al. 2000). Detached coleoptile test showed differential response to NAA in *nec1* plants as the *nec1* coleoptiles showed higher increase in length following application of exogenous auxin (Fig. 20.2c).

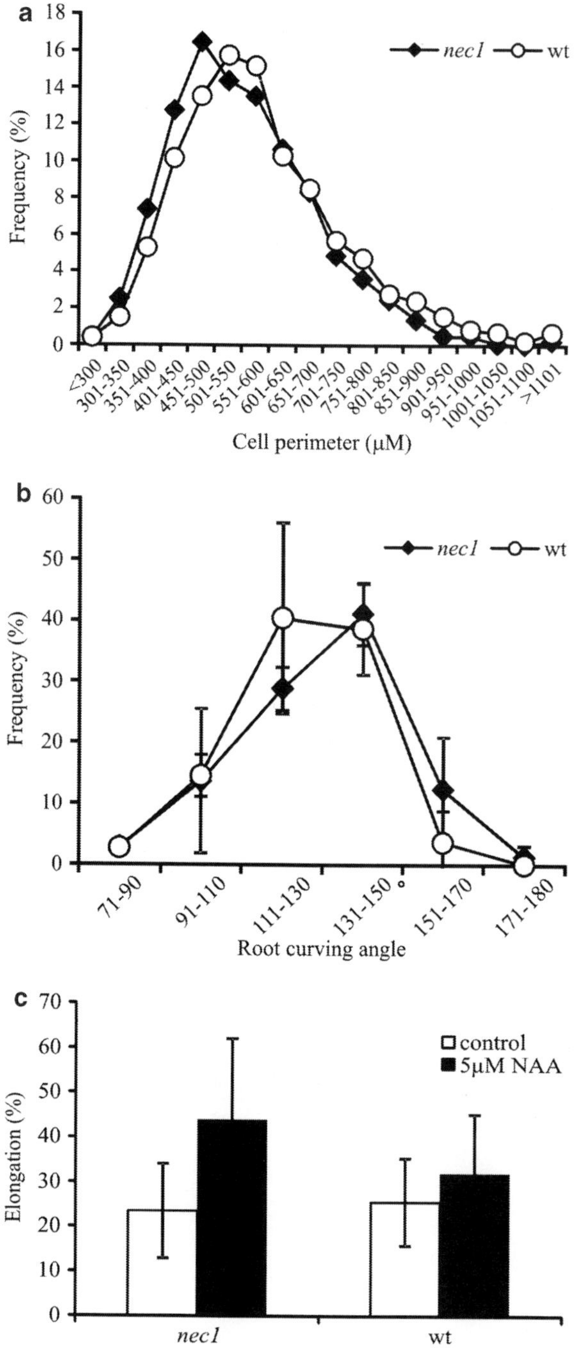

Fig. 20.2 Physiological indicators of auxin response in *nec1*. (**a**) Cell size of *nec1* and wt plants. Perimeter of epidermal cells ($n > 1{,}500$) from 10 plants from *nec1* and wt was measured. (**b**) Angle of root curvature after gravistimulation at 90° to the vertical. Root gravitropic response of *nec1* and wt does not differ significantly. Seedlings were grown for 3 days vertically and then rotated by 90°. Error bars represent the standard deviation of means ($n = 20$ in each – *nec1* and wt). (**c**) Exogenously applied NAA differentially affects coleoptile elongation in *nec1* and wt plants. Results are expressed as percent increase compared to initial segment length. Error bars represent the standard deviation of means

20.3.2 *nec1* Mutation Affects Expression of Auxin Signaling Related Genes

To gain a better understanding of molecular basis of the observed differences in *nec1* physiological auxin response, the expression of several key genes involved in auxin biosynthesis, perception, and downstream signaling was studied using quantitative real-time PCR.

SGT1 genes have been identified as enhancers of the auxin response defect conferred by the *tir1-1* mutation, and the SGT1b protein was shown to be required for the SCF^{TIR1}-mediated degradation of Aux/IAA proteins (Gray et al. 2003). *SGT1* is also required for *MLA*-mediated barley resistance to powdery mildew pathogen *Blumeria graminis* f.sp. *hordei* (Azevedo et al. 2002, 2006). ABP1 (auxin-binding protein 1) was the first protein that was proved to bind auxin directly (Hesse et al. 1989; Teale et al. 2006), although recent discovery of the TIR1 as the auxin receptor (Dharmasiri et al. 2005; Kepinski and Leyser 2005) suggests that it may not be its primary function. Nonetheless, several recent studies suggest it is essential for auxin perception and response (Braun et al. 2008; David et al. 2007; Effendi et al. 2011; Tromas et al. 2009). Moreover, ABP1 has been implicated in stomatal regulation (Gehring et al. 1998). *H^+-ATPase* encodes a plasma membrane proton pump that is dephosphorylated through an ABA-dependent mechanism, in order to cause stomatal closure (Zhang et al. 2004). Salt and osmotic stress has been shown to increase expression of the *H^+-ATPase* (Gaxiola et al. 2007). H^+-ATPase is a central enzyme in the acid growth theory that postulates activation of the enzyme by auxin resulting in extrusion of H^+, lowering of the extracellular pH, and loosening of the cell wall that permits cell expansion (Kutschera 2006; Lüthen et al. 1990; Schenck et al. 2010). *TIR1* encodes a protein that is a part of SCF^{TIR1} ubiquitin-ligase complex involved in degradation of Aux/IAA proteins and serves as an auxin receptor (Dharmasiri et al. 2005; Kepinski and Leyser 2005; Tan et al. 2007). *HVP1* encodes a vacuolar H^+-inorganic pyrophosphatase that has been shown to be strongly induced by application of exogenous synthetic auxin 2,4- dichlorophenoxyacetic acid (2,4-D) (Fukuda and Tanaka 2006). *YUC1, VT2*, and *NIT2* were chosen as genes representing different auxin biosynthesis pathways. *YUCCA* genes encode a family of flavin monooxygenases that are rate-limiting enzymes in the auxin biosynthesis pathway through tryptamine in *A. thaliana* (Cheng et al. 2006). In addition, expression of *YUCCA* genes has been correlated with endogenous auxin levels in *A. thaliana* and barley (Sakata et al. 2010). *VT2* gene in maize is an orthologue of *A. thaliana* genes belonging to *TAA* gene family. *ZmVT2* was shown to be involved in auxin biosynthesis through indole-3-pyruvic acid (IPA) pathway (Phillips et al. 2011). *NIT2* gene has been implicated in conversion of indole-3-acetonitrile (IAN) to IAA in the indole-3-acetaldoxime (IAOx) auxin biosynthesis pathway in maize (Kriechbaumer et al. 2007).

The expression of *SGT1, ABP1*, and *H^+-ATPase* genes was not changed in *nec1*; however, *TIR1, HVP1, VT2*, and *YUC1* genes were significantly repressed (Fig. 20.3) suggesting that some of the auxin responses are altered in *nec1* mutant possibly due to the altered endogenous auxin level caused by changes in auxin biosynthesis.

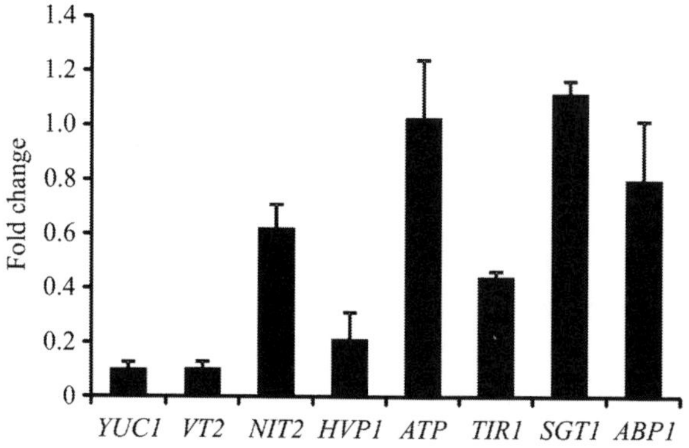

Fig. 20.3 *nec1* mutation affects expression of auxin-related genes in barley. Fold change of transcript abundance in *nec1* relative to wt plants. Error bars represent the standard deviation of means

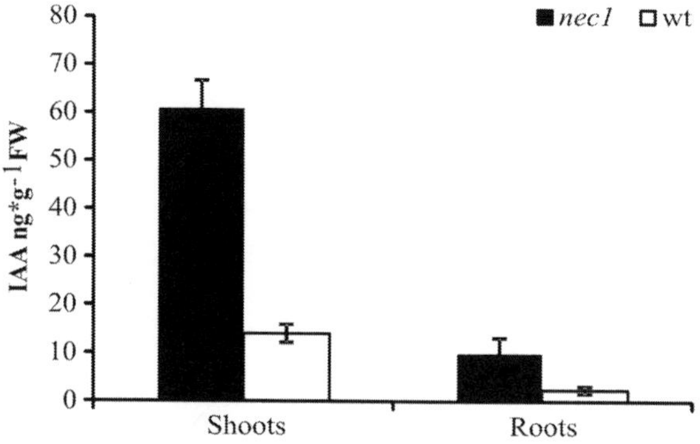

Fig. 20.4 *nec1* exhibits significantly increased endogenous IAA level as quantified by HPLC. Error bars represent the standard deviation of means

20.3.3 IAA Concentration in nec1 Is Significantly Increased

Since significant changes in auxin biosynthesis gene expression were identified in *nec1* mutant, we tested if there are any changes in phytohormone concentration in the *nec1* mutant; therefore, we developed a high-performance liquid chromatography (HPLC) method to simultaneously study the auxin and abscisic acid content in barley leaves and roots (Nakurte et al. unpublished). Highly significant increase of the auxin was observed in both shoots and roots of the *nec1* mutant, while the shoot to root ratio was similar in mutant and wt plants (Fig. 20.4).

Similar increase in auxin concentration was observed in allelic mutants of *nec1* – GSHO989 and FN085 – as well as in the orthologous *A. thaliana dnd2* mutant (data not shown), supporting the hypothesis that mutation in the *CNGC4* gene affects auxin concentration.

20.4 Discussion

Although the role of auxin as a plant disease resistance affecting agent has been suspected for almost two decades, molecular mechanisms linking auxin signaling and plant immunity are largely unknown (Kazan and Manners 2009). Antagonistic interaction with SA-mediated signaling is one of the proposed mechanisms explaining the negative effect of auxin on plant disease resistance (Chen et al. 2007; Wang et al. 2007; Iglesias et al. 2011). Interestingly, several SA-overaccumulating mutants are also impaired in auxin signal perception and vice versa – mutations impairing auxin-related signaling also affect disease resistance. For example, *A. thaliana* SA-overaccumulating mutants *cpr6* and *snc1* produce less IAA, display reduced apical dominance, and are less sensitive to exogenous auxin (Wang et al. 2007). Besides, over-expression of auxin-related genes *GH3* and *WRKY31* enhanced disease resistance of rice to *Xanthomonas oryzae* and *Magnaporthe grisea,* respectively (Ding et al. 2008; Zhang et al. 2009a).

CNGC4 has previously been shown to affect SA-mediated disease resistance in barley (Keisa et al. 2011) and *A. thaliana* (Balague et al. 2003; Jurkowski et al. 2004). In this study, we show that SA-overaccumulating barley mutant *nec1* comprising nonfunctional CNGC4 displays physiological responses resembling altered auxin signaling. *nec1* showed altered sensitivity to exogenous auxin as suggested by coleoptile elongation and stomatal conductance. Physiological indicators representing effects of endogenous auxin did not reveal any significant difference between *nec1* and wt plants. *nec1* exhibited somewhat smaller size of epidermal cells and did not show any change in root gravitropic response. Uncertain physiological data prompted us to analyze expression of auxin-elated genes in *nec1*. Expression analysis of auxin-related genes in *nec1* revealed significant repression of *VT2* (*AtTAA1* homologue) and *YUC1* – genes involved in auxin biosynthesis. Although previously, *TAAs* and *YUCCAs* have been suggested to represent two different auxin biosynthesis pathways (IPA and TAM pathway, respectively) (McSteen 2010), recently it has been shown that both *TAAs* and *YUCCAs* are involved in IPA auxin biosynthesis pathway (Won et al. 2011; Mashiguchi et al. 2011). Therefore, significant repression of both *YUC1* and *VT2* genes in *nec1* suggests that IPA pathway of auxin biosynthesis is likely repressed in *nec1* mutant. Low endogenous auxin concentration could explain altered *nec1* sensitivity to low-dose exogenous auxin application and decreased cell size of *nec1*; therefore, we analyzed IAA content of *nec1*. Surprisingly, HPLC analysis identified significantly increased IAA content in *nec1* shoots as well as roots. In view of contradictory results from gene expression and HPLC analyses, further experiments elucidating the exact effect of *nec1* mutation on auxin biosynthesis in plants will be needed. Elevated auxin concentration throughout *nec1* plants suggests that increased sensitivity to auxin is unlikely caused by changes in auxin transport, but rather might result from alterations in auxin biosynthesis or signaling.

Although it still remains to be elucidated how the lack of cyclic nucleotide-gated ion channel 4 in *nec1* leads to increased level of SA and IAA, it might be anticipated that CNGC4 contributes to both pathways through Ca^{2+}/calmodulin regulatory mechanisms.

Although Ca^{2+} permeability of CNGC4 has not been clearly demonstrated (Balague et al. 2003), a related ion channel CNGC2 has been shown to ensure rise of cytosolic Ca^{2+} in response to PAMPs (Ma et al. 2009). Plant CNGCs are negatively regulated by CaM (Kaplan et al. 2007), and HvCNGC4 comprises calmodulin (CaM) binding domain (Rostoks et al. 2006) suggesting that CNGC4 might be involved in Ca^{2+}/CaM-related signaling. Ca^{2+}/CaM have been implicated as secondary messengers in auxin (Yang and Poovaiah 2000), as well as SA signal transduction (Du et al. 2009). Interestingly, another Ca^{2+} permeable channel, calcium exchanger 1 CAX1, has also been shown to affect auxin-related signaling since *cax1* mutants exhibit reduced apical dominance, inhibited lateral root development, and reduced sensitivity to exogenous auxin treatment (Cheng et al. 2003). Moreover, mutations in the barley *HvCAX1* gene provide broad-spectrum resistance to barley stem rust (Zhang et al. 2009b). Taking into account above mentioned evidence, *nec1* may serve as a model system revealing role of SA-auxin-Ca^{2+} crosstalk in plant disease resistance.

Acknowledgments The study was funded by European Social Fund project 2009/0224/1DP/1.1.1.2.0/09/APIA/VIAA/055 and Latvian Council of Science grant Z-6142-090.

References

Azevedo, C., Sadanandom, A., Kitagawa, K., Freialdenhoven, A., Shirasu, K., & Schulze-Lefert, P. (2002). The RAR1 interactor SGT1, an essential component of *R* gene-triggered disease resistance. *Science, 295*, 2073–2076.

Azevedo, C., Betsuyaku, S., Peart, J., Takahashi, A., Noël, L., Sadanandom, A., Casais, C., Parker, J., & Shirasu, K. (2006). Role of SGT1 in resistance protein accumulation in plant immunity. *EMBO Journal, 25*, 2007–2016.

Balague, C., Lin, B., Alcon, C., Flottes, G., Malmstrom, S., Kohler, C., Neuhaus, G., Pelletier, G., Gaymard, F., & Roby, D. (2003). HLM1, an essential signaling component in the hypersensitive response, is a member of the cyclic nucleotide-gated channel ion channel family. *The Plant Cell, 15*, 365–379.

Braun, N., Wyrzykowska, J., Muller, P., David, K., Couch, D., Perrot-Rechenmann, C., & Fleming, A. J. (2008). Conditional repression of AUXIN BINDING PROTEIN1 reveals that it coordinates cell division and cell expansion during postembryonic shoot development in *Arabidopsis* and tobacco. *The Plant Cell, 20*, 2746–2762.

Bright, J., Desikan, R., Hancock, J. T., Weir, I. S., & Neill, S. J. (2006). ABA-induced NO generation and stomatal closure in *Arabidopsis* are dependent on H_2O_2 synthesis. *The Plant Journal, 45*, 113–122.

Chen, Z., Agnew, J. L., Cohen, J. D., He, P., Shan, L., Sheen, J., & Kunkel, B. N. (2007). *Pseudomonas syringae* type III effector AvrRpt2 alters *Arabidopsis thaliana* auxin physiology. *Proceedings of the National Academy of Sciences of the United States of America, 104*, 20131–20136.

Cheng, N. H., Pittman, J. K., Barkla, B. J., Shigaki, T., & Hirschi, K. D. (2003). The *Arabidopsis cax1* mutant exhibits impaired ion homeostasis, development, and hormonal responses and reveals interplay among vacuolar transporters. *The Plant Cell, 15*, 347–364.

Cheng, Y., Dai, X., & Zhao, Y. (2006). Auxin biosynthesis by the YUCCA flavin monooxygenases controls the formation of floral organs and vascular tissues in *Arabidopsis*. *Genes & Development, 20*, 1790–1799.

David, K. M., Couch, D., Braun, N., Brown, S., Grosclaude, J., & Perrot-Rechenmann, C. (2007). The auxin-binding protein 1 is essential for the control of cell cycle. *The Plant Journal, 50*, 197–206.

Dharmasiri, N., Dharmasiri, S., & Estelle, M. (2005). The F-box protein TIR1 is an auxin receptor. *Nature, 435*, 441–445.

Ding, X., Cao, Y., Huang, L., Zhao, J., Xu, C., Li, X., & Wang, S. (2008). Activation of the indole-3-acetic acid–amido synthetase GH3-8 suppresses expansin expression and promotes salicylate- and jasmonate-independent basal immunity in rice. *The Plant Cell, 20*, 228–240.

Dobrev, P. I., & Kaminek, M. (2002). Fast and efficient separation of cytokinins from auxin and abscisic acid and their purification using mixed-mode solid-phase extraction. *Journal of Chromatography A, 950*, 21–29.

Du, L., Ali, G. S., Simons, K. A., Hou, J., Yang, T., Reddy, A. S. N., & Poovaiah, B. W. (2009). Ca^{2+}/calmodulin regulates salicylic-acid-mediated plant immunity. *Nature, 457*, 1154–1158.

Effendi, Y., Rietz, S., Fischer, U., & Scherer, G. F. E. (2011). The heterozygous *abp1/ABP1* insertional mutant has defects in functions requiring polar auxin transport and in regulation of early auxin-regulated genes. *The Plant Journal, 65*, 282–294.

Fukuda, A., & Tanaka, Y. (2006). Effects of ABA, auxin, and gibberellin on the expression of genes for vacuolar H^+ -inorganic pyrophosphatase, H^+ -ATPase subunit A, and Na^+/H^+ antiporter in barley. *Plant Physiology and Biochemistry, 44*, 351–358.

Gaxiola, R. A., Palmgren, M. G., & Schumacher, K. (2007). Plant proton pumps. *FEBS Letters, 581*, 2204–2214.

Gehring, C. A., McConchie, R. M., Venis, M. A., & Parish, R. W. (1998). Auxin-binding-protein antibodies and peptides influence stomatal opening and alter cytoplasmic pH. *Planta, 205*, 581–586.

Glazebrook, J. (2005). Contrasting mechanisms of defense against biotrophic and necrotrophic pathogens. *Annual Review of Phytopathology, 43*, 205–227.

Grant, M. R., & Jones, J. D. G. (2009). Hormone (dis)harmony moulds plant health and disease. *Science, 324*, 750–752.

Gray, W. M., Muskett, P. R., Chuang, H., & Parker, J. E. (2003). Arabidopsis SGT1b is required for SCF^{TIR1}-mediated auxin response. *The Plant Cell, 15*, 1310–1319.

Hesse, T., Feldwisch, J., Balshusemann, D., Bauw, G., Puype, M., Vandekerckhove, J., Lobler, M., Klambt, D., Schell, J., & Palme, K. (1989). Molecular cloning and structural analysis of a gene from *Zea mays* (L.) coding for a putative receptor for the plant hormone auxin. *EMBO Journal, 8*, 2453–2461.

Iglesias, M. J., Terrile, M. C., & Casalongué, C. A. (2011). Auxin and salicylic acid signalings counteract during the adaptive response to stress. *Plant Signaling & Behavior, 6*, 452–454.

Jurkowski, G. I., Smith, R. K., Jr., Yu, I. C., Ham, J. H., Sharma, S. B., Klessig, D. F., Fengler, K. A., & Bent, A. F. (2004). *Arabidopsis DND2*, a second cyclic nucleotide-gated ion channel gene for which mutation causes the "defense, no death" phenotype. *Molecular Plant-Microbe Interactions, 17*, 511–520.

Kaplan, B., Sherman, T., & Fromm, H. (2007). Cyclic nucleotide-gated channels in plants. *FEBS Letters, 581*, 2237–2246.

Kazan, K., & Manners, J. M. (2009). Linking development to defense: auxin in plant–pathogen interactions. *Trends in Plant Science, 14*, 373–382.

Keisa, A., Kanberga-Silina, K., Nakurte, I., Kunga, L., & Rostoks, N. (2011). Differential disease resistance response in the barley necrotic mutant *nec1*. *BMC Plant Biology, 11*, 66.

Kepinski, S., & Leyser, O. (2005). The *Arabidopsis* F-box protein TIR1 is an auxin receptor. *Nature, 435*, 446–451.

Khazaie, H., Mohammady, S., Monneveux, P., & Stoddard, F. (2011). The determination of direct effect of carbon isotope discrimination (Δ), stomatal characteristics and water use efficiency on grain yield in wheat using sequential path analysis. *Australian Journal of Crop Science, 5*, 466–472.

Kotake, T., Nakagawa, N., Takeda, K., & Sakurai, N. (2000). Auxin-induced elongation growth and expressions of cell wall-bound exo and endo-β-glucanases in barley coleoptiles. *Plant & Cell Physiology, 41*, 1272–1278.

Kriechbaumer, V., Park, W. J., Piotrowski, M., Meeley, R. B., Gierl, A., & Glawischnig, E. (2007). Maize nitrilases have a dual role in auxin homeostasis and b-cyanoalanine hydrolysis. *Journal of Experimental Botany, 58*, 4225–4233.

Kutschera, U. (2006). Acid growth and plant development. *Science, 311*, 952–954.

Livak, K. J., & Schmittgen, T. D. (2001). Analysis of relative gene expression data using real-time quantitative PCR and the $2^{-\Delta\Delta C_t}$ method. *Methods, 25*, 402–408.

Lüthen, H., Bigdon, M., & Böttger, M. (1990). Reexamination of the acid growth theory of auxin action. *Plant Physiology, 93*, 931–939.

Ma, W., Qi, Z., Smigel, A., Walker, R. K., Verma, R., & Berkowitz, G. A. (2009). Ca^{2+}, cAMP, and transduction of non-self perception during plant immune responses. *Proceedings of the National Academy of Sciences of the United States of America, 106*, 20995–21000.

Mashiguchi, K., Tanaka, K., Sakai, T., Sugawara, S., Kawaide, H., Natsume, M., Hanada, A., Yaeno, T., Shirasu, K., Yao, H., McSteen, P., Zhao, Y., Hayashi, K., Kamiya, Y., & Kasahar, H. (2011). The main auxin biosynthesis pathway in *Arabidopsis. Proceedings of the National Academy of Sciences of the United States of America, 108*, 18512–18517.

McSteen, P. (2010). Auxin and monocot development. *Cold Spring Harbor Perspectives in Biology, 2*, a001479.

Navarro, L., Dunoyer, P., Jay, F., Arnold, B., Dharmasiri, N., Estelle, M., Voinnet, O., & Jones, J. D. G. (2006). A plant miRNA contributes to antibacterial resistance by repressing auxin signaling. *Science, 312*, 436–439.

Perrot-Rechenmann, C. (2010). Cellular responses to auxin: Division versus expansion. *Cold Spring Harbor Perspectives in Biology, 2*, a001446.

Phillips, K. A., Skirpan, A. L., Liu, X., Christensen, A., Slewinski, T. L., HudsonC, Barazesh S., Cohen, J. D., Malcomber, S., & McSteen, P. (2011). *vanishing tassel2* encodes a grass-specific tryptophan aminotransferase required for vegetative and reproductive development in maize. *The Plant Cell, 23*, 550–566.

Pieterse, C. M. J., Leon-Reyes, A., Van der Ent, S., & Van Wees, S. C. M. (2009). Networking by small-molecule hormones in plant immunity. *Nature Chemical Biology, 5*, 308–316.

Rostoks, N., Schmierer, D., Mudie, S., Drader, T., Brueggeman, R., Caldwell, D. G., Waugh, R., & Kleinhofs, A. (2006). Barley necrotic locus *nec1* encodes the cyclic nucleotide-gated ion channel 4 homologous to the *Arabidopsis HLM1. Molecular and General Genetics, 275*, 159–168.

Sakata, T., Oshino, T., Miura, S., Tomabechi, M., Tsunaga, Y., Higashitani, N., Miyazawa, Y., Takahashi, H., Watanabe, M., & Higashitani, A. (2010). Auxin reverse plant male sterility caused by high temperatures. *Proceedings of the National Academy of Sciences of the United States of America, 107*, 8569–8574.

Schenck, D., Christian, M., Jones, A., & Luthen, H. (2010). Rapid auxin-induced cell expansion and gene expression: a four-decade-old question revisited. *Plant Physiology, 152*, 1183–1185.

Spaepen, S., & Vanderleyden, J. (2011). Auxin and plant-microbe interactions. *Cold Spring Harbor Perspectives in Biology, 3*, a001438.

Spoel, S. H., & Dong, X. (2008). Making sense of hormone crosstalk during plant immune responses. *Cell Host & Microbe, 3*, 348–351.

Tan, X., Calderon-Villalobos, L. I., Sharon, M., Zheng, C., Robinson, C. V., Estelle, M., & Zheng, N. (2007). Mechanism of auxin perception by the TIR1 ubiquitin ligase. *Nature, 446*, 640–645.

Tanaka, Y., Sano, T., Tamaoki, M., Nakajima, N., Kondo, N., & Hasezawa, S. (2006). Cytokinin and auxin inhibit abscisic acid-induced stomatal closure by enhancing ethylene production in Arabidopsis. *Journal of Experimental Botany, 57*, 2259–2266.

Teale, W. D., Paponov, I. A., & Palme, K. (2006). Auxin in action: Signalling, transport and the control of plant growth and development. *Nature Reviews Molecular Cell Biology, 7*, 847–859.

Tromas, A., Braun, N., Muller, P., Khodus, T., Paponov, I. A., Palme, K., Ljung, K., Lee, J. Y., Benfey, P., Murray, J. A., Scheres, B., & Perrot-Rechenmann, C. (2009). The AUXIN BINDING PROTEIN 1 is required for differential auxin responses mediating root growth. *PLoS One, 4*, e6648.

Truman, W. M., Bennett, M. H., Turnbull, C. G. N., & Grant, M. R. (2010). *Arabidopsis* auxin mutants are compromised in systemic acquired resistance and exhibit aberrant accumulation of various indolic compounds. *Plant Physiology, 152*, 1562–1573.

Verhage, A., van Wees, S. C. M., & Pieterse, C. M. J. (2010). Plant immunity: It's the hormones talking, but what do they say? *Plant Physiology, 154*, 536–540.
Wang, D., Pajerowska-Mukhtar, K., Hendrickson Culler, A., & Dong, X. (2007). Salicylic acid inhibits pathogen growth in plants through repression of the auxin signaling pathway. *Current Biology, 17*, 1784–1790.
Wang, Y., Gao, M., Li, Q., Wang, L., Wang, J., Jeon, J.-S., Qu, N., Zhang, Y., & He, Z. (2008). OsRAR1 and OsSGT1 physically interact and function in rice basal disease resistance. *Molecular Plant-Microbe Interactions, 21*, 294–303.
Won, C., Shen, X., Mashiguchi, K., Zheng, Z., Dai, X., Cheng, Y., Kasahara, H., Kamiya, Y., Choryc, J., & Zhao, Y. (2011). Conversion of tryptophan to indole-3-acetic acid by tryptophan aminotransferases of *Arabidopsis* and YUCCAs in *Arabidopsis*. *Proceedings of the National Academy of Sciences of the United States of America, 108*, 18518–18523.
Yang, T., & Poovaiah, B. W. (2000). Molecular and biochemical evidence for the involvement of calcium/calmodulin in auxin action. *Journal of Biological Chemistry, 275*, 3137–3143.
Zhang, X., Wang, H., Takemiya, A., Song, C., Kinoshita, T., & Shimazaki, K. (2004). Inhibition of blue light-dependent H^+ pumping by abscisic acid through hydrogen peroxide-induced dephosphorylation of the plasma membrane H^+-ATPase in guard cell protoplasts. *Plant Physiology, 136*, 4150–4158.
Zhang, J., Peng, Y., & Guo, Z. (2009a). Constitutive expression of pathogen-inducible OsWRKY31 enhances disease resistance and affects root growth and auxin response in transgenic rice plants. *Cell Research, 18*, 508–521.
Zhang, L., Lavery, L., Gill, U., Gill, K., Steffenson, B., Yan, G., Chen, X., & Kleinhofs, A. (2009b). A cation/proton-exchanging protein is a candidate for the barley *NecS1* gene controlling necrosis and enhanced defense response to stem rust. *Theoretical and Applied Genetics, 118*, 385–397.

Chapter 21
Vulnerability of Cultivated and Wild Barley to African Stem Rust Race TTKSK

Brian J. Steffenson, Hao Zhou, Yuan Chai, and Stefania Grando

Abstract Stem rust is one of the most important diseases of wheat and barley worldwide. Highly virulent stem rust races such as TTKSK (aka isolate Ug99) from Africa represent one of the greatest threats to stable cereal production in 50 years, but little is known regarding the vulnerability of barley. This study was undertaken to determine the reaction of a large and diverse collection of cultivated and wild barley germplasm to race TTKSK at the seedling stage. Of the 1,902 cultivated barley accessions evaluated, only 43 (2.3%) consistently exhibited resistant to moderately resistant reactions across experiments. Similarly, only 16 (1.7%) of the 935 wild barley accessions consistently exhibited resistant to moderately resistant reactions. Pedigree analyses revealed that some of the resistant cultivated accessions were derived from the same parental sources. These results underscore the extreme vulnerability of barley to stem rust race TTKSK and the need to identify additional sources of resistance.

Keywords Stem rust • Wild barley • Disease resistance

21.1 Introduction

Stem rust (caused by *Puccinia graminis* Pers.:Pers. f. sp. *tritici* Eriks. & E. Henn.) is one of the most important diseases of wheat and barley. It can be a serious problem of barley in the major production areas of the Great Plains in North America (Steffenson 1992) and also in northeastern Australia (Dill-Macky et al. 1991). Stem

B.J. Steffenson (✉) • H. Zhou • Y. Chai
Department of Plant Pathology, University of Minnesota, St Paul, MN 55108, USA
e-mail: bsteffen@umn.edu

S. Grando
International Center for Agricultural Research in the Dry Areas, Aleppo, Syria

G. Zhang et al. (eds.), *Advance in Barley Sciences: Proceedings of 11th International Barley Genetics Symposium*, DOI 10.1007/978-94-007-4682-4_21,

rust also can infect barley crops in East Africa, Yemen, Iran, Kazakhstan, Tajikistan, and Georgia (Kumarse Nazari, David Hodson, and Mahbubjon Rahmatov, personal communication). Since the mid-1940s, barley stem rust in North America has been kept in check through the widespread use of cultivars carrying the resistance gene *Rpg1* (Steffenson 1992). However, a race with virulence for *Rpg1* appeared in 1988 and a few years later (1990–1991) caused scattered losses on barley in both the Upper Midwest region of the United States and eastern prairie provinces of Canada (Steffenson 1992). Since the mid-1990s, no significant losses have been reported due to stem rust on barley in the United States. However, the world is now faced with the greatest threat to stable cereal production in over 50 years: highly virulent stem rust races from Africa (Singh et al. 2008). In 1998, heavy stem rust infections were observed on wheat lines carrying the widely effective resistance gene *Sr31* in a field nursery in southwest Uganda. The stem rust isolate collected from this nursery was designated "Ug99" (an abbreviation for the country of origin and year it was received for analysis) (Pretorius et al. 2000) and was subsequently assayed for its virulence phenotype on 16 differential lines of wheat (Jin et al. 2008). Isolate Ug99 was initially keyed to race TTKS (Wanyera et al. 2006). Subsequent virulence typing on an additional set of four wheat differential lines (with *Sr* genes of *Sr24, Sr31, Sr38,* and *SrMcN)* led to the expanded race designation of TTKSK (Jin et al. 2008).

Since it was first discovered in Uganda in 1998 (Pretorius et al. 2000), race TTKSK has been detected in Kenya, Ethiopia, Sudan, Yemen, and Iran (Singh et al. 2008; Nazari et al. 2009). It is certain to spread to other countries given the ease by which rust urediniospores can be disseminated over long distances by wind and also international travelers. Race TTKSK is virulent on over 70% of the world's wheat cultivars (Singh et al. 2006) and is therefore a major threat to the world food supply. Although not a major food crop, barley is nonetheless an important staple for some of the most destitute people living in the highlands of Ethiopia, Eritrea, Yemen, Tibet, Nepal, Ecuador, and Peru. Moreover, the crop is an important component in the agricultural economy of many developed countries for its use as malt in brewing, animal feed, and also specialty foods. Stem rust can significantly decrease both the yield and quality of barley (Dill-Macky et al. 1991; Harder and Legge 2000); therefore, it is urgent that an investigation be made to assess the potential vulnerability of barley to race TTKSK. This study was undertaken to determine the reaction of a large and diverse collection of cultivated and wild barley germplasm to wheat stem rust race TTKSK at the seedling stage.

21.2 Materials and Methods

21.2.1 Plant Materials

For the stem rust evaluations, a large collection of barley cultivars, breeding lines, and landraces was obtained from various genebanks, research institutions, universities, private companies, and individual collaborators around the world. This germplasm

(1,902 accessions in total) represents a broad sampling of the genetic diversity in cultivated barley. In addition to cultivated barley, accessions of wild barley (*Hordeum vulgare* subsp. *spontaneum*) also were evaluated for their stem rust phenotype. The *H. vulgare* subsp. *spontaneum* germplasm (935 accessions in total) included accessions chiefly from the Fertile Crescent, but also other regions across the geographic range of the subspecies.

To assess whether previously identified resistance genes in barley are effective against race TTKSK, we also tested the *Rpg1* sources of Chevron (CIho 1111) and Morex (CIho 15773), the *Rpg2* source of Hietpas 5 (CIho 7124), the *Rpg3* source of PI 382313, the *rpg4*/*Rpg5* sources of Q21861 (carries *rpg4*, *Rpg5*, and *Rpg1*) and Q/SM20 (a Q21861-derived doubled haploid progeny carrying *rpg4* and *Rpg5* only), and the *rpg6* source of 212Y1(Steffenson et al. 2009). Susceptible controls included the barley cultivars of Steptoe (CIho 15229) and Hiproly (PI 60693) and wheat accessions of Line E (PI 357308) and McNair 701 (CItr 15288). Previous research revealed that accessions carrying genes at the complex *rpg4*/*Rpg5* locus are resistant to race TTKSK (Steffenson et al. 2009); thus, Q21861 and Q/SM20 were considered as resistant controls in the experiments.

21.2.2 Plant Growth Conditions, Inoculation Protocol, and Infection/Incubation Period

Stem rust evaluations were done at the Minnesota Department of Agriculture/University of Minnesota Agricultural Experiment Station Biosafety Level-3 containment facility on the St. Paul campus. The experiments were conducted during the winter months of 2009–2010 and 2010–2011. A single pustule isolate (04KEN156/04) of race TTKSK was used in all experiments. Urediniospores were applied to 9-day-old plants using special atomizers pressured by a pump. An inoculum concentration of 14 mg urediniospores/0.7 ml oil was applied at a rate of approximately 0.14 mg/plant. Thereafter, the plants were incubated according to the conditions previously described by Steffenson et al. (2009).

21.2.3 Disease Assessment

Twelve to fourteen days after inoculation, the infection types (ITs) on each accession were scored using a 0–4 scale. The IT scale used for barley is a modification of the one developed for wheat and is based primarily on uredinial size (Steffenson et al. 2009). Barley frequently exhibits two or more ITs on a single leaf when infected with stem rust (Sun and Steffenson 2005). All of the observed ITs were recorded in order of their frequency on the leaves; however, only the two most common ones (i.e., the IT mode) were considered since they usually comprise more than 85% of all those observed on individual accessions (Steffenson et al. 2009). ITs were classified into five general categories as follows: 0 or 0 as highly resistant

(HR); 1 as resistant (R); 2 as moderately resistant (MR); 3^- as moderately susceptible (MS); and 3, 3^+, or 4 as susceptible (S).

The experiment was conducted in a completely randomized design with one replicate and was repeated once over time. Accessions exhibiting resistant to moderately resistant reactions were repeated at least one additional time (in most, every case) to confirm the phenotype. Some cultivated and wild barley accessions were only tested once due to space or seed quantity limitations, but all of the plants within these accessions exhibited only or predominantly susceptible reactions to race TTKSK. The resistant (Q21861 and Q/SM20) and susceptible controls (Steptoe, Hiproly, Line E, and McNair 701) were replicated every 100 test entries to monitor both the infection levels and ITs in each experiment. Only accessions consistently exhibiting HR, R, or MR reactions across all experiments were considered worthy of further investigation for TTKSK resistance and are listed herein. Data on the reaction of all other accessions are available upon request from the authors. To trace the possible sources of TTKSK resistance in cultivated barley accessions, pedigree analyses were conducted whenever possible.

21.3 Results

21.3.1 Reaction of Susceptible Controls

Susceptible controls were included in multiple replicates in all experiments to monitor the infection level and virulence phenotype of race TTKSK. Moderate to high infection levels were observed in all experiments, allowing for reliable scoring of ITs on accessions. Barley (Hiproly and Steptoe) and wheat (McNair 701 and Line E) controls all exhibited the expected compatible ITs of 3 and 4, respectively (Table 21.1). The IT difference between species reflects the lower compatibility of barley as compared to wheat to stem rust.

21.3.2 Reaction of Accessions with Recognized Stem Rust Resistance Genes

Barley accessions carrying *Rpg1* only (Chevron and Morex) exhibited mostly susceptible to moderately susceptible reactions (Table 21.1). Chevron, one of the original sources of *Rpg1*, exhibited slightly lower reactions (IT mode of 3–2) than the derived source of Morex (IT mode of 3). Hietpas 5, the source of *Rpg2*, also was susceptible, exhibiting IT 3. The sources of *Rpg3* (PI 382313) and *rpg6* (212Y1) exhibited predominantly moderately susceptible but also some moderately resistant ITs (3–2). The sources carrying the gene complex of *rpg4*/*Rpg5* (Q21861 and Q/SM20) were resistant to highly resistant (0;1 to 10) to race TTKSK.

Table 21.1 Infection type mode, range, and general reaction of controls and accessions with recognized stem rust resistance genes to *Puccinia graminis* f. sp. tritici pathotype TTKSK at the seedling stage

Accession	Other designator/ notation	Description	IT mode[a]	General reaction[b]	Range[c]
Chevron	PI 38061	Source of *Rpg1*	3–2	MS–MR	2 to 3
Morex	CIho 15773	Source of *Rpg1*	3	S	3– to 3
Hietpas 5	CIho 7124	Source of *Rpg2*	3	S	3
PI 382313	–	Source of *Rpg3*	3–2	MS–MR	2 to 3–
Q21861	PI 584766	Source of *rpg4* & *Rpg5* plus *Rpg1*	0; 1	HR–R	0; to 2
Q/SM20	–	Source of *rpg4* & *Rpg5*	10;	R–HR	0; to 2
212Y1	–	Source of *rpg6*	3–2	MS–MR	2 to 3–
Hiproly	PI 60693	Susceptible barely control	3	S	3– to 3
Steptoe	CIho 15229	Susceptible barley control	3	S	3 to 3+
McNair 701	CItr 15288	Susceptible wheat control	4	S	3+ to 4
Line E	CItr 357308	Susceptible wheat control	4	S	3 to 4

[a]Infection types (ITs) were based on a 0–4 scale. The IT mode represents the one or two most common ITs observed in order of frequency on accessions over all experiments. Symbols + and – denote more or less sporulation of classically described uredinia, respectively

[b]A general reaction was assigned to the ITs where 0 or 0; is highly resistant (HR), 1 is resistant (R), 2 is moderately resistant (MR), 3– is moderately susceptible (MS), and 3, 3+, or 4 is susceptible (S)

[c]The IT range is the lowest and highest types observed on accessions over all experiments

21.3.3 Reaction of Cultivated Barley Accessions from Around the World

Of the 1,902 cultivated barley accessions evaluated to stem rust, only 43 (2.3%) consistently exhibited resistant to moderately resistant reactions (IT modes ranging from 0;1 to 2) across experiments (Table 21.2). An IT mode of 21 was most common among these accessions. Nearly all of the resistant accessions identified were cultivars or breeding lines from different barley improvement programs around the world.

Complete pedigree information was not available or not known for many of the resistant accessions identified; thus, it was not possible to reliably trace the original sources of stem rust resistance. It was, however, possible to trace or at least strongly speculate on the origin of resistance in several well documented cases. For example, all of the resistant lines from Saskatchewan and most of those from Manitoba in Canada trace to line Q21861 (Brian Rossnagel and William Legge, personal communication), which carries the *rpg4*/*Rpg5* complex and also *Rpg1* (Table 21.1). In addition to Q21861, PI 382313 (with *Rpg3*) also was used as a parent for a few lines (BM 8923–30, BM 9238–15, and BM 9723–53) in the Manitoba program, but may not have contributed any resistance to race TTKSK because it exhibited mostly moderately susceptible reactions at both the seedling (Table 21.1) and adult plant stages (B. Steffenson, unpublished). From Uruguay, both CLE 202 INIA Ceibo

Table 21.2 List of cultivated barley accessions exhibiting only resistant to moderately resistant infection types to *Puccinia graminis* f. sp. *tritici* race TTKSK at the seedling stage

Accession	Other designator or notation	Germplasm type	Originating institution/location	Country	Seed supplier	IT mode	Gen Rxn	Range
H9403513203/600016	–	Breeding line	Field Crop Development Centre/Lacombe, Canada	Alberta, Canada	James Helm	21	MR–R	1 to 2
TR 02272	–	Breeding line	Agriculture and Agri-Food Canada/Brandon, Canada	Manitoba, Canada	William Legge	0;1	HR–R	0; to 2
BM 8923-30	–	Breeding line	Agriculture and Agri-Food Canada/Brandon, Canada	Manitoba, Canada	William Legge	10;	R–HR	0; to 2
BM 9723-53	–	Breeding line	Agriculture and Agri-Food Canada/Brandon, Canada	Manitoba, Canada	William Legge	10;	R–HR	0; to 1
AC Maple	–	Cultivar	Eastern Cereal and Oilseed Research Centre/Ottawa, Canada	Eastern Canada	Thin Meiw (Alek) Choo	12	R–HR	1 to 2
OAC Laverne	–	Cultivar	University of Guelph/Guelph, Canada	Eastern Canada	Thin Meiw (Alek) Choo	21	MR–R	1 to 2
SB97197	–	Breeding line	Crop Development Centre, Saskatoon, Saskatchewan	Saskatchewan, Canada	Brian Rossnagel	21	MR–R	0; to 2
SH98073	–	Breeding line	Crop Development Centre, Saskatoon, Saskatchewan	Saskatchewan, Canada	Brain Rossnagel	0;1	HR–R	0; to 2
SH98076	-	Breeding line	Crop Development Centre, Saskatoon, Saskatchewan	Saskatchewan, Canada	Brain Rossnagel	0;1	HR-R	0; to 2
MC0181-11	–	Breeding line	Crop Development Centre, Saskatoon, Saskatchewan	Saskatchewan, Canada	Brain Rossnagel	10;	R–HR	0; to 2
MC0181-31	–	Breeding line	Crop Development Centre, Saskatoon, Saskatchewan	Saskatchewan, Canada	Brain Rossnagel	10;	R–HR	0; to 2
6B03-4105	06BA-22	Breeding line	Busch Agricultural Resources Inc./Ft. Collins, CO	Colorado, USA	Blake Cooper	21	MR–R	1 to 2
M04-03	07MN-55	Breeding line	University of Minnes ota/St. Paul, MN	Minnesota, USA	Kevin Smith	21	MR–R	1 to 2

2ND26373	07N2-68	Breeding line	North Dakota State University/ Fargo, ND	North Dakota, USA	Rich Horsley	21	MR–R	1 to 2
ND23821	06N6-87	Breeding line	North Dakota State University/ Fargo, ND	North Dakota, USA	Rich Horsley	21	MR–R	1 to 2
05WA-368.13	07WA-27	Breeding line	Washington State University/ Pullman, WA	Washington, USA	Steve Ullrich	21	MR–R	1 to 2
CLE 202 INIA Ceibo	–	Cultivar	Instituto Nacional de Investigación Agropecuaria (INIA)/Colonia, Uruguay	Uruguay	S. German	21	MR–R	0; to 2
CLE 250	INIA 11	–	Instituto Nacional de Investigación Agro pecuaria (INIA)/Colonia, Uruguay	Uruguay	Les Wright	21	MR–R	1 to 2
Brandham II	LAND 09-52	–	International Center for Agricultural Research in the Dry Areas/Aleppo, Syria	Austria	Stefania Grando	10;	R–HR	0; to 1
Power	–	Cultivar	SE JET Plant Breeding, Horsens, Denmark	Denmark	Dominique Vequaud	21	MR–R	0; to 2
Anakin	–	Cultivar	SE JET Plant Breeding, Horsens, Denmark	Denmark	Dominique Vequaud	0;1	HR–R	0; to 1
Otira	–	Cultivar	SE JET Plant Breeding, Horsens, Denmark	Denmark	Fredrik Ottosoon	10;	R–HR	0; to 2
Fusion	–	Cultivar	SE JET Plant Breeding, Horsens, Denmark	Denmark	Fredrik Ottosson	0;1	HR–R	0; to 1
Magaly	UN 1	Cultivar	U NISIGMA Froissy, France	France	Dominique Vequaud	21	MR–R	1 to 2
Marigold	UN 2	Cultivar	U NISIGMA Froissy, France	France	Dominique Vequaud	21	MR–R	1 to 2
Onyx	LP2/8622	Cultivar	(Lochow Petkus) now KWS LOCHOW Gm bH, Bergen,	Germany	Robbie Waugh	12	R–MR	1 to 2
BIOS 1	K-29634	Cultivar	–	Russia	Olga Kovaleva	21	MR–R	1 to 2
Zad onskij 8	K-30452	Cultivar	–	Russia	Olga Kovaleva	21	MR–R	1 to 2

(continued)

Table 21.2 (continued)

Accession	Other designator or notation	Germplasm type	Originating institution/location	Country	Seed supplier	IT mode	Gen Rxn	Range
County	–	Cultivar-breign introduction	Syngenta Seeds/Fulbourn Cambridge, England	Spain	Ernesto Igartua Arregui	21	MR–R	1 to 2
Dew	CPBT B35	Cultivar	KWS UK Ltd, Hertfordshire, England	United Kingdom	Robbie Waugh	12	R–MR	1 to 2
Macarena	CSBC 2452-14	Cultivar	LS Plant Breeding Ltd, Impington Cambridge, England	United Kingdom	Robbie Waugh	12	R–MR	0; to 2
Paramount	CSBC 4475-9	Cultivar	LS Plant Breeding Ltd, Impington Cambridge, England	United Kingdom	Robbie Waugh	12	R–MR	0; to 2
Giza 130	CHECKS09-61	–	International Center for Agricultural Research in the Dry Areas/Aleppo, Syria	Egypt	Stefania Grando	21	MR–R	0; to 2
BHS 248	572602	–	–	India	Harold Bockelman	21	MR–R	1 to 2
Mano	PI574305	Cultivar	Jiangsu Academy of Agricultural Science	China	Harold Bockelman	12	R–MR	1 to 2

Zang Qing 80	610228	Cultivar	Chinese Academy of Science, Xizang	China	Harold Bockelman	1	R	1 to 2
Zang Qing 148	610230	Cultivar	Chinese Academy of Science, Xizang	China	Harold Bockelman	12	R–MR	0; to 2
Xiu Mai No.3	611489	Cultivar	Zhejiang Agricultural University, Hangzhou, China	China	Harold Bockelman	21	MR–R	1 to 2
Kashima Mugi	467759	Cultivar	Saitama	Japan	Harold Bockelman	21	MR–R	1 to 2
Kangbori	467852	Cultivar	Wheat and Barley Research Institute/Suwon, South Korea	South Korea	Harold Bockelman	21	MR–R	1 to 2
Cantala	483047	Cultivar	Department of Agriculture/ Melbourne, Australia	Australia	Harold Bockelman	21	MR–R	1 to 2
Grimmet	483048	Cultivar	Department of Primary Industries, Brisbane, Australia	Australia	Harold Bockelman	1	R	1 to 2
Tallon	573731	Cultivar	Hermitage Research Station/ Warwick, Australia	Australia	Harold Bockelman	21	MR–R	1 to 2

and CLE 250 were resistant. The origin of resistance in CLE 202 INIA Ceibo is unknown, but it was a parent of CLE 250 and likely contributed the resistance (Silvia German, personal communication). From the Sejet Plant Breeding company in Denmark, four cultivars (Power, Anakin, Otira, and Fusion) exhibited stem rust resistance. Pedigree data revealed that Chariot is a common parent in several of these cultivars (Rasmus Lund Hjortshøj, personal communication). Thus, Chariot could be the possible donor of resistance, although its reaction to stem rust is not known. The donors of resistance in other accessions could not be determined due to incomplete information.

21.3.4 Reaction of Wild Barley

Only 16 (1.7%) of the 935 wild barley accessions consistently exhibited resistant to moderately resistant reactions (IT modes ranging from 0;1 to 2) across experiments (Table 21.3). An IT mode of 21 was most common among these accessions. Resistant accessions were identified from nine different countries, including Afghanistan, Iraq, Israel, Kazakhstan, Lebanon, Syria, Tajikistan, Turkmenistan, and Uzbekistan. Notwithstanding the large differences in sample sizes, the highest percentage of resistance was found in the Central Asian Republics of Tajikistan (2/8 = 25.0%), Uzbekistan (5/20 = 25.0%), and Kazakhstan (1/7 = 14.3%).

21.4 Discussion

Stem rust race TTKSK and other related variants represent one of the most serious threats to world wheat production in nearly 50 years. More than 70% of wheat cultivars worldwide are susceptible to race TTKSK (Singh et al. 2006). Thus, it may only be a matter of time before this race causes widespread epidemics in major wheat-producing regions of the world. Barley is also a host to stem rust, but little was known regarding its vulnerability to race TTKSK. In this study, a large collection of genetically diverse cultivated and wild barley accessions (2,837 in total) was evaluated at the seedling stage to assess the potential vulnerability of the crop to race TTKSK. Only ~2% of the 1,902 cultivated and 935 wild barley accessions tested exhibited consistent resistant reactions to this race. Moreover, preliminary pedigree analyses revealed that some of the resistant cultivated accessions were derived from the same parental sources (i.e., Q21861 and possibly Chariot). These results underscore the extreme vulnerability of barley to race TTKSK. Genetically, barley is far more vulnerable to disease outbreaks than wheat because it is a diploid crop with relatively few accessible relatives for contributing diversity. Bread wheat and durum wheat are hexaploid and tetraploid, respectively, and therefore benefit from the diversity of genes contributed from the different genomes comprising the crops. Moreover, wheat also has been genetically enriched through the transfer of

Table 21.3 List of wild barley accessions exhibiting only resistant to moderately resistant infection types to *Puccinia graminis* f. sp. *tritici* race TTKSK at the seedling stage[a]

Accession	Other accession designator	Province, region, city, or site or origin	Country of origin	Seed supplier	Organization/location of seed supplier	IT mode	Gen Rxn	Range
WBDC 014	38659	Baghlan	Afghanistan	Jan Valkoun	ICARDA/Aleppo, Syria	12	R–MR	1 to 2
WBDC 013	38658	As Sulaymaniyah	Iraq	Jan Valkoun	ICARDA/Aleppo, Syria	2	MR	1 to 2
WBDC 032	38869	Hazafon	Israel	Jan Valkoun	ICARDA/Aleppo, Syria	0; 1	HR–R	0; to 1
ICARDA039	38936	West Bank	Israel	Jan Valkoun	ICARDA/Aleppo, Syria	10;	R–HR	0; to 2
ICARDA040	38947	Yerushalayim	Israel	Jan Valkoun	ICARDA/Aleppo, Syria	10;	R–HR	0; to 2
WBDC 220	131642	Chimkent	Kazakhstan	Jan Valkoun	ICARDA/Aleppo, Syria	12	R–MR	1 to 2
WBDC 138	40183	Hasbaiya	Lebanon	Jan Valkoun	ICARDA/Aleppo, Syria	1	R	0; to 2
WBDC 302	38635	Damascus	Syria	Jan Valkoun	ICARDA/Aleppo, Syria	1	R	0; to 2
WBDC 224	131790	Dushanbe	Tajikistan	Jan Valkoun	ICARDA/Aleppo, Syria	10;	R–HR	0; to 2
WBDC 225	131792	Dushanbe	Tajikistan	Jan Valkoun	ICARDA/Aleppo, Syria	21	MR–R	1 to 2
WBDC 333	135478	Garygalla	Turkmenistan	Jan Valkoun	ICARDA/Aleppo, Syria	21	MR–R	1 to 2
WBDC 119	40108	Dzhizak	Uzbekistan	Jan Valkoun	ICARDA/Aleppo, Syria	0; 1	HR–R	0; to 2
WBDC 209	123972	Dzhizak	Uzbekistan	Jan Valkoun	ICARDA/Aleppo, Syria	21	MR–R	1 to 2
WBDC 213	124035	Samarkand	Uzbekistan	Jan Valkoun	ICARDA/Aleppo, Syria	21	MR–R	0; to 2
WBDC 214	124046	Samarkand	Uzbekistan	Jan Valkoun	ICARDA/Aleppo, Syria	10;	R–HR	0; to 2
WBDC 345	40155	Kashkadar'ya	Uzbekistan	Jan Valkoun	ICARDA/Aleppo, Syria	10;	R–HR	0; to 2

many important resistance genes derived from progenitor and allied species using conventional or specialized cytogenetic techniques. This, coupled with the fact that only a few barley cultivars comprise the bulk of the area planted in any one country, means that large monocultures with virtually no backstop of genetic diversity exist for cultivated barley.

The assessment of barley's vulnerability to race TTKSK was conducted on plants at the seedling stage only. This was done in order to obtain more rigid control over the environmental conditions for precise and reliable phenotyping. Moreover, the space available for adult plant testing in the field in Africa was limited. Since stem rust rarely infects barley in the field until after the heading stage, adult plant resistance is the most critical breeding target for this disease. The resistant accessions identified in this study will be tested in replication in stem rust nurseries in Africa to assess their adult plant reaction under field conditions. Accessions exhibiting broadly effective resistance will be subjected to full genetic analyses to determine the number and chromosomal location of resistance genes. Seedling tests may only identify major effect genes that provide all stage resistance; thus, useful genes conferring adult plant resistance only may go undetected. Indeed, from the evaluation of a large collection of barley breeding lines from the United States, we identified, through an association mapping approach, an adult plant quantitative trait locus that was not detectable at the seedling stage (Zhou 2011). Additional research is therefore being done to identify barley accessions with adult plant resistance in the field.

This study of stem rust resistance in barley was conducted using a single race (TTKSK) from east Africa. Race TTKSK gained worldwide notoriety after it was found infecting wheat lines carrying the widely effective resistance gene *Sr31* in Uganda. Further investigations revealed that this race is virulent on over 70% of the world's wheat cultivars and has spread across east Africa and into the Middle East (Iran) (Nazari et al. 2009; Singh et al. 2006). Spread to other regions is certain to occur because urediniospores are readily disseminated by wind over long distances and can also "hitchhike" on the clothes of international travelers. TTKSK is only one of several dangerous African stem rust races threatening wheat today. In addition to the *Sr31* virulence carried by TTKSK, other recently described races in the lineage possess virulence for *Sr24* (race TTKST) and*Sr36* (race TTKSK). This suite of African stem rust races has rendered wheat even more vulnerable to stem rust epidemics. Race TTKSK is highly virulent on the most widely used stem rust resistance gene in barley, *Rpg1*. It is not known whether the resistance sources identified in this study to race TTKSK are similarly effective against the other described African races, but this aspect is currently being investigated. Regardless, all future barley cultivars must be bred for broad resistance against the spectrum of races present in a region.

References

Dill-Macky, R., Rees, R. G., & Platz, G. J. (1991). Inoculum pressure and the development of stem rust epidemics in barley. *Australian Journal of Agricultural Research, 42*, 769–777.

Harder, D. E., & Legge, W. G. (2000). Effectiveness of different sources of stem rust resistance in barley. *Crop Science, 40*, 826–833.

Jin, Y., Szabo, L. J., Pretorius, Z. A., Singh, R. P., Ward, R., & Fetch, T. J. (2008). Detection of virulence to resistance gene *Sr24* within race TTKS of *Puccinia graminis* f. sp. *tritici*. *Plant Disease, 92*, 923–926.

Nazari, K., Mafi, M., Yahyaoui, A., Singh, R. P., & Park, R. F. (2009). Detection of wheat stem rust (*Puccinia graminis* f. sp. *tritici*) race TTKSK (Ug99) in Iran. *Plant Disease, 93*, 317.

Pretorius, Z. A., Singh, R. P., Wagoire, W. W., & Payne, T. S. (2000). Detection of virulence to wheat stem rust resistance gene *Sr31* in *Puccinia graminis* f. sp. *tritici* in Uganda. *Plant Disease, 84*, 203.

Singh, R. P., Hodson, D. P., Jin, Y., Huerta-Espino, J., Kinyua, M. G., Wanyera, R., Njau, P., & Ward, R. W. (2006). Current status, likely migration and strategies to mitigate the threat to wheat production from race Ug99 (TTKS) of stem rust pathogen. *CAB Reviews, 1*, 1–13.

Singh, R. P., Hodson, D. P., Huerta-Espino, J., Jin, Y., Njau, P., Wanyera, R., Herrata-Foessel, S. A., & Ward, R. W. (2008). Will stem rust destroy the world's wheat crop? *Advances in Agronomy, 98*, 271–309.

Steffenson, B. J. (1992). Analysis of durable resistance to stem rust in barley. *Euphytica, 63*, 153–167.

Steffenson, B. J., Jin, Y., Brueggeman, R. S., Kleinhofs, A., & Sun, Y. (2009). Resistance to stem rust race TTKSK maps to the *rpg4/Rpg5* complex of chromosome 5 H of barley. *Phytopathology, 99*, 1135–1141.

Sun, Y., & Steffenson, B. J. (2005). Reaction of barley seedlings with different stem rust resistance genes to *Puccinia graminis* f. sp. *tritici* and *Puccinia graminis* f. sp. *secalis*. *Canadian Journal of Plant Pathology, 27*, 80–89.

Wanyera, R., Kinyua, M. G., Jin, Y., & Singh, R. P. (2006). The spread of stem rust caused by *Puccinia graminis* f. sp. *tritici*, with virulence on *Sr31* in wheat in Eastern Africa. *Plant Disease, 90*, 113.

Zhou, H. (2011). *Association mapping of multiple disease resistance in US barley breeding germplasm*. Ph.D. dissertation, University of Minnesota, St. Paul.

Chapter 22
Genome-Wide Association Mapping Reveals Genetic Architecture of Durable Spot Blotch Resistance in US Barley Breeding Germplasm

Hao Zhou and Brian J. Steffenson

Abstract Spot blotch, an economically important disease of both barley and wheat, is caused by *Cochliobolus sativus* (anamorph: *Bipolaris sorokiniana*). The disease has been reported in many regions of the world but is particularly severe on barley in the Upper Midwest region of the United States, adjacent areas of Canada, and northeastern Australia. Durable resistance has been attained in Midwest six-rowed malting cultivars for over 50 years and is derived from line NDB112. To identify loci conferring resistance to spot blotch, an association mapping approach was conducted in US breeding germplasm (3,840 lines) genotyped with 3,072 single-nucleotide polymorphism (SNP) markers. Three quantitative trait loci (QTL) *Rcs-qtl-1H-11_10764, Rcs-qtl-3H-11_10565,* and *Rcs-qtl-7H-11_20162* conferring both seedling and adult resistance were identified. Each individual QTL only partially reduced the infection response (IR) (from 0 to 20%) at the seedling stage and severity (from 20 to 29%) at the adult plant stage. However, all three QTL together reduced the seedling IR and adult plant severity by 47 and 83%, respectively, and comprise the Midwest six-rowed durable resistance haplotype (MSDRH). The identified MSDRH will be valuable for marker-assisted selection of spot blotch resistance in breeding programs.

Keywords Spot blotch • Barley • Association mapping • Disease resistance • Haplotype

H. Zhou • B.J. Steffenson (✉)
Department of Plant Pathology, University of Minnesota, St Paul, MN 55108, USA
e-mail: bsteffen@umn.edu

G. Zhang et al. (eds.), *Advance in Barley Sciences: Proceedings of 11th International Barley Genetics Symposium*, DOI 10.1007/978-94-007-4682-4_22,

22.1 Introduction

Spot blotch, an economically important disease of both barley and wheat, is caused by *Cochliobolus sativus* (Ito &Kurib.) Drechs. ex Dastur (anamorph: *Bipolaris sorokiniana* (Sacc.) Shoem.) (Sprague 1950). The disease has been reported in many regions of the world but is particularly severe on barley in the Upper Midwest region of the United States, adjacent areas of Canada (Mathre 1997), and northeastern Australia (Knight et al. 2010).

Although fungicides can be effective in controlling spot blotch, deployment of cultivars with resistance is the most economically and environmentally sound strategy for ameliorating disease loss. This strategy was employed in the Upper Midwest production area, where six-rowed malting cultivars have remained resistant to *C. sativus* for over 50 years (Steffenson et al. 1996) and remain so today. This remarkable example of durable resistance was originally derived from NDB112 (CIho 11531). All of the resistance loci identified previously in NDB112-derived accessions were from small, biparental mapping populations. The main drawback of using such populations is the relatively poor mapping resolution (Flint-Garcia and Thornsberry 2003; Parisseaux and Bernardo 2004). A more powerful approach for quantitative trait loci (QTL) mapping is association mapping (AM), which is based on linkage disequilibrium (LD). With AM, one can utilize diverse germplasm panels (i.e., collections of cultivars, breeding lines, landraces, and/or wild progenitors) and therefore greatly increase the frequency of identifying polymorphic markers, which will ultimately increase map resolution.

Our long-term goal is to develop barley cultivars with durable resistance to important diseases. The specific goals of this research were to identify loci conferring spot blotch resistance in US barley breeding germplasm using an AM approach and elucidate the genetic architecture of durable spot blotch resistance in Midwest six-rowed germplasm.

22.2 Materials and Methods

The panel used in this AM study was developed by the Barley Coordinated Agricultural Project (BCAP) and consists of advanced breeding lines from ten US barley improvement programs: eight spring and two winter or winter/facultative type programs (Table 22.1). Two barley oligonucleotide pool assays (BOPA1 and BOPA2) (Close et al. 2009) containing allele-specific oligos for a set of 3,072 SNPs were used to genotype the barley lines.

AM analyses for seedling resistance were conducted in the complete panel (3,840 lines) designated hereafter as CAP. AM analyses for adult plant resistance were conducted in the complete spring panel (3,072 lines) hereafter designated as CAP-S. Seedling evaluations were conducted in the greenhouse during the autumn and winter months of 2006–2009. Isolate ND85F (pathotype 1) of *C. sativus* (Valjavec-Gratian

Table 22.1 Breeding programs of the Barley Coordinated Agricultural Project contributing germplasm to the association mapping panel for spot blotch resistance

Breeding program	Breeder	No. of lines	Growth habit	Row type	Primary use
University of Minnesota (MN)	Kevin Smith	384	Spring	Six	Malting
North Dakota State University (N6)	Richard Horsley	384	Spring	Six	Malting
USDA-ARS Aberdeen (AB)	Don Obert	384	Spring	Six/two	Malting/feed
Utah State University (UT)	David Hole	384	Spring	Six/two	Feed
Busch Agricultural Resources Inc. (BA)	Blake Cooper	384	Spring	Six/two	Malting
North Dakota State University (N2)	Richard Horsley	384	Spring	Two	Malting
Washington State University (WA)	Steve Ullrich	384	Spring	Six/two	Malting/feed/food
Montana State University (MT)	Tom Blake	384	Spring	Two	Malting/feed/food
Oregon State University (OR)	Patrick Hayes	384	Winter/facultative	Six/two	Malting/feed/food
Virginia Tech. University (VT)	Carl Griffey	384	Winter	Six	Feed

and Steffenson 1997) was used in all experiments. Adult disease evaluations were conducted at the Minnesota Agricultural Experiment Station in St. Paul with the same isolate.

A mixed linear model (MLM), implemented in software TASSEL (version 3.0) (www.maizegenetics.net), was used to detect associations between SNP markers and spot blotch resistance, both at the seedling and adult plant stages. The Benjamini-Hochberg false discovery rate (BH-FDR) (Benjamini and Hochberg 1995) of q-value = 0.05 was used to correct for multiple comparisons using program QVALUE (Storey 2002).

22.3 Results

The lowest overall seedling disease levels found across breeding programs were in six-rowed lines from MN and N6 (Fig. 22.1), whose mean infection responses (IRs) were 2.6 (median = 2.5) and 2.9 (median = 3.0), respectively. Nearly all of the lines from these two programs had very low disease reactions as only six and nine lines exhibited IRs above 4.0, respectively. Mean and median IRs for most of the other breeding programs were higher than those for MN and N6, and they also varied greater in their range (Fig. 22.1).

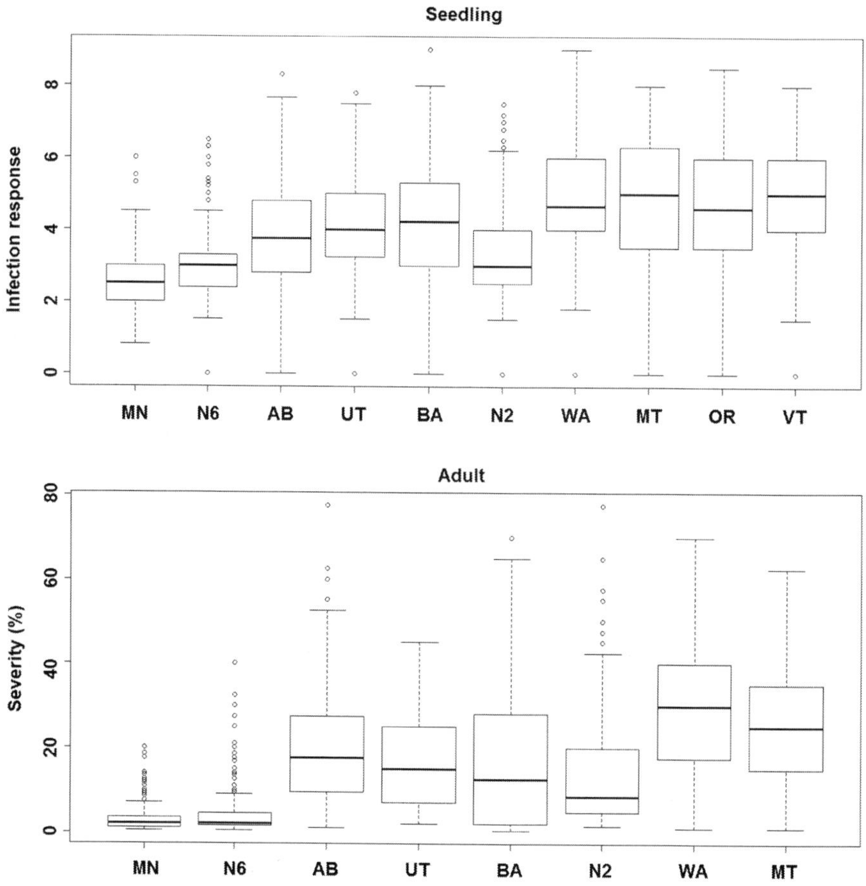

Fig. 22.1 Boxplots of spot blotch seedling and adult plant phenotypic data of US barley breeding germplasm: University of Minnesota (*MN*), North Dakota State University (*N6*), USDA-ARS Aberdeen (*AB*), Utah State University (*UT*), Busch Agricultural Resources Inc. (*BA*), North Dakota State University (*N2*), Washington State University (*WA*), Montana State University (*MT*), Oregon State University (*OR*), and Virginia Polytechnic Institute and State University (*VT*)

As in the seedling tests, six-rowed lines from MN and N6 were highly resistant at the adult plant stage, exhibiting the lowest and narrowest range of severities (Fig. 22.1). Mean severities for MN and N6 were 2.9 (median = 2.0) and 4.0 (median = 2.0), respectively. The mean and median severities for most of the other breeding programs were markedly higher than those for MN and N6, and they also varied greater in their range (Fig. 22.1).

Three common genomic regions (Fig. 22.2) on chromosomes 1H, 3H, and 7H were significantly associated with both seedling and adult plant resistance. The most significant SNP markers found were 11_10764, 11_10565, and 11_20162, respectively, on these three regions (Fig. 22.2); thus, the resistance QTL were named as *Rcs-qtl-1H-11_10764, Rcs-qtl-3H-11_10565,* and *Rcs-qtl-7H-11_20162*, respectively.

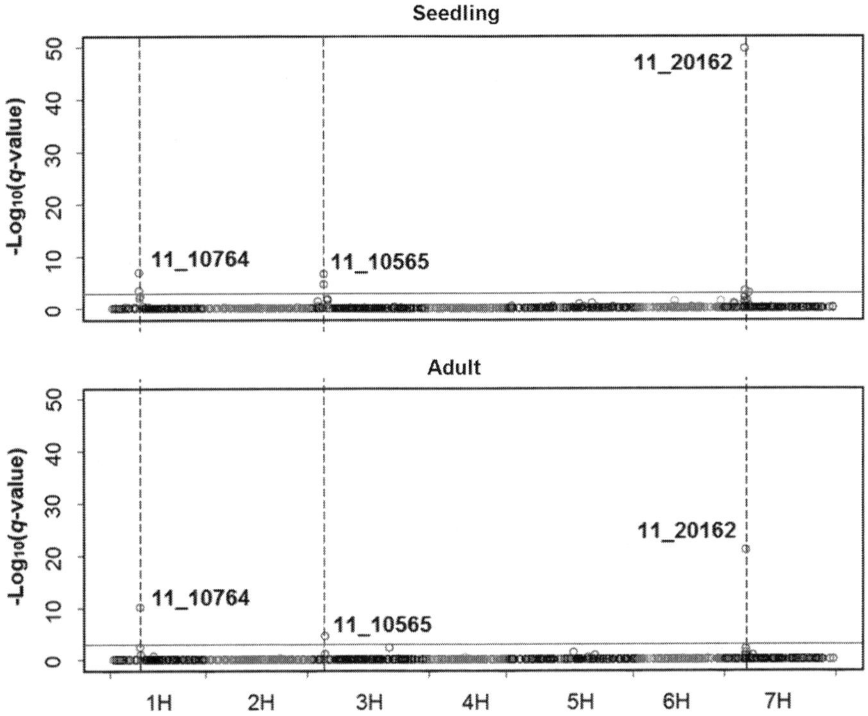

Fig. 22.2 Genome-wide association scan of spot blotch seedling and adult plant resistance of US barley breeding germplasm. The *vertical dotted lines* indicate the genomic regions that were found significantly associated with both seedling and adult plant spot blotch resistance with the most significantly associated markers for the respective QTL labeled. The *horizontal lines* indicate the threshold of significance (false discovery rate = 0.05)

To assess the effect of *Rcs-qtl-1H-11_10764*, *Rcs-qtl-3H-11_10565*, and *Rcs-qtl-7H-11_20162*, lines carrying different combinations of the resistance alleles or no resistance alleles were selected and assessed for their mean IR and disease severity calculated within each breeding program and also across all ten breeding programs (Table 22.2). The results showed that compared to lines lacking any resistance alleles, a single resistance QTL had no to a moderate effect in reducing the IR (from 0 to 20%) and disease severity (20–29%). Lines carrying two QTL exhibited 5–31% lower IRs and 52–56% lower disease severities. Finally, lines carrying all three QTL had 47% lower IRs and 83% lower severities compared to lines without any resistance alleles.

Haplotypes were investigated within the most resistant (IR $\leq$2 and severity $\leq$15%) class of lines (312 lines) from the complete panel and also in ten additional six-rowed cultivars or lines (NDB112, Morex, Excel, Drummond, Foster, Legacy, Quest, M123, Robust, Stander) that have been well characterized for their resistance in the Midwest over a period of years. All of these highly resistant lines carry the same resistant haplotype of "A" at *Rcs-qtl-1H-11_10764*, "B" at *Rcs-qtl-3H-11_10565*,

Table 22.2 Summary of allele effects for different combinations of spot blotch resistance QTL in US barley breeding germplasm

	Rcs-qtl-1H-11_10764			*Rcs-qtl-3H-11_10565*			*Rcs-qtl-7H-11_20162*			None		
Germplasm	No.[a]	IR[b]	Severity[c]	No.	IR	Severity	No.	IR	Severity	No.	IR	Severity
MN	0	–	–	0	–	–	0	–	–	0	–	–
N6	0	–	–	0	–	–	0	–	–	0	–	–
AB	16	6.3	27	0	–	–	15	5.6	30.7	5	6.8	44.5
UT	1	4.5	12	3	6.9	34.6	57	5	24.3	16	6.4	28.2
BA	21	5.2	39.7	13	7.2	38.9	11	6	32.8	4	7.1	45.2
N2	0	–	–	0			5	4.9	25.9	0	–	–
WA	13	5.4	39.2	14	6.3	39.6	106	5.6	35.6	19	6.7	44.7
MT	11	5.2	46	6	7	23	7	4.8	33	12	6.7	51
OR	24	6.1	–	5	6.5	–	30	4.8	–	26	5.7	–
VT	48	5.6	–	2	5.9	–	94	4.2	–	172	5.3	–
All programs	134	5.5 (14%)	32.8 (23%)	43	6.7 (0%)	34 (20%)	325	5.1 (20%)	30.4 (29%)	254	6.4	42.7

Germplasm	*Rcs-qtl-3H-11_10565* *Rcs-qtl-7H-11_20162*			*Rcs-qtl-1H-11_10764* *Rcs-qtl-7H-11_20162*			*Rcs-qtl-1H-11_10764* *Rcs-qtl-3H-11_10565*			All three QTL		
	No	IR	Severity	No	IR	Severity	No.	IR	Severity	No.	IR	Severity
MN	12	3.6	3.1	0	–	–	8	4.6	4.3	362		3
N6	57	3.6	7.8	4	3.4	2.6	5	5.4	9.9	310	3.2	2.5
AB	18	5.1	22.7	75	4.2	22.5	43	5.1	25.7	197	3	6.3
UT	50	4.3	23	38	4.3	13.9	34	5.1	17.8	129	2.9	6
BA	27	5.6	31.1	103	4.7	16.6	39	4.7	19.5	169	3.2	1.8

N2	2	3.9	33.8	15	4.5	23.9	21	5.7	19.2	333	2.7	8.1
WA	46	4.8	33.4	64	4.3	21.9	28	5.1	28.4	77	3.8	16.7
MT	2	4.7	10.8	88	4.8	29.7	81	6	25.5	159	3.3	12.8
OR	4	4.5	–	220	4.9	–	10	6	–	47	4.5	–
VT	0	–	–	33	5	–	0	–	–	1	3.8	–
All programs	218	4.5 (30%)	20.7 (52%)	640	4.4 (31%)	18.7 (56%)	269	5.3 (5%)	18.8 (56%)	1,784	3.7	7.1 (83%)
											3.4 (47%)	

[a] Number of lines carrying the resistance allele(*s*) for the respective QTL

[b] Mean infection response (IR) at seedling stage. Percentage in brackets is the disease reduction as assessed by IR at seedling stage compared to barley lines lacking any resistance QTL

[c] Mean disease severity (%) at adult plant stage. Percentage in brackets is the disease reduction as assessed by severity at adult stage compared to barley lines lacking any resistance QTL

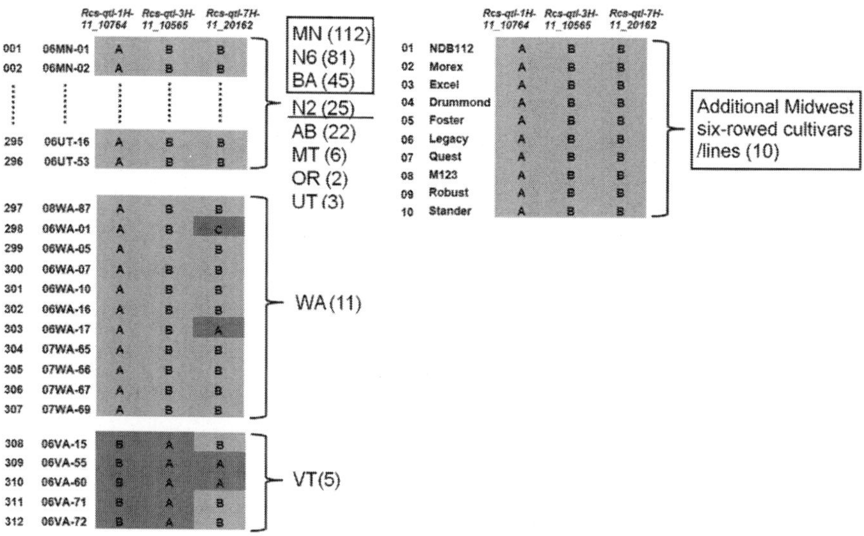

Fig. 22.3 Haplotypes of the most spot blotch resistant lines (312 lines, seedling infection response ≤2, and adult severity ≤15%) in US barley breeding germplasm and also ten well-characterized highly resistant six-rowed cultivars/lines from the Midwest at three QTL, *Rcs-qtl-1H-11_10764*, *Rcs-qtl-3H-11_10565*, and *Rcs-qtl-7H-11_20162*. Midwest six-rowed cultivars or lines are listed in the *boxes*, and Midwest two-rowed lines are *underlined*. Numbers of resistant lines in each breeding program are given in parentheses at *right*

and "B" at *Rcs-qtl-7H-11_20162*, except two from WA and five from VT (Fig. 22.3). Since all of the most widely grown Midwest cultivars over the past 33 years (and presumably those before this era) carry this haplotype, it was designated as the Midwest six-rowed durable resistance haplotype (MSDRH).

22.4 Discussion

Six-rowed malting barley cultivars grown in the Midwest have remained resistant to spot blotch for over 50 years, a remarkable record given the large plantings of cultivars with the resistance, the widespread distribution of the pathogen, and the consistently favorable weather conditions for disease development in the region (Steffenson 2000). This durable resistance was derived from line NDB112 (Wilcoxson et al. 1990) and has been bred into every six-rowed malting cultivar released since the late 1960s in the region. Using the mapping population Steptoe/Morex, Steffenson et al. (1996) found that adult plant resistance in the NDB112-derived cultivar Morex was due to a major effect QTL on chromosome 1H and a minor effect QTL (possibly *Rcs5*) on chromosome 7H. Bilgic et al. (2005) corroborated the results of Steffenson et al. (1996) in the same Steptoe/Morex population naming the respective QTL *Rcs-qtl-1H*-6-7 and *Rcs-qtl-7H*-2-4 but also identified two additional ones contributed by Morex in the population, namely, *Rcs-qtl-2H-3-5* and *Rcs-qtl-3H-11-12*. Using an AM approach with a large germplasm panel, three QTL (*Rcs-qtl-1H-11_10764*, *Rcs-qtl-3H-11_10565,* and *Rcs-qtl-7H-11_20162*) were identified for both seedling and adult plant resistance in US barley breeding germplasm and are likely coincident with the three QTL (*Rcs-qtl-1H-6-7*, *Rcs-qtl-3H-2-4*, and *Rcs-qtl-7H*-2-4 [presumably *Rcs5*], respectively) previously described from the biparental populations based on the comparative barley BIN (barleygenomics.wsu.edu) and SNP marker maps (Close et al. 2009).

To retain the durable spot blotch resistance in Midwest six-rowed germplasm, lines from the MN and N6 programs have been routinely screened for resistance in the field. The low disease phenotype statistics (Fig. 22.1) exhibited by this germplasm and the absence of any significant spot blotch outbreaks on widely grown cultivars in the field validate the successful selection and retention of this durable resistance. The three QTL found for spot blotch resistance by genome-wide AM of CAP and CAP-S were subsequently used for haplotype analysis of the most resistant CAP lines (312 lines) as well as ten other well-characterized resistant six-rowed cultivars or lines from the Midwest.

This analysis revealed the MSDRH of "A" at *Rcs-qtl-1H-11_10764,* "B"at *Rcs-qtl-3H-11_10565,* and "B" at *Rcs-qtl-7H-11* (Fig. 22.3). One of these other resistant lines included NDB112, the original source of the durable resistance. Even though NDB112 exhibits an even higher level of resistance than derived cultivars (B. Steffenson, unpublished data) and may therefore possess additional resistance loci, it appears that its essential loci for conferring durable resistance have been successfully transferred into all of these resistant Midwest six-rowed barleys and comprise the MSDRH (Fig. 22.3).

The allelic effects in reducing spot blotch disease levels were greater in lines with combinations of QTL compared to single QTL. A single QTL reduced the seedling IR and adult plant severity by 0–20% and 20–29%; two QTL by 5–31% and 47–56%; and three QTL (i.e., the MSDRH) by 47 and 83%, respectively (Table 22.2). The additive effects of multiple QTL contributing to higher levels of disease resistance also were reported for stripe rust of barley by Castro et al. (2003). To retain both the level and longevity of resistance, it is essential that all three QTL for spot blotch resistance be maintained in cultivars.

The Midwest two-rowed program of N2 is an interesting contrast to the six-rowed programs with respect to durable spot blotch resistance. As was the case for the Midwest six-rowed programs, lines from the two-rowed program have been routinely selected for spot blotch resistance for over 20 years (R. Horsley, personal communication). However, the mean seedling IR and adult plant severity of N2 lines were consistently higher than for Midwest six-rowed lines (Fig. 22.1). Yet haplotype analysis revealed that 88% (333 of 376) of N2 lines had the three QTL (i.e., MSDRH) (Table 22.2). This result suggests that the spot blotch resistance conferred by the MSDRH is not as effective in this two-rowed genetic background. The case of cultivar Bowman offers further evidence for this hypothesis. It was the first two-rowed cultivar released from the North Dakota program and carries the MSDRH identified in this study. In many spot blotch screening nurseries conducted with *C. sativus* pathotype 1 (ND85F) over the past 22 years, Bowman, like other N2 lines from CAP, consistently exhibited higher disease phenotype statistics than comparable six-rowed lines with the MSDRH (data not shown). Moreover, Bowman succumbed to a new virulence type (pathotype 2) of *C. sativus* after only 6 years of cultivation, while a succession of different six-rowed cultivars has remained highly resistant (Fetch and Steffenson 1994; Valjavec-Gratian and Steffenson 1997). The genetic architecture responsible for Bowman having a lower resistance level to pathotype 1 and susceptibility to pathotype 2, compared to Midwest six-rowed germplasm, is not known. This example demonstrates that even when a well-characterized haplotype conferring durable resistance has been successfully transferred to a line, there is no guarantee that it will confer the same level or longevity of resistance.

In summary, the spot blotch resistance bred into Midwest six-rowed barley cultivars has remained effective for over 50 years and is a remarkable example of durable resistance. Using a genome-wide AM approach, the genetic architecture of this durable spot blotch resistance was clearly resolved and the contributing haplotype (MSDRH) identified. The MSDRH will be valuable for the marker-assisted selection of spot blotch resistance in breeding lines or as a component of a larger genomic selection effort in barley breeding. However, field testing should still be done to confirm full expression of the MSDRH given the example with Midwest two-rowed barley.

References

Benjamini, Y., & Hochberg, Y. (1995). Controlling the false discovery rate: A practical and powerful approach to multiple testing. *Journal of Royal Statistical Society, 57*, 289–300.

Bilgic, H., Steffenson, B. J., & Hayes, P. (2005). Comprehensive genetic analyses reveal differential expression of spot blotch resistance in four populations of barley. *Theoretical and Applied Genetics, 111*(7), 1238–1250.

Castro, A., Chen, X., Hayes, P. M., & Johnston, M. (2003). Pyramiding quantitative trait locus (QTL) alleles determining resistance to barley stripe rust: Effects on resistance at the seedling stage. *Crop Science, 43*, 651–659.

Close, T. J., Bhat, P. R., Lonardi, S., Wu, Y., Rostoks, N., Ramsay, L., Druka, A., Stein, N., Svensson, J. T., Wanamaker, S., Bozdag, S., Roose, M. L., Moscou, M. J., Chao, S., Varshney, R. K., Szucs, P., Sato, K., Hayes, P. M., Matthews, D. E., Kleinhofs, A., Muehlbauer, G. J., DeYoung, J., Marshall, D. F., Madishetty, K., Fenton, R. D., Condamine, P., Graner, A., & Waugh, R. (2009). Development and implementation of high-throughput SNP genotyping in barley. *BMC Genomics, 10*, 582.

Fetch, T. G., & Steffenson, B. J. (1994). Identification of *Cochliobolus sativus* isolates expressing differential virulence on two-row barley genotypes from North Dakota. *Canadian Journal of Plant Pathology, 16*, 202–206.

Flint-Garcia, S. A., & Thornsberry, J. M. (2003). Structure of linkage disequilibrium in plants. *Annual Review of Plant Biology, 54*, 357–374.

Knight, N. L., Platz, G. J., Lehmensiek, A., & Sutherland, M. W. (2010). An investigation of genetic variation among Australian isolates of *Bipolaris sorokiniana* from different cereal tissues and comparison of their abilities to cause spot blotch on barley. *Australasian Plant Pathology, 39*, 207–216.

Mathre, D. E. (1997). *Compendium of barley diseases* (2nd ed.). St. Paul: The American Phytopathological Society Press.

Parisseaux, B., & Bernardo, R. (2004). In silico mapping of quantitative trait loci in maize. *Theoretical and Applied Genetics, 109*, 508–514.

Sprague, R. (1950). *Diseases of cereals and grasses in North America: Fungi, except smuts and rusts*. New York: Ronald Press.

Steffenson, B. J. (2000). Durable resistance to spot blotch and stem rust in barley. In *Proceedings of the 8th International Barley Genetics Symposium*, Glen Osmond (Barley Genetics VIII, Vol. I, pp. 39–44).

Steffenson, B. J., Hayes, P. M., & Kleinhofs, A. (1996). Genetics of seedling and adult plant resistance to net blotch (*Pyrenophora teres f. teres*) and spot blotch (*Cochliobolus sativus*) in barley. *Theoretical and Applied Genetics, 92*, 552–558.

Storey, J. D. (2002). A direct approach to false discovery rates. *Journal of the Royal Statistical Society, 64*, 479–498.

Valjavec-Gratian, M., & Steffenson, B. J. (1997). Pathotypes of *Cochliobolus sativus* on barley in North Dakota. *Plant Disease, 81*, 1275–1278.

Wilcoxson, R. D., Rasmusson, D. C., & Miles, M. R. (1990). Development of barley resistant to spot blotch and genetics of resistance. *Plant Disease, 74*, 207–210.

Chapter 23
Genetic Fine Mapping of a Novel Leaf Rust Resistance Gene and a *Barley Yellow Dwarf Virus* Tolerance (BYDV) Introgressed from *Hordeum bulbosum* by the Use of the 9K iSelect Chip

Perovic Dragan, Doris Kopahnke, Brian J. Steffenson, Jutta Förster, Janine König, Benjamin Kilian, Jörg Plieske, Gregor Durstewitz, Viktor Korzun, Ilona Kraemer, Antje Habekuss, Paul Johnston, Richrad Pickering, and Frank Ordon

Abstract Leaf rust and barley yellow dwarf, caused by Puccinia hordei Otth and barley yellow dwarf virus (BYDV)/cereal yellow dwarf virus (CYDV), are important diseases of barley (Hordeum vulgare L.) worldwide. Screening of spring barley landraces from Serbia led to the identification of the accession 'MBR1012' carrying resistance to the most widespread virulent leaf rust pathotypes in Europe, while a barley line carrying an introgression derived from H. bulbosum on chromosome 2HL was found to be highly tolerant to BYDV-PAV. In a population of 91 doubled haploid lines, derived from the cross MBR1012 (R) × Scarlett (S), the resistance gene against leaf rust was mapped in the telomeric region of chromosome 1HS by using simple sequence repeats (SSR). In parallel, the population was genotyped on the newly developed Illumina iSelect custom 9K BeadChipQ00231, resulting in the identification of closer linked markers. To exploit BYDV tolerance, DH lines derived

P. Dragan (✉) • D. Kopahnke • J. König • I. Kraemer • A. Habekuss • F. Ordon
Julius Kuehn-Institute (JKI), Federal Research Institute for Cultivated Plants,
Institute for Resistance Research and Stress Tolerance,
Erwin-Baur-Str. 27, 06484 Quedlinburg, Germany
e-mail: dragan.perovic@jki.bund.de

B.J. Steffenson
Department of Plant Pathology, University of Minnesota, 495 Borlaug Hall,
1991 Upper Buford Circle, St. Paul, MN 55108-6030, USA

J. Förster
Saaten-Union Biotec GmbH, Hovedisser Str. 92, 33818, Leopoldshöhe, Germany

B. Kilian
Leibniz Institute of Plant Genetics and Crop Plant Research (IPK),
Corrensstr. 3, 06466 Gatersleben, Germany

J. Plieske • G. Durstewitz
TraitGenetics GmbH, Am Schwabeplan 1b, 06466 Gatersleben, Germany

G. Zhang et al. (eds.), *Advance in Barley Sciences: Proceedings of 11th International Barley Genetics Symposium*, DOI 10.1007/978-94-007-4682-4_23,

from the cross (Emir x H. bulbosum) x Emir have been analysed in three steps. In a first step, out of 221 DH lines, 27 plants carrying a recombination event in the H. bulbosum fragment were selected based on nine markers specific for chromosome 2HL. In a next step, selected recombinant plants were analysed on a custom-made Illumina BeadXpress Array (384 SNPs) and on the 9k iSelect BeadChip. Finally, artificially inoculated DH lines carrying introgressions of different sizes will be screened for BYDV virus tolerance by artificial inoculation in order to map this tolerance. Results obtained revealed the presence of novel resistance/tolerance genes and in parallel provide the tools for their efficient deployment in barley breeding

Keyword Barley • *Puccinia hordei* • *Barley yellow dwarf virus* • Resistance • *Hordeum bulbosum* introgression • SSRs • SNPs • iSelect

23.1 Introduction

Leaf rust, caused by *Puccinia hordei* Otth, and barley yellow dwarf, caused by different aphid transmitted virus species belonging to *barley yellow dwarf virus* (BYDV) and *cereal yellow dwarf virus* (CYDV), are important diseases of barley (*Hordeum vulgare* L.) worldwide. Under experimental conditions, yield losses caused by leaf rust can be up to 60% in susceptible barley cultivars, but losses about half that level are common in practice (Cotterill et al. 1992; Das et al. 2007). Barley yellow dwarf (BYD), which is the most prevalent and economically important virus disease of cereals (Miller and Rasochová 1997), causes yield losses in barley up to 40% (Lister and Ranieri 1995). In order to prevent these yield losses, resistance is the most cost-effective and environmentally friendly approach. Although many resistance genes are known in barley (Ordon 2009), most of these resistances have been overcome by virulent races, raising the need for new and durable sources of resistance against the most important fungal and viral pathogens. This is a prerequisite to ensure a sustainable barley production in the future, especially as some pathogens like rust and BYDV may gain more importance due to global warming in specific regions of the world.

In cultivated barley (*Hordeum vulgare* ssp. *vulgare)* and wild barley (*Hordeum vulgare* ssp. *spontaneum)*, 20 major leaf rust resistance genes (*Rph1–Rph20*) have been described (Golegaonkar et al. 2009; Hickey et al. 2011). These major genes are effective against different pathotypes of *P. hordei* but when deployed singly

V. Korzun
KWS LOCHOW GmbH, Grimsehl str. 31, 37574, Einbeck, Germany

P. Johnston • R. Pickering
New Zealand Institute for Plant and Food Research Limited, Private Bag 4704, Christchurch 8140, New Zealand

have often been overcome by new pathotypes. The leaf rust resistance genes *Rph7, Rph15* and *Rph16* are still effective in Europe (Niks et al. 2000; Perovic et al. 2004) but are potentially vulnerable because pathotypes with virulence against *Rph7* are known in Israel (Golan et al. 1978), Morocco (Parlevliet 1981) and the USA (Steffenson et al. 1993) already. Therefore, it may be just a matter of time until these resistance genes are overcome in Europe in the same way as *Rph3* and *Rph10* (Dreiseitl 1990; Fetch et al. 1998). Within the primary gene pool of barley, which includes cultivars, landraces and wild barley accessions, the latter two are of particular importance regarding breeding for disease resistance. Incorporation of new leaf rust resistance genes from landraces was reported only for the genes *Rph3* and *Rph7* (Walther and Lehmann 1980; Jin et al. 1995), while a much larger number (*Rph2, Rph10, Rph11, Rph12, Rph13, Rph15 and Rph16*) were identified in the wild progenitor *H. vulgare* ssp. *spontaneum* (Franckowiak et al. 1996; Ivandic et al. 1998). In this regard, although the diversity present in barley landraces represents a good base for search and transfer of important traits into elite cultivars, they are scarcely used in breeding programmes (Silvar et al. 2010). In order to identify new sources for leaf rust resistance, barley germplasm from eco-geographically diverse collection sites in Serbia was evaluated (Perovic et al. 2001). One of the landraces from this collection ('MBR1012') exhibited a hypersensitive resistance reaction to *P. hordei* isolate I-80, which is virulent to all known major resistance genes in the European barley gene pool, except *Rph7, Rph15* and *Rph16* (Weerasena et al. 2004). Therefore, studies were conducted (1) to investigate the genetics of leaf rust resistance in 'MBR1012', (2) to develop molecular markers for this resistance gene facilitating efficient marker-based selection procedures, (3) to saturate the resistance locus through the use of the newly developed Illumina iSelect custom 9K BeadChip and (4) to develop resources towards the isolation of the corresponding resistance gene.

Regarding BYDV tolerance, *ryd1* detected in the spring barley cultivar 'Rojo' (Suneson 1955) as well as *Ryd2* and *Ryd3* with similar effects against BYDV-PAV and –MAV are known. Both genes were detected in Ethiopian landraces (Schaller et al. 1964; Niks et al. 2004). Besides this, several QTL for BYDV tolerance have been identified (Toojinda et al. 2000; Scheurer et al. 2001; Kraakman et al. 2006). Depending on the virus isolate used and the genetic background, these genes confer tolerance to BYDV, but pyramiding of *Ryd2* located on chromosome 3H (Collins et al. 1996) and *Ryd3* located on chromosome 6H (Niks et al. 2004) resulted in quantitative resistance to BYDV-PAV (Riedel et al. 2011). In contrast to this, a gene called *Ryd4*^{Hb} conferring complete resistance to BYDV-PAV has been transferred from *Hordeum bulbosum* L., which is the sole member of the secondary gene pool of *Hordeum vulgare* (von Bothmer et al. 1995), into cultivated barley. *Hordeum bulbosum* has proven to be a valuable source of genetic diversity for barley crop improvement already. Despite crossing difficulties, successful introgressions of agronomical useful genes, e.g. resistance to powdery mildew (Pohler and Szigat 1982; Xu and Kasha 1992; Pickering et al. 1995), leaf rust (Szigat et al. 1997; Pickering et al. 1998), the BaMMV/BaYMV complex (Ruge et al. 2003; Ruge-Wehling et al. 2006) and BYDV (Scholz et al. 2009), have been achieved. During previous screenings of barley lines carrying an introgression derived from *H. bulbosum* (Johnston et al. 2009), a line highly tolerant to BYDV carrying an introgression on chromosome 2HL was identified. Therefore, studies were conducted on DH lines carrying this introgression using SSRs, a custom-made Illumina GoldenGate Array and the newly developed Illumina iSelect custom 9K BeadChip in order to identify lines with introgressions of different sizes. In a next step, these lines will be artificially inoculated with BYDV and further analysed towards the identification of a candidate gene and the development of perfect markers to facilitate use of this BYDV tolerance in breeding programmes.

23.2 Materials and Methods

23.2.1 Plant Material

An F_1-derived doubled haploid (DH) population, produced via anther culture, comprising 91 lines of a cross between the leaf-rust-resistant landrace 'MBR1012' and the susceptible German cultivar 'Scarlett', was used for mapping (König et al. 2012). A diverse collection of 51 barley accessions was evaluated for resistance to determine the diagnostic value of the closest linked molecular markers. For construction of a high-resolution mapping population for the respective leaf rust resistance gene, F_2 plants of the cross 'MBR1012'×'Scarlett' have been analysed with flanking markers in order to generate segmental recombinant inbreed lines (RILs). Also, a cross between landrace 'MBR1012' and the *Rph4*-resistant cultivar 'Gold' was made in order to test for allelism of two resistant genes that are mapped at a similar chromosomal position.

A barley line carrying an introgression derived from *H. bulbosum* on chromosome 2H (Johnston et al. 2009) that was highly tolerant to BYDV was backcrossed twice to the spring barley cultivar 'Emir', and the $BC2F_1$ plants were used for DH-line production.

23.2.2 Resistance Tests

Regarding leaf rust, five to ten plants of each DH line and the parents were inoculated with the leaf rust isolate I-80 in the greenhouse. Plants were inoculated at the seedling stage, according to Ivandic et al. (1998). Infection types were recorded between 10 and 12 days after inoculation according to the 0–4 scale of Levine and Cherewick (1952). Plants exhibiting infection types from 0, 0_{nc} (hypersensitive reactions with necrotic/chlorotic 'flecks'), 1, 2- or 0-2- were considered as resistant, while those exhibiting infection types 2+, 3, 3-4 and 4 were considered as susceptible. The chi-square test was used to assess segregation ratios for goodness of fit to expected ratios. Additionally, a collection of 13 European leaf rust isolates

(Walther 1987; Kopahnke unpublished data) was tested on the parental lines and accessions of 'Gold' (carrying leaf rust resistance gene *Rph4*), 'Cebada Capa' (*Rph7*), NIL Bowman-*Rph15* (Rph15), 'Hsp680' (*Rph16*) and 'L94' (susceptible control). Inoculation and scoring were performed as described for isolate I-80. The leaf rust isolates were collected from different parts of Europe and selected from single-spore progenies (Walther 1979; Walther 1987; Kopahnke unpublished data). The isolates are maintained at the Julius Kuehn Institute, Federal Research Institute for Cultivated Plants (JKI) and Institute for Resistance Research and Stress Tolerance, in Quedlinburg, Germany.

23.2.3 DNA Extraction and Marker Analyses

Genomic DNA of DH lines was extracted from leaves of 14-day-old plants according to Stein et al. (2001), while DNA extraction of F_2 plants for the construction of a high-resolution mapping population was performed according to Dorokhov and Klocke (1997), after modification for a semi-high-throughput procedure. The concentration and quality of DNA of DH lines were determined using the NanoDrop ND-100 spectrophotometer (PeQLab, Erlangen, Germany) and gel electrophoresis. All DNA of DH lines were adjusted to a final concentration of 20 ng/μl, while the DNA extracted from F_2 plants was used without previous adjustment. For bulk segregant analysis (BSA) (Michelmore et al. 1991), equal aliquots (10 μl) of DNA from nine resistant and nine susceptible DH lines were pooled. DNA for the iSelect analysis was adjusted to a final concentration of 50 ng/μl.

In the leaf rust project, a total of 175 SSRs and 73 SNPs were screened for polymorphism between the parents and bulks. The sequences of the SSR primer pairs and amplification protocols were obtained from Struss and Plieske (1998), Ramsay et al. (2000), Maccaulay et al. (2001), Thiel et al. (2003) and Varshney et al. (2007), while the sequences of the SNP primer pairs, amplification protocol and restriction sites were obtained from Kota et al. (2008). The sequences of the SSR primer pairs and CAPS markers for determining the size of the *H. bulbosum* introgression were obtained from Johnston et al. (2009) and Varshney et al. (2007).

PCR was performed in a volume of 10 μl, containing 1 μl of 10× buffer, 1 μl of 25 mM $MgCl_2$, 0.2 μl of each 10 mM dNTPs, forward primer (1.0 pmol/μl) and reverse primer (10.0 pmol/μl), 0.08 μl 5U Hot FIREPol®DNA polymerase (Solis BioDyne, Tartu, Estonia), 6.12 μl HPLC gradient grade water (Carl Roth, Karlsruhe, Germany) and 1 μl template DNA. For SSR amplification, M13-tailed forward primers were used so that 0.1 μl of 'M13' primer (10.0 pmol/μl) (5′-CACGACGTTGTAAAACGAC – 3′) labelled with 5′ fluorescent dyes was added to the reaction mix (Boutin-Ganache et al. 2001). DNA amplification was performed in a Gene Amp® PCR System 9700 (Applied Biosystems, Darmstadt, Germany). The following PCR conditions were used for all primers: 94°C for 5 min followed by a touchdown PCR (–0.5°C/cycle) with 12 cycles of 30 s at

94°C, 30 s at 62°C, 30 s at 72°C followed by 35 cycles with 30 s at 94°C, 30 s at 56°C, 30 s at 72°C and a final extension at 72°C for 10 min. Detection of allele sizes for the SSR marker was conducted using a capillary electrophoresis ABI PRISM® 3100 genetic analyser (Applied Biosystems, Darmstadt, Germany) or a CEQ™ 8000 Genetic Analysis System (Beckman). SNP markers (Kota et al. 2008) were amplified in a volume of 20 µl, containing 2.0 µl of 10x PCR buffer with $MgCl_2$ (25 mM), 0.4 µl of each 10 mM dNTPs, 0.5 µl forward primer (1.0 pmol/µl) and reverse primer (10.0 pmol/µl), 0.16 µl 5U Taq DNA polymerase (Qiagen, Hilden, Germany), 6.12 µl HPLC gradient grade water (Carl Roth, Karlsruhe, Germany) and 2.0 µl template DNA using the same PCR conditions as described for the SSRs. The applied SNP markers were converted to cleaved amplified polymorphic sequence (CAPS) markers by digesting PCR products with corresponding restriction endonucleases. The PCR products were digested with corresponding restriction endonucleases to generate CAPS in a total volume of 20 µl with 10 µl PCR product, 2 µl of the relevant NE Buffer, 7,9 µl HPLC gradient grade water (Carl Roth, Karlsruhe, Germany) and 0,1 µl enzyme (New England Biolabs). The cleaved DNA fragments were separated on 1.8% agarose gels as described before.

Both DH populations, i.e. 91 DH lines for leaf rust resistance mapping and 27*H. bulbosum* introgression DH lines showing recombination of an original *H. bulbosum* fragment, were genotyped on the newly developed 9K Infinium iSelect high-density custom genotyping BeadChip. Additionally, *H. bulbosum* introgression DH lines were genotyped on an Illumina custom-made BeadXpress Array (384 SNPs) (Korzun and Stein, personal com.).

23.2.4 Map Construction and Statistical Tests

The genetic map of DH lines was constructed using JoinMap 4.0 (van Ooijen 2006) applying the Kosambi function (Kosambi 1944), while the map of the iSelect SNPs was constructed by using a maximum likelihood algorithm of JoinMap 4. Only markers with a LOD score of 3 were integrated into the map.

23.3 Results

23.3.1 Leaf Rust Phenotypic Analysis and Infection Patterns

Infection type of the susceptible parent Scarlett was 2–3, whereas the corresponding one of the resistant parent 'MBR1012' was 0-2-. The 91 DH lines derived from the cross of the resistant landrace 'MBR1012' and the susceptible German cultivar 'Scarlett' segregated 48 resistant to 43 susceptible ($\chi^2=0.29$). This segregation pattern fits well to the expected 1 resistant: 1 susceptible ratio, indicating that leaf rust

Table 23.1 Differential reactions of six barley cultivars/lines possessing known leaf rust resistance genes and the susceptible standard line L94 after inoculation with 13 *Puccinia hordei* isolates (König et al. 2012)

Cultivar	Gold	Cebada Capa	Bowmann	Scarlett	H.sp 680	MBR1012	L 94
Isolat/gene	*Rph4*	*Rph7*	*Rph15*	*Rph3/Rph9/Rph12*	*Rph 16*	$Rph_{MBR1012}$	susc.
R8-1	3-4	0_C	0	0	0_N	0	3-4
R8-2	3-4	0_C	0	0	0_N	0	3-4
R 14-1	3-4	0_C	0	0	0_N	0	3-4
R14-2	3-4	0_C	0	0	0_N	0	3-4
R 16-1	3-4	0_C	0	0	0_N	0	3-4
R34-3	3-4	0_C	0	0	0_N	0	3-4
R54-3	3-4	0_C	0	0	0_N	0	3-4
I 80	3-4	0_C	0-2-	3	0_N	0-2-	3-4
30-1*	3-4	0_C	0-2-	0	0_N	0, 0-1	3-4
30-1/2*	3-4	0_C	0	0, 0-2-	0_N	0, 0-2-	3-4
23-3	3-4	0_C	0	0	0_N	0	3-4
23-1/2/3*	3-4	0_C	0	0	0_N	2-	3-4
23-1+3*	3-4	0_C	0	0	0_N	0, 0-2-	3-4

resistance against isolate I-80 in 'MBR1012' is inherited in a monogenic manner (König et al. 2012).

In order to compare the resistance of 'MBR1012' and previously described resistance genes, a collection of 13 European leaf rust isolates (Walther 1987; Kopahnke unpublished data) was tested on the parental lines and accessions carrying related and still effective resistant genes. The seven barley accessions analysed displayed very different reaction patterns in response to the collection of 13 European leaf rust isolates (Table 23.1).

Susceptible 'L94' and 'Gold' carrying *-Rph*4 were highly susceptible to the entire collection of leaf rust isolates, while our susceptible parental cultivar 'Scarlett' exhibited resistance to 12 isolates and was only susceptible to isolate I-80. Although 'Cebada Capa' (*Rph*7), 'H.sp.680' (*Rph*16) and 'MBR1012' turned out to be resistant to all isolates, their reaction patterns varied from 0c (chlorosis), 0n (necrosis) to 0 -2-, respectively.

23.3.2 *Linkage Analysis of Leaf Rust Resistance Gene from Landrace MBR1012*

The parents and the two bulks of the population 'MBR1012' × 'Scarlett' were analysed with a set of 248 co-dominant molecular markers, which were evenly distributed along the seven chromosomes of barley. From these, 89 SSR (51%) and 21 SNP markers (29%) were polymorphic. Polymorphisms between the resistant and susceptible bulk were detected on chromosome 1H (König et al. 2012). Overall, out of the 32 SSR and 11 SNP markers localised on chromosome

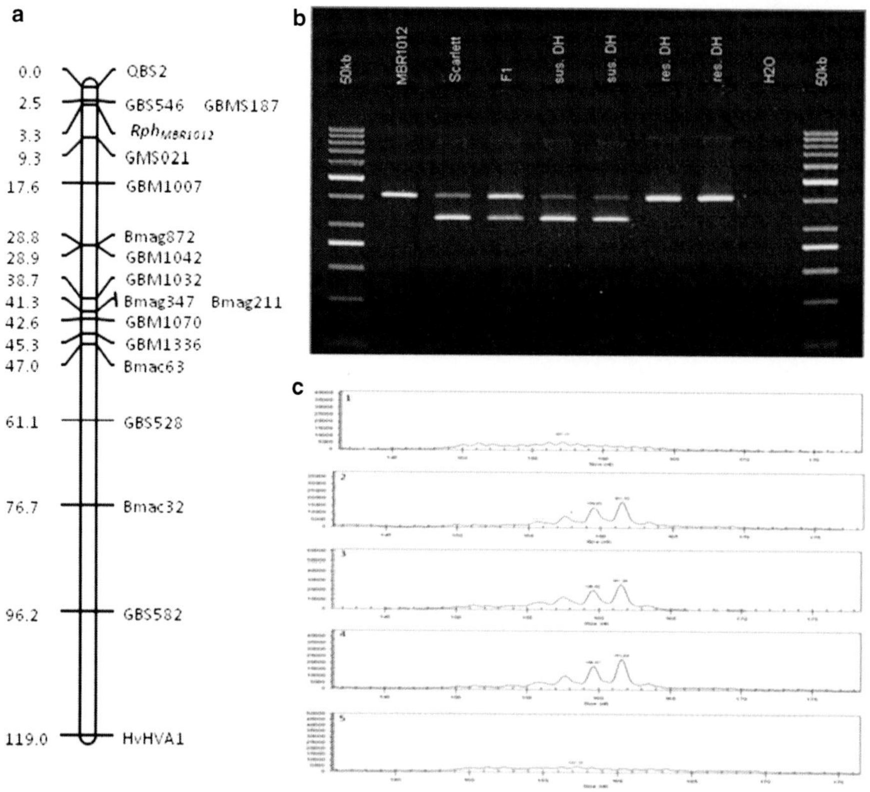

Fig. 23.1 (**a**) Genetic map of chromosome 1H including the resistance locus $Rph_{MBR1012}$; (**b**) Pattern of the CAPS marker GBS546 after digestion with *Hha*I closely linked to resistance locus Rph_{MBR101}; (**c**) Chromatograms of the SSR marker GBMS187 closely linked to the resistance gene $Rph_{MBR1012}$. The order of genotypes are resistant parent 'MBR1012' (1), susceptible parent 'Scarlett' (2), F1-plant (3), susceptible DH line (4) and resistant DH line (5) (König et al. 2012)

1H, 14 and 3 were polymorphic, respectively. Based on these markers, a final genetic map of 119 cM of chromosome 1H was constructed (Fig. 23.1a). The leaf rust resistance gene in 'MBR1012' was mapped in the telomeric region of the short arm of chromosome 1H. The closest linked markers are GBMS187 and GBS546, which map 0.8 cM distal to the resistance gene. On the proximal side, the closest marker identified is GMS21, which maps 6.0 cM from the resistance locus (König et al. 2012).

For construction of a high-resolution mapping population, flanking markers QBS2 and GMS0021 were used. A set of about 2400 F_2 plants were screened. Out of these, about 350 plants showing a single recombination and 25 plants being homozygous recombinant between flanking markers were selected. The single recombinant plants will be screened in F_3 generation in order to obtain homozygous recombinant inbreed lines.

23.3.3 Diagnostic Value of Markers Linked to $Rph_{MBR1012}$

To assess the diagnostic value of the four closest linked markers to the leaf rust resistance gene in 'MBR1012', a set of 51 barley accessions were assayed (König et al. 2012). This barley collection consisted of the original sources of described *Rph* genes, near-isogenic lines of cultivar Bowman with introgressed *Rph* genes, 2 lines with partial leaf rust resistance and 10 susceptible lines with no known *Rph* genes. The number of detected alleles per marker in this analysed collection varied from three alleles for GBS546 and 6 alleles for GMS21 to 9 and 11 alleles for QBS2 and GBMS187, respectively. Restriction pattern B of SNP marker GBS546 (Fig. 23.1b), characteristic for the resistant parent MBR1012, was also observed for 10 other genotypes. Similarly, for the closest proximal marker SSR GMS21, 10 genotypes showed the same allele pattern as the resistant parent MBR1012. Accuracy in prediction of $Rph_{MBR1012}$ was 74.5% for QBS2, 78.5% for GMS21, 80.5% for GBS546 and 100% for GBMS187. Therefore, QBS2, GBS546 and GMS21 have no diagnostic value for tagging the resistance gene described in MBR1012. The closely linked marker GBMS187 (Fig. 23.1c) detected a null allele in the resistant parent 'MBR1012', whereas in the other examined genotypes, eight different alleles were identified. Although the null allele of the GBMS187 in the parent 'MBR1012' is unique in the investigated germplasm, the use of this marker in breeding is of limited value due to its dominant mode of inheritance.

23.3.4 Selection of H. bulbosum Introgression Lines

Barley doubled haploid (DH) lines produced from $BC2F_1$ plants of the cross (Emir x *H. bulbosum*) x 'Emir' by anther culture technique were planted in the greenhouse. Out of 406 regenerated plants, 221 were diploid while the rest turned out to be haploid. Initially, a set of five CAPS (cleaved amplified polymorphism) markers and one SSR marker, which had turned out to be polymorphic in previous experiments, were used for genotyping of these 221 DH lines. In order to more precisely estimate the size of the *H. bulbosum* introgression, a set of 12 additional SSR markers from chromosome 2HL were selected, and three turned out to be polymorphic. In total, a set of 27 recombinant introgression lines were detected by using 9 markers and further analysed. Out of 221 DHs, four lines showed heterozygous signals, 118 DH lines carried the original *H. bulbosum* fragment, 72 DH lines the fragment of cv. Emir and 27 DH lines showed recombination (Table 23.2).

23.3.5 The 9K Infinium iSelect Genotyping

The 91 DH lines of the cross between the leaf-rust-resistant landrace 'MBR1012' and the susceptible cultivar 'Scarlett' and the 27 plants carrying *H. bulbosum*

Table 23.2 Graphical overview of some barley lines carrying *H. bulbosum* introgressions of different sizes

Marker	SSR1	SSR2	CAPS1	CAPS2	SSR3	CAPS3	CAPS4	SSR4	CAPS5
DH line	Bin12	Bin12	Bin12	Bin13	Bin13	Bin14	Bin14	Bin14	Bin15
DH-Pick14	HV	HV	HV	HV	HB	HB	HB	HB	HB
DH-Pick15	HV	HV	HV	HV	HB	HB	HB	HB	HB
DH-Pick22	HB	–	HV	HV	HV	HV	HB	–	HB
DH-Pick167	HB	HB	HB	HB	HB	HB	HB	HV	HB
DH-Pick168	HB	HB	HB	HB	HB	HB	HB	HV	HV
DH-Pick200	HV	HV	HV	HB	HB	HB	HB	HB	HB
DH-Pick204	HV	HV	HV	–	HV	HV	HV	HV	HB
DH-Pick214	HV	HV	HV	–	HV	HV	HV	HV	HB
DH-Pick245	HV	HV	HV	HV	HV	HV	HV	HB	HB
DH-Pick267	HV	HV	HV	HV	HV	HV	HB	HB	HB
DH-Pick272	HB	HB	HB	HB	HB	HB	HB	HV	HV
DH-Pick275	HV	HB	HB	HB	HB	HB	HB	HB	HB
DH-Pick276	HB*	HV	HV	HV	HV	HV	HV	HV	HV
DH-Pick279	HB	HB	HB	HB	HB	HB	HB	HV	HV
DH-Pick287	HB	HB	HV	HV	HV	HV	HV	HB	HB
DH-Pick307	HB	HB	HB	HB	HV	HV	HV	HV	HV
DH-Pick311	HB	HB	HB	HB	HB	HB	HB	HB	HV
DH-Pick313	HB	HV	–	HV	HB	HB	HB	HB	HB
DH-Pick323	HB	HB	HB	HB	HB	HB	HV	HV	HV
DH-Pick364	HV	HV	HV	HV	HV	HV	HV	HB	HB
DH-Pick383	HV	HV	HV	HV	HB	HB	HB	HB	HB
DH-Pick391	HB	HB	HB	HB	HB	HB	HB	HV	HV
DH-Pick403	HV	HV	HV	HV	HB	HB	HB	HB	HB

fragment of different sizes, including four parental lines, were genotyped on the newly developed Illumina iSelect custom 9 K BeadChip (Comadran et al. in prep). Based on these results, a set of 2,806 SNP markers have been placed onto the 'MBR1012' × 'Scarlett' map, which after including 111 SSRs and SNPs consists of 2,917 markers covering 2,326 cM. Regarding chromosomes 1H and 2H, sets of 313 and 520 markers, respectively, were polymorphic and could be used for construction of corresponding high-density chromosomal maps. At the $Rph_{MBR1012}$ locus, 32 new SNP markers were placed between flanking markers, allowing fine mapping and marker saturation of this resistance locus (data not shown).

To exploit BYDV tolerance, 27 DH lines carrying a recombination event in the *H. bulbosum* fragment, which were selected based on nine markers specific for chromosome 2HL, were further analysed using Illumina SNP platforms. Since genotyping on an Illumina BeadXpress Array (384 SNPs) confirmed homogeneity of the background and presence of *H. bulbosum* introgression on the long arm of chromosome 2H at 27 selected lines, genotyping of these lines was performed on the 9k iSelect BeadChip. In a targeted region on chromosome 2HL, more than 80

SNP markers revealed orthologue positions at the 'MBR1012'×'Scarlett' map, and about 50 SNP markers displayed polymorphism, allowing fine mapping of the *H. bulbosum* introgression.

23.4 Discussion

The appearance of new virulent races of fungal and viral pathogens and the depletion of the gene pool of cultivated barley for major resistance genes demand a continuous supply of new sources of disease resistance. Regarding breeding for disease resistance, particular importance has to be given to landraces and wild barley accessions (*Hordeum vulgare* ssp. *spontaneum*) as well as to the secondary breeding pool that has only one species, i.e. *Hordeum bulbosum*. Within the current projects, loci conferring resistance or tolerance, respectively, from the primary and secondary gene pool of barley were detected and characterised.

23.4.1 *Leaf Rust Resistance Gene*

The leaf rust resistance in landrace 'MBR1012' is conferred by a major gene, which exhibits a hypersensitive reaction to many of the most widespread virulent European *P. hordei* isolates and is located on the distal portion of the short arm of chromosome 1H (König et al. 2012). The leaf rust resistance locus $Rph_{MBR1012}$ was mapped in a 6.8 cM interval between the markers GBMS187/GBS546 and GMS21. GBMS187 and GBS546 co-segregate and are the closest linked markers (0.8 cM distal) to $Rph_{MBR1012}$. Although they co-segregated, an assessment of germplasm revealed different accuracy in the prediction of the MBR1012 resistance gene, i.e. 80.5% for GBS546 and 100% for GBMS187. It is important to note that GBMS187 (Fig. 23.1c) is a dominant marker in our population; hence, it may not be the best marker for MAS of this leaf rust resistance gene.

Other dominant leaf rust resistance genes have been mapped to nearly all barley chromosomes (see overview by Golegaonkar et al. 2009). Only the dominant gene *Rph4*, derived from cultivar 'Gold', was localised on chromosome 1HS through linkage (~17 cM) with the powdery mildew resistance locus *Mla* (McDaniel and Hathcock 1969). The fact that the leaf rust resistance gene in 'MBR1012' exhibits a completely different resistance spectrum than *Rph4* to the common European leaf rust isolates (including I-80, Table 23.1) provides strong evidence that the two genes are different. Before it can be unequivocally stated that the leaf rust resistance gene in 'MBR1012' is different from all other previously described genes in barley, a provisional locus designation of $Rph_{MBR1012}$ is assigned.

Genotyping at the high-density SNP platform revealed that plenty of new markers could be mapped at the $Rph_{MBR1012}$ resistance locus, enabling very efficient marker saturation. Regarding construction of a high-resolution mapping population, genetic distance between flanking markers was estimated at 8.3 cM based on the 2400 F_2 plants in contrast to 9.3 cM in the DH population of 91 lines. However, the distances estimated in both populations are in good accordance.

23.4.2 *BYDV Tolerance*

Selection procedure of *H. bulbosum* introgression lines with different fragment sizes via PCR based SSR and CAPS markers and by genotyping on an Illumina BeadXpress Array turned out to be efficient. Out of 221 DH lines, 27 were recombinant and homogeneous in the genetic background. In case the BYDV tolerance carrying fragment can be further delimited thereby reducing the linkage drag, this tolerance may be efficiently used in barley breeding, thereby broadening the genetic base of resistance/tolerance to BYDV in cultivated barley, which up to now is quite limited. This procedure may be also used to make $Ryd4^{Hb}$ available for practical barley breeding (Scholz et al. 2009). Besides this, respective closely linked markers are of special importance with respect to breeding for BYDV tolerance as an efficient phenotypic selection for tolerance requires artificial inoculation procedures based on viruliferous aphids, which is difficult to integrate into applied selection schemes.

23.5 Conclusions (and Future Work)

A new resistance gene to *P. hordei* was identified in a Serbian landrace. Results of the allelism test will answer whether this gene is novel or it is a new allele of the already known *Rph4* gene. Although further research on this gene is needed, the results of this work open the opportunity to broaden the genetic base of resistance to *P. hordei* in barley and provide breeding programmes with genetic markers that will facilitate the faster incorporation of this resistance into elite cultivars. The iSelect platform identified plenty of polymorphic markers, allowing efficient marker saturation of the resistance locus, which enable the effective screening for recombinant inbreed lines that will be derived from a high-resolution mapping population after screening of 5000 F_2 plants.

H. bulbosum introgression lines carrying the shortest fragment and being tolerant together with a non-tolerant line will be subjected to expression profiling. By this approach, genes involved in BYDV tolerance may be identified, and based on these genes, SNP markers will be developed, facilitating the introgression of BYDV tolerance with a minimum of linkage drag. Also, the conversion of the iSelect markers in, e.g. pyrosequencing markers (Silvar et al. 2011), will allow a rapid introgression of this tolerance into cultivated barley.

Acknowledgements Results were obtained within the project ExpResBar funded by the German Federal Ministry of Education and Research (BMBF) under the grant number 0315702B within the KBBE-II call.

References

Boutin-Ganache, I., Raposo, M., Raymond, M., & Deschepper, C. F. (2001). M13-tailed primers improve the readability and usability of microsatellite analyses performed with two different allele sizing methods. *Biotechniques, 31*, 24–28.

Collins, N. C., Paltridge, N. G., Ford, C. M., & Symons, R. H. (1996). The *Yd2* gene for *Barley yellow dwarf virus* resistance maps close to the centromere on the long arm of barley chromosome 3. *Theoretical and Applied Genetics, 92*, 858–864.

Cotterill, P. J., Rees, R. G., Platz, G. J., & Dill-Macky, R. (1992). Effects of leaf rust on selected Australian barley. *Australian Journal of Experimental Agriculture, 32*, 747–751.

Das, M. K., Griffey, C. A., Baldwin, R. E., Waldenmaier, C. M., Vaughn, M. E., Price, A. M., & Brooks, W. S. (2007). Host resistance and fungicide control of leaf rust (*Puccinia hordei*) in barley (*Hordeum vulgare*) and effects on grain yield and yield components. *Crop Protection, 26*, 1422–1430.

Dorokhov, D. B., & Klocke, E. (1997). A rapid and economic technique for RAPD analysis of plant genomes. *Genetika, 33*(4), 443–450.

Dreiseitl, A. (1990). Overcoming the resistance of barley conferred by gene *Pa*3 against leaf rust (*Puccinia hordei* Otth.). *Genetics Selection, 26*, 159–160.

Fetch, T. G., Steffenson, B. J., & Jin, Y. (1998). Worldwide virulence of *P. hordei* on barley. *Phytopathology, 88*(Suppl), 28–34.

Franckowiak, J. D., Jin, Y., & Steffenson, B. J. (1996). Recommended allele symbols for leaf rust resistance genes in barley. *Barley Genetics Newsletter, 27*, 36–44.

Golan, T., Anikster, Y., Moseman, J. G., & Wahl, I. (1978). A new virulent strain of *Puccinia hordei. Euphytica, 27*, 185–189.

Golegaonkar, P. G., Karaoglu, H., & Park, R. F. (2009). Molecular mapping of leaf rust resistance gene *Rph*14 in *Hordeum vulgare. Theoretical and Applied Genetics, 119*, 1281–1288.

Hickey, L. T., Lawson, W., Platz, G. J., Dieters, M., Arief, V. N., Germán, S., Fletcher, S., Park, R. F., Singh, D., Pereyra, S., & Franckowiak, J. (2011). Mapping *Rph20*: A gene conferring adult plant resistance to *Puccinia hordei* in barley. *Theoretical and Applied Genetics, 123*, 55–68.

Ivandic, V., Walther, U., & Graner, A. (1998). Molecular mapping of a new gene in wild barley conferring complete resistance to leaf rust (*Puccinia hordei* Otth). *Theoretical and Applied Genetics, 97*, 1235–1239.

Jin, Y., Steffenson, B. J., & Bockelman, H. E. (1995). Evolution of cultivated and wild barley for resistance to pathotypes of *Puccinia hordei* with wide virulence. *Genetic Resources and Crop Evolution, 42*, 1–6.

Johnston, A. P., Gail, M., Timmerman-Vaughan, K. J., Farnden, F., & Pickering, R. (2009). Marker development and characterization of *Hordeum b bulbosum* introgression lines: A resource for barley improvement. *Theoretical and Applied Genetics, 118*, 1429–1437.

König, J., Steffenson, B., Kopahnke, D., Przulj, N., Romeis, T., Ordon, F., & Perovic, D. (2012). Genetic mapping of novel leaf rust (*Puccinia hordei* Otth) resistance in barley landrace MBR1012. *Molecular breeding*. doi:10.1007/s11032-012-9712-0.

Kosambi, D. D. (1944). The estimation of map distances from recombination values. *Annals of Eugenics, 12*, 172–175.

Kota, R., Varshney, R. K., Prasad, M., Zhang, H., Stein, N., & Graner, A. (2008). EST-derived single nucleotide polymorphism markers for assembling genetic and physical maps of the barley genome. *Functional & Integrative Genomics, 8*, 223–233.

Kraakman, A. T. W., Martinez, F., Mussiraliev, B., van Eeuwijk, F. A., & Niks, R. E. (2006). Linkage disequilibrium mapping of morphological, resistance, and other agronomically relevant traits in modern spring barley cultivars. *Molecular Breeding, 17*(1), 41–58.

Levine, M. N., & Cherewick, W. J. (1952). Studies on dwarf leaf rust of barley. *US Department of Agriculture Technical Bulletin, 1056*, 1–17.

Lister, R. M., & Ranieri, R. (1995). Distribution and economic importance of *Barley yellow dwarf.* In C. J. D'Arcy & P. A. Burnett (Eds.), *Barley yellow dwarf: 40 years of progress* (pp. 29–53). St. Paul: APS Press.

Maccaulay, M., Ramsay, L., Powell, W., & Waugh, R. (2001). A representative, highly informative 'genotyping set' of barley SSR's. *Theoretical and Applied Genetics, 102*, 801–809.

McDaniel, M. E., & Hathcock, B. R. (1969). Linkage of *Pa4* and *Mla* Loci in Barley. *Crop Science, 9*, 822–823.

Michelmore, R. W., Paran, I., & Kesseli, R. V. (1991). Identification of markers linked to disease – Resistance genes by bulked segregant analysis: A rapid method to detect markers in specific genomic regions by using segregating populations. *Proceedings of the National Academy of Sciences, 88*, 9828–9832.

Miller, W. A., & Rasochová, L. (1997). Barley yellow dwarf viruses. *Annual Review of Phytopathology, 35*, 167–190.

Niks, R. E., Walther, U., Jaiser, H., Martinez, F., Rubiales, D., Anderson, O., Flath, K., Gymer, P., Heinrichs, F., Jonsson, R., Kuntze, L., Rasmussen, M., & Richter, E. (2000). Resistance against barley leaf rust (*Puccinia hordei*) in West European spring barley germplasm. *Agronomie, 20*, 769–782.

Niks, R. E., Habekuss, A., Bekele, B., & Ordon, F. (2004). A novel major gene on chromosome 6H for resistance of barley against the *Barley yellow dwarf virus. Theoretical and Applied Genetics, 109*, 1536–1543.

Ordon, F. (2009). Coordinator's report: Disease and pest resistance genes. *Barley Genetics Newsletter, 39*, 58–68.

Parlevliet, J. E. (1981). Race-non-specific disease resistance. In J. F. Jenkyn & R. T. Plumb (Eds.), *Strategies for the control of cereal disease* (pp. 47–54). Oxford: Blackwell Scientific Publications.

Perovic, D., Przulj, N., Milovanovic, M., Prodanovic, S., Perovic, J., Kopahnke, D., Ordon, F., & Graner, A. (2001). Characterisation of spring barley genetic resources in Yugoslavia. In *Proceedings of a Symposium Dedicated to the 100th Birthday of Rudolf Mansfeld, Band 22*, 301–306

Perovic, D., Stein, N., Zhang, H., Drescher, A., Prasad, M., Kota, R., Kopahnke, D., & Graner, A. (2004). An integrated approach for comparative mapping in rice and barley with special reference to *Rph*16. *Functional & Integrative Genomics, 4*, 74–83.

Pickering, R., Hill, A., Michel, M., & Timmerman-Vaughan, G. (1995). The transfer of a powdery mildew resistance gene from *Hordeum bulbosum* L. to barley (*H. vulgare* L.) chromosome 2 (2I). *Theoretical and Applied Genetics, 91*, 1288–1292.

Pickering, R. A., Steffenson, B. J., Hill, A. M., & Borovkova, I. (1998). Association of leaf rust and powdery mildew resistance in a recombinant derived from a Hordeum vulgare 9 Hordeum bulbosum hybrid. *Plant Breeding, 117*, 83–84.

Pohler, W., & Szigat, G. (1982). Versuche zur rekombinativen Genübertragung von der Wildgerste Hordeum bulbosum auf die Kulturgerste H. vulgare. 1. Mitt. Die Rückkreuzung VV 9 BBVV. *Arch Züchtungsforsch Berlin, 12*(8), 7–100.

Ramsay, L., Macaulay, M., degli Ivanissevich, S., McLean, K., Cardle, L., Fuller, J., Edwards, K. J., Tuvesson, S., Morgante, M., Massari, A., Maestri, E., Marmiroli, N., Sjakste, T., Ganal, M., Powell, W., & Waugh, R. (2000). A simple sequence repeat – Based linkage map of barley. *Genetics, 156*, 1997–2005.

Riedel, C., Habekuß, A., Schliephake, E., Niks, R., Broer, I., & Ordon, F. (2011). Pyramiding of *Ryd2* and *Ryd3* conferring tolerance to a German isolate of *Barley yellow dwarf virus*-PAV

(BYDV-PAV-ASL-1) leads to quantitative resistance against this isolate. *Theoretical and Applied Genetics, 123*, 69–76.

Ruge, B., Linz, A., Pickering, R., Proeseler, G., Greif, P., & Wehling, P. (2003). Mapping of Rym14Hb, a gene introgressed from Hordeum bulbosum and conferring resistance to BaMMV and BaYMV in barley. *Theoretical and Applied Genetics, 107*, 965–971.

Ruge-Wehling, B., Linz, A., Habekuß, A., & Wehling, P. (2006). Mapping of Rym16Hb, the second soil-borne virus-resistance gene introgressed from Hordeum bulbosum. *Theoretical and Applied Genetics, 113*, 867–873.

Schaller, C. W., Qualset, C. O., & Rutger, J. N. (1964). Inheritance and linkage of the Yd2 gene conditioning resistance to the Barley yellow dwarf disease in barley. *Crop Science, 4*, 544–548.

Scheurer, K. S., Friedt, W., Huth, W., Waugh, R., & Ordon, F. (2001). QTL analysis of tolerance to a German strain of BYDV-PAV in barley (*Hordeum vulgare* L.). *Theoretical and Applied Genetics, 103*, 1074–1083.

Scholz, M., Ruge-Wehling, B., Habekuss, A., Schrader, O., Pendinen, G., Fischer, K., & Wehling, P. (2009). Ryd4*Hb*: A novel resistance gene introgressed from *Hordeum bulbosum* into barley and conferring complete and dominant resistance to the *Barley yellow dwarf virus*. *Theoretical and Applied Genetics, 119*, 837–849.

Silvar, C., Casas, A. M., Kopahnke, D., Habekuss, A., Schweizer, G., Gracia, M. P., Lasa, J. M., Ciudad, F. J., Molina-Cano, J. L., Igartua, E., & Ordon, F. (2010). Screening the Spanish Barley Core Collection for disease resistance. *Plant Breeding, 129*, 45–52.

Silvar, C., Perovic, D., Casas, A. M., Igartua, E., & Ordon, F. (2011). Development of a cost-effective pyrosequencing approach for SNP genotyping in barley. *Plant Breeding, 130*, 394–397.

Steffenson, B. J., Jin, Y., & Griffey, C. A. (1993). Pathotypes of *Puccinia hordei* with virulence for the barley leaf rust resistance gene *Rph7* in the United States. *Plant Disease, 77*, 867–869.

Stein, N., Herren, G., & Keller, B. (2001). A new DNA extraction method for high – Throughput marker in a large – Genome species such as *Triticum aestivum*. *Plant Breeding, 120*, 354–356.

Struss, D., & Plieske, J. (1998). The use of microsatellite markers for detection of genetic diversity in barley populations. *Theoretical and Applied Genetics, 97*, 308–315.

Suneson, C. A. (1955). Breeding for resistance to yellow dwarf virus in barley. *Agronomy Journal, 47*, 283.

Szigat, G., Herrmann, M., & Rapke, H. (1997). Integration von Bastardpflanzen mit der Wildgerste Hordeum bulbosum in den Zuchtprozeß von Wintergerste. In F. Begemann (Ed.), *Zuüchterische Nutzung Pflanzengenetischer Ressourcen*. Ergebnisse und Forschungsbedarf. *Schriften zu Genetischen Ressourcen 8*, pp. 267–270 (In German)

Thiel, T., Michalek, W., Varshney, R. K., & Graner, A. (2003). Exploiting EST databases for the development and characterization of gene derived SSR markers in barley. *Theoretical and Applied Genetics, 103*, 411–422.

Toojinda, T., Broers, L. H., Chen, X. M., Hayes, P. M., Kleinhofs, A., Korte, J., Kudrna, D., Leung, H., Line, R. F., Powell, W., Ramsay, L., Vivar, H., & Waugh, R. (2000). Mapping quantitative and qualitative disease resistance genes in a doubled haploid population of barley (*Hordeum vulgare*). *Theoretical and Applied Genetics, 101*, 580–589.

Van Ooijen, J. W. (2006). *Join Map®4.0 software for the calculation of genetic linkage maps in experimental populations*. Wageningen: Kyazma BV.

Varshney, R. K., Marcel, T. C., Ramsay, L., Russell, J., Röder, M. S., Stein, N., Waugh, R., Langridge, P., Niks, R. E., & Graner, A. (2007). A high density barley microsatellite consensus map with 775 SSR loci. *Theoretical and Applied Genetics, 114*, 1091–1103. doi:10.1007/s00122-007-0503-7.

von Bothmer, R., Jacobsen, N., Baden, C., Jorgensen, R., & Linde-Laursen, I. (1995). *An ecogeographical study of the genus Hordeum* (2nd ed.). Rome: International Plant Genetic Resources Institute (IPGRI).

Walther, U. (1979). Die Virulenz- und Resistenzgensituation bei *Puccinia hordei* Otth. Arch. Züchtungsforsch., Berlin 9:49–54 Zwergrost (*Puccinia hordei* Otth.). *Kulturpflanze, 28*, 227–228.

Walther, U. (1987). Inheritance of resistance of *Puccinia hordei* Otth. in the spring barley variety Trumpf. *Cereal Rusts Bulletin 15*, 20–26; *Otth Theoretical and Applied Genetics 108*, 712–719.

Walther, U., & Lehmann, C. O. (1980). Resistenzeigenschaften im Gersten- und Weizensortiment Gatersleben. *24. Prüfung von Sommer- und Wintergersten auf Verhalten gegenüber*

Weerasena, J. S., Steffenson, B. J., & Falk, A. B. (2004). Conversion of an amplified fragment length polymorphism marker into a co-dominant marker in mapping of the *Rph*15 gene conferring resistance to barley leaf rust, *Puccinia hordei. TAG Theoretical and Applied Genetics 108*(4), 712–719. doi:10.1007/s00122-003-1470-2.

Xu, J., & Kasha, K. J. (1992). Transfer of a dominant gene for powdery mildew resistance and DNA from *Hordeum bulbosum* into cultivated barley (*H. vulgare*). *Theoretical and Applied Genetics, 84*, 771–777.

Chapter 24
A Major QTL Controlling Adult Plant Resistance for Barley Leaf Rust

Chengdao Li, Sanjiv Gupta, Xiao-Qi Zhang, Sharon Westcott, Jian Yang, Robert Park, Greg Platz, Robert Loughman, and Reg Lance

Abstract Race-specific resistance genes (*Rph*) for leaf rust (*Puccinia hordei*) are often overcome by new pathotypes with matching virulence. Adult plant resistance (APR) is considered potentially more durable for controlling barley leaf rust. Previous studies established that the cultivar Pompadour carried APR to leaf rust. A doubled haploid population (DH) of 200 lines developed from a cross Pompadour/Stirling, and the parents were phenotyped for leaf rust resistance at five field experimental sites in three agricultural zones in Australia. Using a linkage map of SSR and DArT molecular markers, a major QTL associated with the leaf rust resistance was identified on the short arm of chromosome 5H. This QTL explained between 31% and 86% of the phenotypic variation for the APR at different sites. A PCR-based molecular

C. Li (✉)
Department of Agriculture and Food,
3 Baron-Hay Court, South Perth, WA 6151, Australia

State Agricultural Biotechnology Centre, Division of Science and Engineering,
Murdoch University, Murdoch, WA 6150, Australia
e-mail: chengdao.li@agric.wa.gov.au

S. Gupta • X.-Q. Zhang
State Agricultural Biotechnology Centre, Division of Science and Engineering,
Murdoch University, Murdoch, WA 6150, Australia

S. Westcott • J. Yang • R. Loughman • R. Lance
Department of Agriculture and Food,
3 Baron-Hay Court, South Perth, WA 6151, Australia

R. Park
Plant Breeding Institute Cobbitty, University of Sydney,
Private Mail Bag 11, Camden, NSW 2570, Australia

G. Platz
Hermitage Research Station, Department of Employment, Economic Development
and Innovation, Warwick, QLD 4370, Australia

G. Zhang et al. (eds.), *Advance in Barley Sciences: Proceedings of 11th International Barley Genetics Symposium*, DOI 10.1007/978-94-007-4682-4_24,

marker was developed and mapped at 1.6 cM to the APR gene. The present study provides new genetic material and a molecular tool for breeding new varieties with adult plant leaf rust resistance using marker-assisted selection.

Keywords *Hordeum vulgare* • Adult plant resistance • Molecular markers • *Puccinia hordei*

24.1 Introduction

Leaf rust is one of the most important barley diseases around the world, resulting in yield losses up to 30% (Park and Karakousis 2002). Deployment of leaf rust resistant gene is regarded as the most cost-effective and environmentally safe means for controlling the disease. To date, at least 19 isolate-specific *Rph* loci conferring seedling resistance to *P. hordei* have been characterised. Some of the resistant genes have been successfully used to develop commercial varieties (e.g. *Rph12*, *Rph3*, *Rphx*). However, deployment of single race-specific *Rph* genes allows the pathogen to evolve and accumulate new virulence (Park 2003). This creates the need to identify new effective genes and can lead to an ongoing cycle of breeding, deployment and subsequent loss of resistance.

The *Rph12* resistance has been successfully used to develop commercial barley varieties in Australia. However, the new virulent pathotypes have rendered them of little value in conferring leaf rust resistance (Cotterill et al. 1994; Park 2003). The *Rph3* gene has been utilised widely in Europe (Clifford 1985). Significant effort has been made to transfer the *Rph3* gene into Australian commercial barley varieties in Barley Breeding Australia breeding programmes through marker-assisted selection. However, an isolate with virulence for this resistance was detected in early 2009. In late 2009, several nurseries in Southern Queensland were infected with *Puccinia hordei* virulent on *Rph3* (R Park, G Platz and J Franckowiak, unpublished data). Indeed, pathotypes with virulence on genes *Rph1* to *Rph15* and *Rph19* already exist in nature (Fetch et al. 1998; Park and Karakousis 2002). Given a background of continual breakdown of major gene resistance (seedling), deployment of adult plant resistance (APR) was suggested as the way to increase the durability of host resistance (Park 2003; Woldeab et al. 2007).

In contrast to hypersensitive response of the major *Rph* genes, quantitative or partial resistance, including both seedling and adult plant resistance that expresses only during postseedling growth stages, was non race-specific, incomplete and controlled by multiple genes (Marcel et al. 2007). It reduces the rate of epidemic development through its effect on latent period, infection frequency, pustule size, infection period or spore production and thus was considered a more durable source of resistance (Qi et al. 1999, 2000). So far, 21 QTLs have been mapped for partial resistance (Qi et al. 1998, 1999, 2000; Marcel et al. 2007, 2008). In the present study, we investigated the genetic basis of adult plant resistance to leaf rust and identified a major QTL for which PCR-based molecular markers were developed for use in breeding adult plant resistance to leaf rust in barley.

24.2 Materials and Methods

24.2.1 Plant Materials

A population with 200 doubled haploid (DH) lines was produced through another culture from F_1 plants of the cross Pompadour/Stirling. This population was used to map QTLs controlling resistance to net blotch (caused by Pyrenophora teres f. teres) with partial linkage maps of chromosomes 3H and 6H (Gupta et al. 2010). Stirling was bred from a cross of 'Dampier'//(A14)'Prior'/'Ymer'/3/'Piroline' and has been a major malting barley cultivar in Western Australia since the 1980s. Pompadour was derived from a cross of 'FDO'/'Patty'. Using Australian isolates of *P. hordei*, Golegaonkar et al. (2008) demonstrated that Pompadour also carries excellent APR to leaf rust. It has additional resistances to net type net blotch, stripe rust and powdery mildew (Gupta et al. 2010; Gupta unpublished).

24.2.2 Field Experiment for Leaf Rust Phenotype

The DH population was phenotyped for APR to leaf rust at five field sites in the Western, Southern and Northern cropping regions in Australia. The Southern region experiences a temperate climate that is characterised by a predominantly spring rainfall pattern. The Western region experiences a Mediterranean climate, where yields depend upon good winter rains. The Northern region is characterised by a tropical/subtropical climate, where yields often depend upon conservation of soil moisture from summer rainfall. The Western region is also geographically isolated from the other regions. Climatic and varietal differences have resulted in different virulences in crop diseases across the three cropping regions (Gupta et al. 2010). There is significant variability in the leaf rust population in Australia, and differences in pathogen virulence often occur across the three crop-growing regions (Cotterill et al. 1992, 1994; Park 2003).

The Perth and Esperance sites were located in the coast areas of Western region and are approximately 800 km apart. The Perth site has higher temperature than Esperance. The Lansdowne and Karalee sites were located in the Southern region. Barley materials have shown different resistant response to leaf rust at the two sites (Golegaonkar et al. 2009). The Toowoomba site was located in the Northern region. The distance between the Western and the Southern or Northern sites was over 3,500 km.

DH lines and their parents were each planted in a 2-m row in the disease nurseries. The leaf rust susceptible cultivar 'Gus' was used as a control and disease spreader for uniform inoculum increase. Randomised block designs were used in the experiment with two or three replications at different sites.

24.2.2.1 Disease Rating

Adult plant responses were assessed up to three times following anthesis in the Western and Southern regions when disease severity on the susceptible check was 60S or higher. A modified Cobb scale (Peterson et al. 1948) was used to assess disease severity (percent leaf area affected). Pathotypes 5453P– and 5652P+ were used to inoculate the disease nurseries in the Western and Southern regions, respectively.

In the Northern region, adult plant responses to leaf rust pathotype 4610P+ were assessed at Toowoomba (Queensland) in 2004 and 2005. Hill plots of the DH lines, parents and controls were sown in paired rows, parallel and adjacent to spreader rows of susceptible cultivar Gus. Rows were 0.75 m apart with 0.5 m between each plots. The epidemic was initiated in spreader rows at mid-tillering by artificial inoculation and promoted with sprinkler irrigation as required. The epidemic in 2004 was variable, and assessments were taken later than ideal, thus was not used for QTL analysis. In 2005, a uniform, heavy epidemic developed which gave good discrimination within the population. Leaf rust was scored on a 0–9 scale with 9 being very susceptible.

24.2.2.2 Genotype Analysis of the DH Population

Genomic DNA was extracted from approximately 2 g of young leaf tissue from the parents – 'Pompadour' and 'Stirling' – and the 200 DHL following the method of Rogowsky et al. (1991). Over 300 SSR markers from the barley linkage maps (Liu et al. 1996; Ramsay et al. 2000) were used to screen the parents to identify polymorphic markers. Bulk-segregant analysis (BSA) (Michelmore et al. 1991) was carried out by grouping, ten lines each into resistant and susceptible categories based on the phenotypic data at the Perth site. However, the BSA failed to identify molecular markers associated with the leaf rust resistance. In addition, a set of SSR markers was genotyped using MRT™ (Multiplex-Ready Technology) genotyping platform (Hayden et al. 2008). In total, 56 SSR markers were mapped in the 200 DHL. In addition, 42 DNA samples were selected randomly from the original 200 DH lines for DArT analysis which was conducted by Diversity Arrays Technology Pty Ltd (http://www.diversityarrays.com) using the barley version 2.0 array.

24.2.2.3 Construction of Linkage Map

A total of 567 DArT markers were assayed for the subpopulation of 40 DH lines, as two other DH lines were eliminated for analysis due to poor quality data. DArT markers which either did not have polymorphism within the subpopulation or missed the genotypes of both parents were removed from the analysis. The markers with identical genetic information were considered as one haplotype block. Thus, we chose one marker in the haplotype block and removed the remaining markers. In this way, 153 informative DArT markers were selected. In addition, the full population of 200 DH lines was genotyped utilising 56 SSR markers.

Two molecular linkage maps were constructed: one based on the 56 SSR markers in the 200 DH lines and the second map included both SSR and DArT markers. Because only a subgroup of the DH lines (40 DH) were genotyped by DArT, the genotypic data of the remaining 160 DH lines for DArT markers were treated as missing data. The MapManager programme (Manly et al. 2001) was employed to construct the linkage map. The default settings of MapManager were used to order the combined markers on a chromosome by chromosome basis to create a consensus linkage map.

24.2.2.4 QTL Analysis

The two maps were used for comparison of the QTL analyses using the subpopulation and the full population. The software WinQTL cartographer 2.0 was first used for QTL analysis based on the results from individual sites. A QTL was identified using composite interval mapping with threshold values LOD>3. QTLNetwork 2.0 was further adopted to detect if there was epistasis of QTL, QTL-by-environment (QE) interaction and epistasis-by-environment interaction by combining analysis of the data from the five sites (Yang et al. 2007, 2008). One thousand permutation tests were computed to control the genome-wise type I error rate at 0.05. Variation, correlation and distribution analyses were performed using the software JMP (SAS). The broad-sense heritability for the single QTL was estimated using the five sites as replications.

24.2.2.5 Development of Simple PCR Markers Based on the DArT Markers

Based on the QTL mapping result from the subpopulation with 40 DH lines, ten DArT marker sequences were retrieved from the Diversity Arrays Technology Pty Ltd database (http://www.diversityarrays.com) around the chromosome 5H QTL based on the consensus map (Wenzl et al. 2006). After removing the redundant sequences, the first round PCR primers were designed to amplify DNA fragments from Pompadour and Stirling. The DNA bands were cut from the agarose gel and cloned in T-vector using a TA-cloning kit from Promega. The inserts of the plasmid DNA were sequenced using the BigDye™ Terminator method on an Applied Biosystems 3,730 DNA Sequencer (SABC, Murdoch University, Western Australia) to detect the presence of SNPs. The second-round PCR primers were designed based on the sequence polymorphisms. Three simple PCR markers were developed based on the DArT markers. The simple PCR markers were integrated into the map of the full population using the RECORD programme (Van et al. 2005).

24.3 Results

24.3.1 Phenotyping Response to P. hordei

Variation in disease severity (percent leaf area affected) was observed from resistant (0) to susceptible (>60%) at the five experimental sites. Pompadour showed resistance,

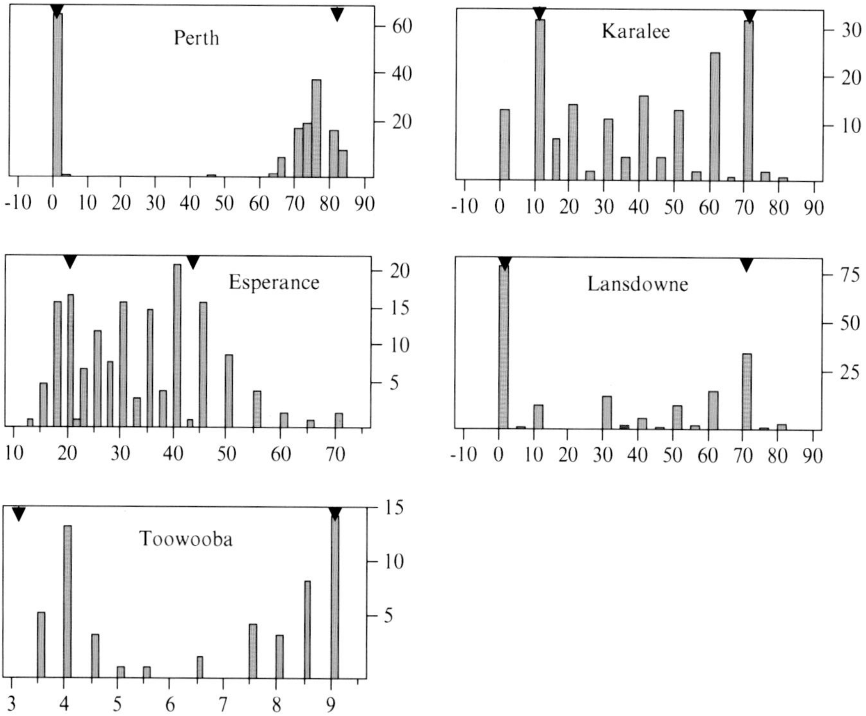

Fig. 24.1 Distribution of disease severity of doubled haploid lines in the Pompadour/Stirling population against barley leaf rust at five sites across Australia. The *arrows* represented the parental response, and the *Y* axis indicated the number of lines

Table 24.1 Correlation coefficients (*R*) of disease resistance for adult plant leaf rust resistance at the five sites in the Pompadour/Stirling population ($p<0.0001$)

Sites	Esperance	Karalee	Lansdowne	Toowoomba
Perth	0.64	0.57	0.73	0.95
Esperance		0.51	0.60	0.71
Karalee			0.80	0.50
Lansdowne				0.76

and Stirling was fully susceptible. Transgressive segregation was only observed at the Esperance site in the Western region (Fig. 24.1). A clear bimodal distribution was apparent at the Perth site in the Western region and the Toowoomba site in the Northern region despite the two sites being over 3,500 km apart. Continuous variation was observed in other three sites (Fig. 24.1).

Correlation analysis demonstrated that phenotypic variation was significantly correlated between the trial sites ($P<0.0001$), which indicated a common gene effect among the five trial sites (Table 24.1). The Perth site in the Western region

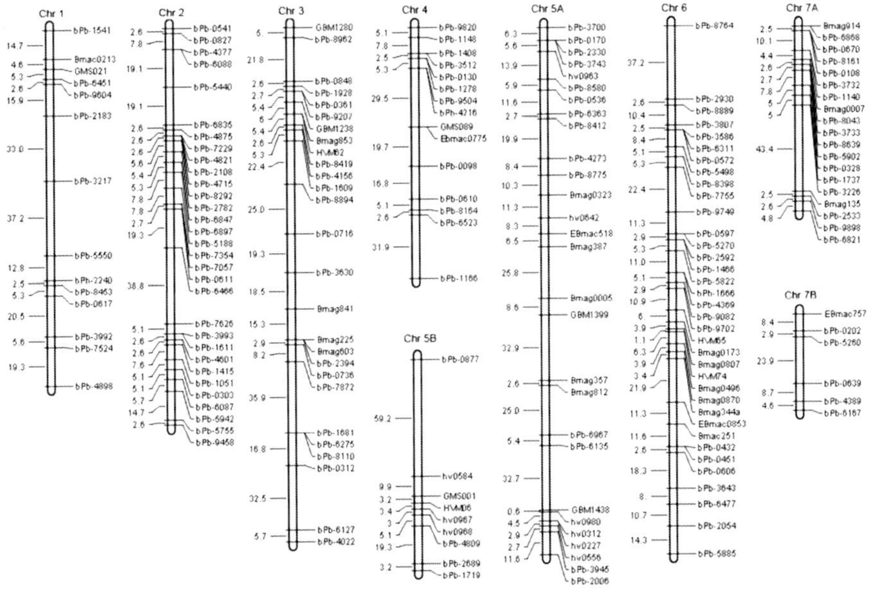

Fig. 24.2 The combined linkage map comprised of DArT and SSR markers. The linkage groups were constructed by MapManager and visualised by MapDraw. Chromosomes 5 and 7 were separated into two linkage groups, denoted by Chr 5A and Chr 5B and Chr 7A and Chr 7B, respectively. The chr1 to chr7 in figure is corresponding to chromosome 1 H to 7 H

had the best correlation with the Toowoomba site in the Northern region ($R = 0.95^{***}$). The Karalee site in the Southern region showed the least correlation with the other sites in the Northern and Western region, but the correlation between the two sites in the Southern region was moderately high ($R = 0.80^{***}$).

24.3.2 The Linkage Map

A combined linkage map consisting of a total number of 185 markers included 142 DArT markers and 43 SSR markers (Fig. 24.2). The redundant markers were removed from the map. All markers within each individual chromosome (Chr) were grouped into a single linkage group except chromosomes 5H and 7H, both of which were separated into two linkage groups denoted by Chr 5HA and Chr 5HB, and Chr 7HA and Chr 7HB, respectively. The combined linkage map had a total coverage of 1,545.7 cM with an average distance of 8.36 cM. The linkage map was visualised by a microsoft excel macro MapDraw (Liu and Meng 2003) and presented in Fig. 24.2. After removing the DArT markers from the combined linkage map, the SSR linkage map covered 562.5 cM with an average marker distance of 14.07 cM.

24.3.3 Mapping QTL Conferring Adult Plant Resistance to Leaf Rust

A single QTL was mapped to the short arm of chromosome 5H in both the sub- and full populations. Marker interval, heritability, LOD score and additive effect of the chromosomal 5H QTL for adult plant leaf rust resistance at five sites are reported in Table 24.2. In the subpopulation, the QTL was mapped to the marker interval of bpb0170 and bpb0536. The total genetic distance was about 31 cM. However, the QTL peak shifted slightly at the different sites. The QTL was located between markers Hv0963-bpb8580 at Perth and Lansdowne and located between markers Bpb8580-bpb0536 at Esperance and Toowoomba. The QTL was mapped further towards the telomere region at the Karalee site (Fig. 24.3; Table 24.2). The Karalee site also had the lowest heritability. When the five sites were treated as environmental factor, a single QTL was mapped to the 10-cM marker interval of Hv0963-Bpb-8580 (Fig. 24.3). The QTL explained 58% of the phenotypic variation, and the QTL x environment interaction accounted for 12% of phenotypic variation.

24.3.4 Conversion of the DArT Markers into Simple PCR Markers

Based on comparison of the QTL location in the above analysis with the DArT consensus map (Wenzl et al. 2006), ten DArT marker sequences were initially selected for development of simple PCR markers. Further analysis identified two duplicated sequences, and three other DArT markers had limited sequence information. Five sequences were eventually used to develop simple PCR markers.

Primers (5′CAGGAAATTGCTAGTGTCATGG and 5′AGCTGCAACAATAGGTGCAGA) based on bPb-2872 sequence generated a 1.3-kb DNA fragment from Stirling but not from Pompadour. It seemed that there was distinct difference in the sequences between the two parents at this locus. Subsequently, another primer pair designed within the fragment were used to amplify DNA from both parents. The results confirmed that there was a DNA sequence difference in Pompadour.

Two SNPs were identified between the parents based on the primers (5′AGATGTCCGTTTCAGACCTAG and 5′CTCATACCCGTATTGTCACTC) from bPb-9562. An allele-specific marker was developed based on one SNP.

Another allele-specific marker was developed based on bPb-6485. There was one SNP identified between Stirling and Pompadour using primers designed from bPb-6485 (5′AGCCTCACTAAACTACCAGC and 5′GGGTATAAGAGAAGGATAAGC). To differentiate with the original DArT markers, the simple PCR marker was named Bxxx, e.g. a simple PCR marker developed from bPb-9562 was named B9562.

The three PCR markers were used to analyse the 200 DH lines, and two mapped to the QTL region. B9562 was 2.2 cM away from the SSR marker Hv0963 towards the telomere, and B2872 was 5.2 cM away from Hv0963 towards the centromere (Figs. 24.3 and 24.4).

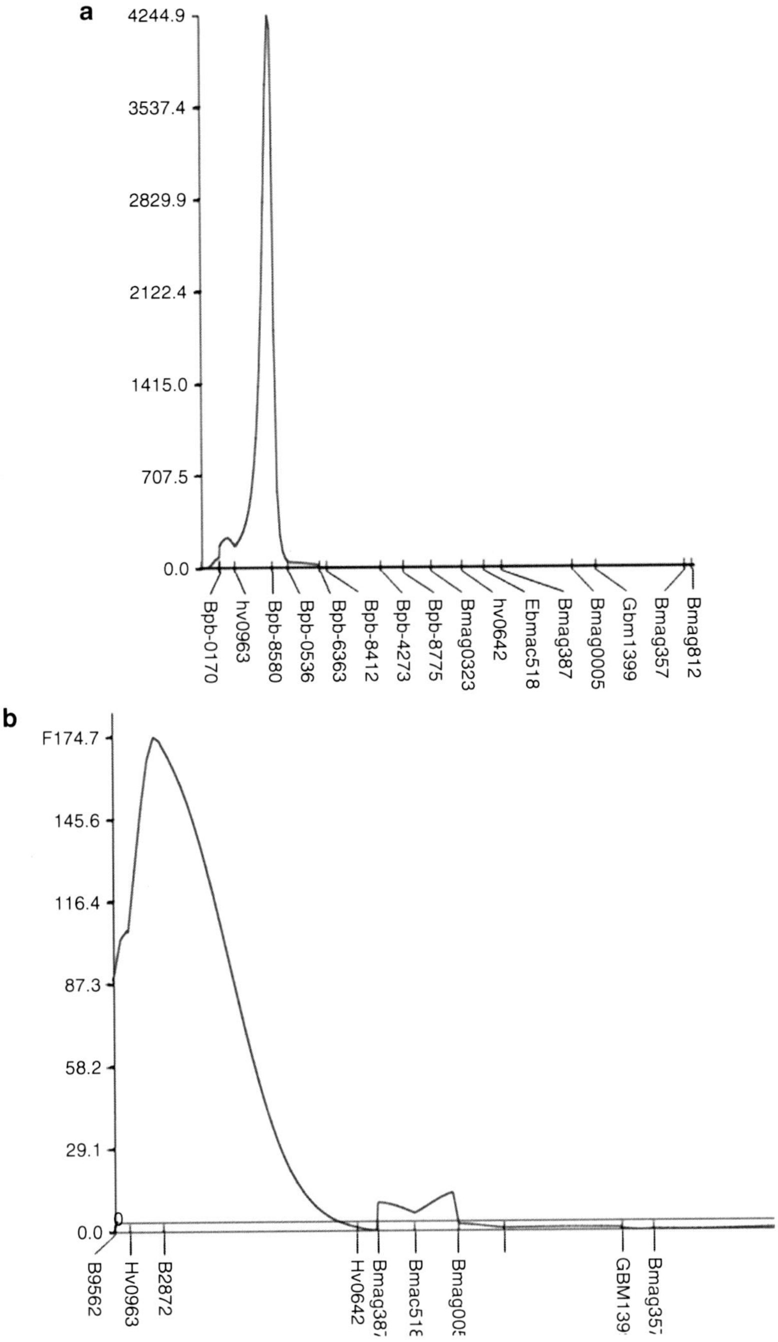

Fig. 24.3 Graphic visualised QTL mapping results to combining the results at five sites by QTLNetwork software. The QTL peak on chromosome 5H was detected either by the subpopulation with the combined linkage map (**a**) or by the full population with the SSR and simple PCR marker linkage map (**b**). The *Y* axis is the *F*-value

Table 24.2 Marker interval, broad-sense heritability and additive effect of the chromosomal 5H QTL for adult plant leaf rust resistance at five sites

Trial sites	Pop	Marker interval	Heritability	Additive effect	LOD
Esperance	Full	Hv0963-B2872	0.36	11.8 ± 1.7	10.89
	Sub	Bpb8580-bpb0536	0.60	10.6 ± 1.4	7.86
Perth	Full	B2872-Hv0642	0.86	49.2 ± 3.0	155.5
	Sub	Hv0963-bpb8580	0.99	38.1 ± 0.6	40.86
Lansdowne	Full	B2872-Hv0642	0.46	31.4 ± 3.3	31.78
	Sub	Hv0963-bpb8580	0.70	24.5 ± 2.6	10.31
Karalee	Full	B2872-Hv0642	0.31	23.9 ± 3.1	13.27
	Sub	Bpb0170-bpb02330	0.41	16.4 ± 3.2	4.95
Toowoomba	Full	B2872-Hv0642	0.82	3.26 ± 0.3	26.4
	Sub	Bpb8580-bpb0536	0.96	2.2 ± 0.1	9.84

24.3.5 *Mapping the QTL for APR to Leaf Rust in the Whole Population*

In the initial mapping with only the SSR markers, a single QTL was mapped between the marker interval of Hv0963 and Hv0642 at all sites (data not shown). The marker interval was about 35.7 cM. QTL mapping results for both marker intervals, and heritability improved significantly with addition of the two DArT-converted markers B2872 and B9562. As found in the subpopulation mapping, a single QTL was mapped for the five sites (Table 24.2; Fig. 24.3). The Karalee site had the lowest heritability of 0.31, and the Perth site had the highest heritability of 0.86. The QTL was mapped between the 5.2-cM marker interval of Hv0963-B2872 at the Perth site, while it was mapped between B2872-Hv0642 at the other four sites. When the five sites were treated as an environmental factor, a single QTL was mapped to the marker interval of Hv0963-B2872 (Fig. 24.3). This QTL explained 42% of the phenotypic variation, and the QTL x environment interaction accounted for 12% of the phenotypic variation.

24.3.6 *Mapping the Adult Plant Resistance as a Single Gene*

As the phenotypic results at the Perth and Toowoomba sites could be clearly differentiated as resistant and susceptible (Fig. 24.1), we utilised data from these sites to map the APR as a single gene. In the subpopulation, the resistance gene was mapped between the markers Hv0963 and bPb-8580 where it was 8.2 cM away from the SSR marker Hv0963. In the full population, the resistance gene was mapped between the markers B2872 and Hv0642. It was about 1.6 cM away from the DArT-converted simple PCR marker B2872 and 6.8 cM away from the SSR marker Hv0963 (Fig. 24.4)

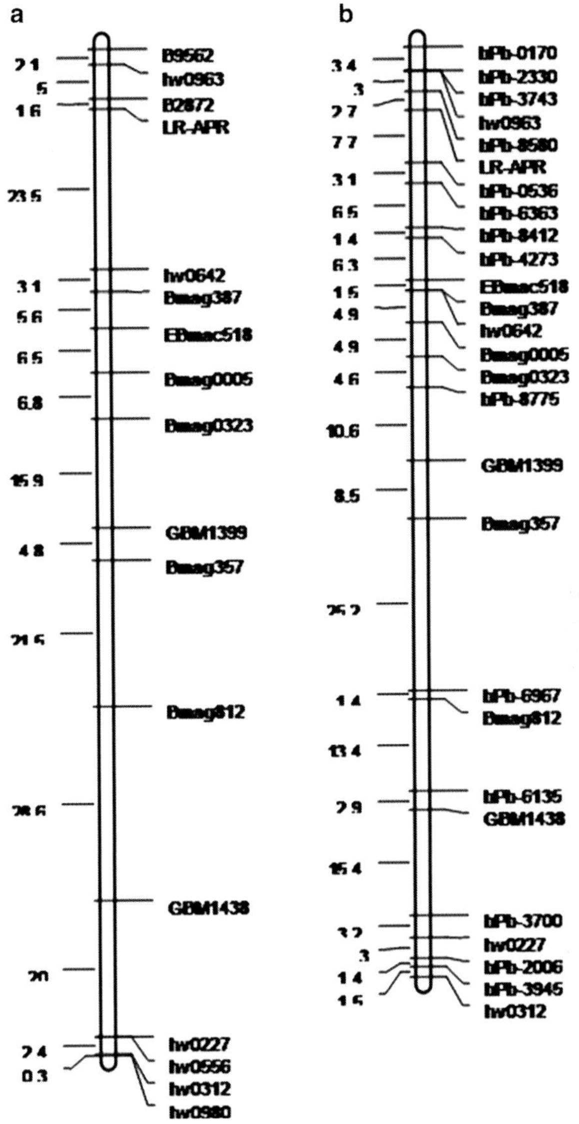

Fig. 24.4 Mapping the QTL as a single gene on chromosome 5H either by the full population with the SSR and simple PCR marker linkage map (**a**) or the subpopulation with the combined linkage map (**b**). The maps were generated by the programme RECORD

24.4 Discussion

In the search for improved genetic solutions for leaf rust resistance, efforts have been made to identify quantitative trait loci (QTL) controlling leaf rust resistance in the last decade, which may provide durable resistance to leaf rust. So far, 21 QTLs controlling partial resistance to barley leaf rust have been mapped to the 7 barley chromosomes in different genetic population (Qi et al. 1998, 1999, 2000; Marcel et al. 2007, 2008). Different genetic populations segregate for different sets of QTLs with only few QTLs shared by any pair of cultivars (Marcel et al. 2007). In a recent study, two QTLs were identified on chromosomes 3H and 7H for leaf rust resistance from barley cultivar Baronesse (Rossi et al. 2006). These results reflected that there were abundant partial resistant QTLs for leaf rust in the barley germplasm (Qi et al. 2000; Marcel et al. 2007). On the other hand, this also presented a challenge to barley breeders on how to effectively combine a large number of QTLs with small effect in the breeding programmes. In a previous study, the barley cultivar Pompadour was identified as an effective APR source to leaf rust (Golegaonkar et al. 2009). The present study showed that a single gene conferring APR in Pompadour was effective in five different environments, and it explained up to 86% of phenotypic variation. This is in contrast to previous studies with large numbers of small effect QTLs for partial resistance to leaf rust in barley (Qi et al. 1998, 1999, 2000; Marcel et al. 2007, 2008).

Among the 23 QTLs for partial resistance and the 19 *Rph* genes for seedling resistance, only the *Rphq4* was mapped to the short arm of chromosome 5H (Qi et al. 1998, 1999, 2000; Marcel et al. 2007, 2008; Rossi et al. 2006). Due to a lack of molecular markers in common between the current and the previous studies, we could not judge if the APR QTL in the present study was mapped to the same location as *Rphq4* from the barley cultivar Vada. It was reported that *Rphq4* was effective in seedlings and in adult plants against two isolates of *P. hordei* but ineffective in both development stages against a third isolate. Thus, *Rphq4* is an isolate-specific resistance QTL for both seedling and adult plant resistance (Marcel et al. 2008). In contrast, Pompadour lacks seedling resistance to Australian isolates of *P. hordei* (Golegaonkar et al. 2009), although there was significant difference for the method to measure the seedling resistance in the two studies. In the present study, a single QTL for the resistance to three different isolates was detected in diverse environments across three barley production regions around Australia. This result further supported that the adult plant resistance from Pompadour is non-isolate specific.

In a previous study, eight barley varieties/lines Athos, Gilbert, Patty, Pompadour, Nagrad, RAH1995, Vada and WI3407 demonstrated APR to *P. hordei* in multi-pathotype tests and field observations (Golegaonkar et al. 2009). Different levels of APR were observed between Vada and Pompadour in the same study. To further test if the APR QTL from Pompadour is the same as *Rphq4*, two SSR markers flanked the APR QTL were used to fingerprint four different barley varieties with APR from the above study (Fig. 24.5). Vada showed different alleles for both SSR markers. This is not surprising as the partial resistance to *P. hordei* in Vada was derived from *H. laevigatum* (Neervoort and Parlevliet 1978). Based on the above evidences, the APR for leaf rust from Pompadour is different from the *Rphq4* in Vada. Pompadour (TR) demonstrated higher levels of APR than Vada (10MR to TR) in the previous study (Golegaonkar et al. 2009). As multiple QTLs conferred the APR in Vada (Qi et al. 1998, 1999, 2000; Marcel et al. 2007, 2008), the APR QTL from Pompadour

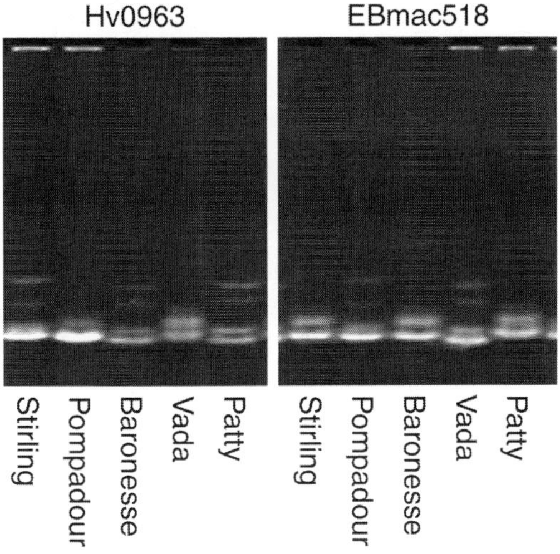

Fig. 24.5 DNA fingerprints of four barley varieties with APR for leaf rust using the two SSR markers flanked the APR gene from Pompadour

should offer better tolerance than *Rphq4* from Vada. Further research is required to clarify if the APR QTL from Pompadour and *Rphq4* from Vada are different genes or alleles. It is well known that barley cultivars which include *H. laevigatum* in their pedigree have harder endosperms, higher β-glucan and lower α-amylase levels, thus resulted in poor malting quality (Swanston 1987). The APR QTL from Pompadour will provide a new genetic source to develop leaf rust resistance malting barley cultivar.

Genes with large genetic effect for adult plant resistance of leaf rust have been identified in wheat such as Lr12, Lr22a, Lr22b, Lr35, Lr48 and Lr49 (Bansal et al. 2008; Hiebert et al. 2007), which have become the major breeding targets in many breeding programmes throughout the world. Identification of the adult plant resistance QTL in the present study provides the basis to implement a similar strategy as wheat for genetic control of barley leaf rust using an alternative genetic resource. Furthermore, we developed a simple PCR-based molecular marker, estimated to be about 1.6 cM away from the APR gene, which will provide an efficient tool for integrating the gene into breeding programmes through marker-assisted selection.

Although some sources of APR are known to be durable, it is not known whether or not the gene identified in the current study will prove to be durable. Thus, caution is needed in deploying this gene as it may be vulnerable to pathogen variation. One approach is to combine this gene with other effective major resistance genes such as *Rph7*.

Bulk-segregant analysis has been used successfully to identify molecular markers associated with various genes. In the initial attempt, we failed to identify SSR markers associated with the APR gene using BSA. This was due to the lack of SSR marker coverage in the telomere region where the resistance gene is located. Instead, a subpopulation of 40 DH lines was genotyped using array-based molecular marker DArT. Compared to the full population mapping, the subpopulation provided reasonable estimate of the QTL/gene location and effect (Figs. 24.3 and 24.4; Table 24.2).

Array-based molecular marker systems can generate large numbers of polymorphic markers for each individual, and in so doing, the cost of each genetic data point is reduced significantly. On the other hand, the genotyping cost for each individual in a population is still high. Thus, an optimum and appropriate population size is important for efficient QTL mapping using high-throughput marker systems. The present study demonstrated that combining high-throughput molecular markers with a small population size can provide an efficient approach to identify molecular markers linked with the target genes.

Our results demonstrated that the QTL had the lowest heritability at the Karalee site which showed the lowest correlation with other sites (Tables 24.1 and 24.2). The previous genetic study also indicated that there is a second resistant gene at the Karalee site (Golegaonkar et al. 2009). There are several different options to search for the second resistant gene. One possibility is to construct a subpopulation without the resistant allele of the chromosome 5H QTL and to map the subpopulation using different molecular markers for identification of the second adult plant resistance gene at the Karalee site in the future.

References

Bansal, U. K., Hayden, M. J., Venkata, B. P., Khanna, R., Saini, R. G., & Bariana, H. S. (2008). Genetic mapping of adult plant leaf rust resistance genes *Lr48* and *Lr49* in common wheat. *Theoretical and Applied Genetics, 117*, 307–312.

Clifford, B. C. (1985). Barley leaf rust, disease, distribution, epidemiology, and control. In A. P. Roelfs & W. R. Bushnell (Eds.), *The cereal rusts* (Vol. 11, pp. 173–205). New York: Academic.

Cotterill, P. J., Rees, R. G., & Vertigan, W. A. (1992). Detection of *Puccinia hordei* virulent on the *Pa9* and Triumph resistance genes in barley in Australia. *Australasian Plant Pathology, 21*, 32–34.

Cotterill, P. J., Rees, R. G., & Platz, G. J. (1994). Response of Australian barley cultivars to leaf rust (*Puccinia hordei*). *Australian Journal of Experimental Agriculture, 34*, 783–788.

Fetch, T. G., Steffenson, B. J., & Jin, Y. (1998). Worldwide virulence of *Puccinia hordei* on barley. *Phytopathology, 88*, S28.

Golegaonkar, P. G., Park, R. F., & Singh, D. (2008). Evaluation of seedling and adult plant resistance to *Puccinia hordei* in barley. *Euphytica, 166*, 183–197.

Golegaonkar, P. G., Park, R. F., & Singh, D. (2009). Genetic analysis of adult plant resistance *Puccinia hordei* in barley. *Plant Breeding, 129*, 162–166.

Gupta, S., Li, C. D., Loughman, R., Cakir, M., Platz, G., Westcott, S., Bradley, J., Broughton, S., Appels, R., & Lance, R. C. M. (2010). Quantitative trait loci and epistatic interactions in barley conferring resistance to net type net blotch (*Pyrenophora teres* f. *teres*) pathotypes. *Plant Breeding, 129*, 362–368.

Hayden, M. J., Nguyen, T. M., Waterman, A., McMichael, G. L., & Chalmers, K. J. (2008). Application of multiplex-ready PCR for fluorescence-based SSR genotyping in barley and wheat. *Molecular Breeding, 21*, 271–281.

Hiebert, C. W., Thomas, J. B., Somers, D. J., McCallum, B. D., & Fox, S. L. (2007). Microsatellite mapping of adult-plant leaf rust resistance gene Lr22a in wheat. *Theoretical and Applied Genetics, 115*, 877–884.

Liu, R. H., & Meng, J. L. (2003). MapDraw: A Microsoft Excel macro for drawing genetic linkage maps based on given genetic linkage data. *HEREDITAS (Beijing), 25*, 317–321.

Liu, Z. W., Biyashev, R. M., & Saghai Maroof, M. A. (1996). Development of simple sequence repeat DNA markers and their integration into a barley linkage map. *Theoretical and Applied Genetics, 93*, 869–876.

Manly, K. F., Cudmore, R. H., Jr., & Meer, J. M. (2001). Map Manager QTX, cross-platform software for genetic mapping. *Mammalian Genome, 12*, 930–932.

Marcel, T. C., Varshney, R. K., Barbieri, M., Jafary, H., de Kock, J. D., Graner, A., & Niks, R. E. (2007). A high-density consensus map of barley to compare the distribution of QTLs for partial resistance to *Puccinia hordei* and of defence gene homologues. *Theoretical and Applied Genetics, 114*, 487–500.

Marcel, T. C., Gorguet, C. B., Ta, M. T., Kohutova, Z., Vels, A., & Niks, R. E. (2008). Isolate specificity of quantitative trait loci for partial resistance of barley to *Puccinia hordei* confirmed in mapping populations and near-isogenic lines. *New Phytologist, 177*, 743–755.

Michelmore, R. W., Paran, I., & Kesseli, R. V. (1991). Identification of markers linked to disease resistance genes by bulked segregant analysis: A rapid method to detect markers in specific genomic regions by using segregating populations. *Proceedings of the National Academy of Sciences of the United States of America, 88*, 9828–9832.

Neervoort, W. J., & Parlevliet, J. E. (1978). Partial resistance of barley to leaf rust, *Puccinia hordei*. Analysis of components of partial resistance in eight barley cultivars. *Euphytica, 27*, 33–39.

Park, R. F. (2003). Pathogenic specialization and pathotype distribution of *Puccinia hordei* in Australia, 1992 to 2001. *Plant Disease, 87*, 1311–1316.

Park, R. F., & Karakousis, A. (2002). Characterization and mapping of gene *Rph19* conferring resistance to *Puccinia hordei* in the cultivar 'Reka 1' and several Australian barleys. *Plant Breeding, 121*, 232–236.

Peterson, R. F., Campbell, A. B., & Hannah, A. E. (1948). A diagrammatic scale for estimating rust intensity on leaves and stems of cereals. *Canadian Journal of Research, 26*, 496–500.

Qi, X., Niks, R. E., & Stam PLindhout, P. (1998). Identification of QTLs for partial resistance to leaf rust (*Puccinia hordei*) in barley. *Theoretical and Applied Genetics, 96*, 1205–1215.

Qi, X., Jiang, G., Chen, W., Niks, R. E., Stam, P., & Lindhout, P. (1999). Isolate-specific QTLs for partial resistance to *Puccinia hordei* in barley. *Theoretical and Applied Genetics, 99*, 877–884.

Qi, X., Fufa, Q. F., Sijtsma, D., Niks, R. E., Lindhout, P., & Stam, P. (2000). The evidence for abundance of QTLs for partial resistance to *Puccinia hordei* on the barley genome. *Molecular Breeding, 6*, 1–9.

Ramsay, L. M., Macaulay, K., Ivanissevich, S. D., MacLean, K. L., Cardle, L., Fuller, J., Edwards, K. J., Tuvesson, S., Morgante, M., Massari, A., Maestri, E., Marmiroli, N., Sjakste, T., Ganal, M., Powell, W., & Waugh, R. (2000). A simple sequence repeat-based linkage map of barley. *Genetics, 156*, 1997–2005.

Rogowsky, P. M., Guidett, F. L., Langridge, P., Shepherd, K. W., & Koebner, R. M. D. (1991). Isolation and characterization of wheat-rye recombinants involving chromosome arm 1DS of wheat. *Theoretical and Applied Genetics, 82*, 537–544.

Rossi, C., Cuesta-Marcos, A., Vales, I., Gomez-Pando, L., Orjeda, G., Wise, R., Sato, K., Hori, K., Capettini, F., Vivar, H., Chen, X., & Hayes, P. (2006). Mapping multiple disease resistance genes using a barley mapping population evaluated in Peru, Mexico, and the USA. *Molecular Breeding, 18*, 355–366.

Swanston, J. S. (1987). The consequences, for malting quality of *Hordeum laevigatum* as a source of mildew resistance in barley breeding. *Annals of Applied Biology, 110*, 351–355.

Van, Os, Stam, H. P., Visser, R. G. F., & Van Eck, H. J. (2005). RECORD: A novel method for ordering loci on a genetic linkage map. *Theoretical and Applied Genetics, 112*, 30–40.

Wenzl, P., Li, H., Carling, J., Zhou, M., Raman, H., Paul, E., Hearnden, P., Maier, C., Xia, L., Caig, V., Ovesna, J., Cakir, M., Poulsen, D., Wang, J., Raman, R., Smith, K. P., Muehlbauer, G., Chalmers, K. J., Kleinhofs, A., Huttner, E., & Kilian, A. (2006). A high-density consensus map of barley linking DArT markers to SSR, RFLP and STS loci and agricultural traits. *BMC Genomics, 7*, 206.

Woldeab, G. C., Fininsa, H., Singh, D., Yuen, J., & Crossa, J. (2007). Variation in partial resistance to barley leaf rust (*Puccinia hordei*) and agronomic characters of Ethiopian landrace lines. *Euphytica, 158*, 139–151.

Yang, J., Zhu, J., & Williams, R. W. (2007). Mapping the genetic architecture of complex traits in experimental populations. *Bioinformatics, 23*, 1527–1536.

Yang, J. C., Hu, H., Yu, R., Xia, Z., Ye, X., & Zhu, J. (2008). QTLNetwork: Mapping and visualizing genetic architecture of complex traits in experimental populations. *Bioinformatics, 24*, 721–723.

Chapter 25
Large Population with Low Marker Density Verse Small Population with High Marker Density for QTL Mapping: A Case Study for Mapping QTL Controlling Barley Net Blotch Resistance

Jian Yang, Chengdao Li, Xue Gong, Sanjiv Gupta, Reg Lance, Guoping Zhang, Rob Loughman, and Jun Zhu

Abstract The development of array-based high-throughput genotyping methods created significant opportunities to increase the number of genetic populations for linkage analysis. In the present study, a strategy was proposed for mapping QTLs (quantitative trait loci) based on DArT (diversity arrays technology) genotyping system. A consensus linkage map was constructed with both DArT and SSR markers by utilizing a subgroup DH population, and a second linkage map was constructed with SSR markers alone and a more extensive full DH population. Resistance to

J. Yang
College of Agriculture and Biotechnology, Zhejiang University,
Hangzhou 310029, China

Department of Agriculture and Food Western Australia,
3 Baron-Hay Court, South Perth, WA 6151, Australia

C. Li (✉) • X. Gong
Department of Agriculture and Food Western Australia,
3 Baron-Hay Court, South Perth, WA 6151, Australia

Western Australia State Agricultural Biotechnology Centre,
Murdoch University, Perth, WA 6150, Australia
e-mail: chengdao.li@agric.wa.gov.au

S. Gupta
Western Australia State Agricultural Biotechnology Centre,
Murdoch University, Perth, WA 6150, Australia

R. Lance • R. Loughman
Department of Agriculture and Food Western Australia,
3 Baron-Hay Court, South Perth, WA 6151, Australia

G. Zhang • J. Zhu (✉)
College of Agriculture and Biotechnology, Zhejiang University,
Hangzhou 310029, China
e-mail: jzhu@zju.edu.cn

G. Zhang et al. (eds.), *Advance in Barley Sciences: Proceedings of 11th International Barley Genetics Symposium*, DOI 10.1007/978-94-007-4682-4_25,

barley net-type net blotch disease was analyzed using the subpopulation data with the high-density consensus linkage map and the full-population data with the low-density SSR linkage map, respectively. Two interactive QTLs were detected either by the sub- or full population. Simulation studies were conducted to validate the strategy presented in this chapter. In addition, a computer program written in C++ is freely available on the web to deal with the data files. Based on both real data analysis and simulation studies, we concluded that high-density molecular markers, small population size, and precise phenotyping can improve the precision of mapping major-effect QTL and the efficiency of conducting QTL mapping experiment.

Keywords Quantitative trait locus (QTL) • Diversity arrays technology (DArT) • Population size • Marker density

25.1 Introduction

In the past two decades, numerous gel-based molecular marker analysis methods have been developed, such as the restriction fragment length polymorphism (RFLP) (Botstein et al. 1980), simple sequence repeats (SSR) (Weber and May 1989), random amplified polymorphic DNA (RAPD) (Williams et al. 1990), and amplified fragment length polymorphism (AFLP) (Vos et al. 1995). These genotyping methods have greatly facilitated characterization of germplasm, assessment of biodiversity, mapping of quantitative trait loci (QTLs), and marker assisted selection (MAS) but are constrained by their dependence on gel electrophoresis methods, resulting in low throughput. In barley, particularly, it is extremely difficult to develop such kind of gel-based marker due to the lack of reference genome sequence.

Recently, a novel molecular marker technique, called diversity arrays technology (DArT), was developed using the rice genome (Jaccoud et al. 2001) and has been applied in barley (Wenzl et al. 2004), Arabidopsis (Wittenberg et al. 2005), cassava (Xia et al. 2005), pigeon pea (Yang et al. 2006), and wheat (Akbari et al. 2006). This technology is genome sequence-independent and deployed on a microarray platform that allows high-throughput screening of hundreds of molecular markers simultaneously. It is especially suitable for generating genome-wide markers for genetic linkage mapping and identifying markers closely linked to genes for MAS applications (Wittenberg et al. 2005).

With the development of molecular marker technology, quantitative trait loci (QTL) mapping has been used routinely as a tool for detecting genetic loci underlying the traits of agricultural and medical importance. At present, most QTL analysis experiments use only one or a few mapping populations due to the limitations of the genotyping capability of gel-based markers. Linkage disequilibrium (LD) was proposed as a solution to identify marker-trait associations in a large set of germplasm, but it is restricted by the low levels of LD in barley (Morrell et al. 2005). The development of DArT marker techniques provides an opportunity for simultaneously genotyping a large number of genetic individuals from various structured populations. These high-throughput techniques can generate large numbers of polymorphic markers for each individual; thereby, the cost of each genetic data point was significantly reduced. On the other hand, the genotyping cost for each individual in a population is still high. Thus, an optimum and appropriate population size is important for efficient QTL mapping using high-throughput marker systems. In the present study, we proposed a strategy for analysis of quantitative trait loci based on the DArT genotyping technique. We demonstrated this strategy using a barley dataset and conducted a simulation study to examine the validity of the presented strategy.

25.2 Material and Methods

25.2.1 Barley Population and Net-Type Net Blotch Phenotyping

Net-type net blotch (NTNB) caused by *Pyrenophora teres* f. *teres* Drechs. (anamorph: *Drechslera teres* f. *teres* (Sacc.) Shoemaker) is a major disease for barley around the world. A population of 200 doubled-haploid lines (DHLs) was derived from across between two barley cultivars, Pompadour and Stirling, using another culture. The population was inoculated using five NTNB pathotypes: 97NB1, 95NB100, NB50, NB81, and NB52B. As these pathotypes were collected from different environments, they were also treated as different environmental factors in analysis. The details for disease inoculation have been reported elsewhere (Gupta et al. 2009). The pathogenic behaviors of the pathotypes 97NB1, 95NB100, and NB81 are correlated with one another. Similarly, the pathogenic behaviors of the pathotypes NB50 and NB52B are correlated with each other but are dissimilar from the other three.

25.2.2 Marker Analysis

Genomic DNA was extracted from approximately 2 g of young leaf tissue following the method of Rogowsky et al. (Rogowsky et al. 1991). SSR markers were selected from published barley maps (Becker and Heun 1995; Liu et al. 1996; Ramsay et al. 2000) based on their distribution along the barley chromosomes. In addition, a set of SSR markers was genotyped using MRT™ (Multiplex-Ready Technology) genotyping platform according to Hayden et al. (2008). In total, 56 SSR markers have been mapped previously in this population. The genotyping and phenotyping details of the population have been reported in a previous paper (Gupta et al. 2008).

25.2.3 DArT Analysis for a Subpopulation

A commercial service for genotyping was provided by Diversity Arrays Technology Pty Ltd (http://www.diversityarrays.com) using the barley version 2.0 array. The DArT is set on a 96-well standard format, and the cost is charged on per plate basis. In the present study, we explored the possibility of genotyping two genetic populations on one plate for QTL mapping. In this case, the maximum number of samples for each population is 48, which included 2 internal controls, 2 parental lines with two replicates, and 42 DNA samples from the population. The 42 DNA samples were randomly selected for DArT analysis from the original 200 DH population.

25.2.4 Data Analysis

In Fig. 25.1, we illustrated the procedure of constructing a consensus linkage map consisting of both DArT and SSR markers and a second linkage map only with SSR markers. A total number of 567 DArT markers were assayed for the subpopulation of 40 DHLs, as two other DH lines were eliminated for analysis due to poor quality data. Those DArT markers that either did not have polymorphism within the subpopulation or missed the genotypes of both parents were removed from the analysis. Some of the adjacent DArT markers, which had identical strain distribution patterns, provide identical genetic information, and, thus, only one of these markers was chosen to represent the genetic information. All the markers in one haplotype block had the identical genetic information. Thus, we chose one marker in the haplotype block to represent the genetic information of that haplotype block and removed the remaining markers. Ultimately, we obtained 153 informative DArT markers. In addition, the full population of 200 DHLs was genotyped utilizing 56 SSR markers selected from the genetic maps in previous studies (Gupta et al. 2008). We then combined the datasets of DArT markers and SSR markers together. Since only a subgroup of the DHLs (40 DHLs) were genotyped by DArT, the genotypic data of the remaining 160 DHLs for DArT markers were treated as missing data. The MapManager program (Manly 1993) was employed to construct the linkage map by combining the genotypic data. Since the chromosome information of all the markers was known, we reordered the markers within each chromosome individually so as to avoid grouping errors. The default settings of MapManager were used to order the combined markers on a chromosome by chromosome basis to create a consensus linkage map. Finally, all the DArT markers on the consensus linkage map were removed to obtain a SSR linkage map. These maps were used for comparison of the QTL analyses using the subpopulation and the full population.

All the 200 DHLs were phenotyped for net-type net blotch disease index by using five different isolates in Australia. The QTL mapping analysis was separated in two parts; the first was using subpopulation data with the consensus linkage map, and the second was using the full-population data with the SSR linkage map. These will be

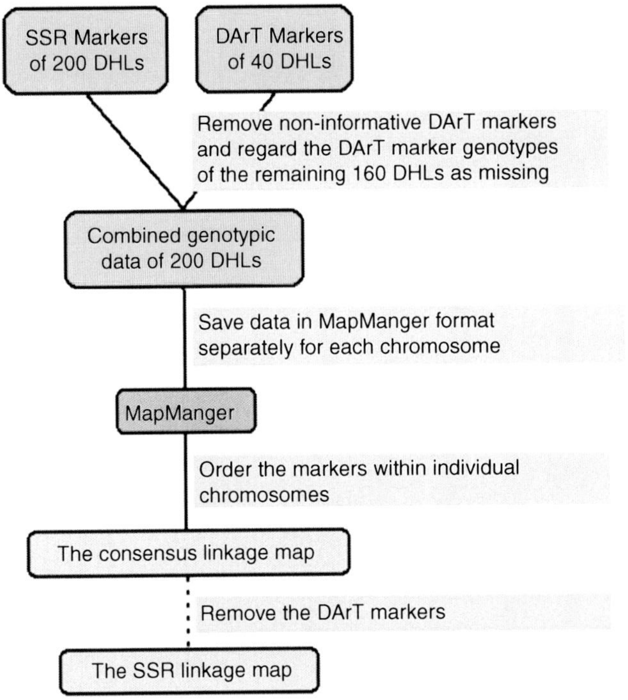

Fig. 25.1 The procedure of constructing linkage maps by DArT and SSR markers

referred as Case I and Case II. The QTL mapping software QTLNetwork 2.0 was adopted to detect QTLs, epistasis, QTL-by-environment (QE) interaction, and epistasis-by-environment interaction (Yang et al. 2007, 2008). One thousand permutation tests were computed to control the genome-wise type I error rate at 0.05. In both cases, we processed the trait data as follows: (1) We analyzed the data for each isolate separately; (2) by comparing the QTL mapping results of the five isolates, we classified the five isolates into two groups (97NB1, 95NB100, and NB81 being group 1 and NB50 and NB52B being group 2) according to the similarity of the QTL mapping results; (3) we treated the two groups of isolates as two environments and used QTLNetwork 2.0 to simultaneously analyze the multiple-environment data.

25.3 Results

25.3.1 The Linkage Map

A consensus linkage map consisting of a total number of 185 markers including 142 DArT markers and 43 SSR markers was constructed by MapManager. All the markers within each individual chromosome (Chr) were grouped into a single linkage group

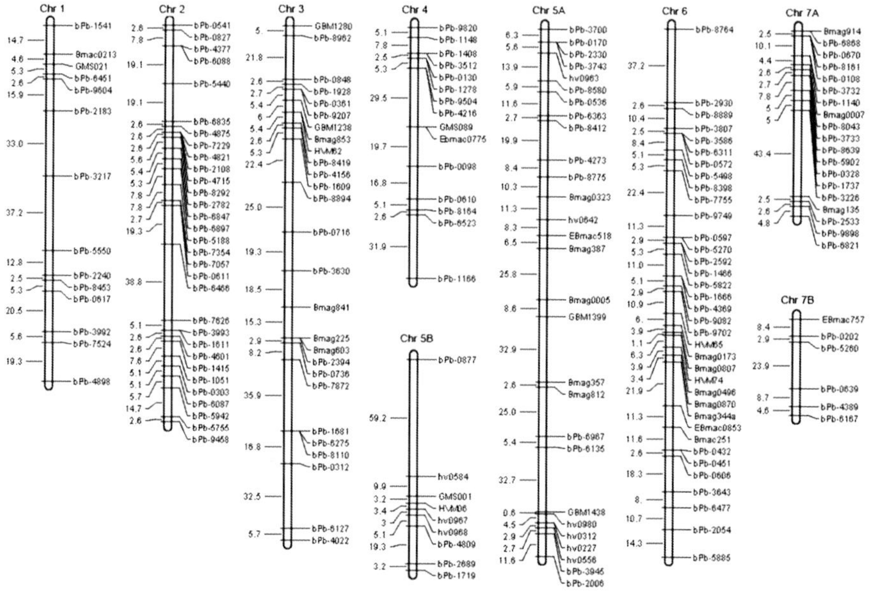

Fig. 25.2 The consensus linkage map comprised of DArT and SSR markers. The marker data were prepared by the approach presented in Fig. 25.1. The linkage groups were constructed by MapManager and visualized by MapDraw. Chromosomes 5 and 7 were separated into two linkage groups, denoted by Chr 5A and Chr 5B, Chr 7A and Chr 7B, respectively. The SSR linkage map can be obtained by removing all the DArT markers

except chromosomes 5 and 7, both of which were separated into two linkage groups denoted by Chr 5A and Chr 5B, and Chr 7A and Chr 7B, respectively. The consensus linkage map had a total coverage of 1,545.7 cM with an average spacing of 8.36 cM. The linkage map was visualized by a microsoft excel macro MapDraw (Liu and Meng 2003) and presented in Fig. 25.2. When we removed the DArT markers on the consensus linkage map, we obtained an SSR linkage map having the coverage of 562.5 cM with the density of 14.07 cM.

25.3.2 QTL Mapping Results of the Combined Analysis

The phenotype data has been published previously (Gupta et al. 2008). We reanalyzed the data by QTLNetwork 2.0 using the linkage maps constructed in the present study. In both Cases I and II, two peaks were identified by a one-dimensional genome scan, indicating that there were two QTLs located on Chr 3 and Chr 6, respectively (Fig. 25.3). The QTLs were designated as "NB" with the serial number

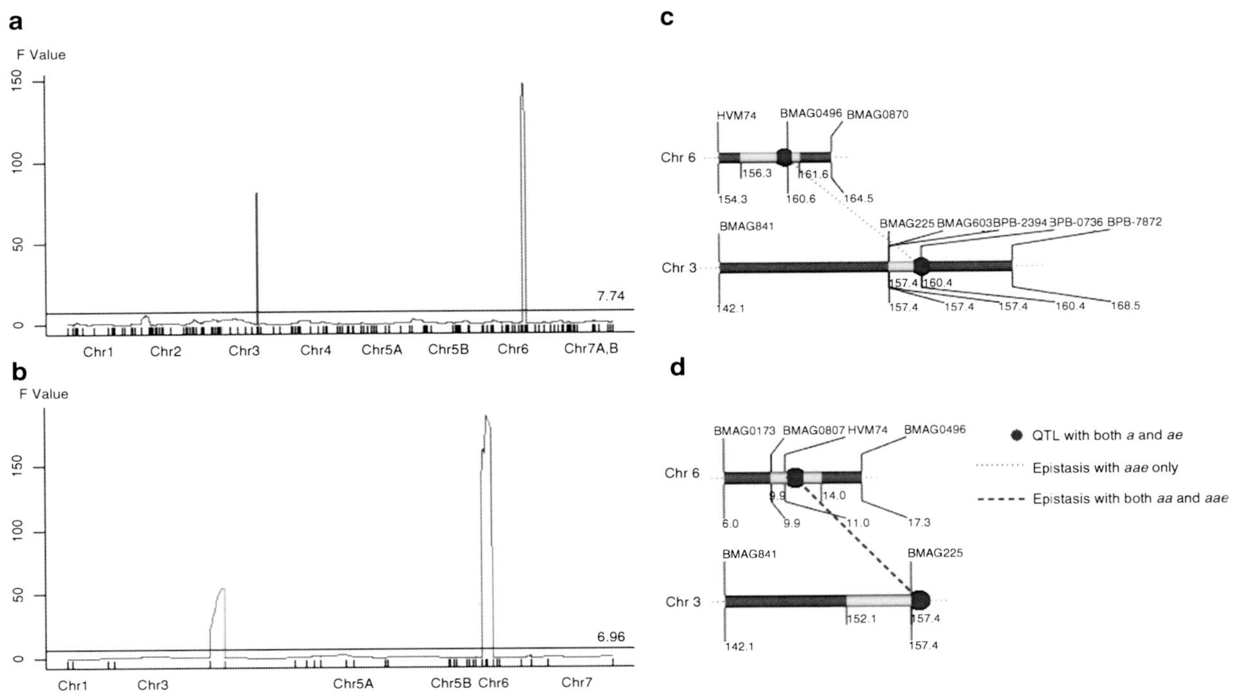

Fig. 25.3 Graphic visualized QTL mapping results by QTLNetwork software. Two peaks were detected on chromosome 3 and chromosome 6 either by the subpopulation with the consensus linkage map (**a**) or by the full population with the SSR linkage map (**b**). The chromosome 2 of the SSR linkage map has only two markers with zero distance and thus is not presented. The QTLNetwork map of the two QTLs by the subpopulation with the consensus linkage map (**c**). The QTLNetwork map of the two QTLs by the full population with the SSR linkage map (**d**). The *blue ball* represents the QTL with both additive (*a*) and additive-by-environment interaction (*ae*) effects. The *blue dashed line* denotes the epistasis between two QTLs with both epistatic (*aa*) and epistasis-by-environment (*aae*) effects. The *green dashed line* denotes the epistasis between the two QTLs with *aae* only

of chromosome and marker interval. It was shown in Fig. 25.3b that the F-value profile of Chr 3 in Case II did not reach to a peak but stopped increasing at the end of the linkage group. With the additional DArT markers, we definitely got a sharp peak at Chr 3 flanked by two DArT markers BPB–2394 and BPB–0736 in Case I (Fig. 25.3a, c). The relative contribution (RC) of the QTL in Case I was almost four times larger than that in Case II (31.62 vs. 8.15%) (Table 25.1) because the position of the QTL in Case II was estimated with bias due to the insufficient coverage of SSR markers. In addition, the QTL on Chr 6 was mapped to nearby the SSR marker HVM74 in both cases, flanked by the markers HVM74 and BMAG0496 in Case I and by the markers BMAG0807 and HVM74 in Case II (Fig. 25.3c, d). The RC of this QTL was 48.12% in Case I which is much higher than that estimated in Case II (25.23%) (Table 25.1). In addition, significant epistatic interaction between these two QTLs was detected in both of the two cases, but the RC of it is small, 1.68% in Case I and 0.99% in Case II, respectively. The epistasis with such a small RC is negligible in a breeding context but may provide important insights into studies of the genetic controls underlying these two QTLs.

There were strong additive-by-environment interaction (*ae*) effects detected in both cases. Take Case I for an instance, the *ae* effects of both of the two QTLs contributed to the phenotypic variation by 28.39 and 42.05%, respectively. For isolates 97NB1, 95NB100, and NB81, the total genetic effects of NB3-19 (chromosome 3H QTL) and NB6-24 (chromosome 6H QTL) were 0.36 and −1.94, while for isolates NB50 and NB52B, the total genetic effect of NB3-19 decreased to −1.94 and that of NB6-24 increased to 1.36 (Table 25.1). Both the two QTLs completely altered their directions for disease resistance responses for different "environments" (pathotype groups). For the isolates 97NB1, 95NB100, and NB81, the Stirling (P2) allele of NB3-19 and the Pompadour (P1) allele of NB6-24 increased the disease resistance, while for the NB50 and NB52B isolates, the P1 allele of NB3-19 and the P2 allele of NB6-24 increased the disease resistance. It was implied that an elite variety that had excellent performance of disease resistance in one "environment" would completely lose its resistance in another "environment." Therefore, different breeding schemes would be required for different environments. Using these QTL information, we predicted isolate-specific superior lines (Yang and Zhu 2005) by substituting alleles between the two parents. Define Q_1Q_1 and q_1q_1 as the allele genotypes of Pompadour and Stirling for NB3-19 and Q_2Q_2 and q_2q_2 for NB6-24. For isolates 97NB1, 95NB100, and NB81, the superior line took the genotype of q_1Q_2/q_1Q_2 with the total genetic effect of −2.63. While for isolates NB50 and NB52B, it took the genotype of Q_1q_2/Q_1q_2 with the total genetic effect of −2.96. As the population mean of this trait was 5.22, both isolate-specific superior lines could increase the disease resistance by more than 50%. It should be noted that the genotypes of the superior lines in the two group isolates were completely reversed, implying that the selected line with best resistance performance in environment 1, however, would be the worst one in environment 2 and *vice versa*.

Table 25.1 QTLs and epistasis identified for net-type net blotch disease in barley

	Case I		Case II			Case I	Case II
QTL	NB3–19	NB6–24	NB3–5	NB6–3	*Epistasis*	(NB3–19, NB6–24)	(NB3–5, NB6–3)
Chromosome	3	6	3	6	*Chromosome*	(3, 6)	(3, 6)
[a]*FMs*	BPB–2394& BPB–0736	HVM74& BMAG0496	BMAG841& BMAG225	BMAG0807& HVM74	*FMs*	(BPB–2394& BPB–0736, HVM74& BMAG0496)	(BMAG841& BMAG225, BMAG0807& HVM74)
Position (cM)	159.4	159.3	157.1	10.9	*Position (cM)*	(159.4, 159.3)	(157.1, 10.9)
[b]*SI (cM)*	157.4–160.4	156.3–161.6	152.1–157.4	9.9–14.0	[b]*SIs (cM)*	(157.4–160.4, 156.3–161.6)	(152.1–157.4, 9.9–14.0)
a (P-value)	−0.76 (0.000)	−0.29 (0.000)	−0.33 (0.000)	−1.00 (0.000)	*aa (P-value)*	−0.14 (0.071)	−0.14 (0.031)
ae_1 *(P-value)*	1.12 (0.000)	−1.65 (0.000)	0.60 (0.000)	−1.00 (0.000)	aae_1 *(P-value)*	0.32 (0.001)	0.18 (0.071)
ae_2 *(P-value)*	−1.18 (0.000)	[illegible].61 (0.000)	−0.61 (0.000)	0.99 (0.000)	aae_2 *(P-value)*	−0.30 (0.010)	−0.17 (0.038)
[c]*RC (%)*	31.62	48.12	8.15	25.23	[c]*RC (%)*	1.68	0.99

The QTLs are designated as "NB" with the serial number of chromosome and marker interval. *a* represents the additive effect of QTL, *ae* Represents the additive-by-environment interaction effect with the subscripted number indicating the number of environments, *aa* Represents the additive-by-additive epistatic effect, *aae* Represents the interaction effect between the additive-by-additive epistasis and the environment with the subscripted number indicating the number of environment

[a]FMs, the flanking markers of QTL separated by the symbol "&"

[b]SI(s), the support interval(s) of QTL position calculated by the odd ratio reduced by a factor 10 (Lander and Botstein 1989)

[c]RC, relative contribution of the QTL or epistasis to phenotype variation

25.3.3 Monte Carlo Simulation

A simulation study was conducted to examine the effectiveness and efficiency of the presented strategy. To make the simulation as close to a real case as possible, we used the real linkage maps constructed in the present study. We separated the simulation study into two parts with different individual numbers and marker densities, referred to as Case I and Case II. In Case I, the consensus linkage map was used to generate a DH population with 46 lines, and in Case II, the SSR linkage map was used to generate a DH population with 184 lines. To make balance the number of observations, we assumed that the 46 DHLs in Case I were investigated in two environments with 4 replications and the 184 DHLs in Case II were investigated in two environments without replication. Three QTLs (*Q*1, *Q*2, and *Q*3) with small, large, and medium effects were assumed on Chr3, Chr5, and Chr7, respectively. Positions and effects of these QTLs can be found in Table 25.2. The RCs of the three QTLs were designated as RC(*Q*1):RC(*Q*2):RC(*Q*3) = 1:4.2:1.7. The heritability of the hypothesized trait was assumed to be 60%. The environmental effects and the random errors explained 30 and 10% of the trait variation, respectively. We regenerated the DH populations according to the hypothesized genetic structure by 200 times, analyzed the simulated datasets by QTLNetwork 2.0, and summarized the results in Table 25.2. In general, the results demonstrated a reasonably accurate estimation of the parameters including the positions and effects of the QTLs. The false discovery rates (FDRs) of detecting QTLs were 0.007 in Case I and 0.018 in Case II, both of which were lower than the predefined significance level of 0.05. The powers of detecting the three QTLs were all more than 95% in both cases (Table 25.2). In general, the standard errors of the estimates in Case I were less than those in Case II, and the positions and effects of the three QTLs were more accurately estimated in Case I than those in Case II. Take *Q*1 for example, the true additive effect of it is −2.50, and this was estimated as −2.45 in Case I and −2.30 in Case II. In Case I, the true position of *Q*1 is 49.0 cM and was estimated as 48.8 with 0.2 cM bias. In Case II, the true position of it is 50.6 cM that was estimated as 48.2 with 2.2 cM bias. The position support interval of this QTL, which was calculated by the odd ratio reduced by a factor 10 (Lander and Botstein 1989), was only 5.2 cM in Case I but was 29.3 cM in Case II.

In addition, we gradually decreased the heritability in Case I from 60 to 10% with 10% decrement and retained the relative contributions of environmental effects at 30% constantly. Thus, the proportion of trait variation attributed by the random errors was increased from 10 to 60% with 10% increment. We performed 200 simulations at each of these heritability levels and calculated the powers and FDRs of detecting QTLs. Generally, the powers of QTLs increased and the FDRs decreased with the increasing of heritability (Fig. 25.4). The power of detecting the QTL with a large effect (*Q*2) was always high; even if the heritability decreased to 10%, it could still stay at about 80%. While for the small-effect QTL (*Q*1), it could only be efficiently detected at a heritability level of more than 40%.

Table 25.2 Summarized simulation results of mapping QTLs using 46 DHLs with the consensus linkage map and 184 DHLs with the SSR linkage map from 200 simulation replicates

	Case I ($h^2=60\%$)			Case II ($h^2=60\%$)		
QTL	1	2	3	1	2	3
[a]RC (%)	8.68	36.51	14.81	8.68	36.51	14.81
Chromosome	3	5	7	3	5	7
Position (cM)	49.0	65.9	154.3	50.6	103.8	13.0
Estimate	48.81 (2.32)	65.67 (1.48)	154.47 (1.34)	48.23 (8.72)	104.48 (2.21)	12.98 (1.21)
[b]SI (cM)	5.2	6.3	4.7	29.3	6.2	4.0
a	−2.5	6.8	−4.7	−2.5	6.8	−4.7
Estimate	−2.45 (0.27)	6.91 (0.29)	−4.68 (0.30)	−2.30 (0.63)	6.62 (0.41)	−4.64 (0.39)
ae_1	−1.83	−2.0	0.0	−1.83	−2	0
Estimate	−1.83 (0.25)	−2.07(0.24)	0.02(0.12)	−1.58(0.46)	−1.90 (0.30)	−0.01 (0.14)
ae_2	1.83	2.0	0.0	1.83	2	0
Estimate	1.79 (0.24)	2.05 (0.23)	−0.02 (0.12)	1.56 (0.46)	1.89 (0.30)	0.01 (0.14)
Power (%)	97.5	100	98.5	97	100	100

The standard errors of the estimates are given in the parentheses.
[a]RC, relative contribution of the QTL to phenotype variation
[b]SI, averaged support interval of QTL position from the 200 simulation replicates

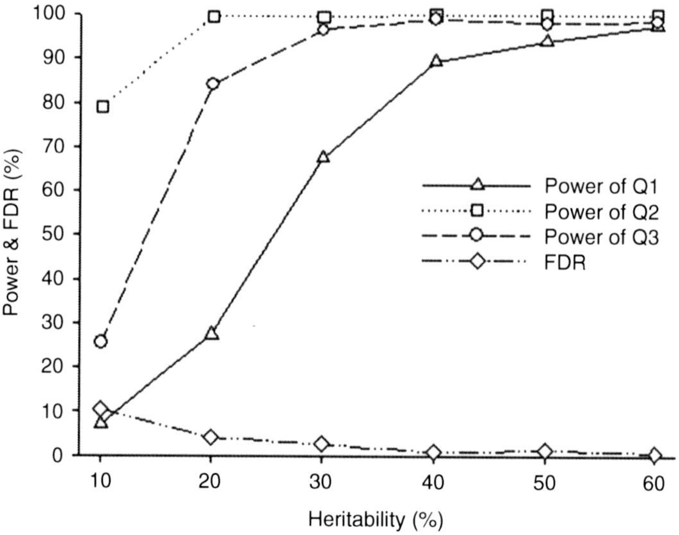

Fig. 25.4 The powers and false discovery rates (FDRs) of detecting QTLs observed by 200 simulations under 6 heritability levels. *Q*1, *Q*2, and *Q*3 are the assumed QTLs with small, large, and media effects, respectively. The ratios of the relative contributions (RCs) of the three QTLs are RC(*Q*1):RC(*Q*2):RC(*Q*3) = 1:4.2:1.7

25.3.4 *Software Development*

A computer program written in the C++ has been developed to deal with the DArT markers, to prepare data files for MapManager for linkage grouping, and these in turn to be used in QTLNetwork 2.0 for QTL mapping. It is able to handle the data with only DArT marker or with both DArT and SSR markers. The program is freely available on the website: http://ibi.zju.edu.cn/software/qtlnetwork/prepdata4qnk/.

25.4 Discussion

Mapping quantitative trait loci underlying agriculturally important traits such as yield, quality, and disease resistance is an important prerequisite to marker-assisted breeding and map-based cloning. In the past, the limitation of genotyping capability of gel-based methods constrained genetic populations to a limited number of crosses involving a few parental varieties. As a result, the large genetic resources available as part of breeding programs have remained a significant untapped resource. This scenario has created a major constraint to the progress of the genetic improvement of crops. The development of high-throughput DArT and SNP genotyping methods created significant opportunities to increase the number of genetic populations for genetic linkage analysis. The current DArT platform has the capacity of simultaneously assaying 96 samples on one slide and can generate a linkage map within 3 days by using a properly formatted genotyping array. It was demonstrated in the present study that a DH population consisting of 46 lines was sufficient for the accurate mapping of QTLs based on DArT markers. Therefore, we can genotype two populations in one slide (the cost for genotyping each population is less than $2,500 USD). In our simulation study, for a trait with a heritability of 60%, we could map a small-effect QTL with a RC of 8.68% using 46 DHLs with 4 replicates in 2 environments. However, to map small-effect QTL underlying the traits with lower heritability (<40%) or to map epistasis, it was suggested that more than 200 DHLs were used (Yang et al. 2007), and therefore, two or three slides are required.

In addition, the combination of both SSR and DArT markers together in the analysis rather than using the DArT marker alone will make the result in the present study comparable with those from the previous studies which were based mostly on SSR markers. However, there is a constraint in constructing the linkage map with both SSR and DArT markers in that the chromosome information for the SSR markers should be known. It is because that without the chromosome information, it will result in a problem of grouping the SSR markers into right chromosomes using such a small population (the subpopulation). Since the chromosome information of DArT markers is available from the Diversity Arrays Technology Pty Ltd, in addition to that of SSR markers, we only need to reorder the markers on each chromosome and thus avoid the aforementioned problem.

Phenotyping in respect to both individual numbers within a population and the precision of the measurement and estimation of a trait is becoming a bottleneck for effective gene mining for breeding programs. With the development of high-throughput genotyping technologies, the genotyping cost of each data point has been significantly reduced, but the phenotyping cost for each individual is still high. The current study and simulation has demonstrated that the enhanced genotyping capabilities will enable the utilization of smaller mapping populations and therefore provide the opportunity to phenotype populations with increased replication and improved precision. This will more precisely and accurately map the traits of interest. In conclusion, high-density molecular markers, small population size, and precision phenotyping will improve the precision of mapping major-effect QTL and the efficiency of conduct QTL mapping experiment.

Acknowledgments The research is supported by Australian Winter Cereal Molecular Marker Program of the Australian Grain Research and Development Corporation, Department of Agriculture and Food Western Australia, the Natural Science Foundation of China, and the National Basic Research Program of China.

References

Akbari, M., Wenzl, P., Caig, V., Carling, J., Xia, L., Yang, S., Uszynski, G., Mohler, V., Lehmensiek, A., Kuchel, H., Hayden, M. J., Howes, N., Sharp, P., Vaughan, P., Rathmell, B., Huttner, E., & Kilian, A. (2006). Diversity arrays technology (DArT) for high-throughput profiling of the hexaploid wheat genome. *Theoretical and Applied Genetics, 113*, 1409–1420.

Becker, J., & Heun, M. (1995). Mapping of digested and undigested random amplified microsatellite polymorphisms in barley. *Genome, 38*, 991–998.

Botstein, D., White, R. L., Skolnick, M., & Davis, R. W. (1980). Construction of a genetic linkage map in man using restriction fragment length polymorphisms. *American Journal of Human Genetics, 32*, 314–331.

Gupta, S., Li, C. D., Loughman, R., Cakir, M., Platz, G., Westcott, S., Bradley, J., Broughton, S., Appels, R., & Lance, R. (2008). Gene identification and epistatic interactions in barley conferring resistance to net type net blotch (*Pyrenophora teres* f. *teres*) pathotypes. *Phytopathology* (in press)

Gupta, S., Li, C. D., Loughman, R., Cakir, M., Platz, G., Westcott, S., Bradley, J., Broughton, S., Appels, R., & Lance, R. (2009). Gene identification and epistatic interactions in barley conferring resistance to net type net blotch (*Pyrenophora teres* f. *teres*) pathotypes. *Phytopathology* (in press)

Hayden, M. J., Nguyen, T. M., Waterman, A., & Chalmers, K. J. (2008). Multiplex-ready PCR: A new method for multiplexed SSR and SNP genotyping. *BMC Genomics, 9*, 80.

Jaccoud, D., Peng, K., Feinstein, D., & Kilian, A. (2001). Diversity arrays: A solid state technology for sequence information independent genotyping. *Nucleic Acids Research, 29*, E25.

Lander, E. S., & Botstein, D. (1989). Mapping Mendelian factors underlying quantitative traits using RFLP linkage maps. *Genetics, 121*, 185–199.

Liu, R. H., & Meng, J. L. (2003). MapDraw: A Microsoft Excel macro for drawing genetic linkage maps based on given genetic linkage data. *HEREDITAS (Beijing), 25*, 317–321.

Liu, Z. W., Biyashev, R. M., & Maroof, M. A. (1996). Development of simple sequence repeat DNA markers and their integration into a barley linkage map. *Theoretical and Applied Genetics, 93*, 869–876.

Manly, K. F. (1993). A Macintosh program for storage and analysis of experimental genetic mapping data. *Mammalian Genome, 4*, 303–313.

Morrell, P. L., Toleno, D. M., Lundy, K. E., & Clegg, M. T. (2005). Low levels of linkage disequilibrium in wild barley (*Hordeum vulgare* ssp. *spontaneum*) despite high rates of self-fertilization. *Proceedings of the National Academy of Sciences of the United States of America, 102*, 2442–2447.

Ramsay, L., Macaulay, M., degli Ivanissevich, S., MacLean, K., Cardle, L., Fuller, J., Edwards, K. J., Tuvesson, S., Morgante, M., Massari, A., Maestri, E., Marmiroli, N., Sjakste, T., Ganal, M., Powell, W., & Waugh, R. (2000). A simple sequence repeat-based linkage map of barley. *Genetics, 156*, 1997–2005.

Rogowsky, P. M., Guidett, F. L. Y., Langridge, P., Shepherd, K. W., & Koebner, R. M. D. (1991). Isolation and characterization of wheat-rye recombinants involving chromosome arm 1DS of wheat. *Theoretical and Applied Genetics, 82*, 537–544.

Vos, P., Hogers, R., Bleeker, M., Reijans, M., van de Lee, T., Hornes, M., Frijters, A., Pot, J., Peleman, J., Kuiper, M., et al. (1995). AFLP: A new technique for DNA fingerprinting. *Nucleic Acids Research, 23*, 4407–4414.

Weber, J. L., & May, P. E. (1989). Abundant class of human DNA polymorphisms which can be typed using the polymerase chain reaction. *American Journal of Human Genetics, 44*, 388–396.

Wenzl, P., Carling, J., Kudrna, D., Jaccoud, D., Huttner, E., Kleinhofs, A., & Kilian, A. (2004). Diversity Arrays Technology (DArT) for whole-genome profiling of barley. *Proceedings of the National Academy of Sciences of the United States of America, 101*, 9915–9920.

Williams, J. G., Kubelik, A. R., Livak, K. J., Rafalski, J. A., & Tingey, S. V. (1990). DNA polymorphisms amplified by arbitrary primers are useful as genetic markers. *Nucleic Acids Research, 18*, 6531–6535.

Wittenberg, A. H., van der Lee, T., Cayla, C., Kilian, A., Visser, R. G., & Schouten, H. J. (2005). Validation of the high-throughput marker technology DArT using the model plant Arabidopsis thaliana. *Molecular Genetics and Genomics, 274*, 30–39.

Xia, L., Peng, K., Yang, S., Wenzl, P., de Vicente, M. C., Fregene, M., & Kilian, A. (2005). DArT for high-throughput genotyping of Cassava (*Manihot esculenta*) and its wild relatives. *Theoretical and Applied Genetics, 110*, 1092–1098.

Yang, J., & Zhu, J. (2005). Methods for predicting superior genotypes under multiple environments based on QTL effects. *Theoretical and Applied Genetics, 110*, 1268–1274.

Yang, S., Pang, W., Ash, G., Harper, J., Carling, J., Wenzl, P., Huttner, E., Zong, X., & Kilian, A. (2006). Low level of genetic diversity in cultivated Pigeon pea compared to its wild relatives is revealed by diversity arrays technology. *Theoretical and Applied Genetics, 113*, 585–595.

Yang, J., Zhu, J., & Williams, R. W. (2007). Mapping the genetic architecture of complex traits in experimental populations. *Bioinformatics, 23*, 1527–1536.

Yang, J., Hu, C., Hu, H., Yu, R., Xia, Z., Ye, X., & Zhu, J. (2008). QTLNetwork: Mapping and visualizing genetic architecture of complex traits in experimental populations. *Bioinformatics, 24*, 721–723.

Chapter 26
"Deep Phenotyping" of Early Plant Response to Abiotic Stress Using Non-invasive Approaches in Barley

Agim Ballvora, Christoph Römer, Mirwaes Wahabzada, Uwe Rascher, Christian Thurau, Christian Bauckhage, Kristian Kersting, Lutz Plümer, and Jens Léon

Abstract The basic mechanisms of yield maintenance under drought conditions are far from being understood. Pre-symptomatic water stress recognition would help to get insides into complex plant mechanistic basis of plant response when confronted to water shortage conditions and is of great relevance in precision plant breeding and production. The plant reactions to drought stress result in spatial, temporal and tissue-specific pattern changes which can be detected using non-invasive sensor techniques, such as hyperspectral imaging. The "response turning time-point" in the temporal curve of plant response to stress rather than the maxima is the most relevant time-point for guided sampling to get insights into mechanistic basis of plant response to drought stress. Comparative hyperspectral image analysis was performed on barley (*Hordeum vulgare*) plants grown under well-watered and water stress conditions in two consecutive years. The obtained massive, high-dimensional data cubes were analysed with a recent matrix factorization technique based on simplex volume maximization of hyperspectral data and compared to several drought-related traits. The results show that it was possible to detect and visualize the accelerated senescence signature in stressed plants earlier than symptoms become visible by the naked eye.

A. Ballvora (✉) • J. Léon
INRES-Plant Breeding and Biotechnology, University of Bonn, Bonn, Germany
e-mail: ballvora@uni-bonn.de

C. Römer • L. Plümer
Institute of Geodesy and Geoinformation, Geoinformation, University of Bonn, Bonn, Germany

M. Wahabzada • C. Thurau • C. Bauckhage • K. Kersting
Fraunhofer Institut für Intelligente Analyse und Informationssysteme, Schloss Birlinghoven, Germany

U. Rascher
ICG-3 Phytosphere: Ecosystem Dynamics, Forschungszentrum Jülich, Jülich, Germany

G. Zhang et al. (eds.), *Advance in Barley Sciences: Proceedings of 11th International Barley Genetics Symposium*, DOI 10.1007/978-94-007-4682-4_26,

Keywords Barley • Phenotyping • Water stress • Hyperspectral image • Response turning point • Non-invasive analysis

26.1 Introduction

Plants are constantly confronted with various kinds of biotic and abiotic stresses, causing serious crop losses in agriculture. Water scarcity is a principle global problem that causes aridity and serious crop losses in agriculture. It has been estimated that drought can cause a depreciation of crop yield up to 70% in conjunction with other abiotic stresses (Boyer 1982; Pinnisi 2008).

Facing a rapidly growing world population, the challenge of "how to feed a hungry world" is becoming more and more imminent. Although progress in plant breeding and agriculture has successfully increased crop production by doubling the yield per hectare in the past decades, the annual yield increase curve has recently been observed to stagnate. Meanwhile, the need for doubling cereal grain yield until 2050 has been identified (Furbank et al. 2009). The slow progress in the development of improving cultivars was mainly due to poor understanding of genetic factors that impart tolerance to drought (Passioura 2002). Although progress has been made in the genetic dissection of drought tolerance mechanisms, additional information on functional traits related to drought stress tolerance is urgently needed. A deep knowledge of the adaptation process and of the genetic mechanisms behind tolerance to the water stress is complex and often showing polygenic inheritance. Understanding the molecular basis underlying the plant response when encountering environmental changes will provide tools for qualitative and quantitative improvements in breeding of new crop cultivars that will fulfil the increasing needs for food, energy and recycling bioproducts. This knowledge will be essential in improving management practices, breeding strategies as well as engineering viable crops for a sustainable agriculture in the coming years.

Barley (*Hordeum vulgare* L.) is a self-pollinating crop with the genome that follows strict diploidy (2n:14), making it a suitable genomic model for *Triticeae*. The genetic diversity of barley underlies a huge variation with respect to drought adaptation and therefore can be cultivated from boreal to equatorial regions of the world (Schulte et al. 2009). Such diverse ecological ranges have established bases for adaptive changes with respect to water availability. This remarkable variation invites us to understand the genetics behind novel fitness of barley genotypes in order to elucidate the underlying mechanisms of plant adaptation under drought. In the last two decades, progress has been made in understanding the genetic basis of drought-related quantitative trait loci (QTL) (Lebreton et al. 1995; Nguyen et al. 2004; Tondelli et al. 2006; McKay et al. 2008). More recently, -omics approaches have offered a direct molecular insight into drought tolerance mechanism (Rabbani et al. 2003; Tran et al. 2004; Ueda et al. 2004; Sakma et al. 2006; Xiong et al. 2006; Talamé et al. 2007; Schreiber et al. 2009; Guo et al. 2009; Abdeen et al. 2010). However, genetic and biochemical approaches are time consuming and still fail to fully predict the performance of new lines in the field. Resistance to water shortage conditions is a web of interactions between the genotype and the environment, leading to phenotypic expressions. The recent rapid developments of destructive and non-destructive technologies offer new opportunities for an effective and high-throughput analysis of plant characteristics. The connectivity and flow of these data towards molecular breeding and farming have been hampered by a bottleneck at the level of phenotyping, the so-called phenomics bottleneck (Tardieu and Schurr 2009; Richards et al. 2010). Effective use of sensors could contribute towards a pinpoint accuracy in phenotyping, reduced experimental requirements and will enable multiple, simultaneous and objective data to be collected.

The hyperspectral camera systems have become affordable and are being widely used in plant sciences. In hyperspectral imaging, the radiative properties of plant leaves or canopies are used for determining structural and physiological traits of vegetation by measuring the portion of radiation that is reflected from plant surfaces and specific absorption properties are calculated (Malenovský et al. 2009; Rascher et al. 2010, 2011; Ustin and Gamon 2010). For instance, the spectral reflectance of vegetation is characterized by a low reflectivity in the visible part of the spectrum (400–700 nm) due to a strong absorption by photosynthetic pigments, while a high reflectivity in the near infrared (700–1,100 nm) is produced by a high scattering of light by the leaf mesophyll tissues (Rascher et al. 2010). In addition, in the short-wave infrared part of the spectrum (1,100–2,500 nm), the reflectance intensity is affected by the water content of plant tissues (Rascher et al. 2010, 2011). Most approaches based on hyperspectral data aim to quantify plant traits by calculating vegetation indices that quantify specific changes in plant structure and composition. The complex physiological effects of drought stress, however, cause changes in the reflectance in most spectral regions (Aldakheel and Danson 1997; Penuelas et al. 1997).

Hyperspectral imaging sensors are an established, sophisticated technology for early stress detection with high spatio-temporal resolution. They gather large, high-dimensional data cubes which pose a significant challenge even to advanced methods of data analysis and machine learning. Production of annotated data sets is costly and difficult, and hence, classical supervised learning algorithms often fail in applied plant sciences. Therefore, new approaches for unsupervised recognition of relevant patterns are needed.

The main objective of the study is the pre-symptomatic water stress recognition and response, considering them of great relevance in precision plant breeding and production. On the other side, this will help to better understand the mechanistic basis of this response.

In this study, we phenotyped three barley cultivars that were grown under water stress conditions in two consecutive years using hyperspectral images. The obtained data were analysed with a matrix factorization technique based on simplex volume maximization to hyperspectral images (Kersting et al. 2012). It could be shown that it was possible to detect and visualize the accelerated senescence signature in stressed plants earlier than the appearance of visible symptoms by the naked eye.

26.2 Materials and Methods

26.2.1 Plant Material

The barley cultivar Scarlett was used in experimental year 2010 and cultivars Wiebke and Barke in 2011. Six plants were sown per pot filled with 17.5 kg of substrate Terrasoil (Cordel & Sohn, Salm, Germany).

26.2.2 Water Treatment of the Plants

The experiments were performed in rain-out shelters at the experimental station of the University of Bonn in Bonn Poppelsdorf, Germany.

In year 2010, the genotype Scarlett was used in two treatments (well-watered and with reduced water) with six pots per treatment. The water stress was started at developmental stage BBCH31 (Biologische Bundesanstalt, Bundessortenamt and Chemical industry) either by reducing the water amount or by withdrawing it completely.

In year 2011, the genotypes Wiebke and Barke were used in pot experiments arranged in a randomized complete block design with three treatments (well-watered and two drought stressed) with four pots per genotype and treatment. The drought stress was induced either by reducing the total amount of water or by the complete withholding of water. By reducing the irrigation, the water potential of substrate remained at the same level as in the well-watered pots for the first 7 days but decreased rapidly in the following 10 days, reaching 40% compared to the control.

26.2.3 Hyperspectral Image Spectroscopy

For the measurements, the plants were transferred into the laboratory. The hyperspectral spectroscopy was performed with a resolution of 640×640 pixels, where each pixel is a vector with 120 recorded wavelengths from the range of 394–891 nm with 4 nm spectral resolution, using the SOC-700 hyperspectral imaging system (Rascher et al. 2007), manufactured by Surface Optics. The camera was mounted at the same level as the lamps in NADIR position. The illumination was provided by six halogen lamps (400 W ECO, OSRAM, Munich, Germany) fixed at distance of 1.6 m from the support where the pots were placed to take the pictures.

26.2.4 Analysis of the Hyperspectral Data

The obtained massive, high-dimensional data cubes were analysed with a recent matrix factorization technique based on simplex volume maximization of hyperspectral (SiVM) data as described by Kersting et al. (2012) and Römer (manuscript in preparation).

26.2.5 Measurement of Water Potential and Biomass

Water potential was measured from plant stems using the Scholander pressure chamber. Above-ground wet and dry biomass was determined destructively by weighting the above-ground vegetative plant material after harvesting and again after drying for 48 h at 45°C.

26.3 Results and Discussion

26.3.1 Hyperspectral Spectroscopy

During the experiments of year 2010, hyperspectral images were taken for each pot (genotype Scarlett) at 10 time-points, twice per week starting from day four of water stress. In year 2011, (genotypes Wiebke and Barke) images were taken every consecutive day starting at the second day of watering reduction: at 11 time-points for the non-watered pots and at 20 time-points for the pots irrigated with reduced water amount (Figs. 26.1 and 26.2). Although the camera used measures in the electromagnetic wavelength range from 394 to 890 nm, the wavelengths below 470 and above 750 nm were discarded because of high levels of noisy signals. The reason for that might be an unstable source of illumination at these light lengths. Therefore, only the images obtained at 470–750 nm were used for data processing and analysis.

26.3.2 Data Processing and Analysis

The separation of the plant from background signals was done using K-means. More than 95% of the background noise was removed, and the plant signals were further analysed to calculate archetypes which differentiate the healthy from the stressed leaves (Römer et al., manuscript in preparation).

The Dirichlet distribution of the spectral data from the year 2010 revealed that at the 14th stress day, the cluster analysis of hyperspectral data could identify with a confidence of higher than 90% the stressed from healthy plants. This was at least 10 days earlier than the visual identification of wilting symptoms by the naked eye. The measurement at day 10 of stress period could not identify the stressed plants from the healthy ones. By increasing the frequency of the measurements in the time-window between day 10 and 14 of stress period, it was possible to identify an earlier time-point of stress response.

The archetypes obtained from the SiVM analysis of the hyperspectral images taken in year 2011 experiments show that starting from day 7,we could differentiate the specific signatures from control plants towards stressed ones (Fig. 26.1) (Römer et al., manuscript in prep.). It should be noted that this coincides with the decrease

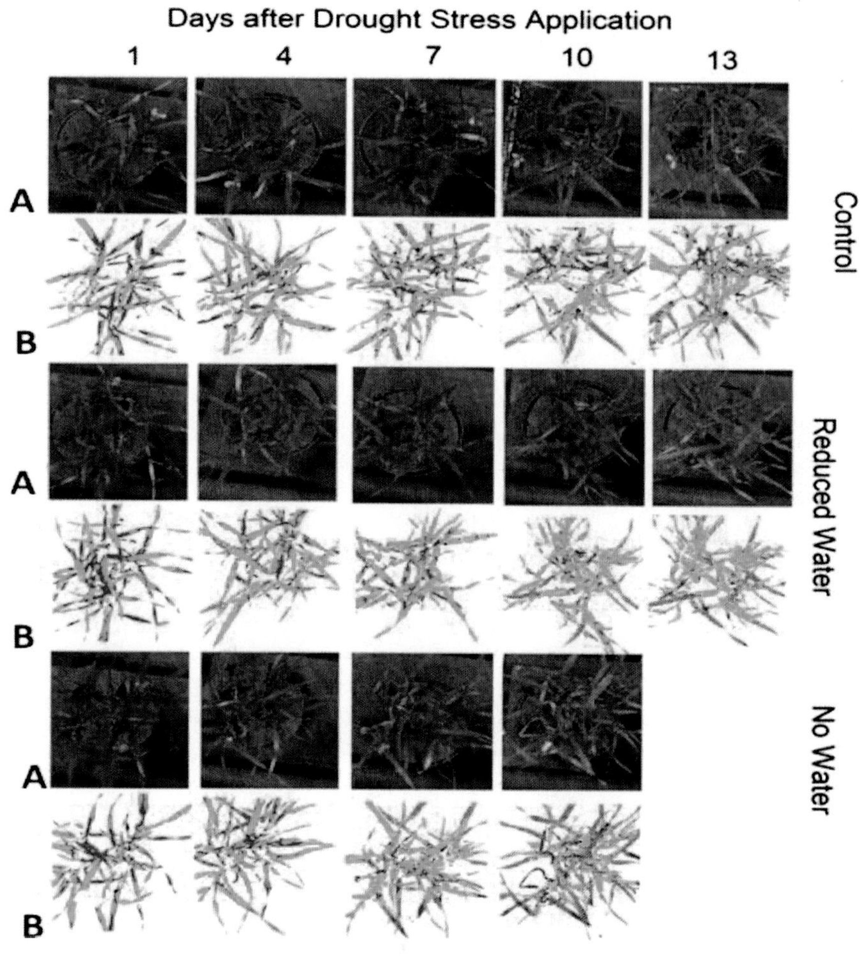

Fig. 26.1 The analysis of the response-progress to drought-progress of barley Barke genotype using hyperspectral images. On the *right*, the growth conditions of the plants are shown . On the *left*, (A) marks the original picture and (B) the false-colour image based on simplex volume maximization (SiVM) analysis. *Green marks* the healthy leaf-tissue and *red* the senescent one. Shown are only the pictures obtained prior to and after the "response turning time-point"

in water capacity of the substrate that started from day 7 too (see Sect. 26.2). As expected, the stress developed much faster after this date in plants without watering than with reduced watering. It could be noticed by the naked eye that at days 10–11, the number of senescent leaves in stressed plants (without watering) was significantly higher than in control ones. Meanwhile in plants where watering was reduced continuously, these symptoms were visible at day 17, earliest. As shown in Römer et al. (manuscript in prep.), this divergence was clearly reflected in archetype curves. The time shift of the divergence of signatures between stressed and healthy plant when comparing the 2010 and 2011 experiments might be a consequence of weather

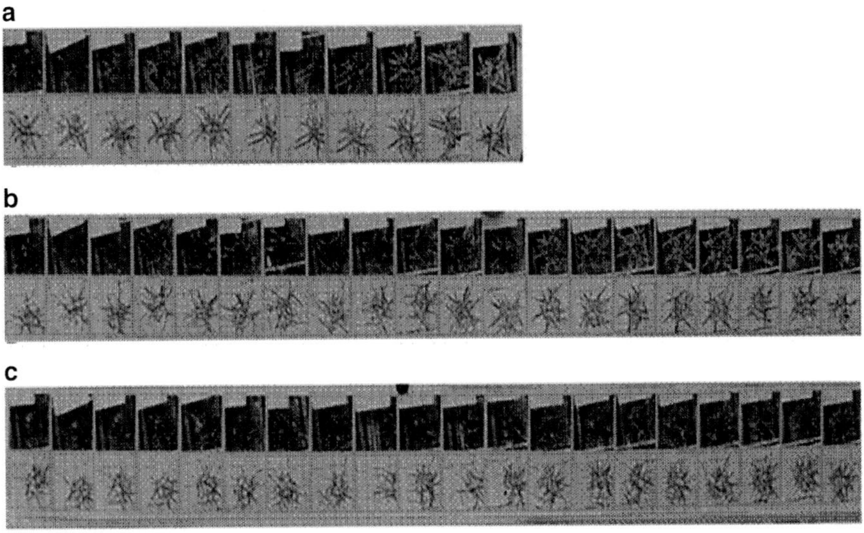

Fig. 26.2 The drought stress distribution in barley plant Wiebke genotype from the 2011 experiments. (**a**) and (**b**) show stressed plants with complete withholding of water and reduced watering, respectively (see Sect. 26.2), and (**c**) the controls with normal watering. The *upper row* of each panel shows the original pictures of the plants, and the lower one is derived from spectral images analysed with matrix factorization technique. *Blue* indicates healthy and *red* senescent leaves

conditions which were temperate than in 2012. The consequence is a longer time until stress conditions occur, and therefore, a delayed response from the plant takes place.

These hyperspectral signatures of stressed and control plants are in correlation to the water potential data (see Fig. 26.3) which indicate that the change in hyperspectral images is in relation to the available water inside the plant.

26.3.3 Water-Related Traits

At the same day of each hyperspectral measurement, the relative chlorophyll content was measured by using SPAD metre (Konica Minolta Sensing, Inc., Osaka, Japan). For a better comparison of the data, the measurements were done on the same leaves which were marked with rubber strips. The data received from the SPAD measurements did not show any significant correlation with the hyperspectral signatures (data not shown).

The water potential and the biomass were determined at midday at time-points 3, 6, 9, 14, 17 and 20 days of water stress period. The data for water potential of two stressed plants (green without water and red with reduced water) and control (the blue bar) are shown in Fig. 26.3. From day 7 of water shortage period, the water

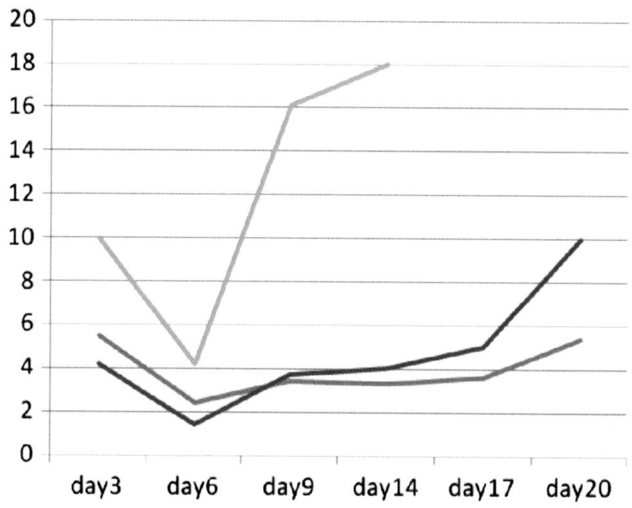

Fig. 26.3 The water potential was measured using a Scholander pressure chamber at six time-points during stress period, genotype Wiebke. *Green* – the fully stressed, *red* – the moderate stress and *blue* – the control

potential of plants under severe stress conditions increases rapidly, whereas that of the plants with moderate stress shows slow increase, which becomes more evident from day 17 and so on.

The curves of the biomass as well as of water content showed a similar time-course (data not shown).

26.4 Conclusions

Pre-symptomatic stress recognition is of great relevance in identification of specific stress-responsive molecules and for the development of effective tools in precision plant breeding and production. We hypothesize that the "response turning point" in the temporal curve of plant response to stress rather than the maxima is the most relevant time-point for a guided sampling to get inside into mechanistic basis of plant response to drought stress. Comparative analysis at level of transcripts, proteins and metabolites between drought stress and control plants at the "response turning time-point" will facilitate the identification of genes acting at the early stages of plant response to drought.

Here we reported the measurement of hyperspectral signatures, which has been applied for recognition of early response of barley plants under water stress conditions in two consecutive years. Comparative hyperspectral image analysis was performed on barley plants grown under well-watered and two regimes of water stress conditions in rain-out shelter. The obtained massive, high-dimensional data cubes were analysed with a recent matrix factorization technique based on simplex volume maximization of hyperspectral data. The results show that it was possible to detect and visualize the accelerated senescence signature in stressed plants earlier than the appearance of visible symptoms by the naked eye.

Acknowledgements This work has been carried in frame of a research programme funded by BMBF with project number 315309/ CROP.SENSe. The authors would like to thank Merle Noschinski for excellent technical assistance, Henrik Schumann, Melanie Herker and Sfefan Teutsch for their support with hyperspectrometry.

References

Abdeen, A., Schnell, J., & Miki, B. (2010). Transcriptome analysis reveals absence of unintended effects in drought-tolerant transgenic plants overexpressing the transcription factor ABF3. *BMC Genomics, 11*, 69. doi:10.1186/1471-2164-11-69.

Aldakheel, Y. Y., & Danson, F. M. (1997). Spectral reflectance of dehydrating leaves: Measurements and modelling. *International Journal of Remote Sensing, 18*(17), 3683–3690.

Boyer, J. S. (1982). Plant productivity and environment. *Science, 218*, 443–448.

Furbank, R. T., Cammerer, S., Sheehy, J., & Edwards, G. (2009). C4Rice: A challenge for plant phenomics. *Functional Plant Biology, 36*(11), 845–865.

Guo, P., Baum, M., Grando, S., Ceccarelli, S., Bai, G., Li, R., von Korff, M., Varshney, R. K., Graner, A., & Valkoun, J. (2009). Differentially expressed genes between drought-tolerant and drought-sensitive barley genotypes in response to drought stress during the reproductive stage. *Journal of Experimental Botany, 60*, 3531–3544.

Kersting, K., Wahabzada, M., Roemer, C., Thurau, C., Ballvora, A., Rascher, U., Leon, J., Bauckhage, C., Pluemer, L. (2012) Simplex distributions for embedding data matrices over time. In I. Davidson & C. Domeniconi (Eds.), *Proceedings of the 12th SIAM International Conference on Data Mining (SDM)*, Anaheim, CA

Lebreton, C., Lazic-Jancic, V., Steed, A., Pekic, S., & Quarrie, S. A. (1995). Identification of QTL for drought responses in maize and their use in testing causal relationships between traits. *Journal of Experimental Botany, 46*, 853–865.

Malenovský, Z., Mishra, K. B., Zemek, F., Rascher, U., & Nedbal, I. (2009). Scientific and technical challenges in remote sensing of plant canopy reflectance and fluorescence. *Journal of Experimental Botany, 60*, 2987–3004.

McKay, J. K., Richards, J. H., Sen, S., Mitchell-Olds, T., Sandra Boles, S., Stahl, E. A., Wayne, T., & Juenger, T. E. (2008). Genetics of drought adaptation in Arabidopsis thaliana II. QTL analysis of a new mapping population, Kas-1x TSU-1. *Evolution, 62*, 3014–3026.

Nguyen, T. T. T., Klueva, N., Chamareck, V., Aarti, A., Magpantay, G., Millena, A. C. M., Pathan, M. S., & Nguyen, H. T. (2004). Saturation mapping of QTL regions and identification of putative candidate genes for drought tolerance in rice. *Molecular Genetic & Genomics, 272*, 35–46.

Passioura, J. B. (2002). Environmental biology and crop improvement. *Functional Plant Biology, 29*, 537–554.

Penuelas, J., Pinol, J., Ogaya, R., & Filella, I. (1997). Photochemical reflectance index and leaf photosynthetic radiation-use-efficiency assessment in Mediterranean trees. *Internat Journal of Remote Sensing, 18*(13), 2869–2875.

Pinnisi, E. (2008). The blue revolution, drop by drop, gene by gene. *Science, 320*, 171–173.

Rabbani, M. A., Maruyama, K., Abe, H., Khan, M. A., Katsura, K., Ito, Y., Yoshiwara, K., Seki, M., Shinozaki, K., & Yamaguchi-Shinozaki, K. (2003). Monitoring expression profiles of rice genes under cold, drought, and high-salinity stresses and abscisic acid application using cDNA microarray and RNA gel-blot analyses. *Plant Physiology, 133*, 1755–1767.

Rascher, U., Nichol, C. L., Small, C., & Hendricks, L. (2007). Monitoring spatio-temporal dynamics of photosynthesis with a portable hyperspectral imaging system. *Photogrammetric Engineering and Remote Sensing, 73*, 45–56.

Rascher, U., Damm, A., van der Linden, S., Okujeni, A., Pieruschka, R., Schickling, A., & Hostert, P. (2010). Sensing of photosynthetic activity of crops. In E.-C. Oerke et al. (Eds.), *Precision crop protection – The challenge and use of heterogeneity*. Dordrecht/Heidelberg/London/New York: Springer.

Rascher, U., Blossfeld, S., Fiorani, F., Jahnke, S., Jansen, M., Kuhn, A. J., Matsubara, S., Märtin, L. L. A., Merchant, A., Metzner, R., Müller-Linow, M., Nagel, K. A., Pieruschka, R., Pinto, F., Schreiber, C. M., Temperton, V. M., Thorpe, M. R., Van Dusschoten, D., Van Volkenburgh, E., Windt, C. W., & Schurr, U. (2011). Non-invasive approaches for phenotyping of enhanced performance traits in bean. *Functional Plant Biology, 38*, 968–983.

Richards, R. A., Rebetzke, G. J., Watt, M., Condon, A. G., Spielmeyer, W., & Dolferus, R. (2010). Breeding for improved water productivity in temperate cereals: Phenotyping, quantitative trait loci, markers and the selection environment. *Functional Plant Biology, 37*(2), 85–97.

Sakma, Y., Maruyama, K., Osakabe, Y., Qin, F., Seki, M., Shinozaki, K., & Yamaguchi-Shinozaki, K. (2006). Functional analysis of an arabidopsis transcription factor, DREB2A, involved in drought-responsive gene expression. *Plant Physiology, 18*, 1292–1309.

Schreiber, A. W., Sutton, T., Caldo, R. A., Kalashyan, E., Lovell, B., Mayo, G., Muehlbauer, G. J., Druka, A., Waugh, R., Wise, R. P., Langridge, P., & Baumann, U. (2009). Comparative transcriptomics in the Triticeae. *BMC Genomics, 10*, 285. doi:10.1186/1471-2164-10-285.

Schulte, D., Close, T. J., Graner, A., Langridge, P., Matsumoto, T., Muehlbauer, G., Sato, K., Schulman, A. H., Waugh, R., Wise, R. P., & Stein, N. (2009). The international barley sequencing consortium–at the threshold of efficient access to the barley genome. *Plant Physiology, 149*, 142–147.

Talamé, V., Ozturk, N. Z., Bohnert, H. J., & Tuberosa, R. (2007). Barley transcript profiles underde hydration shock and drought stress treatments: A comparative analysis. *Journal of Experimental Botany, 58*, 229–240.

Tardieu, T., & Schurr, U. (2009). White paper on plant phenotyping. In *EPSO Workshop on Plant Phenotyping*, Jülich. Germany.

Tondelli, A., Francia, E., Barabaschi, D., Aprile, A., Skinner, J. S., Stockinger, E. J., Stanca, A. M., & Pecchioni, N. (2006). Mapping regulatory genes as candidates for cold and drought stress tolerance in barley. *Theoretical and Applied Genetics, 112*, 445–454.

Tran, L. S. P., Nakashim, K., Sakuma, Y., Simpson, S. D., Fujita, Y., Maruyama, K., Fujita, M., Seki, M., Shinozaki, K., & Yamaguchi-Shinozaki, K. (2004). Isolation and functional analysis of Arabidopsis stress-inducible NAC transcription factors that bind to a drought-responsive cis-element in the early responsive to dehydration stress Promoter. *Plant Cell, 16*, 2481–2498.

Ueda, A., Kathiresan, A., Inada, M., Narita, Y., Nakamura, T., Weiming Shi, W., Tetsuko Takabe, T., & Bennett, J. (2004). Osmotic stress in barley regulates expression of a different set of genes than salt stress does. *Journal of Experimental Botany, 55*, 2213–2218.

Ustin, S., & Gamon, J. A. (2010). Remote sensing of plant functional types. *New Phytologist, 186*, 795–816.

Xiong, L., Wang, R., Mao, G., & Koczan, J. M. (2006). Identification of drought tolerance determinants by genetic analysis of root response to drought stress and abscisic acid. *Plant Physiology, 142*, 1065–1074.

Chapter 27
Barley Adaptation: Teachings from Landraces Will Help to Respond to Climate Change

Ernesto Igartua, Ildikó Karsai, M. Cristina Casao, Otto Veisz, M. Pilar Gracia, and Ana M. Casas

Abstract Adaptation of crops to temperate climates depends to a large extent on plants having the appropriate combination of genes to respond to day length and temperature. Global warming poses new challenges to plant breeding. In many places, current cultivars will no longer be suited for cultivation. To sustain agricultural production, future cultivars must be provided with genes that ensure optimum adaptation to the new conditions. We present several findings on barley adaptation to Mediterranean climates, published over the last years, which resulted from the study of adaptations exhibited by local landraces.

Winter barley is widely grown in the Mediterranean region. We found that winter landraces have some degree of vernalization requirement, tuned to respond to the winter temperatures typical for each region. We present results that demonstrate that the allelic series of the main vernalization gene, *VRNH1*, is essential to determine the length of the cold period needed to promote flowering in barley.

The presence of the photoperiod-responsive allele of the gene *PPDH2* in most Mediterranean landraces is considered as a safety mechanism to promote flowering, which comes into play at least when vernalization conditions are not optimum (rather often in some areas). This mechanism is coordinated with the vernalization pathway through repression by *VRNH2*.

A latitudinal pattern of distribution of *VRNH3* in Spanish barleys suggests a role in day-length adaptation. This gene integrates the photoperiod and vernalization pathways in barley and seems to present an allelic series of at least five functionally

E. Igartua (✉) • M.C. Casao • M.P. Gracia • A.M. Casas
Department of Genetics and Plant Production, Aula Dei Experimental Station, EEAD-CSIC, Avda.Montañana 1005, 50059 Zaragoza, Spain
e-mail: igartua@eead.csic.es; http://www.eead.csic.es

I. Karsai • O. Veisz
Agricultural Research Institute, Hungarian Academy of Sciences, ARI-HAS, 2462 Martonvásár, Brunszvik u. 2, Hungary

G. Zhang et al. (eds.), *Advance in Barley Sciences: Proceedings of 11th International Barley Genetics Symposium*, DOI 10.1007/978-94-007-4682-4_27,

different alleles. We present evidences from several independent sets of materials that demonstrate a distinct phenotypic effect of several of these alleles, in accordance with the latitudinal distribution observed.

Keywords Barley • Adaptation • Vernalization • Photoperiod • Landraces • Breeding • Climate change

27.1 Introduction

Plant breeding must respond to challenges posed by biotic and abiotic factors threatening crop production. Climate change, if anything, will increase these challenges. Winter harshness, length of growth periods, and water availability may suffer changes, and crops will have to undergo adaptive adjustments accordingly. Plant breeders must understand the genetic bases of crop adaptation and should equip themselves with tools to allow breeding new cultivars provided with genes that ensure optimum adaptation to the upcoming conditions and sustain or increase agricultural productions.

Responses to temperature and day length are the main factors underlying adaptation in cereals. These responses determine the onset and the duration of the different growth phases and are responsible for matching the growth cycle of the plants to the prevailing conditions and to the resources available. Among the variables that are used to describe the growth cycle of cereals, flowering time is one of the most critical. This is so because, in seed crops like cereals, it indicates the moment in which the carbohydrate balance of the plant is definitely shifted in favor of the spike, and the resources of the plant are committed toward the economic yield. Therefore, it has to occur at the right time to ensure good agronomic performance.

Many genes have an effect on flowering time of barley. Genes and QTLs have been identified along the seven chromosomes. But several genes have a major role in the control of plant development in barley in response to environmental cues. These are, at least, *VRNH1* and *VRNH2*, involved in the determination of the vernalization requirements, *PPDH1* and *PPDH2*, which have been described as responsive to day length, and *VRNH3*, an integrator of the vernalization and photoperiod pathways toward the promotion of flowering. These genes do not act independently but present complex interactions that are still not fully understood (Laurie et al. 1995; Trevaskis et al. 2007; Distelfeld et al. 2009).

The allelic diversity at these genes can potentially produce a large number of combinations. *PPDH2* and *VRNH2* have just two alleles, either presence or absence. *PPDH1* presents also two alleles, one insensitive to long days, the other one sensitive. The situation for *VRNH1* and *VRNH3* is different. They present allelic series that seem to provide opportunities for fine-tuning the growth cycle to winter temperatures and day length, respectively. The occurrence of mutations at these genes has provided ample genetic variation for phenological traits and has been at the root of the dispersal of the crop and its spread and adaptation into new areas where it was not cultivated before (Cockram et al. 2011).

The Iberian Peninsula is an excellent laboratory to study adaptation of barley. On one hand, barley is grown under a diverse range of climates. Though most of them fall within the broad definition of *Mediterranean climate*, steep gradients of temperature, both annual and seasonal, altitude, and also a wide range of latitudes are present. These geographical features produce a diversity of environments for cereal cultivation. On the other hand, barley has been cultivated in the Iberian Peninsula since at least 7,000 BP, and landraces were maintained in cultivation until the twentieth century, when they were collected by researchers and kept in germplasm banks. Therefore, we have an available complete sampling of barleys, mainly from Spain, that reflects the adaptations accumulated by the crop over a long history of cultivation. A core collection of inbred lines derived from landraces was assembled with the objective of studying diversity and adaptation to the Spanish climates (SBCC, Igartua et al. 1998).

Here, we present a progress report of several findings on barley adaptation, published over the last years, which resulted from the study of Spanish landraces and modern cultivars. We have put together a summary of the conclusions on barley adaptation that we have arrived at after performing experiments at several levels: field trials, experiments under controlled conditions, genetic diversity, agroecology, and gene expression.

27.2 Geographic Patterns, Phenotypic Effects

Mediterranean climates, in the Old World, occur in the region comprising large sections of the countries on the rims of the Mediterranean basin in Europe, North Africa, and East Asia, including several of the most important barley-growing regions of the world. This climate poses several challenges at cereal development. Among them are the risk of frosts, especially late frosts that can damage the reproductive tissues, and the frequent occurrence of terminal heat and drought stress. Temperatures in the late spring tend to rise sharply, and cultivars must complete grain filling before the environment is too harsh.

The main adaptation against winter harshness is the occurrence of "spring" types, which can be sown in the late winter or in the spring and have no vernalization requirement. But winter cultivars are generally preferred in the Mediterranean basin. One reason is the longer growth season attained by a fall sowing, compared to a winter or spring sowing. Also, temperatures in the winter usually do not compromise the survival of the crop.

Under Mediterranean climate, winter cereals must have built-in mechanisms to avoid exposure to terminal abiotic stresses. Besides true tolerance mechanisms, avoidance by tuning the growth cycle to the conditions prevailing at each environment is the most widespread mechanism of plant and crop adaptation. The mechanisms set in place to achieve this are revealed by investigating the relationship between diversity of genes involved in the control of phenological traits and geographic features.

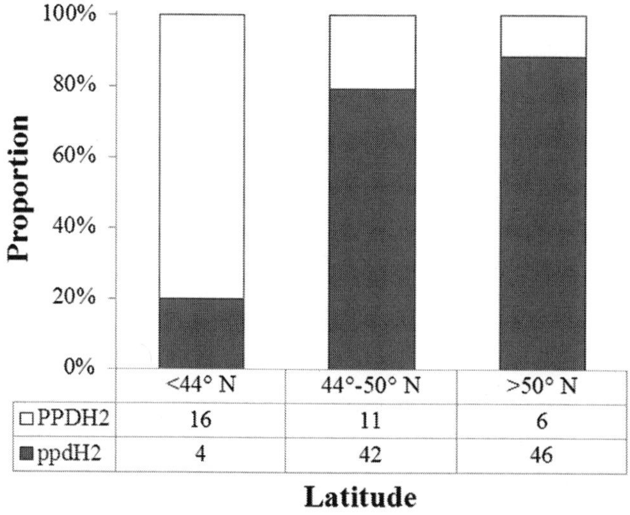

Fig. 27.1 Distribution of *PPDH2* alleles in 125 winter barley cultivars classified according to latitude of origin (*PPDH2*, dominant, functional allele; *ppdH2*, recessive, null allele)

PPDH1 is the main gene governing the response to long photoperiods (Laurie et al. 1994, 1995). The existence of a pattern of geographic distribution for the alleles of *PPDH1* has been acknowledged for sometime (Jones et al. 2008). The sensitive allele occurs much more frequently at lower latitudes. The effect of the sensitive allele is a dramatic advancement of flowering date under long days, thereby avoiding the terminal drought and heat stress frequent at lower latitudes. The presence of the sensitive alleles at lower latitudes produces an induction flowering by long days. In strict winter cultivars, this induction must be preceded by a complete vernalization. Spanish barley landraces presented the sensitive allele in 156 out of 159 lines. Even most of the spring lines had the sensitive *PPDH1* allele (Fig. 27.1).

PPDH2 is widely regarded as a gene responsive to short photoperiods. A latitudinal pattern for the distribution of the alleles of *PPDH2* (as well as a more general role) has been proposed recently (Casao et al. 2011c). This pattern is less evident than in the case of *PPDH1*, as it is revealed only when just winter genotypes are considered. It is widely assumed that winter cultivars do not present the functional version of *PPDH2*. This generalization is valid for winter cultivars from central and northern European countries, but it is less evident for southern European countries. A survey of 125 winter cultivars, mainly from European countries (only 9 non-European) both from the literature (Faure et al. 2007; Cuesta-Marcos et al. 2010) and also our own results, reveals an apparent higher occurrence of the dominant *PPDH2* allele at lower latitudes. This pattern is remarkable because latitudes below 44°N include almost the entire Mediterranean region. In this area, barley is sown predominantly during autumn and, to a large extent, using winter cultivars. The dominant allele was also prevalent in a large set of winter landraces from the SBCC (cultivated between 35 and 44°N). Actually, 129 out of the 143 Spanish winter lines had the dominant allele at *PPDH2*.

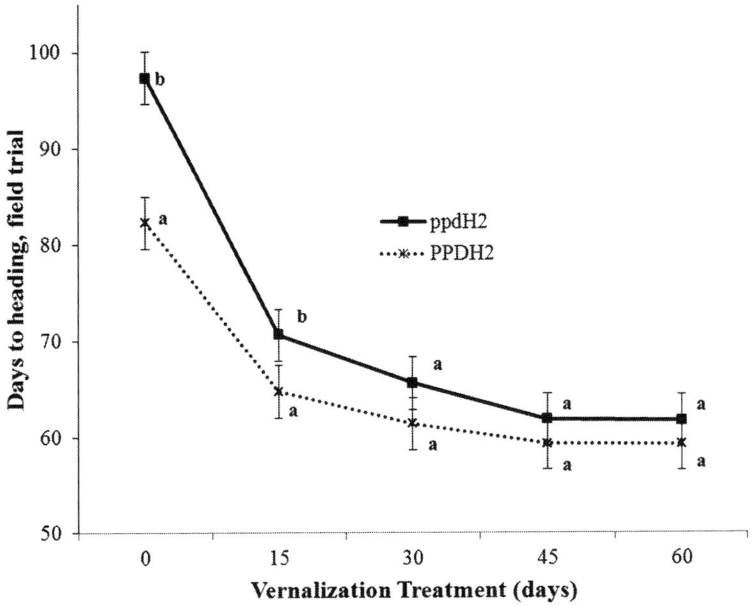

Fig. 27.2 Days to flowering in the field of 70 winter cultivars planted on late March after 0, 15, 30, 45, or 60 days of vernalization with low light intensity. *PPDH2, dominant, functional* allele present; ppdH2, recessive, null allele

The reason for the prevalence of the dominant *PPDH2*allele in winter cultivars at lower latitudes is most likely related to the fact that this allele has been identified as being responsible for the phenomenon known as "short-day vernalization" (Roberts et al. 1988). This misnomer actually stands for a promotion effect caused by the functional allele of *PPDH2* (Casao et al. 2011c), which is evident in winter cultivars when vernalization is not complete. Winter cultivars in higher latitudes tend to have the null allele, *ppdH2*, probably because it has been associated with higher frost tolerance (von Zitzewitz et al. 2011). At lower latitudes, cultivars and landraces carry the responsive allele at *PPDH2*, probably as a backup mechanism to induce flowering even in the absence of complete vernalization (King and Heide 2009). This mechanism makes sense when there is a chance of vernalization not occurring consistently year after year at a certain region. Strong evidence in favor of the existence of this mechanism was delivered by an experiment with 70 winter cultivars carried out in Hungary. Several plants per cultivars were exposed to vernalizing temperatures for increasing periods, with staggered sowings to allow transplantation to the field on March 25th. At that late date, it was expected that the plants would not receive much more additional vernalization. The effect of *PPDH2* on heading date was dramatic, producing a marked advancement of flowering time when vernalization was far from complete (0 and 15 days of vernalization period received prior to transplanting). This effect faded out as the vernalization duration increased (Fig. 27.2).

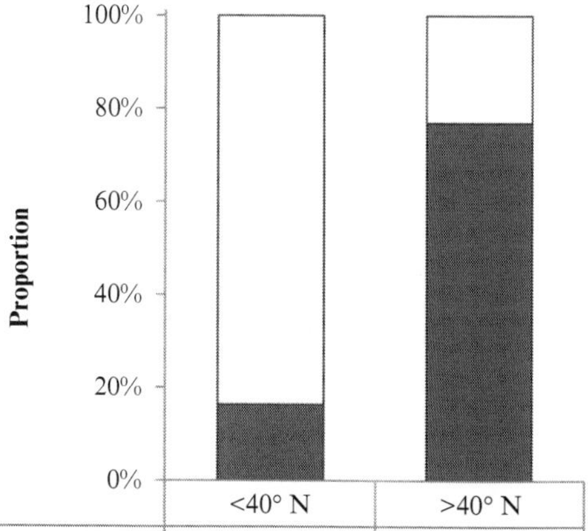

	<40° N	>40° N
□ VRNH3-TC	67	18
■ VRNH3-AG	13	60

Latitude

Fig. 27.3 Latitudinal distribution of polymorphisms described by Yan et al. (2006) at the first intron of *VRNH3*, over a set of 158 inbred lines derived from Spanish landraces (part of the Spanish Barley Core Collection)

A latitudinal pattern was also evident for several *VRNH3* alleles (Casas et al. 2011). The polymorphisms described at the first intron (Yan et al. 2006) were very distinctly associated with latitude (Fig. 27.3). The haplotypes formed by polymorphisms at the promoter and the first intron, first described in that work, suggested an even more staggered latitudinal distribution of the alleles. These haplotypes were also apparently related to effects on flowering time in an association analysis (Fig. 27.4, derived from data in Casas et al. 2011), consistent with the latitudinal distribution, i.e., the "later" the allele, the more northerly its distribution.

Hemming et al. (2009) proposed a classification of the polymorphism at *VRNH1*, which we follow here. There are already several indications of a functional role of the alleles of *VRNH1*. An allelic series at this locus was already proposed by Takahashi and Yasuda (1971). The polymorphisms that define the different alleles (mainly concentrated in the first intron) have been described in several recent works (Fu et al. 2005; von Zitzewitz et al. 2005; Cockram et al. 2007; Sz cs et al. 2007; Hemming et al. 2009), which also provide indication of functional allelic differences. We evaluated the response to vernalization of several winter and spring genotypes, with the most representative *VRNH1* alleles, including the two prevalent in Spanish barleys (*VRNH1*-4 and *VRNH1*-6). The results (Fig. 27.5) indicate a gradient of vernalization requirements, according to the allele present at *VRNH1* (for winter genotypes only) from the strict winter allele *vrnH1*, present in most European winter cultivars, to the spring genotypes, with three intermediate alleles: *VRNH1*-6, related with a vernalization requirement lower than, but close to, strict winter cultivars; *VRNH1*-4, with a low vernalization requirement; and a new allele, here named *Orria*, after the cultivar in which it was found, associated with an even lower vernalization need. These results were partially presented in Casao et al. (2011a).

The geographical distribution of the*VNRH1* alleles of Spanish landraces was not latitudinal, but followed winter harshness patterns (Fig. 27.6). Of the two alleles

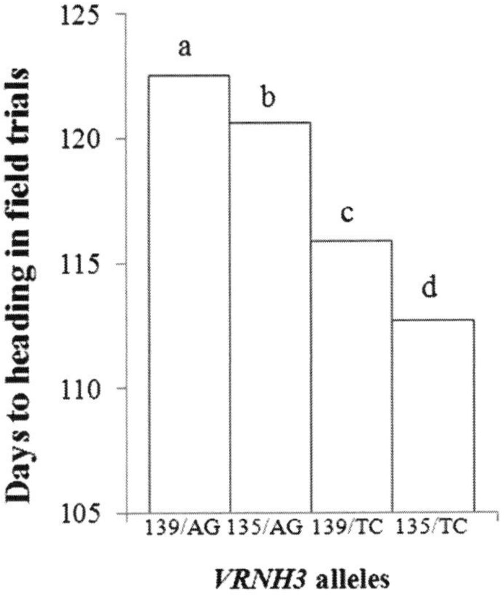

Fig. 27.4 Effect of different *VRNH3* alleles on flowering time, at a set of nine field trials carried out in Spain, under autumn sowing, for a set of 143 inbred lines of winter growth habit derived from Spanish landraces (part of the Spanish Barley Core Collection)

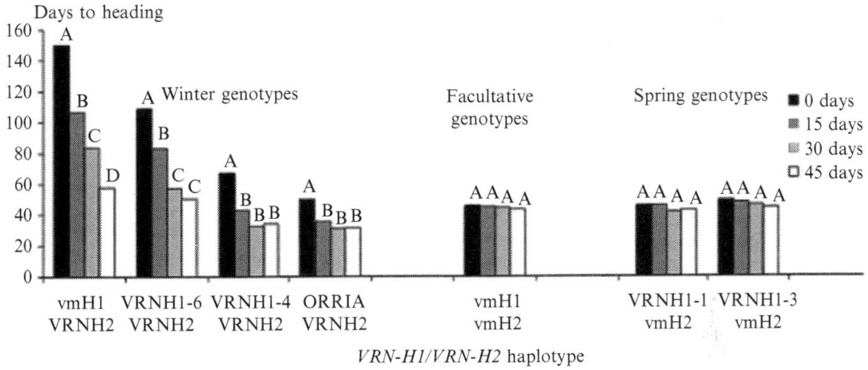

Fig. 27.5 Days to heading (flowering) of several typical barley genotypes (or groups of genotypes) measured in a growth chamber, under long days (16 h), after vernalization treatments of increasing duration (from 0 to 45 days under 4–8 °C, 8 h light), according to the haplotypes determined by the two main vernalization loci, *VRNH1* and *VRNH2*

most frequent among Spanish landraces, the one with the shorter first intron (*VRNH1*-4, according to Hemming et al. 2009) prevailed at regions with warmer winters, according to the climate classification of Papadakis (1975).

The other one, *VRNH1*-6, with a small deletion in the first intron, occurred much more frequently in barleys from regions with colder winters (Fig. 27.6). This allele, quite frequent in Spanish landraces, was very rare in another study of European cultivars (Cockram et al. 2007). There is more evidence pointing at *VRNH1* as the

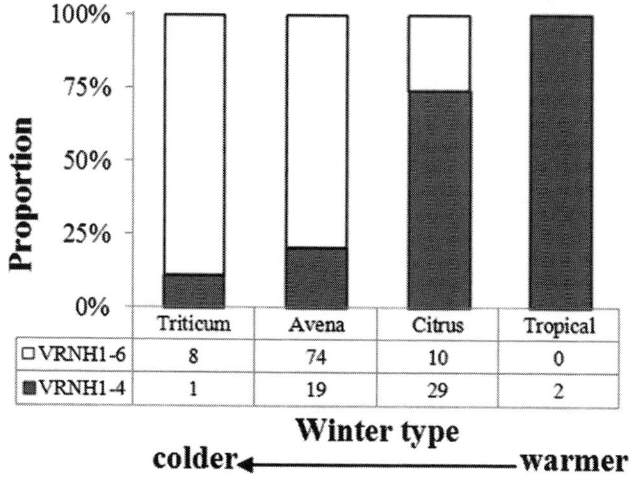

Fig. 27.6 Distribution of *VRNH1* alleles over types of winter, according to the classification of Papadakis (1975), for a set of 143 inbred lines of winter growth habit derived from Spanish landraces (part of the Spanish Barley Core Collection)

main responsible for the differences in vernalization needs of these lines and cultivars. Probably, the most important is an experiment in which the region carrying a *VRNH1*-4 allele was transferred through marker-assisted backcross into a strict winter cultivar (Plaisant, carrying *vrnH1*), used as recurrent parent. Lines with 92–97% recovery of the recurrent genome displayed a vernalization pattern similar to the donor parent. Therefore, there is a good chance that *VRNH1*-4 alone was responsible for that change (Casao et al. 2011b).

Most of the results presented here are based on collections of germplasm and have been derived through association methods. Our group and some collaborators have developed biparental populations of several kinds (RILs, doubled haploids, backcrosses, F2) for several purposes, including the validation of the effects of vernalization and day-length responsive genes. For instance, the distinct effect of the polymorphisms at two regions of *VRNH3*, the first intron and the promoters, has been confirmed with two different populations, an F2 for the intron (Casas et al. 2011) and at a doubled haploid population for the promoter (Ponce et al., submitted). Experiments with populations to investigate the effect of *VRNH1* alleles, *VRNH2*, *PPDH2*, and *PPDH1* alleles, are currently under way, and so far all the effects described in this chapter are being confirmed.

27.3 Gene Expression

Several gene expression experiments (Casao et al. 2011a, c) revealed that the gradual phenotypic effects of the allelic series at VRNH1 had a perfectly matching underlying pattern of gene expression in response to increasing length of vernalization periods.

Another important finding is that PPDH2 seems to have a larger role than previously acknowledged. It is expressed as well under long photoperiods (as reported previously by Kikuchi et al. 2009) and is apparently suppressed by VRNH2. As VRNH2 expression is promoted by long days, its suppressive effect on PPDH2 has been traditionally interpreted as short-day sensitivity.

27.4 Interpretation and Use of Patterns of Adaptation

Only a small percentage, about 5–10%, of the Spanish barleys are spring or facultative types. These are confined to highlands or mountainous regions with cool springs. Most barley landraces from the Iberian Peninsula are winter types. All winter barleys from the SBCC (143) have a functional allele at *VRNH2* (*HvZCCTa-c* are present) and the sensitive allele at *PPDH1*. Then, most of them have the functional *PPDH2* allele (129). Only 14 had the recessive allele, which is prevalent in winter cultivars from more northern countries. These 14 accessions apparently come from colder areas, in general, but the data are too few to conclude anything. The adaptation of barley in this area seems to have been driven first by *VRNH1*. Besides, this gene seems to be causing the genetic structure of the Spanish barleys. Spanish barleys can be divided into four groups (Yahiaoui et al. 2008), the two most numerous corresponding largely to six-row winter barleys, which are differentiated mainly by the *VRNH1* allele carried by most of the accessions that constitute each group (*VRNH1*-4 or *VRNH1*-6). Then, there is a clear latitudinal gradient for the distribution of *VRNH3* alleles. Also, *VRNH3* and *VRNH1* show no significant linkage disequilibrium (which may have been the case if the alleles at both genes were selected concurrently). Therefore, Spanish barleys are, in most cases, winter types with reduced vernalization requirement compared to typical winter cultivars. The size of this requirement is proportional to winter temperatures of the regions of origin of the landraces. This notwithstanding, they add a backup mechanism (a functional *PPDH2* allele) to allow promotion toward flowering even when this reduced vernalization requirement is not fulfilled, or is completed too late. The presence of this backup mechanism suggests two things. First, this mechanism probably takes place before long days that induce the expression of *VRNH2*, which seems to block *PPDH2* expression (Casao et al. 2011a). Second, it indicates that the presence of long-day sensitivity (provided by *PPDH1*) is not a guarantee in itself to induce flowering on time to avoid terminal stress. The issue of what are the day-length thresholds that induce expression of these genes is an open and interesting question.

One of the purposes of this work was to provide guidance and tools for barley breeding. The knowledge of the adaptation traits that were acquired by the cereals in their process of domestication and dispersal can offer excellent clues to breeders, about how to continue breeding for adaptation in a planned manner. Doebley et al. (2006) asked themselves: *can a knowledge of the genes contributing to past crop domestication and improvement guide future breeding efforts?* Their answer was a clear *yes*, for a variety of reasons. We agree with them. Future climate scenarios in

Europe actually suggest that temperatures in central and northern European countries will become closer to those currently experienced in the Mediterranean region. It is clear, therefore, that the adaptation traits found in the Mediterranean can be of direct use for breeders in other regions. But this does not mean that the genetic solutions for adaptation found in extant genotypes are the only ones, neither the best ones possible. Given the genetic richness of barley and its independent diversification at different regions, it seems unlikely that all possible combinations of these major genes (and others) have been tested empirically. Also, the implications of these genes and their combinations on agronomic performance, specifically on grain yield, yield components, and frost tolerance have to be investigated. Some work in this respect is also under way.

Acknowledgements Funding by grants AGL2007-63625, AGL2010-21929, and HH2008-0013 from the Spanish Ministry of Science and Technology (co-funded by the European Regional Development Fund) and by the Hungarian Scientific Research Fund (OTKA NK72913) is acknowledged. Germplasm maintenance of the SBCC was funded by projects RFP2004-00015-00-00 and RFP2009-00005-00-00. MCC was supported by an I3P Predoctoral Fellowship from CSIC.

References

Casao, M. C., Igartua, E., Karsai, I., Lasa, J. M., Gracia, M. P., & Casas, A. M. (2011a). Expression analysis of vernalization and day-length response genes in barley (Hordeumvulgare L.) indicates that VRNH2 is a repressor of PPDH2 (HvFT3) under long days. *Journal of Experimental Botany, 62*, 1939–1949.

Casao, M. C., Igartua, E., Karsai, I., Bhat, P. R., Cuadrado, N., Gracia, M. P., Lasa, J. M., & Casas, A. M. (2011b). Introgression of an intermediate VRNH1 allele in barley (Hordeumvulgare L.) leads to reduced vernalization requirement without affecting freezing tolerance. *Molecular Breeding*. doi:10.1007/s11032-010-9497-y.

Casao, M. C., Karsai, I., Igartua, E., Gracia, M. P., Veisz, O., & Casas, A. M. (2011c). Adaptation of barley to mild winters: A role for PPDH2. *BMC Plant Biology, 11*, 164.

Casas, A. M., Djemel, A., Ciudad, F. J., Yahiaoui, S., Ponce, L. J., Contreras-Moreira, B., Gracia, M. P., Lasa, J. M., & Igartua, E. (2011). HvFT1 (VrnH3) drives latitudinal adaptation in Spanish barleys. *Theoretical and Applied Genetics, 112*, 1293–1304.

Cockram, J., Chiapparino, E., Taylor, S. A., Stamati, K., Donini, P., Laurie, D. A., & O'Sullivan, D. M. (2007). Haplotype analysis of vernalization loci in European barley germplasm reveals novel VRN-H1 alleles and a predominant winter VRN-H1/VRN-H2 multi-locus haplotype. *Theoretical and Applied Genetics, 115*, 993–1001.

Cockram, J., Hones, H., & O'Sullivan, D. M. (2011). Genetic variation at flowering time loci in wild and cultivated barley. *Plant Genetic Resources, 9*, 264–267.

Cuesta-Marcos, A., Szűcs, P., Close, T. J., Filichkin, T., Muehlbauer, G. J., Smith, K. P., & Hayes, P. M. (2010). Genome-wide SNPs and re-sequencing of growth habit and inflorescence genes in barley: Implications for association mapping in germplasm arrays varying in size and structure. *BMC Genomics, 11*, 707.

Distelfeld, A., Li, C., & Dubcovsky, J. (2009). Regulation of flowering in temperate cereals. *Current Opinion in Plant Biology, 12*, 178–184.

Doebley, J. F., Gaut, B. S., & Smith, B. D. (2006). The molecular genetics of crop domestication. *Cell, 127*, 1309–1321.

Faure, S., Higgins, J., Turner, A., & Laurie, D. A. (2007). The FLOWERING LOCUS T-like gene family in barley *Hordeum vulgare*. *Genetics, 176*, 599–609.

Fu, D., Szűcs, P., Yan, L., Helguera, M., Skinner, J., Hayes, P., & Dubcovsky, J. (2005). Large deletions within the first intron of the VRN-1 are associated with spring growth habit in barley and wheat. *Molecular Genetics and Genomics, 273*, 54–65.

Hemming, M. N., Fieg, S., Peacock, W. J., Dennis, E. S., & Trevaskis, B. (2009). Regions associated with repression of the barley (Hordeumvulgare) VERNALIZATION1 gene are not required for cold induction. *Molecular Genetics and Genomics, 282*, 107–117.

Igartua, E., Gracia, M. P., Lasa, J. M., Medina, B., Molina-Cano, J. L., Montoya, J. L., & Romagosa, I. (1998). The Spanish barley core collection. *Genetic Resources and Crop Evolution, 45*, 475–481.

Jones, H., Leigh, F. J., Mackay, I., Bower, M. A., Smith, L. M. J., Charles, M. P., Jones, G., Jones, M. K., Brown, T. A., & Powell, W. (2008). Population-based resequencing reveals that the flowering time adaptation of cultivated barley originated east of the Fertile Crescent. *Molecular Biology and Evolution, 25*, 2211–2219.

Kikuchi, R., Kawahigashi, H., Ando, T., Tonooka, T., & Handa, H. (2009). Molecular and functional characterization of PEBP genes in barley reveal the diversification of their roles in flowering. *Plant Physiology, 149*, 1341–1353.

King, R. W., & Heide, O. M. (2009). Seasonal flowering and evolution: The heritage from Charles Darwin. *Functional Plant Biology, 36*, 1027–1036.

Laurie, D. A., Pratchett, N., Bezant, J. H., & Snape, J. W. (1994). Genetic analysis of a photoperiod response gene on the short arm of chromosome 2 (2H) of *Hordeum vulgare*. *Heredity, 72*, 619–627.

Laurie, D. A., Pratchett, N., Bezant, J. H., & Snape, J. W. (1995). RFLP mapping of five major genes and eight quantitative trait loci controlling flowering time in a winter×spring barley *Hordeum vulgare* L. cross. *Genome, 38*, 575–585.

Papadakis, J. (1975). *Climates of the world and their agricultural potentialities*. Buenos Aires: Edición Argentina. 200 pp.

Roberts, E. H., Summerfield, R. J., Cooper, J. P., & Ellis, R. H. (1988). Environmental control of flowering in barley (Hordeum vulgare L.). I Photoperiod limits to long-day responses, photoperiod-insensitive phases and effects of low temperature and short-day vernalization. *Annals of Botany–London, 62*, 127–144.

Szűcs, P., Skinner, J. S., Karsai, I., Cuesta-Marcos, A., Haggard, K. G., Corey, A. E., Chen, T. H. H., & Hayes, P. M. (2007). Validation of the *VRN-H2/VRN-H1*epistatic model in barley reveals that intron length variation in *VRN-H1* may account for a continuum of vernalization sensitivity. *Molecular Genetics and Genomics, 277*, 249–261.

Takahashi, R., & Yasuda, S. (1971). Genetics of earliness and growth habit in barley. In R. A. Nilan (Ed.), *Barley genetics II* (pp. 388–408). Pullman: Washington State University Press.

Trevaskis, B., Hemming, M. N., Dennis, E. S., & Peacock, W. J. (2007). The molecular basis of vernalization-induced flowering in cereals. *Trends in Plant Science, 12*, 352–357.

von Zitzewitz, J., Szűcs, P., Dubcovsky, J., Yan, L., Pecchioni, N., Francia, E., Casas, A., Chen, T. H. H., Hayes, P. M., & Skinner, J. S. (2005). Molecular and structural characterization of barley vernalization genes. *Plant Molecular Biology, 59*, 449–467.

von Zitzewitz, J., Cuesta-Marcos, A., Condon, F., Castro, A. J., Chao, S., Corey, A., Filichkin, T., Fisk, S. P., Gutierrez, L., Haggard, K., Karsai, I., Muehlbauer, G. J., Smith, K. P., Veisz, O., & Hayes, P. M. (2011). The genetics of winter hardiness in barley: Perspectives from genome-wide association mapping. *Plant Genome, 4*, 76–91.

Yahiaoui, S., Igartua, E., Moralejo, M., Ramsay, L., Molina-Cano, J. L., Ciudad, F. J., Lasa, J. M., Gracia, M. P., & Casas, A. M. (2008). Patterns of genetic and eco-geographical diversity in Spanish barleys. *Theoretical and Applied Genetics, 116*, 271–282.

Yan, L., Fu, D., Li, C., Blechl, A., Tranquilli, G., Bonafede, M., Sanchez, A., Valarik, M., & Dubcovsky, J. (2006). The wheat and barley vernalization gene *VRN3* is an orthologue of FT. *Proceedings of the National Academy of Sciences of the United States of America, 103*, 19581–19586.

Chapter 28
Development of Recombinant Chromosome Substitution Lines for Aluminum Tolerance in Barley

Kazuhiro Sato and Jianfeng Ma

Abstract A series of recombinant chromosome substitution lines (RCSLs) and a near-isogenic line on the isolated aluminum tolerance locus were developed from a cross between elite cultivar Haruna Nijo and aluminum-tolerant cultivar Murasakimochi. High-throughput single nucleotide polymorphism (SNP) genotyping has previously been developed for barley. The parents were genotyped with 1,448 unigene-derived SNPs. Of these 1,448 SNPs, 690 were polymorphic between Haruna Nijo and Murasakimochi, giving an overall polymorphism rate of 47.7%. These 690 SNPs were compared with the consensus map developed by the same marker system, and the numbers of markers per chromosome were as follows: 1H (78), 2H (101), 3H (120), 4H (97), 5H (113), 6H (94), and 7H (87). The SNP markers were well-distributed and reflecting the original distribution of consensus map. The SNPs used to genotype 83 BC_3F_3 (RCSLs). These data were used to create graphical genotypes for each line and thus estimate the location, extent, and total number of introgressions from Murasakimochi in the Haruna Nijo background. The selected 39 RCSLs sample most of the Murasakimochi genome.

Keywords Acid soil tolerance • Aluminum tolerance • Barley • *Hordeum vulgare* • Mapping • Single nucleotide polymorphism

---

Presenting author, Kazuhiro Sato

K. Sato (✉) • J. Ma
Institute of Plant Science and Resources, Okayama University, Chuo, Kurashiki 710-0046, Japan
e-mail: kazsato@rib.okayama-u.ac.jp

G. Zhang et al. (eds.), *Advance in Barley Sciences: Proceedings of 11th International Barley Genetics Symposium*, DOI 10.1007/978-94-007-4682-4_28,

28.1 Introduction

Aluminum (Al) tolerance is one of the key characters of barley (*Hordeum vulgare* L.) to adapt acid soils which comprise 30–40% of the world's arable soils (von Uexküll and Mutert 1995). Barley is most sensitive to Al among cereal crops and shows significant yield loss in acid soils where soluble ionic Al inhibits root growth. However, there is a wide variation in Al resistance among barley cultivars. Zhao et al. (2003) showed that the Al-resistant barley cultivars rapidly secrete citrate from the roots in response to Al and that there is a good correlation between Al resistance and the amount of citrate secretion among different cultivars. Of these cultivars, Al-resistant Murasakimochi and Al-sensitive Morex were used as mapping parents and identified a locus (*HvAACT1*) responsible for the Al-activated citrate secretion by fine mapping combined with microarray expression analysis (Furukawa et al. 2007). This gene belongs to the multidrug and toxic compound extrusion (MATE) family and was constitutively expressed mainly in the roots of the Al-resistant barley cultivar. A good correlation was found between the expression of HvAACT1 and citrate secretion in barley cultivars differing in Al resistance. The results demonstrate that HvAACT1 is an Al-activated citrate transporter responsible for Al resistance in barley.

Murasakimochi is an old Japanese waxy food barley landrace especially used for making local cakes. This haplotype is exotic to other gene pools of present barley cultivars and needs improvements in characters on practical uses and field performances. Even a locus is isolated from this cultivar, commercial production of Al-tolerant barley by transformed HvAACT1 is not practical at present due to technical difficulties of barley transformation and a low public/industrial acceptance of genetically modified cultivars of barley. One of the solutions for transferring cloned Al tolerance locus is backcrossing an elite cultivar (recurrent parent) to develop an introgression line having the Al-tolerant allele. The genome-wide backcross introgression populations (also called recombinant chromosome substitution lines; RCSLs) provide the opportunity to assess unadapted alleles in an adapted genetic background (Pillen et al. 2003, 2004; von Korff et al. 2004; Matus et al. 2003; Schmalenbach et al. 2008; Sato and Takeda 2009; Sato et al. 2011). For example, Sato and Takeda (2009) demonstrated systematic generation of substituted segments of the wild barley H602 into cultivated Haruna Nijo. The present study also aims Haruna Nijo as a target haplotype and introgresses substituted segments of Al locus from Murasakimochi.

To identify substituted segments, whole genome marker genotyping is necessary. The Illumina SNP genotyping system and consensus map developed by Close et al. (2009) have been used to characterize RCSLs for other crosses between EST donors as parents and confirm its efficiency to detect genome-wide marker pormorphisms between EST donors (Sato et al. 2011). However, the number of efficient SNPs between Haruna Nijo (EST donor) and Murasakimochi (non-donor) is unknown. Moreover, the RCSLs of Haruna Nijo/Murasakimochi do not have reference mapping population, e.g., doubled haploid lines or recombinant inbred lines, to map original markers segregating in a mapping population. The consensus map by Close et al. (2009) was derived from four mapping populations of approximately 3,000 EST-derived unigenes. Markers in the consensus map were mapped by the polymorphism of parents in these four mapping populations (Steptoe/Morex, Oregon Wolfe Barley, Morex/Barke, and Haruna Nijo/H602). The number of markers mapped on RCSLs of Haruna Nijo/Murasakimochi is also limited by the markers polymorphic on these pairs of parents in the consensus map.

The present analysis is aiming to (1) identify polymorphic markers between Haruna Nijo and Murasakimochi which are on the consensus map of Close et al. (2009), (2) genotype substituted segment in the RCSLs to develop a set of RCSLs to cover the genome, and (3) develop a near-isogenic line of Al tolerance from Murasakimochi in the background of Haruna Nijo.

28.2 Materials and Methods

28.2.1 Plant Materials

A Japanese malting barley cultivar "Haruna Nijo" was used as a female parent for crossing with a Japanese waxy food barley cultivar "Murasakimochi." Recombinant chromosome substitution lines (RCSLs) were developed from Haruna Nijo/Murasakimochi F_1 plants. The RCSLs were developed in the following manner. The F_1 was crossed with a recurrent parent "Haruna Nijo" to produce BC_1F_1 individuals. Twenty-four BC_1F_1 individuals were crossed with the recurrent parent. Each of 117 BC_2F_1 individuals was again crossed with the recurrent parent to produce BC_3F_1 RCSLs. Each BC_3F_1 individual was self-pollinated to develop a set of 111 BC_3F_2 lines. A subset of 83 RCSLs was randomly selected from the 111 BC_3F_2 lines and advanced to the BC_3F_3 generation at the facilities of Okayama University, Kurashiki, Japan (34°35′N and 133°46′E).

28.2.2 DNA Isolation and Genotyping

Plants were grown in the experimental field in Okayama University and 200–300 mg of seedling leaf tissue was harvested from each plant. Leaf samples were frozen in liquid nitrogen and crushed into fine powder using a multi-bead shocker (Yasui Kikai Co.). Qiagen DNeasy Plant mini kits (QIAGEN Co.) were used to isolate DNA from each sample. DNA concentration was adjusted to 100–200 ng/ul.

Frozen DNA samples were sent to the Southern California Genotyping Consortium, Illumina BeadLab at the University of California – Los Angeles for the OPA-SNP assay with the 1,536-plex detection platform of barley OPA 1 (BOPA1; Close et al. 2009). Genotyping was performed on parents and 83 RCSLs using the

Illumina GoldenGate BeadArray (Fan et al. 2003). Genotype data were manually inspected to correct for excessive emphasis on heterozygote calls using GenCall software (Illumina, San Diego, CA), and only the most reliable calls were retained. The SNP loci are designated by HarvEST (http://harvest.ucr.edu/) unigene assembly #32 numbers.

28.2.3 Map Distance Calculation

SNP calls were compared between Haruna Nijo and Murasakimochi to identify markers segregating in the RCSLs. Genotype calls from the RCSLs were placed in map order as determined from the SNP consensus map (Close et al. 2009). The chromosome segments introgressed into Haruna Nijo from Murasakimochi were estimated from the graphical haplotypes.

28.2.4 Development of Al-Tolerant Isogenic Line and Phenotype Evaluation

Three RCSLs of BC_3F_3 were selected to have Murasakimochi allele by the sequence marker polymorphism within HvAACT1 (Furukawa et al. 2007) and backcrossed with Haruna Nijo to develop BC_4 generation. After selfing the crossed plant, plant having Murasakimochi allele for HvAACT1 was selected and checked the performance in the Al toxic condition.

28.3 Results and Discussion

28.3.1 Application of OPA Markers for Genotyping RCSLs

We used the barley oligonucleotide pooled assay #1 (BOPA1) (1,536 SNPs targeting 1,536 different genes) to find polymorphism between Haruna Nijo and Murasakimochi. Of the 1,448 informative SNPs, 690 were polymorphic between Haruna Nijo and Murasakimochi, giving an overall polymorphism rate of 47.7%. These 690 SNPs were compared with the consensus map of Close et al. (2009). The numbers of markers per chromosome were as follows: 1H (78), 2H (101), 3H (120), 4H (97), 5H (113), 6H (94), and 7H (87). The SNP markers were well-distributed and reflecting the original distribution of consensus map.

Allele calls for these 690 polymorphic SNPs were obtained on 83 RCSLs based on the parental SNP alleles. SNP marker data from the 83 RCSLs were sorted in order according to the consensus map of Close et al. (2009). The RCSLs were homozygous at most loci; the few cases of heterozygote calls were readily recognized, in general defining segments containing tracts of heterozygous genes.

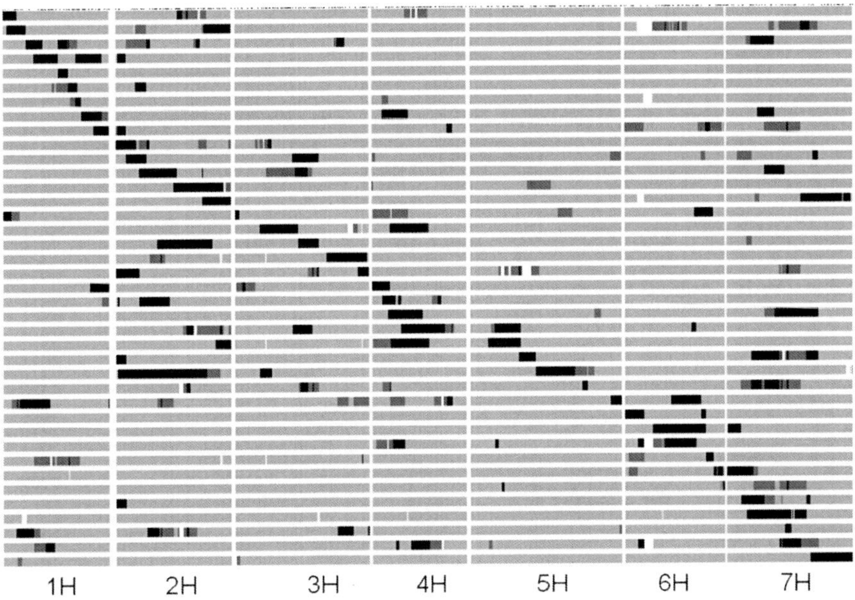

Fig. 28.1 Graphical genotypes of a minimum set of 39 RCSLs representing Murasakimochi substituted segments (*black*) on the background of cv. Haruna Nijo (*gray*). Heterozygous segments are indicated by intermediate *gray* color. Missing markers are indicated by *blank*. Each chromosome is oriented with short arms from *left*

By arranging substituted segments in the RCSLs by chromosome (and from short arm to long arm within each chromosome), a minimum set of RCSLs was selected to represent the unique substituted segments from wild barley. Figure 28.1 shows these 39 selected RCSLs, as illustrated by GGT 2.0 (van Berloo 2008). Some of the substituted segments that overlap are duplicated to achieve maximum genome coverage. The introgressed segments represent the Murasakimochi genome well, except for the chromosome arms of 3HS, 4HL, 5HS, and 5HL. Despite these gaps, substitutions of most of the barley genome, as defined on a genetic rather than a physical basis, were obtained.

28.3.2 *Marker Polymorphism and Mapping*

The variation of 1,301 SNPs (Sato et al. 2011) between Haruna Nijo and Murasakimochi (0.53) was almost comparable with the one between Haruna Nijo and Akashinriki (0.54) (Sato et al. 2011) but still lower compared to other combinations of EST donors, e.g., Barke and Morex (0.61) or Haruna Nijo and H602 (0.57). These differences may be due to ascertainment bias in the SNPs used for the BOPA1 design rather than being true metrics of genetic distance. Moragues et al. (2010) applied the same 1,536 SNPs set for the diversity analysis of 169 barley landraces from Syria and Jordan and 171 European cultivars. They estimated the diversity by using randomly chosen sets of 384 SNPs and the same number of SNPs to maximize the polymorphisms either in landraces and cultivars. They compared the efficiency of these sets and found that preselecting markers from BOPA1 for their diversity in a germplasm set is very worthwhile in terms of the quality of data obtained.

The BOPA1 SNPs were selected based on polymorphisms between EST donors, and the number of useful SNPs from each pair was influenced by the number of ESTs generated from these donors. Most of the ESTs used for the HarvEST:Barley database are from donors of Occidental (non-Asian) origin. Selection of SNPs that have a high minor allele frequency within this subset of barley germplasm, as was the purpose of the BOPA1 design, would be expected to carry ascertainment bias and reduce the general applicability of BOPA1 SNPs to the detection of polymorphisms between Occidental and Oriental (East Asian) barleys. As shown by Sato et al. (2011), the SNPs between Akashinriki and other major donors of EST may be more likely to provide SNPs relevant to distinguishing between Oriental and Occidental haplotypes. Since Akashinriki was the only EST donor of Oriental origin, the polymorphic BOPA1 markers between Haruna Nijo/Akashinriki may be useful in general for mapping and selection in parental combinations of Oriental and Occidental barleys. Actually, of the 1,301 SNPs from BOPA1, 1,193 SNPs agreed between Akashinriki and Murasakimochi, indicating that these two cultivars share the close genetic background with each other.

The opportunity to merge maps that have many markers in common is another considerable advantage of using a standardized marker system by referring to the consensus map (Close et al. 2009). As shown in the expanded cMAP (Fang et al. 2003) web image at http://map.lab.nig.ac.jp:8085/cmap/, the BOPA1 markers tended to cluster within centromeric regions of each chromosome, consistent with the observation reported in Close et al. (2009).

28.3.3 Detection, Alignment, and Selection of Alien Segments Using RCSLs

The current application of SNP markers to RCSLs has allowed us to precisely map substituted segments in barley. The same mapping strategy was used in the Haruna Nijo/H602 population, which gave the same quality of data sets for identifying segmental substitution (Sato and Takeda 2009).

The alignment of 39 BC_3F_3 Haruna Nijo/Murasakimochi RCSLs using high-resolution genotyping detected missing segments only on chromosome arms of 3HL, 4HL, 5HS, and 5HL. This is better genome coverage compared to the Haruna Nijo/H602 (BC_3F_5) population, which did not capture several regions of chromosomes (Sato and Takeda 2009). Since the utility of RCSLs will depend on the genomic region of interest, it may not be necessary to have a complete set of RCSLs. If such a set is deemed necessary, it could be developed by application of high-throughput genotyping in an earlier generation during RCSL development.

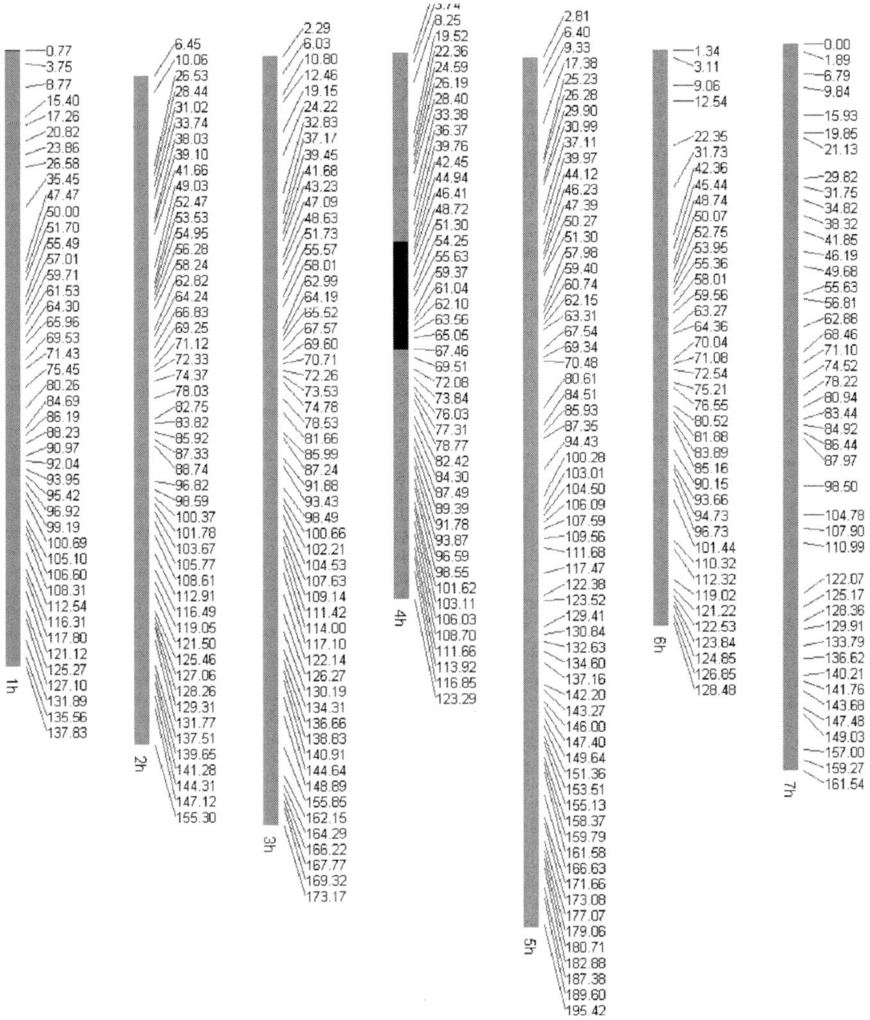

Fig. 28.2 Graphical genotyping of recombinant chromosome substitution line with a segment including Al tolerance locus on chromosome 4H in barley. *Black line* shows the substation segment from Al-tolerant Murasakimochi and *gray lines* show Haruna Nijo genome

We found that a set of 39 RCSLs represents most of the alleles of Murasakimochi in the background of Haruna Nijo. The total length of the consensus map (Close et al. 2009) is 1,076.6 cM, whereas the map length of the RCSLs (1004.9 cM) is 93.3% of the SNP map. This is much higher than the 81.4% of the H602 genome captured in Haruna Nijo in the Haruna Nijo/H602 BC_3F_5 (Sato and Takeda 2009). The graphical genotyping shows clear and reliable identification of substitution segments from Murasakimochi. Therefore, the Haruna Nijo/Murasakimochi RCSLs provide a high level of genome coverage with a high degree of marker saturation.

28.3.4 Development of a Near-Isogenic Line of A1 Tolerance

Due to poor malting quality profiles of hybrids between malting and food barley crosses, it has been difficult to use food barleys in malting barley breeding. RCSL is one approach to exploit exotic germplasm in breeding programs to improve elite malting barley cultivars, e.g., Haruna Nijo. These resources will bring Al tolerance and other new alleles into the malting barley gene pool from food barley to develop improved malting barley cultivars as clearly demonstrated in Fig. 28.2.

28.4 Conclusions (and Future Work)

The efficient system of RCSL development in barley is presented in this study. An RCSL with a segment having a gene of interest is one of the non-transgenic methods to use the cloned gene in the elite germplasm. All EST markers mapped on the substituted segments in the RCSLs can be utilized to analyze genes of interest that are segregating between Haruna Nijo and Murasakimochi. Since BAC libraries are available for Haruna Nijo (Saisho et al. 2007), segregating transcripts on the substituted segments are potential targets of gene isolation, especially unique characters in East Asian food barleys.

Acknowledgments We would like to thank Dr. Joe Deyong, Southern California Genotyping Consortium, Illumina BeadLab, UCLA for the OPA-SNP assay; Drs. Timothy J. Close and Prasanna Bhat, Department of Botany and Plant Sciences, University of California, Riverside for basecalling of OPA assay.

References

Close, T. J., Bhat, P. R., Lonardi, S., Wu, Y., Rostoks, N., Ramsay, L., Druka, A., Stein, N., Svensson, J. T., Wanamaker, S., Bozdag, S., Roose, M. L., Moscou, M. J., Chao, S., Varshney, R. K., Szucs, P., Sato, K., Hayes, P. M., Matthews, D. E., Kleinhofs, A., Muehlbauer, G. J., DeYoung, J., Marshall, D. F., Madishetty, K., Fenton, R. D., Condamine, P., Graner, A., & Waugh, R. (2009). Development and implementation of high-throughput SNP genotyping in barley. *BMC Genomics, 10*, 582.

Fan, J. B., Oliphant, A., Shen, R., Kermani, B. G., Garcia, F., Gunderson, K. L., Hansen, M., Steemers, F., Butler, S. L., Deloukas, P., Galver, L., Hunt, S., McBride, C., Bibikova, M., Rubano, T., Chen, J., Wickham, E., Doucet, D., Chang, W., Campbell, D., Zhang, B., Kruglyak, S., Bentley, D., Haas, J., Rigault, P., Zhou, L., Stuelpnagel, J., & Chee, M. S. (2003). Highly parallel SNP genotyping. *Cold Spring Harbor Symposia on Quantitative Biology, 68*, 69–78.

Fang, Z., Polacco, M., Chen, S., Schroeder, S., Hancock, D., Sanchez, H., & Coe, E. (2003). cMap: The comparative genetic map viewer. *Bioinformatics, 19*, 416–417.

Furukawa, J., Yamaji, N., Wang, H., Mitani, N., Murata, Y., Sato, K., Katsuhara, M., Takeda, K., & Ma, J. F. (2007). An aluminum-activated citrate transporter in barley. *Plant & Cell Physiology, 48*, 1081–1091.

Matus, I., Corey, A., Filichkin, T., Hayes, P. M., Vales, M. I., Kling, J., Riera, O., Sato, K., Powell, W., & Waugh, R. (2003). Development and characterization of recombinant chromosome substitution lines (RCSLs) using *Hordeum vulgare* subsp. *spontaneum* as a source of donor alleles in a *Hordeum vulgare* subsp. *vulgare* background. *Genome, 46*, 1010–1023.

Moragues, M., Comadran, J., Waugh, R., Milne, I., Flavell, A. J., & Russell, J. R. (2010). Effects of ascertainment bias and marker number on estimations of barley diversity from high-throughput SNP genotype data. *Theoretical and Applied Genetics, 120*, 1525–1534.

Pillen, K., Zacharias, A., & Leon, J. (2003). Advanced backcross QTL analysis in barley (*Hordeum vulgare* L.). *Theoretical and Applied Genetics, 107*, 340–352.

Pillen, K., Zacharias, A., & Leon, J. (2004). Comparative AB-QTL analysis in barley using a single exotic donor of *Hordeum vulgare* ssp. *spontaneum*. *Theoretical and Applied Genetics, 108*, 1591–1601.

Saisho, D., Myoraku, E., Kawasaki, S. K. S., & Takeda, K. (2007). Construction and characterization of a bacterial artificial chromosome (BAC) library for Japanese malting barley 'Haruna Nijo'. *Breeding Science, 57*, 29–38.

Sato, K., & Takeda, K. (2009). An application of high-throughput SNP genotyping for barley genome mapping and characterization of recombinant chromosome substitution lines. *Theoretical and Applied Genetics, 119*, 613–619.

Sato, K., Close, T. J., Bhat, P., Muñoz-Amatriaín, M., & Muehlbauer, G. J. (2011). Single nucleotide polymorphism mapping and alignment of recombinant chromosome substitution lines in barley. *Plant & Cell Physiology, 52*, 728–737.

Schmalenbach, I., Korber, N., & Pillen, K. (2008). Selecting a set of wild barley introgression lines and verification of QTL effects for resistance to powdery mildew and leaf rust. *Theoretical and Applied Genetics, 117*, 1093–1106.

van Berloo, R. (2008). GGT 2.0: Versatile software for visualization and analysis of genetic data. *Journal of Heredity, 99*, 232–236.

von Korff, M., Wang, H., Leon, J., & Pillen, K. (2004). Development of candidate introgression lines using an exotic barley accession (*Hordeum vulgare* ssp. *spontaneum*) as donor. *Theoretical and Applied Genetics, 109*, 1736–1745.

von Uexküll, H. R., & Mutert, E. (1995). Global extent, development and economic impact of acid soils. *Plant and Soil, 171*, 1–15.

Zhao, Z., Ma, J. F., Sato, K., & Takeda, K. (2003). Differential Al resistance and citrate secretion in barley (*Hordeum vulgare* L.). *Planta, 217*, 794–800.

Chapter 29
New and Renewed Breeding Methodology

Hayes Patrick and Cuesta-Marcos Alfonso

Abstract Over the past 10,000 years, barley breeders and geneticists have developed increasingly complex breeding methods. The first selectors were incredibly successful at differentiating the major germplasm groups we still recognize: 2-row/6-row, malt/feed/food/forage, and winter/spring/facultative. They also left a legacy of genetic bottlenecks that breeders and pathologists labor to rectify. Over the years, breeding methods have diverged and rejoined, forming a veritable network of oxbows, backwaters, crosscurrents, and channels. Accelerated generation advance – via off-season nurseries and/or doubled haploids – remains a central feature of many programs. The big wave of fashionable marker-assisted selection has crested, leaving significant smaller sets behind for those actively engaged in implementation. Genomic selection is the current draw, with theory in hot pursuit of results. In this section of the conference, key papers highlight the diversity of breeding methods applied around the world.

Keywords Barley • Selection • Marker-assisted selection • QTL • Association mapping • Genomic selection

H. Patrick (✉) • C.-M. Alfonso
Plant Breeding and Genetics Program, Department of Crop and Soil Science, Oregon State University, Corvallis, OR 97331, USA
e-mail: patrick.m.hayes@oregonstate.edu

G. Zhang et al. (eds.), *Advance in Barley Sciences: Proceedings of 11th International Barley Genetics Symposium*, DOI 10.1007/978-94-007-4682-4_29,

The old saying that "there is nothing new under the sun" is not entirely applicable to barley breeding methodology. Some aspects of breeding methodology – for example, visual selection – clearly fall into the "renewed" category. Visual selection based on individual plant phenotype has been, is, and will continue to be effective and essential. At the other end of the spectrum are breeding methods, such as genomic selection (GS), that are truly new to barley breeding. Genomic selection, as applied today, involves a union of computational and genotyping methods that simply did not exist 10 years ago. Between visual selection and genomic selection there lies a fluid continuum of breeding methods: for example, bulk, pedigree, male-sterile-facilitated recurrent selection, etc. These plant breeding methods and selection procedures in barley – from domestication ~10,000 years ago until today (and on until tomorrow) – are addressed in a plethora of books, chapters, reviews, and technical papers (see recent review by Thomas et al. 2011). Therefore, the goal of this presentation will be to frame barley breeding methods – new and renewed – in terms of essential issues and considerations.

In this session of the 11th IBGS, we will learn about using alternative population types and assessment approaches (at the genome, transcriptome, and metabolome levels) to detect QTL. There is a long history of QTL analysis in completely inbred barley mapping populations derived from biparental crosses, leading to hundreds of reports in the past 25 years. Statistical methods for QTL detection in such germplasm have been continuously improved. Details on the many different methods available. The two principal features of this approach are (1) members of the populations trace to a single cross, usually between two lines contrasting for the target trait(s) and (2) there is extensive linkage disequilibrium (LD) between QTL and markers. This allows for the estimation of QTL positions (within a confidence interval), effects, and interactions between QTL without the necessity of high density of markers (Piepho 2000). There are, however, limitations to QTL detection in biparental populations – including bias in the estimation of QTL effects due to reduced sample size (Vales et al. 2005), narrow genetic bases and consequent limited scope of inference (Crepieux et al. 2005), and broad confidence intervals for QTL positions and effects (Darvasi et al. 1993; Hyne et al. 1995) – which can compromise the success of marker-assisted selection (MAS). In this presentation, we will use MAS to refer to selection for specific regions of the region where genes or QTLs are located. Most targets of MAS in the *Triticeae* have been identified through biparental mapping populations. However, often the parents chosen for these populations are genetically and phenotypically diverse in order to facilitate segregation for the trait of interest and insure sufficient marker polymorphism to construct complete linkage maps. Unfortunately, the use of such wide crosses has often lead to the discovery of markers that are not very useful in breeding and resulted in a gap between the discovery of QTL and application of MAS (Bernardo 2008). Genome-wide association mapping (GW-AM) approaches have been applied in crop genetic studies to address some of the limitations of biparental mapping studies (Rafalski 2002; Jannink et al. 2001). In particular, GW-AM allows investigators to assemble arrays of germplasm tailored to the question at hand. In barley, GW-AM panels have been constructed from wild barley for the discovery of novel alleles as well as elite breeding material to characterize useful genetic variation that is more immediately useful in breeding (Roy et al. 2010; Cuesta-Marcos et al. 2010; Massman et al. 2011; Kraakman et al. 2004, 2006; von Zitzewitz et al. 2011). In addition, since germplasm arrays can be assembled from existing genotypes, it is possible to generate populations with sufficient size to overcome problems of lack of power to detect QTL and biased estimates of QTL effects (Vales et al. 2005; Beavis 1994; Melchinger et al. 1998; Schön et al. 2004). The USDA Barley Coordinated Project (CAP) was established to create large phenotypic and SNP marker data sets of elite breeding lines from ten US breeding programs with which to conduct GW-AM (Waugh et al. 2009). The SNP genotyping platform, trait and marker database (The HordeumToolbox; http://hordeumtoolbox.org), and wealth of QTL information generated from the Barley CAP provide a rich framework and set of breeder tools with which to initiate ambitious new breeding efforts.

Regardless of the method used to identify targets for selection, the goal, until recently, has been MAS. This technique has been successful in plant breeding for traits controlled by single genes with large effects. In barley, we have examples of resistance genes for stripe rust and Septoria speckled leaf blotch (Zhong et al. 2006; Castro et al. 2003). The utility of MAS diminishes as the number of selection targets increases (Bernardo 2008). For example, fixing the favorable allele at four independent loci in a single individual in an F2 requires a minimum of 44 or 256 individuals. If one simply enriches the population at each locus, by selecting lines homozygous and heterozygous for the favorable allele, the number of individuals required is four. Hundreds of major QTL or genes have been mapped or cloned, and diagnostic markers exist with which to conduct MAS. In the *Triticeae* in particular, the way to achieve this in most cases has been the use of biparental mapping populations of inbred (or doubled haploid) lines.

Three talks in this session will specifically address breeding methods. One will give insights into the breeding methods used in Europe, a region in which "private breeders" have achieved an enviable record of successful barley variety development. The second will specifically address genomic selection and its applications – a perspective from North America, where the "academic as researcher/breeder" model still holds sway, albeit tenuously. The third talk with this theme will address the coupling of high-throughput genotyping and phenotyping. The final talk in this session will explore the interface of breeding and basic science: once a gene, or genes, underlying a target trait is identified, application and commercialization usually proceed down one path, while understanding structure and function proceeds on another. The paths will inevitably intersect.

The most straightforward of breeding methods – visual selection – was effective 10,000 years ago in selecting for shattering resistance and launching the agricultural revolution (Harlan 1957). Other effective visual selection events occurred in barley that rank nearly as high in importance: for example, six-row inflorescence, spring growth habit, hooded, hull-less, black seed, and dwarfness. A common denominator in all these cases is that the phenotypes usually show complete penetrance and expressivity. Anyone with a good eye and determination can successfully select for these traits, and once the proper alleles are configured and fixed, the phenotype will breed true. The genes determining these major germplasm descriptor traits have all been cloned and characterized, or soon will be. Knowing how things work satisfies fundamental curiosity and leaves one in awe of the level of scientific achievement required to determine, for example, that multiple and independent mutations in *HvHox1* lead to the six-row phenotype, (Komatsuda et al. 2007), a deletion in *PPD-H2* leads to short-day insensitivity (Faure et al. 2007), and deletion of one or more *ZCCT-H* genes leads to spring growth habit (Szűcs et al. 2007).

Value can also be obtained from knowledge of single gene structure and function in terms of breeding method. Although the traits are amenable to visual selection and target alleles can easily be fixed in the self or doubled haploid progeny of the appropriate F1, there are cases when MAS for simply inherited traits may be warranted. For example, in laying the foundation for our food barley doubled haploid genomic selection (DHGS) program at Oregon State University, we are accessing alleles from six-row and two-row germplasm of winter and spring growth habit. Our goal is for all germplasm entering the DHGS pipeline to be two-row and facultative but to vary for other food quality traits (e.g., hulled vs. hull-less, waxy vs. non-waxy vs. resistant starch, aroma, flavor, etc.). Therefore, we will apply MAS (using SNP-based assays) for *ZCCT-H* and *HvHox1* to select for facultative two rows in large segregating populations. Doubled haploids will be produced from selected individuals, and these doubled haploids will be used to initiate GS.

Knowing how things work can allow one to correctly predict outcomes when using simple breeding schemes for traits that involve epistatic interactions. For example, the efficiency of breeding for hooded in germplasm derived from hooded x awned crosses is increased when one properly accounts for the epistatic interaction of *Kap* and *Lks2* (Thomas et al. 2011). Likewise, knowing that vernalization insensitivity can be due to deletion of *ZCCT-H* genes or to deletions in the first intron of *HvBM5A* allows one to correctly predict the recovery of vernalization-sensitive individuals in the progeny derived from the cross of vernalization-insensitive parents (Szűcs et al. 2007). Although 10,000 years old, visual selection is entirely applicable to the current generation of barley transgenics, directed mutagenesis (e.g., CIBUS; http://www.cibus.com/pdfs/Cibus_Brochure.pdf), and efficient detection of induced mutations (e.g., TILLING, Gotwald et al. 2009). For example, in our breeding program, we have worked almost exclusively with six-row winter and facultative malting barley germplasm, based on the directives of research sponsors and the sheer perverse pleasure of doing what no one else does (Hayes 2007). Changes in corporate ownership and brewing technology, however, have led research sponsors to send a very clear signal to cease and desist with six-row and shift to two-row. Shifting a germplasm base from six-row to two-row is a seismic event using conventional breeding methods and a challenging one using complex contemporary methods such as DHGS (as described above). Applying site-specific mutagenesis to our elite and adapted six rows via the approach of CIBUS in order to maintain the desired configuration of alleles at ~29,000 loci while changing a single SNP at *HvHox1* is tempting. Imagine, six-row seed mailed off in year one returns as two-row seed a year later. Breeding could not be simpler. But, alas, like Sisyphus we will only be tempted. Meager research funding could never fund such a technical tour de force. Even if funding were available, the unfortunate chromosome location of *HvHox1* in the centromeric region could prove to be a complicating factor.

Complicating factors are at the root of the next category of traits, and it is in this zone of the continuum – between simple and infinitesimal – where most barley breeders labor away. Envision that "simple" lies at the left and "infinitesimal" at the right of an imaginary line. At the left end lie the traits determined by one or a few genes that show little or no interaction – neither *inter se*, with the environment or with different genetic backgrounds. At the right end lie the traits determined by multiple genes that do interact *inter se*, with the environment and with genetic backgrounds. This simple linear model becomes multidimensional as one progressively adds considerations such as manipulating heritability and developing proxy methods for trait assessment. The choice of appropriate breeding method becomes resource-, case-, time-, and space-dependent. Key considerations are population size, generation advance, how the phenotype will be assessed, and at to what extent which type of "omic" data will be used.

By way of example, single gene disease resistance would lie on the left side of the line, assuming an efficient screening procedure is available. In this situation, the choice of breeding method (i.e., whether or not to use simple visual assessment of disease incidence/severity vs. MAS or to combine the two techniques) will be based on the relative costs of the two procedures. However, if epidemic development is not consistent, the trait shifts to the right as heritability declines, and more complex breeding schemes involving some degree of MAS, if not GS, are indicated. And if one is engaged in purely defensive breeding (i.e., as is currently the case for UG99 stem rust resistance breeding in North America), truly complex breeding schemes involving accelerated gene detection and accelerated approaches to homozygosity, MAS, and phenotyping of selected individuals in exotic locales is necessary. Following the disease resistance theme, quantitative resistance is more likely to be durable than quantitative resistance (Castro et al. 2003). However, effective pyramiding of resistance genes can only be achieved via integrating MAS in the breeding program.

An alternative example of traits in the central zone of the continuum is malting quality. Malting quality is what makes barley a unique and esteemed crop, and it is the versatility of this grain to produce both solid and liquid food that led to its domestication. Very likely, an early hunter-gatherer inadvertently left barley seed in water. The barley germinated, the slurry was colonized by ambient yeast, and the yeast produced ethanol. Not beer as we know beer, but nonetheless a satisfying enough beverage to save civilization (http://store.discovery.com/how-beer-saved-the-world-dvd/detail.php?p=363503). Germinate, inoculate, and sample: what a simple phenotype to measure and breed for! The intervening 10,000 years of breeding led to the development of barley varieties that are in essence finely tuned beer-making machines, optimized purveyors of food for yeast. Malting quality, as it is defined today – not as defined by that first happy breeder/drinker – is a complex phenotype that can be deconstructed into a series of composite traits, each of which may have a complex genetic basis. As a consequence, the literature is replete with QTL reports for each of the components of malting quality (Szűcs et al. 2009; Zhang and Li 2009). Despite the depth and breadth of this knowledge, it remains essentially descriptive rather than predictive. For example, recent work defines the genomic architecture of the founder genotype "Morex" (Muñoz-Amatriaín et al. 2009), and yet in our own breeding program, we are constantly on the prowl for expanded opportunities to empirically measure malting quality traits. Even "simple" components of malting quality, such as beta amylase activity, have turned out to defy initial efforts at detecting and fixing favorable alleles via MAS (Filichkin et al. 2010). Queried as to how we are using MAS for malting quality, we will adroitly turn the subject to "simpler" traits, such as biotic and abiotic stress tolerance (Castro et al. 2003; Von Zitzewitz et al. 2011).

As we approach the right side of our imaginary line, where the complexity level reaches factor 11 (http://www.spinaltapfan.com), it is worth mentioning the tremendous strides that can be made in characterizing and breeding for complex traits using alternative phenotypic assays. Beta glucan, for example, can now be measured easily, cheaply, and quickly using the Megazyme kit with the protocol modification developed by Hu and Burton (2008). Likewise, grain protein can be readily measured using NIRS. The complex pathways underlying synthesis, deposition, and degradation of these compounds are still being elucidated (Burton et al. 2010; Lacerenza et al. 2010). The current *Triticeae* Coordinated Agricultural Project (T-CAP; http://www.triticeaecap.org/) has as a fundamental tenet the discovery of genes determining water use efficiency (WUE) and nitrogen use efficiency (NUE) based on detection via canopy spectral reflectance (CSR). Direct measurement of these phenotypes is not possible in the very large number of genotypes that will be assayed during the life of the project. Optimistically, the results of GW-AM for WUE and NUE will lead to the detection of so many genes and QTL that MAS for WUE and NUE will not be feasible. The indicated breeding method of choice is genomic selection (GS): a method where breeding program complexity is commensurate with genetic and trait complexity.

Genomic selection, strictly speaking a type of MAS, offers several advantages over traditional MAS. While the former is useful for enriching and/or tracking a limited number of single loci in breeding populations, it loses effectiveness as the number of loci increases as is the case for quantitative traits. Another disadvantage of MAS is that it requires the two-step process of identifying markers that exceed a significance threshold in mapping study, followed by performance prediction based on markers. Since the choice of significance threshold is arbitrary, information carried in markers below that threshold is lost. GS avoids this problem by combining the estimate of all marker effects and the prediction of the individual's performance in a single step (Meuwissen et al. 2001). These genomic estimated breeding values (GEBV) are derived from marker effects that are estimated from a training population for which both marker and phenotypic data exist. The primary benefit of GS is that selection can be imposed for quantitative traits very early in the breeding process substantially reducing the breeding cycle. If GS can be coupled with rapid generation advancement, as it is the case of doubled haploid production (Thomas et al. 2003), the breeding cycle can be as short as a single generation. Several recent publications have reviewed and evaluated GS methods through both simulation and empirical studies (Heffner et al. 2009; Lorenz et al. 2011; Iwata and Jannink 2011; Lorenzana and Bernardo 2009). Taken together, these studies indicate that relatively simple statistical models, such as ridge regression BLUP, and training population using 300 individuals and 300 markers were sufficient to generate prediction accuracies that would substantially improve genetic gain per year.

In conclusion, barley breeding appears to be on a track of increasing complexity, higher throughput, acceleration, and efficiency. Barley breeders have at hand an array of tools to simultaneously meet the needs of consumers, processors, and producers. Barley has the track record, and the potential, to meet so many of the world's needs for food, feed, and happiness. It is for all of us to realize this potential. Like Harlan (1957), I trust that each and every one of us can say that "fate smiled on me and gave me a crop to work with that is fascinating in itself and that grows when and where living conditions are ideal."

References

Beavis, W. D. (1994). The power and deceit of QTL experiments: Lessons from comparative QTL studies. In *49th Annual Corn and Sorghum Industry Research Conference* (pp. 250–266). Washington, DC: ASTA.

Bernardo, R. (2008). Molecular markers and selection for complex traits in plants: Learning from the last 20 years. *Crop Science, 48*, 1649–1664.

Burton, R. A., Gidley, M. J., & Fincher, G. B. (2010). Heterogeneity in the chemistry structure and function of plant cell walls. *Nature Chemical Biology, 6*, 724–732.

Castro, A., Chen, X., Hayes, P. M., & Johnston, M. (2003). Pyramiding quantitative trait locus (QTL) alleles determining resistance to barley stripe rust: Effects on resistance at the seedling stage. *Crop Science, 43*, 651–659.

Crepieux, S., Lebreton, C., Flament, P., & Charmet, G. (2005). Application of a new IBD-based QTL mapping method to common wheat breeding population: Analysis of kernel hardness and dough strength. *Theoretical and Applied Genetics, 111*, 1409–1419.

Cuesta-Marcos, A., Szucs, P., Close, T. J., Filichkin, T., Muehlbauer, G. J., Smith, K. P., & Hayes, P. M. (2010). Genome-wide SNPs and re-sequencing of growth habit and inflorescence genes in barley: Implications for association mapping in germplasm arrays varying in size and structure. *BMC Genomics, 11*, 707.

Darvasi, A., Weinreb, A., Minke, V., Weller, J. I., & Soller, M. (1993). Detecting marker-QTL linkage and estimating QTL gene effect and map location using a saturated genetic-map. *Genetics, 134*, 943–951.

Faure, S., Higgins, J., Turner, A., & Laurie, D. A. (2007). The Flowering Locus T-like gene family in barley (*Hordeum vulgare*). *Genetics, 176*, 599–609.

Filichkin, T. P., Vinje, M., Budde, A., Duke, S., Gallagher, L., Helgesson, J., Henson, C., Obert, D., Ohm, O., Petrie, S., Ross, A., & Hayes, P. (2010). Phenotypic variation for diastatic power, β-amylase activity, and β-amylase thermostability vs. allelic variation at the*Bmy1* locus in a sample of North American barley germplasm. *Crop Science, 50*, 826–834.

Gotwald, S., Bauer, P., Komatsuda, T., Lundqvist, U., & Stein, N. S. (2009). TILLING in the two-rowed barley cultivar reveals preferred sites of functional diversity in the gene *HvHox1*. *BMC Research Notes, 2*, 258.

Harlan, J. (1957). *One man's life with barley*. New York: Exposition Press.

Hayes, P. M. (2007). What's your barley? *Brewer's Guardian, 136*, 46.

Heffner, E. L., Sorrells, M. E., & Jannink, J. L. (2009). Genomic selection for crop improvement. *Crop Science, 49*, 1–12.

Hu, G., & Burton, C. (2008). Modification of standard enzymatic protocol to a cost-efficient format for mixed-linkage (1 ->3,1 ->4)-beta-D-glucan measurement. *Cereal Chemistry, 85*, 648–653.

Hyne, V., Kearsey, M. J., Pike, D. J., & Snape, J. W. (1995). QTL analysis: Unreliability and bias in estimation procedures. *Molecular Breeding, 1*, 273–282.

Iwata, H., & Jannink, J. L. (2011). Accuracy of genomic selection prediction in barley breeding programs: A simulation study based on the real single nucleotide polymorphism data of barley breeding lines. *Crop Science, 51*, 1915–1927.

Jannink, J. L., Bink, M. C., & Jansen, R. C. (2001). Using complex plant pedigrees to map valuable genes. *Trends in Plant Science, 6*, 337.

Komatsuda, T., Pourkheirandish, M., He, C., Azhaguvel, P., Kanamori, H., Perovic, D., Stein, N., Graner, A., Wicker, T., Tagiri, A., Lundqvist, U., Fujimura, T., Matsuoka, M., Matsumoto, T., & Yano, M. (2007). Six-rowed barley originated from a mutation in a homeodomain-leucine zipper I-class homeobox gene. *Proceedings of the National Academy of Sciences of the United States of America, 104*, 1424–1429.

Kraakman, A. T. W., Niks, R. E., van den Berg, P. M., Stam, P., & van Eeuwijk, F. A. (2004). Linkage disequilibrium mapping of yield and yield stability in modern spring barley cultivars. *Genetics, 168*, 435–446.

Kraakman, A. T. W., Martinez, F., Mussiraliev, B., van Eeuwijk, F. A., & Niks, R. E. (2006). Linkage disequilibrium mapping of morphological, resistance, and other agronomically relevant traits in modern spring barley cultivars. *Molecular Breeding, 17*, 41–58.

Lacerenza, J. A., Parrott, D. L., & Fischer, A. M. (2010). A major grain protein content locus on barley (*Hordeum vulgare* L.) chromosome six influences flowering time and sequential leaf senescence. *Journal of Experimental Botany, 61*, 3137–3149.

Lorenz, A. J., Chao, S., Asoro, F. G., Heffner, E. L., Hayashi, T., Iwata, H., Smith, K. P., Sorrells, M. E., & Jannink, J. L. (2011). Genomic selection inplant breeding: Knowledge and prospects. *Advances in Agronomy, 110*, 77–123.

Lorenzana, R. E., & Bernardo, R. (2009). Accuracy of genotypic value predictions for marker-based selection in biparental plant populations. *Theoretical and Applied Genetics, 120*, 151–161.

Massman, J., Cooper, B., Horsley, R., Neate, S., Macky, R. D., Chao, S., Dong, Y., Schwarz, P., Muehlbauer, G. J., & Smith, K. P. (2011). Genome-wide association mapping of Fusarium head blight resistance in contemporary barley breeding germplasm. *Molecular Breeding, 27*, 439–454.

Melchinger, A. E., Utz, H. F., & Schön, C. C. (1998). Quantitative trait locus (QTL) mapping using different testers and independent population samples in maize reveals low power of QTL detection and large bias in estimates of QTL effects. *Genetics, 149*, 383–403.

Meuwissen, T. H., Hayes, B. J., & Goddard, M. E. (2001). Prediction of total genetic value using genome-wide dense marker maps. *Genetics, 157*, 1819–1829.

Muñoz-Amatriaín, M., Cistué, L., Xiong, Y., Bilgic, H., Budde, A. D., Schmitt, M. R., Smith, K. P., Hayes, P. M., & Muehlbauer, G. J. (2009). Structural and functional characterization of a winter malting barley. *Theoretical and Applied Genetics, 120*, 971–984.

Piepho, H. (2000). Optimal marker density for interval mapping in a backcross population. *Heredity, 84*, 437–440.

Rafalski, A. (2002). Applications of single nucleotide polymorphisms in crop genetics. *Current Opinion in Plant Biology, 5*, 94–100.

Roy, J. K., Smith, K. P., Muehlbauer, G. J., Chao, S., Close, T. J., & Steffenson, B. J. (2010). Association mapping of spot blotch resistance in wild barley. *Molecular Breeding, 26*, 243–256.

Schön, C. C., Utz, H. F., Groh, S., Truberg, B., Openshaw, S., & Melchinger, A. E. (2004). Quantitative trait locus mapping based on resampling in a vast maize testcross experiment and its relevance to quantitative genetics for complex traits. *Genetics, 167*, 485–498.

Szűcs, P., Skinner, J., Karsai, I., Cuesta-Marcos, A., Haggard, K. G., Corey, A. E., Chen, T. H. H., & Hayes, P. M. (2007). Validation of the *VRN-H2/VRN-H1* epistatic model in barley reveals that intron length variation in *VRN-H1* may account for a continuum of vernalization sensitivity. *Molecular Genetics and Genomics, 277*, 249–261.

Szűcs, P., Blake, V., Bhat, P. R., Chao, S., Close, T. J., Cuesta-Marcos, A., Muehlbauer, G. J., Ramsay, L., Waugh, R., & Hayes, P. M. (2009). An integrated resource for barley linkage map and malting quality QTL alignment. *Plant Genome, 2*, 134–140.

Thomas, W. T. B., Gertson, B., & Forster, B. P. (2003). Doubled haploids in breeding. In M. Maluszynski, K. J. Kasha, B. P. Forster, & I. Szarejko (Eds.), *Doubled haploid production in crop plants: A Manual* (pp. 337–350). Dordrecht/Boston/London: Kluwer Academic Publications.

Thomas, W. T. B., Hayes, P. M., & Dahleen, L. S. (2011). Application of molecular genetics and transformation to barley improvement. In S. E. Ullrich (Ed.), *Barley: Production, improvement, and uses*. West Sussex: Wiley-Blackwell.

Vales, M. I., Schön, C. C., Capettini, F., Chen, X. M., Corey, A. E., Mather, D. E., Mundt, C. C., Richardson, K. L., Sandoval-Islas, J. S., Ut, H. F., & Hayes, P. M. (2005). Effect of population size on the estimation of QTL: A test using resistance to barley stripe rust. *Theoretical and Applied Genetics, 111*, 1260–1270.

Von Zitzewitz, J., Condon, F., Corey, A., Cuesta-Marcos, A., Filichkina, T., Haggard, K., Fisk, S. P., Smith, K. P., Muehlbauer, G. J., Karsai, I., & Hayes, P. M. (2011). The genetics of winter hardiness in barley: Perspectives from genome-wide association mapping. *Plant Genome, 4*, 76–91.

Waugh, R., Jannink, J. L., Muehlbauer, G. J., & Ramsay, L. (2009). The emergence of whole genome association scans in barley. *Current Opinion in Plant Biology, 12*, 218–222.

Zhang, G., & Li, C. (2009). *Genetics and improvement of barley malt quality*. New York: Springer.

Zhong, S., Toubia-Rahme, H., Steffenson, B., & Smith, K. P. (2006). Molecular mapping and marker-assisted selection of genes for Septoria speckled leaf blotch resistance in barley. *Phytopathology, 96*, 993–999.

Chapter 30
Barley in Tropical Areas: The Brazilian Experience

Euclydes Minella

Abstract Domesticated barley, *Hordeum vulgare ssp. vulgare*, is poorly adapted to acid, aluminum toxic soils, and warm and wet springs, conditions that normally occur in the world subtropical/tropical areas. Tailoring barley to fit these adverse conditions has challenged researchers and farmers in Brazil since early 1920s. Increasing beer consumption has pushed domestic malt and malting barley production since 1930s. The selection of soil acidity/aluminum toxicity more tolerant varieties in the late 1950s was a major achievement in establishing barley production. The release of the net blotch resistant, high-yielding variety "Cevada BR 2" by Embrapa in 1990 was a cornerstone in the consolidation of a malting barley industry. The adoption of the "Zero Tillage" by mid-1990s in the double-cropping grain production system of the subtropical has significantly improved the yield potential and, consequently, the competitiveness of barley compared to wheat and other competitor cereals. The increased soil production capacity under "no-till" demanded for varieties more adapted to this technology. The release of the disease-resistant, short-strawed, high-yielding varieties BRS 195 (2000), BRS Cauê, BRS Elis (2006), and BRS Brau (2010) by Embrapa is the third major event in the crop history, revolutionizing barley production. The widespread use of these improved genotypes has boosted yield and malting quality, making barley even a more competitive crop in this century. Since then, farm productivities over 6,000 kg/ha have been harvested in favorable seasons, increasing the average farm yield from 1,500 kg/ha in the 1980s to 3,500 kg/ha in the 2000s. Varieties BRS 180 (1999), BRS 195 (2000), BRS Sampa (2009), and BRS Manduri (2011) made malting barley production economically feasible also in more tropical environments of the southeast and central-west (cerrados) regions, under irrigation, where yields over 7,000 kg/ha are not difficult to obtain. However, the breakthroughs in yield, quality, and disease resistance have not

E. Minella (✉)
Embrapa Trigo, BR 285 Km 294, cx. Postal 451, 99001-970, Passo Fundo, RS, Brazil
e-mail: eminella@cnpt.embrapa.br

G. Zhang et al. (eds.), *Advance in Barley Sciences: Proceedings of 11th International Barley Genetics Symposium*, DOI 10.1007/978-94-007-4682-4_30,

reduced to a satisfactory level yet, the production/quality instability due to excessive rainfall during harvesting, particularly in ENSO years. On the average, the volume of harvested grain that does not meet the required quality for malting is close to 20%. Besides preharvesting sprouting, increased losses to Fusarium head blight and/or DON contamination, possibly associated with the no-till practice, have been observed and are of concern and need to be addressed for the country's sustainability of the malting barley industry.

Keywords Barley • Tropical production • NTPS • Breeding • Soil acidity

30.1 Introduction

Barley was introduced in Brazil several times before becoming a crop in the twentieth century, with commercial production starting in 1930. Since the onset, only the production for malting has been economically competitive. The only barley used for feed is the amount produced annually rejected by malting industry. Production evolved and is still concentrated in southern Brazil in the subtropical region where cool springs favor malting quality. Major growing areas are between latitudes 24° and 31°S, in the highland plateaus (400–1,100 m asl) of Rio Grande do Sul, Santa Catarina, and Parana states. The regional climate is classified as a Koppen's Cfa (humid subtropical), with rainfall well distributed throughout the year and a mean temperature of the warmest month above 22 °C. Barley is sown from late fall to early winter (May–July), following a summer crop of either soybean or corn, in the traditional double-cropping system area. The crop cycle is completed in 110–140 days, depending on the variety, location, planting date and seasonal temperature, and rainfall regimes. Harvesting takes place from late October to early December. Subsoil acidity/aluminum toxicity is a recurrent problem; the originally acid soils restrict root growth beyond the limed layer, making barley more vulnerable to moisture stress than the more acid-soil tolerant crops such as wheat and oats. Rainfall during the growing season averages 700 mm. High seasonal and regional variability in the amount and distribution of rainfall is the major yield and/or quality limiting factor. Excessive rainfall during the reproductive stage, particularly in El Niño Southern Oscillation (ENSO) years, is the most unfavorable environment for malting quality. Seasonal drought and/or temperature stresses (heat, frost) at critical stages and fungal and virus diseases are other major limiting factors. Since 2000, a small amount of malting barley has been produced under irrigation in tropical Brazil in the states of São Paulo, Goiás, and Minas Gerais.

The objective of this chapter is to present the Brazilian experience and the results of research aiming at tailoring barley for production in acid soil in both high-rainfall subtropical and hot and dry tropical regions, respectively.

30.2 Production

Since the onset, barley is produced under contract between farmers and the malting-brewing industry. The malting industry provides seed of the varieties they want to malt, buying the production meeting the quality standards established in the contract. In the beginning and for some time, the seed distributed by the industry was imported mainly from Argentina. Local production was small and confined to the Rio Grande do Sul, Santa Catarina, and Parana states in areas where wood forest was removed for agricultural production by settlers, immigrants of European origins (Germans, Italians, and Russians). Until the end of 1950s, barley was produced manually and/or animal driven only in high pH, fertile soils, while wheat and rye were already grown also in the poor acid soil lands favorable to mechanization. Locally selected varieties more tolerant to soil acidity/aluminum toxicity allowed barley to compete in the mechanized areas, boosting production during the 1960s, when acreages up to 40,000 ha were reported. Average grain yield at the farm level was then around 1,000 kg/ha. Production almost disappeared early in 1970s due to production/quality instability and the competition of the international market.

The federal government barley and malt self-sufficiency plan launched in 1976 boosted both the capacity of the malting industry and barley production. As part of the plan, Embrapa, the Ministry of Agriculture Agricultural and Livestock Research Branch founded in 1973, was assigned the task to provide the required scientific and technical support. Embrapa through its National Wheat Research Unit started working with barley in 1976, with the mission of expanding, diversifying, and coordinating all the country's research and development efforts on the crop, then provided only by the industry.

The public support for research and production really helped establishing malting barley as a commercial crop in southern Brazil, giving farmers an alternative winter crop to grow on the stubble of a summer cash crop (soybean or corn) harvest. Rio Grande do Sul and Paraná have, since being the major barley- and malt-producing states, the second being the current leader with an average of 60% of the production. Since 2000, barley is also being produced in small scale under irrigation in Goiás, Minas Gerais and Federal District, and in São Paulo, in the southeast.

Public support for barley production ended in late 1980s, leaving the local production up to the humor of the malting industry. The high production cost relative to other winter crops mainly due to the susceptibility of varieties to net blotch, then the major disease, together with a lower price of foreigner barley reduced local acreage to the levels of late 1970s. Domestic production started recovering in 1995 due to the release the variety BR 2, by Embrapa, the first Brazilian variety resistant to net blotch, widely adapted and highly competitive in grain yield, combined with the beginning of production under the "no-till production system" (NTPS) being adopted in the major producing states. The increased competitiveness of barley under the NTPS, as well as the price increase in the international market, made domestic production interesting again to the brewing industry. Malting barley production in the last decade averages 264,000 Mt, supplying 48% of current malting demand for barley (560,000 Mt) by the industry which in turn supply 40% of the country's needs for brewing malt. Brazil is a major malt importer, and the domestic brewing consumption has steadily increased in the last years, indicating an increased demand for both malt and barley. Current trend indicates a significant increase in the local malting capacity in a few years and, consequently, increasing the demand for malting barley. Considering the current agronomic and quality competitiveness of barley, it can be postulated a steady growth in both acreage and production of malting barley in Brazil.

30.3 Research

Public barley research was initiated together with that of wheat in 1920. During the 1930–1975 period, research was carried by the brewing industry interested in malting barley production, with little or no support from government institutions. Major efforts then were directed to breeding, soil improvement, and crop agronomy. Under the leadership of Embrapa, the amount of research efforts were significantly increased after 1976 with inputs from state research systems, universities, and farmer organizations. Besides breeding, during the official support to domestic production, barley received a substantial amount of research on crop production and crop protection. Unfortunately, the cut in official incentives to the crop negatively affected the resources allocated to barley research mainly from public institutions. Since 1990, Embrapa is practically the sole public institution with significant research inputs in barley. On the private sector, AmBev and Cooperativa Agraria Agroindustrial run individual research programs with major efforts in variety development and crop agronomy. Academic studies on barley are still insignificant today. Embrapa Trigo runs a malting barley breeding program through a formal technical and financial support of the malting industry (currently AmBev, Cooperativa Agraria Agroindustrial, and Malteria do Vale). Since its onset, Embrapa's breeding program has been integrated with the research and quality analysis capabilities of the malting industry. The varieties obtained from the program, after intensive agronomic and malting quality testing, are incorporated into commercial production through the partners' seed production and farmer technical assistance apparatus.

Embrapa's barley improvement program was implemented in 1976 at the National Wheat Research Center located in Passo Fundo, RS (28°15′ S, 52° 24″ W, 687 masl), one of the country's major barley production regions. Since its beginning, the major goals of the program have been assembling and enhancing a diverse gene pool (pre-breeding) and developing new barley cultivars adapted to the local soil and climate with competitive yield, malting quality, and disease resistance. Getting hold of viable seeds from the local developed germplasm and from exotic genetic material was the first task of the program. Cultivars and breeding lines made available by the brewing companies Brahma and Antarctica and by the state institutions IAPAR (Paraná) and IAC (São Paulo) became the germplasm base. Crossing work started in 1978, combining the local adapted malting lines with known sources of disease resistance and malting quality introduced from Canada, Australia, Europe, South America, and the United States.

The program has target the idea that to be competitive under local conditions, a phenotype should combine the largest number of the following traits: malting quality, yield potential (5,000 and 7,000 kg/ha for rain fed and irrigated conditions, respectively), kernel plumpness of 85%, early maturity (less 130 and 110 days for rain fed and irrigated, respectively), short and stiff straw, resistance to lodging and to preharvest sprouting (PHS), and resistance to net blotch, powdery mildew, leaf rust, BYDV, spot blotch, and head scab. The methodology applied has evolved from conventional breeding to a combination of the conventional and modern methods. Nowadays, 5% of the F_1's are advanced through single-seed descent (SSD) and doubled-haploid (DH) methods and 95% through the selected bulk method through the F_4. Double haploids are still produced through anther culture. By the SSD, the materials are advanced three generations per year. The selection for adaptation to soil and environmental conditions and for disease resistance is made in the field, under natural and artificial pathogen infection. Selection based on malting quality is started only at advanced generations of inbreeding or inbred lines, based on the industry's quality analysis output.

30.4 Results

Barley research in Brazil is over 90 years old now, whereas Embrapa's program has been around for 35 years, having the germplasm and the knowledge developed so far made substantial contributions to the consolidation of a malting barley industry. Furthermore, important knowledge on the performance of the crop in acid soil subtropical/tropical areas (Minella and Sorrells 1992, 2002, 1997; Peruzzo and Arias 1996) has also been accumulated. As a result, during the last 35 years barley production has progressed significantly. Current average acreage, production, and yield are, respectively, three, six, and two times larger than those of the 1970s, when Embrapa started working with barley (Table 30.1).

During the 1980s and early 1990s, the average farm yield increased by 30% mainly due to improved agronomy and plant protection practices developed by Embrapa and other institutions. Significant improvements were obtained in the geographical distribution of the production, seeding time and density, soil fertility increase through liming and P and K amendments, N fertilization, and in disease (mainly net blotch) and insect (mainly aphids) control methods. The aphid population that frequently caused severe crop losses was controlled biologically, through a massive release of exotic aphid natural enemies.

In the 1990s, the yield increase in farmer field was even greater than that of the 1980s, being the difference mainly due to the wide use of Embrapa's varieties available to farmers since 1992 (Tables 30.1 and 30.2) and the NTPS (Silva and Minella 1996; Minella 2000, 2010), adopted by most of the producers by the end of the decade.

Table 30.1 Average acreage, production, and grain yield and % relative to 1970s of malting barley in Brazil

	Acreage		Production		Grain yield	
Decade	Ha	%	Mt	%	Kg/ha	%[a]
1970	42,962	100	50,018	100	1,124	100
1980	102,947	239	146,884	293	1,459	130
1990	93,812	218	197,230	394	2,036	181
2000	10,198	257	264,072	528	2,550	227

[a]Percent change over 1970s

Table 30.2 Evolution of the malting barley crop in Brazil in the twenty-first century: average acreage, production, and grain yield

	Acreage	Production	Yield	Embrapa
Year	Ha	Mt	Kg/ha	%[a]
2001	135,640	274,888	2,027	60.0
2002	145,156	224,403	1,546	51.0
2003	136,971	381,220	2,783	63.0
2004	146,803	395,277	2,692	73.0
2005	127,961	282,245	2,207	72.7
2006	90,661	250,291	2,761	57.6
2007	101,414	187,165	1,846	67.1
2008	65,285	194,263	2,976	65.3
2009	71,920	192,518	2,673	71.5
2010	80,172	258,451	3,220	67.0
Decade	110,198	264,072	2,550	65.0

[a]Proportion of the acreage seeded with Embrapa varieties

The release of BR 2, the first Embrapa's true malting barley, released in 1989 was cornerstone in both barley breeding and production. Due to its earliness, short straw, wide adaptation, high yield, and resistance to net blotch, BR 2 was the leading variety for 12 years (1994–2003). It was sown to over 80% of the acreage, peaking at 91% in 1997. Its wide use allowed also a reduction of at least one fungicide spray against net blotch, lowering production cost and the impact on the environment. The net blotch resistance of BR 2 and its derivatives continues to be as effective as it was 20 years ago. It is believed that BR 2 represented a 40% yield increase, bringing back the competitiveness of the local crop in comparison with home-grown wheat and imported barley. Barley production was resumed with Embrapa's genetics in 1995 and is currently stabilized at around 264,000 Mt/year, supplying on the average 48% of the malting capacity. Before Embrapa's varieties and the NTPS technology, the country was importing an average of 70% of the barley used for malting.

In this century, the average yield is 25% higher than of that of the 1990s, and the increase has been attributed to the wide use since 2003, of the high-yield, short-strawed, and lodging-resistant variety BRS 195 (Minella et al. 2005) and to the higher use of nitrogen with it. Because its superior yield potential and acceptable lodging resistance

Table 30.3 Evolution of malting barley in Brazil since late 1970s

Parameter	1970s	2000s	%[a]
Grain yield (kg/ha)	1,124	2,550	227
Kernel plumpness (%)	65.0	90.0	38
Barley protein (%)	12.5	10.0	76
Malt extract (%)	79.0	82.0	104

[a]2000's/1970's (%)

in areas of several years under NTPS (chemical, physical, and biologically improved soils), BRS 195 released in 2002 was the leading variety in both rainfed and irrigated areas until 2009. In general it yielded 1,000 kg/ha more than the regular (non-dwarf) varieties. Lodging has become a major problem for regular type varieties in soils with a long history under the NTPS.

The average malting quality of the produced barley has also been improved substantially (Table 30.3) and varieties Embrapa 127, BRS Borema (Minella et al. 2006), BRS Elis, BRS Cauê, BRS Brau, and BRS Sampa usually produce a malting quality profile competitive with those of the world top quality varieties such as Scarlett.

The release of BRS 180 (six-rowed barley) in 1999 (Silva et al. 2000) was the starting point for commercial production of barley with malting in the "Cerrado" region (Goiás and Minas Gerais states) under central pivot irrigation. Because of the large distances from the nearest malting plant (Taubaté, São Paulo), the high transportation cost still makes production in those regions not economically competitive with barley imported from the Mercosul. Since 2005, production for malting is under way in irrigated areas of the São Paulo estate, where BRS 195 is more competitive than BRS 180 in both yield and quality. Since 2010, BRS Sampa is replacing BRS 195 for being more competitive in kernel plumpness and yield. Major limiting production factors in the region so far have been lodging, occasional occurrence of PHS and rice blast, and dry heat at the end of grain-filling period.

Soil improved production capacity under NTPS, and modern varieties doubled the yield potential of barley in the subtropical areas where nowadays productivities above 6,000 kg/ha at the farmers fields are already common in favorable seasons. The yield potential of the new varieties BRS Sampa and BRS Manduri under irrigation is over 8,000 kg/ha. In areas under NTPS for more than a decade, barley has been also much less affected by acidity and/or aluminum toxicity of the subsoil than when the soil was tilled. A production threat possible associated with the NTPS in southern Brazil is the increased frequency of FHB and spot blotch epidemics. Therefore, reducing the losses associated with these diseases to acceptable levels is currently the major challenge for the barley scientific and technological communities, including plant pathologists, breeders, and agronomists. Unfortunately, the level of resistance to the causal pathogens in current varieties and breeding lines is far below the desired one, and the variability level in the germplasm used is not promising for a genetic solution of the problem.

The high yield potential already achieved in association with the increased value of corn may turn the crop economically competitive also as feed what can boost barley production through the expansion of acreage outside of the current production of malting. Research, mainly in breeding, is under way to assemble and develop germplasm for feed, food, and unmalted adjunct for beer making.

Currently the major effort for the irrigated areas is breeding to reduce the crop cycle to 100, 20 days less than current cultivars. Inbred lines with this phenotype are already in yield trials with preliminary promising results. Research is also under way to reduce losses to rice blast, an emerging problem in the irrigated areas.

Finalizing it can be stated that the future of barley in tropical and subtropical areas in Brazil looks very promising.

References

Minella, E. (2000). Adapting barley to unfavorable environments: Results from Brazil. *International Barley Genetics Symposium, 8 (2000) Proceedings* (Vol. 3, pp. 267–268). Adelaide: Adelaide University, Department of Plant Science/Grains Research & Development Corporation.

Minella, E. (2010). The evolution of barley in Brazil: Contributions of Embrapa in 30 years of research and development. In S. Ceccarelli & S. Grando (Eds.), *Proceedings of the 10th International Barley Genetics Symposium*, April 5–10, 2008, Alexandria, Egypt ((pp. 774–779). Aleppo, Syria: ICARDA.

Minella, E., & Sorrells, M. E. (1992). Aluminum tolerance in barley: Genetic relationships among genotypes of diverse origin. *Crop Science, 32*, 503–598.

Minella, E., & Sorrells, M. E. (1997). Inheritance and chromosome location of Alp, a gene controlling aluminum tolerance in 'Dayton' barley. *Plant Breeding, 116*, 465–469.

Minella, E., & Sorrells, M. E. (2002). Genetic analysis of aluminum tolerance in Brazilian barleys. *Pesquisa Agropecuária Brasileira, 37*(8), 1099–1103.

Minella, E., Silva, M. S., Arias, G., & Linhares, G. (2005). A BRS 195 Malting Barley cultivar. *Crop Breeding and Applied Biotechnology, 2*, 321–322.

Minella, E., Silva, M. S., Arias, G., & Linhares, G. A. (2006). Malting Barley BRS Borema. *Crop Breeding and Applied Biotechnology, 6*, 322–324.

Peruzzo, G., & Arias, G. (1996). Barley and other cereals root development in a Brazilian acid soil. In *International Oat Conference, 5 and International Barley Genetics Symposium,7, Saskatoon, SK, Canada. Proceedings*, Saskatoon-SK, Canada, University of Saskatchewan.

Silva, M. S., & Minella, E. (1996). Sterility in Brazilian malting barley cultivar and lines. In Slinkard, A., Scoles, G., & Rossnagel, B. (Eds.), *Proceedings of the 5th International Oat Conference and 7th International Barley Genetics Symposium* (pp. 577–579)

Silva, D. B., Guerra, F., Minella, E., & Arias, G. (2000). BRS 180 – Cevada cervejeira para cultivo irrigado. *Pesquisa Agropecuária Brasileira, 35*(8), 1689–1694.

Chapter 31
Performance and Yield Components of Forage Barley Grown Under Harsh Environmental Conditions of Kuwait

Habibah S. Al-Menaie, Hayam S. Mahgoub, Ouhoud Al-Ragam, Noor Al-Dosery, Meena Mathew, and Nisha Suresh

Abstract Kuwait being an arid country faces several climatic challenges which impose constraints on sustainable agricultural development. Hence, it is essential to manage the natural resources in an efficient manner to meet the increasing demand for food and fodder in the country. The growth performance of forage barley was studied under the harsh environmental conditions of Kuwait. Six promising lines of barley genotypes were sown at the rate of 400 seeds/m^2, and a randomized complete block design with three replications was used. All the recommended horticultural practices were adopted. This chapter will focus on the performance and yield components of barley lines successfully grown under harsh and a biotic environmental condition of Kuwait.

Keywords Arid country • Sustainable agriculture • Horticultural practices

31.1 Introduction

Farming is being practiced under irrigated conditions in an area of 46,965 ha in Kuwait which includes both open field and greenhouse cultivation. The climate of Kuwait is characterized by very low annual rainfall and high degree of aridity. In addition, there are problems resulting from the degradation of natural resources due to salinity, water logging, and desertification. Since the production technologies are being adopted without considering the soil and water characteristics of Kuwait, rise

H.S. Al-Menaie (✉) • H.S. Mahgoub • O. Al-Ragam • N. Al-Dosery • M. Mathew • N. Suresh
Arid Land Agriculture and Greenery Department, Kuwait Institute for Scientific Research, P.O. Box 24885, Safat 13109, State of Kuwait
e-mail: hmanaie@kisr.edu.kw

G. Zhang et al. (eds.), *Advance in Barley Sciences: Proceedings of 11th International Barley Genetics Symposium*, DOI 10.1007/978-94-007-4682-4_31,

in groundwater table, groundwater pollution, and degradation in the productive capacity of soil occurs which in turn affects plant productivity. Such practices present a threat to the environment and sustainability of agricultural production in Kuwait. Hence, managing the natural resources in an efficient manner is one of the most important criteria for food production in the country.

The combined effects of aridity and soil salinity limit the range of crops that can be cultivated in Kuwait, which resulted in the import of over 96% of its food (Encyclopedia of the Nations 2005). Due to global financial crisis, the world was engrossed with food shortages and staple food prices rose enormously. Forage prices were also increased due to the elevated fertilizer, energy, and transportation costs. Only few agricultural crops are found to be tolerant to both salinity and drought, and barley (*Hordeum vulgare* L.) is one among them. The shortage of forage production in Kuwait has resulted in the inclusion of barley as a forage crop in the farming plans of farmers. Barley is considered to be relatively tolerant to alkaline soils, drought, and extremes of temperature and also is classified as one of the most salt-tolerant crop species (ICARDA 2000). It is an important versatile food and forage crop in dry areas and is also included in the list of priority crops of Kuwait (KISR 1996). According to Al-Menaie et al. (2004) and Al-Menaie (2003), sustainable barley cultivation is feasible in Kuwait by careful consideration of soil type, local ground water, soil rejuvenation, water table control, and genotype.

Recently, the Kuwait Institute for Scientific Research (KISR) outlined the strategic policy framework for promoting new methods and technologies, aiming to achieve sustainability in the production of food and forage in the coming decade. Exploitation of forage barley directly helps in reducing the share in import expenditure. The project has national relevance by proposing the screening of barley genotypes for forage production under Kuwait's environmental conditions. The limited availability of traditional water resources necessitates the use of brackish water resources, and evaluating the potential of forage cultivation under brackish water is a positive strategy in managing the available water resources of the country. Hence, a study was initiated by KISR to identify the most suitable barley genotypes tolerant to the climatic conditions in Kuwait.

31.2 Materials and Methods

Two preliminary experiments were conducted for 2 years to evaluate the yield performance for 40 drought-tolerant barley lines obtained from the Arab Centre for the Studies of Arid Zones and Dry Lands (ACSAD) under Kuwait's environment. Six promising lines were selected as the most adapted lines. This study focused on the performance and yield components of these lines under two sources of water irrigation. The pedigrees of these promising lines are detailed in Table 31.1. The Agricultural Research Station (ARS), Sulaibiya, was selected as the planting site. The experimental site had a hot climate with a moderate winter, and dry and hot summer. Daily maximum, minimum, and mean temperatures were recorded by a

Table 31.1 The pedigree of barley promising lines

1	ACSAD	ER/Apm//AC 253/3/Ath/Lignee 686 ACS-B-11198-1IZ-2IZ-2IZ-0IZ
2	ACSAD	ER/Apm//AC 253/3/Ath/Lignee 686 ACS-B-11198-18IZ-1IZ-1IZ-0IZ
3	ACSAD	Rihane-03/4Alanda//Lignee527/Arar/3/Centinela/2* Calicuchima ACS-B-11245-9IZ-2IZ-1IZ-0IZ
4	ACSAD	ACSAD 1182/4/Alanda//Lignee 527/Arar/3/Centinela/2* Calicuchima CS-B-B-11265-5IZ-2IZ-1IZ-0IZ
5	ACSAD	ACSAD 1182/4/Alanda//Lignee 527/Arar/3/Centinela/2* Calicuchima ACS-B-11265-13IZ-2IZ-1IZ-0IZ
6	ACSAD	Moroc 9-75//WI 2291/CI 01387/3/H.Spont-41-1/Tadmor/7/11012-2/Impala//Birence/3/Arabi Abiad/4/5604/1025/5/SB73358-B-104-16-1-3//ER/Apm/6 W12291/Bgs//Hml-02 ACS-B-11294-2IZ-1IZ-2IZ-0IZ

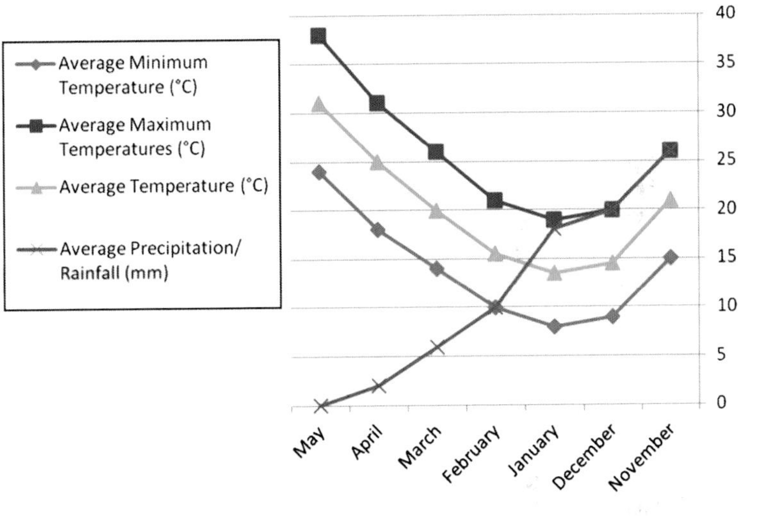

Fig. 31.1 Average temperature (°C) and rainfall status of experimental site

meteorological station placed in the experimental site (Fig. 31.1). A split-plot design with three replications was adopted. Fresh and brackish water were assigned to the main strips. Meanwhile, in each strip the six promising lines were randomly allocated to the subplots. The subplot area was 3 m^2, consisting of six rows (2.5 m long and 20 cm apart). Plant density was 400 plants/m^2. The fertilization rate was 60 kg ha^{-1} N plus 50 kg ha^{-1} P_2O_5. The plots were irrigated whenever necessary and maintained at field capacity to avoid water stress.

Phenological Traits and Yield Evaluation: The phenological characteristics of different genotypes were determined, taking into account the number of days observed from sowing until the upper most spikes appeared beyond the auricles of flag leaf sheath (50% heading on plant basis), i.e., days to heading (DHE). The number of days between the sowing date and time at which 50% of the spikes had matured was counted. When the crops were matured, five spikes were randomly selected from each plot, cut, and threshed to obtain the number of grains/spike. The plant height was measured from ground level to the end of the main spike. At harvest, two external rows from each plot were eliminated to avoid border effect. Thus, four rows were harvested, weighed, and threshed, and their grain yields were weighed and adjusted to tons/hectare to indicate the biological and grain yield. The thousand-grain weight (1,000 KW) was obtained by counting 1,000 grains in each plot with the help of a grain counter and weighing them. Then the averages of these weights were calculated.

Tolerance and Susceptibility Index: Genotypes were compared for their yield potentials under fresh and brackish water. Stress tolerance index according to Ramirez-Vallejo and Kelly (1998) was used.

Statistical Analysis: Combined analyses of variance for grain yield and its related characters were performed over trails after verifying the homogeneity of trial variance errors using Bartlett's test. Least significant difference (LSD) values were calculated at the 5% probability level. The microcomputer statistical program Statistical Package for the Social Sciences (SPSS) was used.

31.3 Results and Discussion

31.3.1 Days to Heading

Under normal condition days to heading ranged from 80 to 85 days. Salinity stress caused a reduction in the number of days to heading by about 8%. This decrease ranged from 2% for line 4–12% for line 5. The results clearly showed that brackish water had a significant effect in reducing the number of days to heading. Hence, under stress environment early heading lines may perform better (Fig. 31.2).

31.3.2 Plant Height

Brackish irrigation water reduced plant height by about 17% as compared to fresh water. This reduction ranged from 2% (Line 2) to 20% (Line 3). Results revealed that plants with medium height were more tolerant to salinity stress. On the other hand, salinity-susceptible plants have fast growth, more plant height, and larger leaf area as compared to the tolerant ones under salinity conditions (Fig. 31.3).

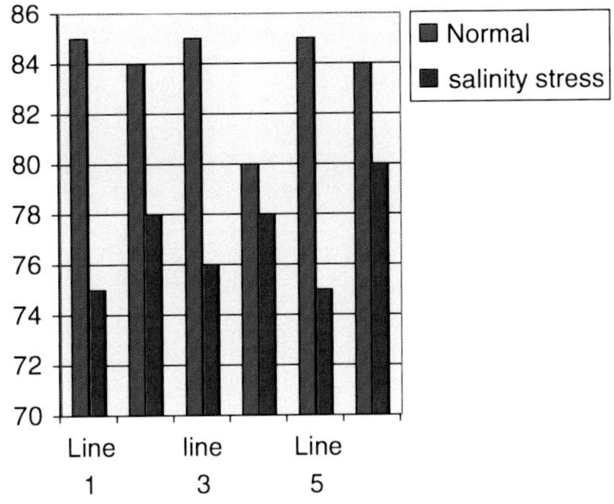

Fig. 31.2 Means of days to heading as affected by irrigation sources

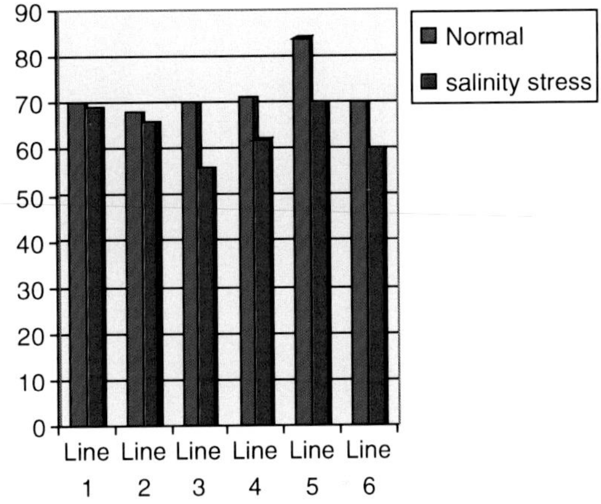

Fig. 31.3 Means of plant height as affected by irrigation sources

31.3.3 Number of Spikes per Square Meter

It is one of the major characteristics that affect grain yield. Salinity stress during vegetative growth decreased number of spike per square meter. Average reduction in number of spikes per square meter caused by salinity stress was 9%. This reduction was only about 2% for line 6, while it is about 16% for line 5. This may due to the increased salt concentration in root medium, combined with reduced plant absorption of essential mineral nutrients which resulted in reduced plant growth (Fig. 31.4).

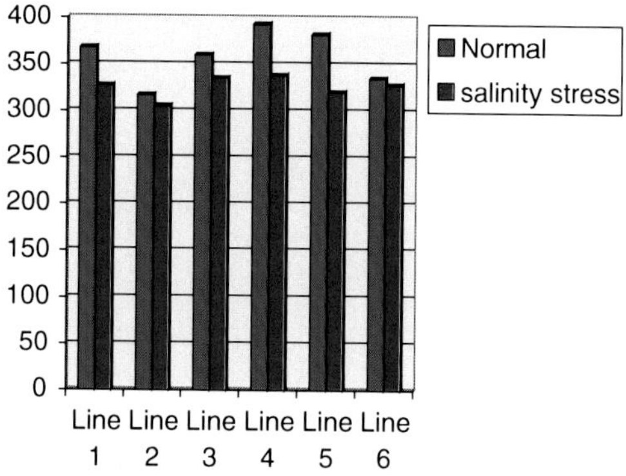

Fig. 31.4 Means of number of spikes per square meter as affected by irrigation sources

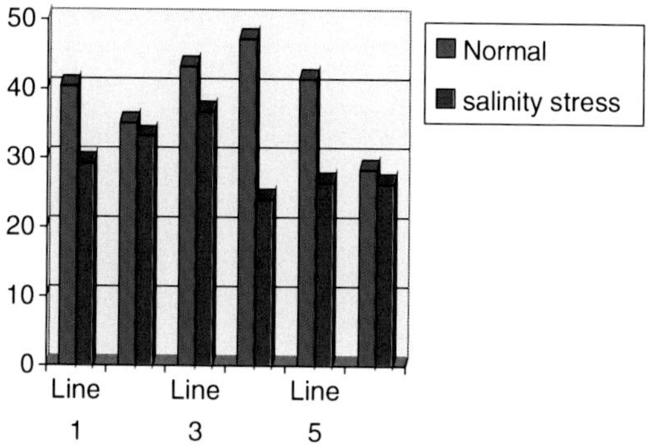

Fig. 31.5 Means of thousand-grain weight as affected by irrigation sources

31.3.4 Thousand-Grain Weight

Salinity stress had significantly reduced thousand-grain weight. The average reduction was about 23%. This reduction was due to the decrease in the period of grain formation. The lowest reduction was observed by line 3 (5%), whereas for line 4, it was about 49%. This reduction of grain weight was not only caused by salinity stress but also due to the combined effect between salinity and high temperature. This combined effect during grain-filling stage may affect the grain maturity due to the reduction of grain-filling period (Fig. 31.5).

31.3.5 Grain Yield

Use of brackish water for irrigation reduced grain yield by about 41% as compared to freshwater. Results obtained showed that maximum grain yield was obtained under freshwater irrigation by line 3 (2.85 tons/ha). Statistical analysis showed significant

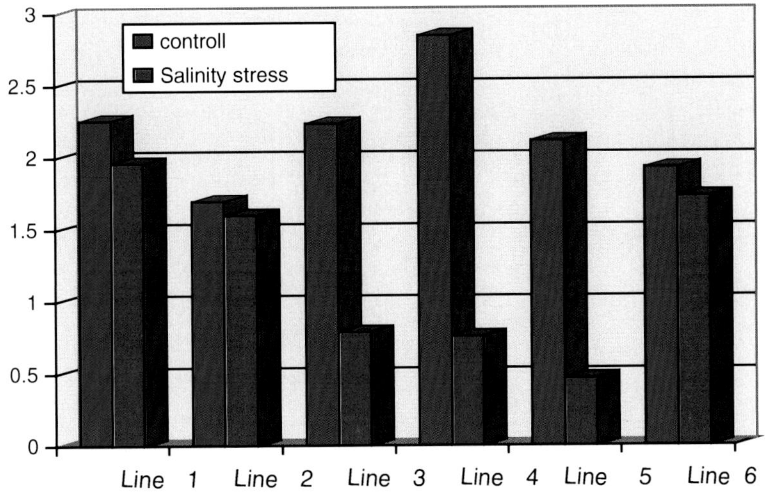

Fig. 31.6 Stress index

effect of interaction between brackish water and genotypes. All promising lines showed significant reduction in grain yield when it was irrigated with brackish water. However, lines 1, 3, and 5 produced higher yield than lines 2 and 3. This reduction may be due to the combined stress effect between high temperature and salinity. This had a significant influence on agronomic and morphological characters which in turn is reflected on the average yield of genotypes.

31.3.6 Stress Indices

Highest stress tolerance index was recorded by lines 6 and 1. Meanwhile, the lowest value was obtained from lines 3, 4, and 5. By comparing grain yields of new promising lines and using stress indices as a basis, the genotype with high grain yield both under stress and optimum conditions such as line 6 and line 1 can be selected as tolerant genotypes (Fig. 31.6).

31.4 Summary and Conclusion

The growth performance of six lines of forage barley was studied under the harsh environmental conditions of Kuwait. The seeds were sown at the rate of 400 seeds/m^2, and a randomized complete block design with three replications was adopted.

The phenological characteristics of different genotypes including days to heading, plant height, number of spikes per square meter, grain yield, and 1,000-grain weight were recorded and statistically analyzed. Results showed that salinity stress significantly affected the number of days to heading, plant height, number of spikes per square meter, 1,000-grain weight, and grain yield. Lines 1 and 2 recorded the highest grain yield both under stress and optimum conditions. The promising lines need to be retested before they can be recommended for cultivation under Kuwait's environment.

Acknowledgements The project team would like to thank the Kuwait Institute for Scientific Research for their financial support and interest in this study.

References

Al-Menaie, H.S. (2003). *Matching soil, water and genotype for rarely cultivation in Kuwait*. Ph.D. thesis, University of Reading, Reading, UK.

Al-Menaie, H. S., Caligari, P. D. S., & Foster, B. P. (2004). Matching soil, water and genotype for barley cultivation in Kuwait. In *Proceedings of the 9th International barley Genetics Symposium*, Prague.

Encyclopedia of the Nations, Britannica online Science and Technology. (2005). Available at http://www.britannica.com

ICARDA. (2000). *Germplasm program- Barley germplasm improvement for increased productivity and yield stability* (Annual Report, p. 126). Aleppo: ICARDA.

KISR. (1996). *Agricultural Master Plan of the State of Kuwait (1995–2015), plan overview*. Prepared by KISR, AAD/FRD, Kuwait.

Ramirez-Vallejo, P., & Kelly, J. D. (1998). Traits related to drought resistance in common bean. *Euphytica, 99*, 127–136.

Chapter 32
Variability of Spring Barley Traits Essential for Organic Farming in Association Mapping Population

Linda Legzdina, Ieva Mezaka, Indra Beinarovica, Aina Kokare, Guna Usele, Dace Piliksere, and Nils Rostoks

Abstract Association mapping population consisting of 154 Latvian and foreign spring barley genotypes contrasting for traits that are important for organic agriculture was established with the aim to develop molecular markers useful in breeding for organic farming. The mapping population was genotyped at 3072 single-nucleotide polymorphism loci using Illumina GoldenGate platform to provide marker data for association mapping. Field trials in two organically and two conventionally managed locations are being carried out during three seasons. The following traits essential for organic farming were phenotyped: plant morphological traits ensuring competitive ability against weeds, grain yield in organic farming, yield stability/ adaptability to organic conditions, nutrient use efficiency (measured as ability to form acceptable grain yield and accumulate protein in grain under conditions of organic farming) and prevalence of diseases. This chapter gives an overview on preliminary phenotyping results. ANOVA showed that genotype and location significantly influenced most of analysed traits ($p<0.01$). The average yield reduction in organic trials, when compared to conventional, was 1.2 $t\,ha^{-1}$, and it ranged from 4.2 $t\,ha^{-1}$ reduction to 1.2 $t\,ha^{-1}$ increase. In respect to morphological traits related to competitive ability against weeds (canopy height in beginning of plant elongation, plant ground cover in tillering, length and width of flag leaves, tillering capacity, plant height before harvest), there was a tendency that average trait values were higher in conventional farming locations, but the coefficients of variation were higher in organic locations in most of the cases. The differences in protein content between

L. Legzdina (✉) • I. Mezaka • I. Beinarovica • A. Kokare • G. Usele • D. Piliksere
State Priekuli Plant Breeding Institute, Zinatnes street 1a, Priekuli, LV-4126, Latvia
e-mail: lindaleg@navigator.lv

N. Rostoks
Faculty of Biology, University of Latvia, 4 Kronvalda Blvd., Riga, LV-1586, Latvia

G. Zhang et al. (eds.), *Advance in Barley Sciences: Proceedings of 11th International Barley Genetics Symposium*, DOI 10.1007/978-94-007-4682-4_32,

conventional and organic trials correlated significantly (r=0.732), and its variation was significantly effected by genotype. The average reduction of protein content in organic fields, if compared to conventional, was 2%, and it ranged 0–6.5%.

Keywords Phenotyping • Competitive ability against weeds • Nutrient use efficiency

32.1 Introduction

Development of organic agriculture in general has been much slower in comparison to conventional agriculture; one of the reasons is considerably greater environmental variability wherewith the need for adaptation of varieties to varied environmental conditions is important issue in organic agriculture (Wolfe et al. 2008).

Backes and Østergård (2008) concluded that there might be a high potential for marker assisted selection (MAS) if markers that contain information about GxE interaction in plant breeding for organic farming would be used. For certain traits, QTL effect may differ between environments: there is significant effect in one environment and no effect in another. In this case, the gene responds only to the specific environmental factors. A possible solution can be inclusion of GxE interaction in QTL analysis (Backes and Østergård 2008). Standards of organic agriculture do not exclude use of molecular markers as such; one of the reservations of organic sector is the use of harmful chemicals and enzymes produced from genetically modified organisms in marker development. MAS could be used efficiently in breeding for organic farming besides to phenotypic selection for particular traits if markers for relevant traits are available. There is a need to discuss this issue between molecular scientists and representatives of organic sector (Lammerts van Bueren et al. 2010).

Primarily quantitative trait mapping has been conducted in bi-parental populations. Recently, association mapping approach has emerged as high-throughput genotyping and sequencing technologies have become available (Zhu et al. 2008). Association mapping in barley has resulted in significant findings of marker-trait associations of morphological traits (Cockram et al. 2008) and agronomical traits, including yield and yield stability (Kraakman et al. 2004, 2006), grain quality (Beattie et al. 2010) and disease resistance (Maasman et al. 2011; Roy et al. 2010).

Several research results regarding differences in variety performance between organic and conventional farming systems have been reported. High genetic correlations between both systems for most traits have been found; however, possibility of rankings to be coincident has been estimated as moderate, and combining of information from both organic and conventional trials has been suggested (Przystalski et al. 2008). Differences in lentil variety ranking according to the yield in organic and conventional environments were demonstrated by Vlachostergios and Roupakias (2008). There are only a few reports regarding a comparison of selection results in organically and conventionally managed trials. Reid et al. (2009) concluded that indirect selection of spring wheat in conventionally managed trials would not give the best possible result for organically managed production. However, prediction of potential gains from selection in organically managed fields is difficult because of greater variability for certain traits.

Plant characters needed for organic agriculture include efficient use of nutrients, weed competition, disease and pest resistance, yield, yield stability and quality. It may be useful to identify pleiotropic characters with a positive value for a wide range of physiological needs and concentrate on major characters that integrate many minor and variable ones (Wolfe et al. 2008).

Nutrient use and uptake efficiency is of particular importance in breeding for organic farming. Regarding nitrogen, higher potential for grain protein production is needed in organic agriculture than in conventional (Wolfe et al. 2008). Insufficiency of N uptake may limit leaf area growth during the period around the end of stem elongation and beginning of flowering and consequently reduce grain number per unit area and yield via a reduction in radiation interception and use efficiency (Dreccer 2006). Baresel et al. (2005) concluded that varieties adapted to conventional farming are unlikely to be successful under organic farming, especially in environments providing low productivity; genotypes with a high early uptake of nitrogen and a high ability for translocation to the grain should be more adapted to organic farming, whereas a late uptake of N is probably less important in this farming system because there is usually a shortage of nitrogen in the later season, whereas in conventional farming, addition of nitrogen late in the season can be used to increase protein content in the grain. Plant root characteristics can be used as a trait that indicates differences in nutrient uptake by different varieties. In organic and lower input farming, root system should be able to explore deeper soil layers and be more active than in conventional farming systems; interaction with beneficial soil microorganisms that promote nutrient uptake is more essential (Lammerts van Bueren 2002).

Baresel et al. (2005) predict that MAS for nitrogen uptake and nitrogen use efficiency will be difficult, because those are very complex traits which depend on many single factors, whose importance vary according to environmental characteristic. In rice, seven QTLs associated with N-deficiency tolerance have been identified by measuring six related traits in seedling stage (Feng et al. 2010).

Traits related to competitive ability against weeds for the most part have not been of high priority in conventional plant breeding (Wolfe et al. 2008), but importance arose with a development of organic farming. The previous findings on crop competitive ability against weeds are inconsistent. Several studies with winter wheat (Wicks et al. 1986; Challaiah et al. 1986) show that emergence rate, early growth rate, number of tillers and maximum canopy height were characteristics of major importance, while study on spring barley (Christensen 1995) does not confirm similar results on emergence rate and tillering. More recent studies (Hoad et al. 2008; Østergård et al. 2008; Wolfe et al. 2008) conclude that competitive ability is usually not attributed to a single trait, but the interaction among a series of desirable characteristics is important. Traits associated with high ground cover and shading ability, such as rapid early growth rate, high tillering ability, planophile leaf habit and plant height, are widely reported as important characteristics for increasing crop competitiveness. However, both crop performance and weed growth are affected by many other factors such as composition of weed community, methods of crop management and various soil and weather conditions (Hoad et al. 2008). In addition, less well understood are physiological traits for desirable below-ground competition that therefore is with less practical value to plant breeders (Wolfe et al. 2008).

Sinclair and Vadez (2002) reported that horizontal-leaf crops are likely to be more desirable in low-nutrient environment. Loss of leaf area is one of the main consequences of inadequate N and P availability. Crops with more horizontal leaves are more likely to shade competing weeds and lessen the negative impact of the weeds.

Seed-borne diseases are of great importance in organic farming (e.g. loose smut and leaf stripe for barley) because no practical and effective permissible seed treatments are available; diseases effected by nitrogen supply and plant density like powdery mildew are generally less important than in conventional farming (Wolfe et al. 2008).

The aim of the investigation is to develop molecular markers useful in breeding for organic farming and to clarify if significant marker-trait associations differ under organic and conventional farming systems. Additional aim is to identify genotypes useful in breeding for organic farming. This chapter gives an overview on preliminary phenotyping results.

32.2 Material and Methods

Association mapping population consisting of 154 Latvian and foreign spring barley genotypes contrasting for traits that are important for organic farming was established with the aim to find significant marker-trait associations using genome-wide association mapping approach and to develop molecular markers useful in breeding for organic farming. Population includes 19 hulless barley accessions and 4 six-row barleys; 107 genotypes are with Latvian origin (27 varieties and 80 breeding lines).

Field trials in two organically and two conventionally managed locations are being carried out during three seasons (2010–2012). In this chapter, we present the phenotyping results of season 2010. The first organic location (O1) was situated in a research field of the plant breeding institute; the second organic location (O2) included an organic farmer's field within 5 km distance from the institute. In both locations, fields were for more than 5 years certified for organic farming. The first conventional location (C1) was located in barley breeding field (with a medium level of mineral fertiliser input) and the second conventional location (C2) in a seed production field of the institute.

In all locations, there was sod-podzolic loamy sand soil. In O1, green manure was used as precrop (peas, around 20 $t\,ha^{-1}$), and in O2, stable manure (around 40 $t\,ha^{-1}$) was applied in 2009. In C1 and C2, a complex mineral fertiliser was applied before cultivation of the soil: N-81, P_2O_5–48, K_2O–84 kg ha^{-1}. Harrowing at tillering stage in O1 and spraying with herbicide Secator 0.15 $L\,ha^{-1}$ in C1 and C2 were used for weed management. Insecticide Karate 0.2 $L\,ha^{-1}$ was applied in C1 and C2 to control leaf aphids.

Table 32.1 Description of the soil characteristics and crop management systems of the trials under organic (O1, O2) and conventional (C1, C2) farming systems)

Location	pH KCl	P_2O_5, mg kg^{-1}	K_2O, mg kg^{-1}	Humus content, g kg^{-1}	N content in soil in spring, kg ha^{-1}	Precrop
O1	5.7	111	144	28	10.41	Peas for green manure
O2	6.5	265	173	35	7.29	Perennial grasses
C1	5.5	100	132	26	12.51	Potatoes
C2	5.6	115	159	28	8.43	Potatoes

The summer of 2010 was the warmest in the history of Latvian meteorological observations and was characterised by thermal stability. Starting from the 24th of June, average daily air temperature for more than 2 months constantly kept above normal. In C1 location, unusually high for Latvia infection with barley yellow dwarf virus (BYDV) was observed (Table 32.1).

The following traits were assessed: plant growth habit (growth stage according to Zadoks decimal scale (GS) 25–29), 1 – erect, 9 – planophile; canopy height (GS 31–32); cereal plant ground cover and weed ground cover (GS 31–32, visually estimated percentage of plant covered area in the plot); length and width of flag leaf (GS 47–51), five measurements per plot; and plant height (GS 90). Productive tillering was calculated per 1 m^2 from plant emergence (GS 13, registered number of plants per 0.05 m^2) and number of productive tillers registered after harvest in the same plot area. Visual assessment of plant density in tillering stage (scores 0–5) was done. Natural infection with powdery mildew (*Blumeria graminis* f. sp. *hordei*) and net blotch (*Drechslera teres*) was assessed according to scale 0–9; for loose smut (*Ustilago nuda*) and leaf stripe (*Drechslera graminea*), infected plants per plot were counted. Protein content in grain was measured by Infratec 1241 Analyser, and protein yield was calculated.

32.3 Results and Discussion

There was a tendency that average trait values were higher under conventional farming locations whereas the variation among the genotypes was higher under organic locations (in agreement with Reid et al. (2009)) in most of the cases (Table 32.3). Lower variation was observed in C2 location where the growing conditions were more close to optimum, but variation was higher in C1 (especially for grain yield, crop ground cover, plant height and protein yield) what can be explained by the effect of BYDV infection. The coefficients of variation of trait values between the genotypes ranged from 11.4% for plant height in C2 to 46% for tillering in O1.

ANOVA showed that most of the analysed traits were significantly influenced by genotype and location (Table 32.2). The influence of genotype surpassed that of location for grain yield and traits related to competitive ability against weeds except

Table 32.2 Partitioning of sum of squares and correlations between the trait expressions in two organic (O1, O2) and two conventional (C1, C2) locations

	Partitioning of sum of squares, %		Coefficient of correlation					
Traits	Genotype	Location	O1–O2	O1–C1	O2–C1	O1–C2	O2–C2	C1–C2
Grain yield	31.3	29.8	0.208*	0.181*	0.148	0.458**	0.249**	0.330**
Growth habit	69.7	2.9	0.541**	0.600**	0.660**	0.608**	0.628**	0.726**
Canopy height	57.5	16.2	0.563**	0.594**	0.541**	0.630**	0.516**	0.665**
Crop ground cover	29.5	39.0	0.287**	0.249**	0.168*	0.455**	0.290**	0.423**
Length of flag leave	37.9	30.5	0.369**	0.277**	0.382**	0.421**	0.506**	0.423**
Width of flag leave	64.1	15.3	0.717**	0.575**	0.699**	0.655**	0.744**	0.689**
Productive tillering	28.6	15.2	0.135	0.143	–0.021	–0.034	0.255**	0.218**
Plant height	55.6	21.2	0.732**	0.569**	0.580**	0.671**	0.626**	0.546**
Protein yield	20.6	47.5	0.143	0.105	0.114	0.339**	0.139	0.280**

Effect of genotype and location significant ($p<0.01$); * $p<0.05$, ** $p<0.01$

Table 32.3 Mean, range and coefficient of variation (V%) for selected traits under organic (O1, O2) and conventional (C1, C2) farming systems

Traits		O1	O2	Mean (O1, O2)	C1	C2	Mean (C1, C2)
Grain yield, t ha^{-1}	Mean	3.01 ab*	2.78 b	2.90	2.66 b	4.12 a	3.39
	Min	0.76	0.30	1.20	0.27	1.22	0.86
	Max	5.16	4.51	4.39	6.03	6.89	5.89
	V%	25.6	26.5	26.1	38.6	24.4	31.5
Plant growth habit, points 1–9	Mean	5.5	4.7	5.1	4.8	5.1	4.9
	Min	1.0	1.0	1.0	1.0	1.0	1.0
	Max	9.0	9.0	9.0	9.0	9.0	9.0
	V%	30.4	40.3	35.3	37.2	32.5	34.9
Canopy height, cm	Mean	25.4 ab	21.5 b	23.4	26.2 ab	27.6 a	26.9
	Min	10.0	12.0	11.0	10.0	16.0	13.0
	Max	35.0	33.0	32.0	39.0	39.0	37.5
	V%	22.0	20.7	21.3	22.0	17.9	19.9
Crop ground cover, %	Mean	33.1 b	25.6 b	29.4	38.2 ab	46.3 a	42.2
	Min	8.0	10.0	11.5	10.0	18.0	16.5
	Max	60.0	48.0	50.0	75.0	75.0	67.5
	V%	27.4	25.0	26.2	28.2	23.3	25.8
Length of flag leave, cm	Mean	11.7 b	11.9 b	11.8	11.8 b	15.1 a	13.4
	Min	6.0	8.6	7.3	6.8	9.8	9.1
	Max	17.8	17.2	17.5	18.0	21.8	18.4
	V%	17.2	15.9	16.5	17.8	17.2	17.5
Width of flag leave, mm	Mean	7.7 b	8.9 ab	8.3	8.4 ab	9.7 a	9.1
	Min	4.0	5.6	4.8	5.0	6.6	5.9
	Max	18.2	16.6	17.3	14.6	16.8	15.4
	V%	24.4	19.2	21.8	18.7	18.8	18.8

(continued)

Table 32.3 (continued)

Traits		O1	O2	Mean (O1, O2)	C1	C2	Mean (C1, C2)
Productive tillering coefficient	Mean	1.7	1.6	1.7	2.3	2.5	2.4
	Min	0.4	0.8	0.7	0.2	1.0	1.1
	Max	4.8	5.6	4.1	4.5	7.0	5.2
	V%	46.0	42.2	44.1	37.8	42.6	40.2
Plant height, cm	Mean	77.3 b	75.6 b	76.4	80.6 ab	92.9 a	86.7
	Min	50.0	51.0	52.0	49.0	63.0	60.5
	Max	131.0	110.0	120.0	129.0	122.0	125.5
	V%	17.9	16.9	17.4	18.4	11.4	14.9
Protein yield, t ha^{-1}	Mean	0.42 b	0.36 b	0.39	0.36 b	0.63 a	0.49
	Min	0.15	0.04	0.19	0.04	0.28	0.18
	Max	0.72	0.61	0.62	0.74	0.99	0.79
	V%	25.2	27.4	26.3	36.2	21.0	28.6

*Trait means followed by different letters within a row differ significantly ($p<0.01$)

plant ground cover. Correlations were significant for all the measured traits only between both conventional locations (Table 32.2).

32.3.1 Traits Related to Competitive Ability Against Weeds

In respect to morphological traits related to competitive ability against weeds, the correlations between the locations were significant (in agreement with findings of Przystalski et al. 2008) except for productive tillering. Mean values of plant growth habit did not significantly differ between the locations; correlations between all the locations were significant. Canopy height in beginning of stem elongation was significantly lower in O2 if compared to C2 location. For crop ground cover, the correlations between the locations were comparatively weaker; mean trait values were significantly higher in C2 than in both organic locations. The influence of genotype on the variation of flag leaf width was larger if compared to flag leaf length. The higher values of both parameters in C2 are in agreement with Dreccer (2000) in respect of N shortage limiting effect on leaf area growth. The values of productive tillering correlated significantly between the locations only in two cases, and the influence of genotype was comparatively lower for this trait. It can be explained by the influence of poor plant density in individual cases and loss of plants in the period between registration of field emergence and productive tillers (reason for values below 0). The methodology is planned to be improved in the following trial years. For weed ground cover, which was estimated under organic conditions only (data not shown), there was no significant effect of genotype found, and therefore, the possibility to use this trait for mapping is doubtful.

32.3.2 Infection with Diseases

Infection with powdery mildew, loose smut and leaf stripe was significantly influenced by genotype and location, and correlations between the locations were significant ($p<0.01$); for net blotch, there was no significant effect of genotype found (data not shown). The level of net blotch infection in 2010 was comparatively high, and the heavy prevalence of BYDV in C1 location influenced development and scoring possibility for net blotch and powdery mildew as well. The highest infection level with powdery mildew was in C2 location (average score 4); in organic locations, it was slightly lower (3 and 2.9 in O1 and O2, respectively), but in C1 the lowest (1.2).

Loose smut and leaf stripe are seed-borne diseases, and even though the same seed material was used for all locations, the effect of location was significant. Differences might be caused by variations of the local environmental factors favouring the development of disease and in the case of loose smut by the influence of plant density. The loose smut infection varied from 0 to 147 infected plants per plot with the highest level in C2 location and lowest level in O2. The number of plants infected with leaf stripe ranged from 0 to 78; the level was slightly higher under organic growing conditions if compared to conventional ones.

32.3.3 Traits Related to Nutrient Use Efficiency

Phenotyping of traits related to nutrient use efficiency in field trials with large amount of genotypes is complicated; no direct trait measurements (e.g. for root characteristics) were possible in this study because of lack of appropriate methods commensurable to our resources. Nutrient use efficiency can be characterised by genotype ability to form sufficient grain yield under organic growing conditions in comparison to conventional ones where easily available for plants mineral fertiliser is applied. The use of secondary traits in respect of selection for grain yield was suggested by Bänziger et al. (2000); under stress conditions, the heritability of grain yield usually decreases, whereas the heritability of some secondary traits remains high, while at the same time, the genetic correlation between grain yield and those traits increases. Trait 'ears per plant' is recommended as useful secondary trait having high relationship with grain yield under N stress conditions.

The differences in yield between conventional location C2 and both organic locations were calculated (Fig. 32.1). One of the reasons for lower yield under organic locations can be insufficient amount of available nitrogen (Dreccer 2000). The data from location C1 was not used for calculations because of the strong influence of BYDV infection; the mean yield in this location did not significantly differ from the organic locations and was even lower than in O1 (Table 32.3). The average yield reduction was 1.2 $t\,ha^{-1}$, and it ranged from 4.2 $t\,ha^{-1}$ reduction to 1.2 $t\,ha^{-1}$ increase. Mean yield increase or non-significant yield reduction ($p<0.05$) was found for 83 genotypes; two varieties ('Camila' and 'Klinta') and two Latvian breeding lines had increased yield in both organic locations.

It is suggested to measure the nutrient use efficiency indirectly by protein yield (Löschenberger et al. 2008). We calculated protein yield (Table 32.3); the effect of location on the variability of this trait surpassed that of genotype, and the correlations between the locations were not significant in most of the cases (Table 32.2). It suggests that mapping of this trait could be with less success if compared to other investigated traits.

We also analysed differences in protein content between conventional (C2) and organic (O1 and O2) trials (Fig. 32.1); they both correlated significantly ($r=0.732$, $p<0.01$), and there was significant influence of genotype found (partitioning of mean squares was 75.5% for genotype and 12.8% for location, $p<0.01$). The average reduction of protein content was 2%, and it ranged 0–6.5%. The difference was insignificant for 33 genotypes; Latvian variety 'Druvis' had the best result with no reduction on average. There was a significant negative correlation between protein reduction and grain yield ($r=-0.412$ and -0.354 for O1 and O2 locations, respectively, $p<0.01$). Genotypes with smaller protein reduction were with a tendency to have a higher yield. It might suggest the possibility to use protein reduction between conventional and organic trials as indirect selection criteria in breeding for improved nutrient use efficiency without negative effect on grain yield. The average reduction of protein yield was 0.24 $t\,ha^{-1}$, and it ranged from 0.63 $t\,ha^{-1}$ reduction to 0.11 $t\,ha^{-1}$ increase.

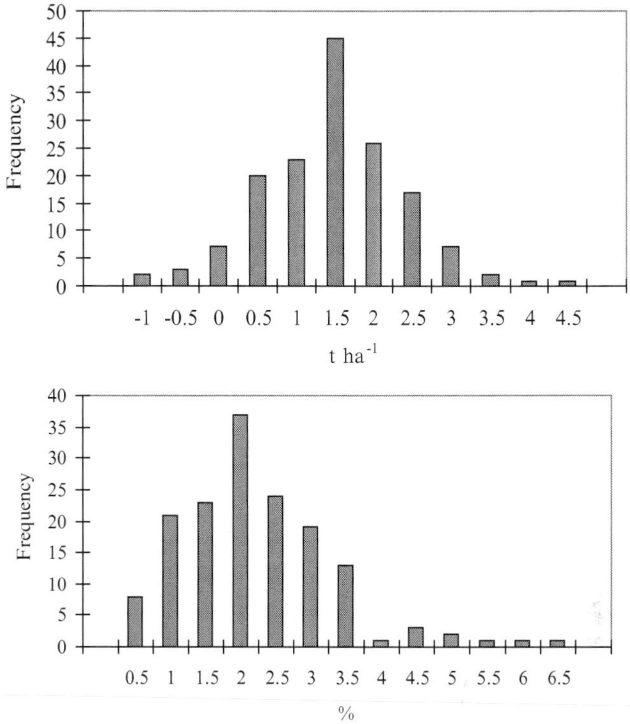

Fig. 32.1 Distribution of mean yield difference (*top*) and mean protein content difference (*bottom*) values between conventional location C2 and both organic locations (O1 and O2)

To summarise, the coefficients of variation of trait values between the genotypes ranged from 11.4 to 46%; weed ground cover, productive tillering and protein yield can be traits for which mapping is problematic; data from two more seasons has to be added to make more sound conclusions and to reach the aims of investigation.

Acknowledgements This study is performed with financial support of European Social Fund co-financed project 2009/0218/1DP/1.1.1.2.0/09/APIA/VIAA/099. Thanks to Dr. E. Lammerts van Bueren for help with ideas during initialisation of this research.

References

Backes, G., & Østergård, H. (2008). Molecular markers to exploit genotype–environment interactions of relevance in organic growing systems. *Euphytica, 163*, 523–531.

Bänziger, M., Edmeades, G. O., Beck, D., & Bellon, M. (2000). *Breeding for Drought and Nitrogen Stress Tolerance in Maize: From Theory to Practice*. Mexico: CIMMYT. 68.

Baresel, J., Reents, H. J., & Zimmermann, G. (2005, January 17–19). Field evaluation criteria for nitrogen uptake and nitrogen use efficiency. In *Proceedings of the COST SUSVAR/ECO-PB workshop on organic plant breeding strategies and the use of molecular markers*, Driebergen, The Netherlands, pp. 49–51.

Beattie, A. D., Edney, M. J., Scoles, G. J., & Rossnagel, B. G. (2010). Association mapping of malting quality data from western Canadian two-row barley cooperative trials. *Crop Science, 50*, 1649–1663.

Challaiah, Burnside O. C., Wicks, G. A., & Johnson, V. A. (1986). Competition between winter wheat (*Triticum aestivum*) cultivars and Downy brome (*Bromus tectorum*). *Weed Science, 34*, 689–693.

Christensen, S. (1995). Weed suppression ability of spring barley varieties. *Weed Research, 35*, 241–247.

Cockram, J., White, J., Leight, F. J., Lea, V. J., Chiapparino, E., Laurie, D. A., Mackay, I. J., Powell, W., & O'Sullivan, D. M. (2008). Association mapping of partitioning loci in barley (*Hordeum vulgare* ssp. vulgare L.). *BMC Genetics, 9*, 16.

Dreccer, M. F. (2006). Nitrogen use at the leaf and canopy level: A framework to improve crop N use efficiency. In S. Sham et al. (Eds.), *Enhancing the efficiency of nitrogen utilization in plants* (pp. 97–101). Binghamton: Food Products Press.

Feng, Y., Cao, L. Y., Wu, W. M., Shen, X. H., Zhan, X. D., Zhai, R. R., Wang, R. C., Chen, D. B., & Cheng, S. H. (2010). Mapping QTLs for nitrogen-deficiency tolerance at seedling stage in rice (*Oryza sativa* L.). *Plant Breeding, 129*, 652–656.

Hoad, S., Topp, C., & Davies, K. (2008). Selection of cereals for weed suppression in organic agriculture: A method based on cultivar sensitivity to weed growth. *Euphytica, 163*, 355–366.

Kraakman, A. T. W., Niks, R. E., Van den Berg, P. M. M. M., Stam, P., & Van Eeuwijk, F. A. (2004). Linkage disequilibrium mapping of yield and yield stability in modern spring barley cultivars. *Genetics, 168*, 435–446.

Kraakman, A. T. W., Martinez, F., Mussiraliev, B., van Eeuwijk, F. A., & Niks, R. E. (2006). Linkage disequilibrium mapping of morphological, resistance and other agronomically relevant traits in modern spring barley cultivars. *Molecular Breeding, 17*, 41–58.

Lammerts van Bueren, E. T., Backes, G., de Vriend, H., & Ostergaard, H. (2010). The role of molecular markers and marker assisted selection in breeding for organic agriculture. *Euphytica, 175*, 51–64.

Löschenberger, F., Fleck, A., Grausgruber, H., Hetzendorfer, H., Hof, G., Lafferty, J., Marn, M., Neumayer, A., Pfaffinger, G., & Birschitzky, J. (2008). Breeding for organic agriculture: The example of winter wheat in Austria. *Euphytica, 163*, 469–481.

Maasman, J., Cooper, B., Horsley, R., Neate, S., Dill-Macky, R., Chao, S., Dong, Y., Schwarz, P., Muehlbauer, J., & Smith, K. P. (2011). Genome-wide association mapping of Fusarium head blight resistance in contemporary barley breeding germplasm. *Molecular Breeding, 27*(4), 439–454.

Østergård, H., Kristensen, K., Pinnschmidt, H. O., Klarskov Hansen, P., & Hovmøller, M. S. (2008). Predicting spring barley yield from variety-specific yield potential, disease resistance and straw length, and from environment-specific disease loads and weed pressure. *Euphytica, 163*, 391–408.

Przystalski, M., Osman, A., Thiemt, E. M., Rolland, B., Ericson, L., Østergård, H., Levy, L., Wolfe, M. A., Buchse, A., Piepho, H. P., & Krajewski, P. (2008). Comparing the performance of cereal varieties in organic and non-organic cropping systems in different European countries. *Euphytica, 163*, 417–433.

Reid, T. A., Yang, R. C., Salmon, D. F., & Spaner, D. (2009). Should spring wheat breeding for organically managed systems be conducted on organically managed land? *Euphytica, 169*, 239–252.

Roy, J. K., Smith, K. P., Muehlbauer, G. J., Chao, S., Close, T. J., & Steffenson, B. J. (2010). Association mapping of spot blotch resistance in wild barley. *Molecular Breeding, 26*, 243–256.

Sinclair, T. R., & Vadez, V. (2002). Physiological traits for crop yield improvement in low N and P environments. *Plant and Soil, 245*, 1–15.

Lammerts van Bueren, E. T. (2002). *Organic plant breeding and propagation: concepts and strategies*. PhD thesis, Wageningen University, The Netherlands, p. 207.

Vlachostergios, D. N., & Roupakias, D. G. (2008). Response to conventional and organic environment of thirty-six lentil (*Lens culinaris* Medik.) varieties. *Euphytica, 163*, 449–457.
Wicks, G. A., Ramsel, R. E., Nordquist, P. T., & Schmidt, J. W. (1986). Impact of wheat cultivars on establishment and suppression of summer annual weeds. *Agronomy Journal, 78*, 59–62.
Wolfe, M. S., Baresel, J. P., Desclaux, D., Goldringer, I., Hoad, S., Kovacs, G., Löschenberger, F., Miedaner, T., Østergård, H., & Lammerts van Bueren, E. T. (2008). Developments in breeding cereals for organic agriculture. *Euphytica, 163*, 323–346.
Zhu, C., Gore, M., Buckler, E. S., & Yu, J. (2008). Status and prospects of association mapping in plants. *The Plant Genome, 1*, 5–20.

Chapter 33
Barley Production and Breeding in Europe: Modern Cultivars Combine Disease Resistance, Malting Quality and High Yield

Wolfgang Friedt and Frank Ordon

Abstract Barley is still one of the major agricultural crops in Europe although cultivation area and grain production are obviously declining. While winter barley is clearly outyielding spring barley, the latter is generally characterised by superior malting and brewing quality. Nonetheless, the use of winter barley is continuously replacing spring barley not only as animal feed but also for malt production. In major barley-growing countries like Germany, grain yield of classical winter barley cultivars, i.e. line varieties, has been increased by ca. 70 kg ha^{-1} $year^{-1}$ in the last decades. The implementation of "haploidy steps" and use of doubled haploids in line breeding have caused an acceleration of breeding progress. More recently, the application of marker-assisted selection allowed an efficient combination of effective genes or loci (pyramiding) for enhancing resistance against fungal and viral diseases and improving product quality. Today, the development of a suitable hybrid system has formed the basis for an exploitation of heterosis in barley via hybrid breeding. New hybrid varieties tend to have in general a higher yield potential than line varieties. On this basis, it is feasible to enhance the performance of barley and to improve its competitiveness with other major crops such as wheat, maize or oilseed rape.

Keywords Genetic gain • Winter versus spring barley • Line varieties • Doubled haploids • Marker-assisted breeding • Disease and stress resistance • Malting/brewing quality • Grain yield • Restored hybrids • New traits

---

W. Friedt (✉)
Department of Plant Breeding, IFZ Research Centre, Justus-Liebig-University, Giessen, Germany
e-mail: wolfgang.friedt@agrar.uni-giessen.de

F. Ordon
Institute for Resistance Research and Stress Tolerance, Julius Kuehn-Institute (JKI), Quedlinburg, Germany
e-mail: frank.ordon@jki.bund.de

G. Zhang et al. (eds.), *Advance in Barley Sciences: Proceedings of 11th International Barley Genetics Symposium*, DOI 10.1007/978-94-007-4682-4_33,

33.1 Introduction

On a global scale, barley (*Hordeum vulgare* L.) ranks fifth among crop plants regarding acreage. But, the barley area harvested worldwide has declined from a maximum of more than 80 million ha (Mha) in the 1970s to nowadays less than 60 Mha. In Europe, barley is number two, next to wheat, with an acreage of about 23 Mha in 2010 (http://faostat.fao.org), while it was about 57 Mha in the 1980s (Fig. 33.1). At the same time, yield in Europe increased from an average of 1.5 $t\,ha^{-1}$ at the beginning of the 1960s to about 3.5 $t\,ha^{-1}$ today (Fig. 33.1). However, in the main barley-growing countries of Western Europe, average crop yield today is nearly 7.0 $t\,ha^{-1}$ (http://faostat.fao.org/).

The major European barley-growing countries, France, Germany, Russia, Spain, Ukraine and the UK, account for about three quarters of the total annual barley grain production of about 105 million tonnes (Mt), representing 67% of the world production (ca. 157 Mt in 2008). Almost two-thirds (62%) of the total European barley produce come from the EU member states (EU 27) covering 50% of the cultivation area in Europe. The average grain yield per unit area in the EU (27) is 25% higher than the European overall mean (4.54 vs. 3.62 $t\,ha^{-1}$, cf. http://faostat.fao.org, Fig. 33.1). These figures demonstrate the outstanding importance of barley for Europe and of Europe for the global barley grain production.

During the last decades, the productivity of barley has been increased by an annual rate of approximately 1–2%, which is due to (1) genetic breeding progress in terms of more productive and stable cultivars, (2) more efficient disease and pest control, (3) improved fertilisation schemes and (4) improved agricultural plant production technology, e.g. sowing, cultivation, harvest, storage, etc.

In this chapter, the genetic gain contributing to yield increase, improved resistance and quality and the combination of all three trait complexes achieved in Europe is briefly reviewed.

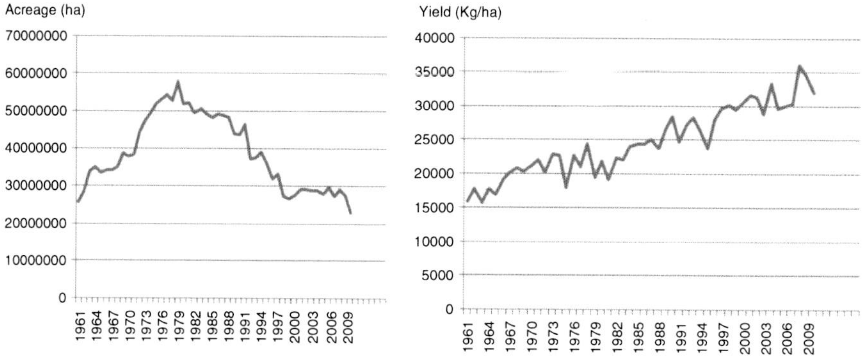

Fig. 33.1 Development of barley acreage (*left*) and grain yield (*right*) in Europe (http://faostat.fao.org)

33.2 Enhancement of Grain Yield

Large differences can be observed concerning the genetic gain contributing to yield increase depending on the country and the time period analysed (Abeledo et al. 2002). For example, in Austria, an analysis of 24 two-rowed spring barley cultivars covering a period of 160 years revealed a slight yield increase until the 1960s, followed by a yield increase of about 60 kg ha^{-1} $year^{-1}$ in the following decades (Grausgruber et al. 2002). In Spain, the genetic gain for yield was estimated at 40 kg ha^{-1} $year^{-1}$ in two-rowed and 33 kg ha^{-1} $year^{-1}$ in six-rowed cultivars in productive environments (Munoz et al. 1998). But in very poor-growing areas, the yield of six-rowed types has declined by 15 kg ha^{-1} $year^{-1}$. In Sweden, the average genetic gain for two-rowed spring barley was estimated at 31% since the end of the nineteenth century (Mac Key 1993), while Peltonen-Sainlo and Karjalainen (1991) came to the conclusion that barley yield did not significantly increase in Finland during 1920–1988. For Nordic spring barley cultivars, an average genetic gain of 13 kg ha^{-1} $year^{-1}$ was estimated for two-rowed cultivars released in the period 1942–1988 and 22 kg ha^{-1} $year^{-1}$ for six-rowed cultivars (Ortiz et al. 2002).

In Germany, winter barley yield increased on average by about 70 kg ha^{-1} $year^{-1}$ since the 1960s (Ahlemeyer et al. 2008; Ahlemeyer 2009). The latter authors carried out extensive field trials with a set of 113 cultivars being particularly important in their decade of release at 12 locations over three successive years. The results reveal that about 40 kg ha^{-1} $year^{-1}$, i.e. 50% of the yield increase of two-rowed and six-rowed winter barley cultivars, were due to breeding progress (Fig. 33.2). In the two-rowed varieties, the number of ears per square meter increased in this period by 2.27 m^{-2} a^{-1}, while in the six-rowed cultivars, an increase of 1,000-kernel weight by 0.21 g a^{-1} was observed. Interestingly, a significant negative correlation between the year of release and the lodging score was observed so that yield stability of winter

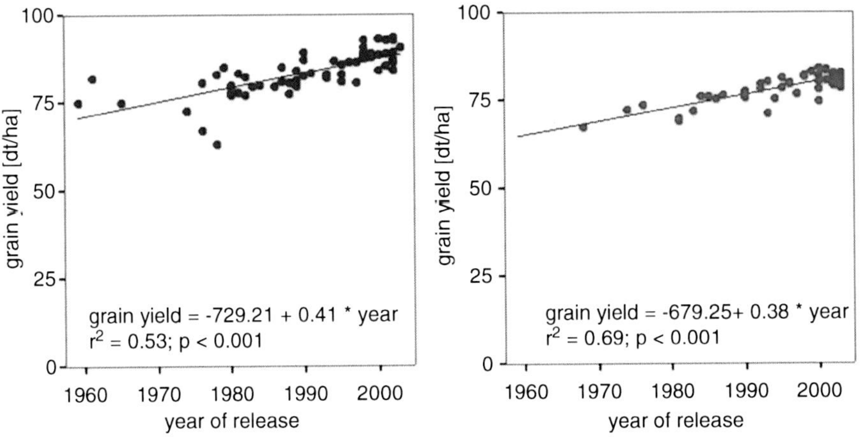

Fig. 33.2 Genetic gain and breeding progress in six-rowed (*left*) and in two-rowed (*right*) winter barley cultivars in Germany since the 1960s (Ahlemeyer 2009)

barley has been significantly increased via improved lodging resistance during the last decades (J. Ahlemeyer, pers. comm.).

Genomic regions involved in this yield increase and the improvement of other agronomic traits were identified in a genome-wide association mapping approach (GWAS) using 109 German winter barley cultivars out of the set used by Ahlemeyer (2009) including 6-rowed and 2-rowed types representing the period 1959–2003 (Rode et al. 2011). Based on the phenotypic data obtained for calculating the genetic gain, i.e. 12 locations for 3 years (see above) and 833 SNPs with an allele frequency higher than 5% obtained via the Illumina Golden Gate Bead Array technology, linkage disequilibrium (LD) was calculated at 7.35 cM, and 91 significant marker-trait associations were detected corresponding to 48 different genomic regions (Table 33.1). Two regions associated with grain yield each were identified on chromosomes 3H and 5H, but many additional regions were detected for traits contributing to grain yield and yield stability underlining the complexity of agronomic traits such as yield. A majority of relevant regions for different traits was found on chromosome 5H (15). Whole genome selection strategies may be useful to improve such traits more efficiently in the future (cf. Waugh et al. 2010).

33.3 Invention of Hybrid Cultivars Based on Cytoplasmic Male Sterility

The vast majority of barley cultivars grown in Europe still represent pure lines which are either developed by conventional methods, i.e. pedigree and progeny selection, or via haploidy steps leading to homozygous doubled haploid lines (Friedt et al. 2000; Anonymous 2011). Such variety types are homozygous and homogeneous producing high yields of uniform grain quality. Alternatively, the systematic use of heterozygosity and heterosis would basically allow an enhanced yield potential of the barley crop.

After Ahokas (1979, 1982) had described the cytoplasmic male sterility (CMS) systems msm1 and msm2 and the restorer gene *Rfm1* in barley, it has taken more than two decades until the first CMS-based hybrid cultivars appeared on the European market. Today, a number of registered winter barley hybrids are available for cultivation in Europe. Examples are cvs. Hobbit, SY Boogy, SY Wahoo, SY Bamboo, Tatoo, Yoole and Zzoom. In Germany, Zzoom was listed in 2008, and Hobbit has been registered in 2010. In different official trials in Germany with variable production intensities, this hybrid has shown to combine maximal yield with resistance against major diseases such as powdery mildew or soil-borne viruses (Anonymous 2011).

The dominant fertility restoration gene *Rfm1* is able to restore fertility in the presence of both msm1 and msm2 male-sterile cytoplasms of barley. The *Rfm1* locus has been mapped to the short arm of chromosome 6H (Matsui et al. 2001). More recently, closely linked AFLP-based STS markers in a distance of ca. 1 cM to *Rfm1* have been described (Murakami et al. 2005). These diagnostic genetic markers

Table 33.1 Summary of all marker-trait associations found in a collection of 109 barley cultivars: Number of genomic regions associated with one of the analysed traits on the seven chromosomes of barley (Rode et al. 2011)

	Chromosome							
Trait	1H	2H	3H	4H	5H	6H	7H	Total regions
Yield			2		2			4
TKW		2			1	2		5
Grains	1							1
Spikes	1				1	1		3
HLW	2							2
Height	2	1	2	1	4			10
Lodge1			1	1	3	1		6
Lodge2		1		1	2	1		5
Head.	1		1	1				3
Brack.	1			1	1	1		4
Neck.		1						1
Sieve		1			1	2		4
Total	8	6	6	5	15	8		48

Yield grain yield, *TKW* 1,000 kernel weight, *Grains* grains per spike, *Spikes* spikes/m^2, *HLW* hectolitre weight, *Height* plant height, *Lodge1* lodging at flowering, *Lodge2* lodging before harvest, *Head.* heading date, *Brack.* brackling, *Neck.* necking, *Sieve* sieve fraction

are expected to accelerate the selection of new improved restorer lines by marker-assisted breeding and facilitate the development of superior new hybrids.

However, up to now, lines still represent the dominant variety type cultivated in Europe, although especially in France, hybrids are recently gaining more importance. An overview on the most important cultivars in Europe is given below.

Recent leading winter barley varieties in France include cvs. Alinghi, Arturio, Azurel, Cartel, Cervoise, Escadre, Estérel, Gigga, KWS Cassia, Limpid, Metaxa, Salamandre and Touareg. The cultivation of hybrid varieties has recently been strongly expanded in France, e.g. Tatoo, Hobbit, SY Boogy, SY Wahoo, and SY Bamboo (http://www.semences-cereales-certifiees.fr/economie/). Major winter barley varieties preferred by the maltsters are Vanessa (2-rowed), Arturio, Azurel, Cartel and Estérel (all 6-rowed). Recommended spring barley cultivars include Beatrix, Bellini, Concerto, Henley, NFC Tipple, Pewter, Prestige, Sebastian and others (www.lafranceagricole.fr/actualite-agricole/).

Major 6-rowed winter barley cultivars in Germany according to the actual descriptive variety list (Anonymous 2011) are Christelle (cross: Laverda × SCOB2251), Fridericus (Carola × LP6-564), Highlight (LEU5033 × Cornelia × Cartola), Hobbit (9737 × RE15), Laverda ((Ludmilla × GW1836) × Merlot), Lomerit ((Askanova × Grete) × Ouana × 1332-29), Naomie ((Julia × NS90517/16) × Carola) and Souleyka (Laverda × Pelikan). Leading 2-rowed types are Anisette (Opal × Reni), Campanile, Finesse (N89510 × ZE90.1896), Metaxa (Sunbeam × Clara) and Sandra (Artist × Carat)). Leading spring barley varieties comprise Braemar, Grace, Marthe and Quench (all 2-rowed).

A couple of spring barley varieties were registered in Germany and recommended in the United Kingdom as well. For example, spring barley cvs. Concerto, NFC Tipple, Propino, Publican, Quench (all for malting) and Westminster (feed variety). Correspondingly, winter barley cultivar Pelican (6-rowed) is registered in Germany and other countries, too. On the contrary, the two-rowed varieties are obviously more specifically adapted to regions so that distinct types are prevalent in different countries.

33.4 Combining High Grain Yield with Improved Disease Resistance

Modern winter barley cultivars such as the hybrid Hobbit combine a high-yield potential with complete yellow mosaic virus resistance and reasonable mildew (*Blumeria graminis* f. sp. *hordei*) plus Rhynchosporium resistance (Anonymous 2011).

The major disease of barley and other cereals is still powdery mildew. Many monogenic resistance genes, e.g. the *Mla* locus, have been identified and exploited throughout the last decades. Most of the resistance genes or alleles have sooner or later been overcome by new mildew strains, except *mlo*, which in European barley breeding up to now is particularly employed in spring types. Therefore, breeders and growers are highly interested in new and stable resistance sources even if it only provides partial protection against fungal damage. Such basal resistance of barley to powdery mildew is a quantitative trait based on non-hypersensitive defence mechanisms.

A recent functional genomics approach by Aghnoum et al. (2010) indicates that many candidate genes are involved in the defence reaction. Since it is not known which of these candidate genes have allelic variation contributing to the natural diversity in mildew resistance, a number of EST or cDNA clone sequences expected to play a role in the barley–*Blumeria* interaction were mapped and considered as candidate genes for basal resistance. The said authors have mapped QTL for powdery mildew resistance in six mapping populations at the seedling and adult plant stages and developed a high-density integrated genetic map containing almost 7,000 markers for comparing QTL and candidate gene positions over mapping populations. At the seedling stage, 12 QTL were mapped and 13 QTL at the adult plant stage; four common QTL in the two stages were identified. Six of the candidate genes are mapped in the same positions as QTL for basal mildew resistance. The authors therefore concluded that these six candidate genes may be responsible for the phenotypic effects of the QTL for basal resistance of barley against *B. graminis* f. sp. *hordei* (Aghnoum et al. 2010). Combining as many resistance-related QTL as possible by marker-assisted breeding (pyramiding) is expected to enhance the level of quantitative pathogen resistance.

Like mildew, soil-borne barley mosaic viruses, i.e. *barley mild mosaic virus* (BaMMV) and *barley yellow mosaic virus* (BaYMV), are recognised as very important disease-causing agents of winter barley in Europe today. Due to a constant

Table 33.2 Grain yield of BaMMV/BaYMV resistant and susceptible winter barley cultivars released in Germany, 1986–2011 (Anonymous 1986, 1995, 2005, 2011)

	No. of cultivars		Average grain yield[a]	
Year	Resistant	Susceptible	Resistant	Susceptible
1986	6	37	4.3[a]	5.6
1995	24	41	6.5	6.3
2005	52	23	6.7	6.1
2011	55	9	6.9	6.4

[a]Scale: 1 = minimum, 9 = maximum score

spread of the area infested and yield losses of up to 50%, barley yellow mosaic disease (caused by different virus strains) has become one of the most important diseases of winter barley. Chemical measures to prevent yield losses are not effective due to transmission of these viruses by the soil-borne plasmodiophorid *Polymyxa graminis*. Therefore, breeding for resistance is the only way to ensure winter barley cultivation in the growing areas of infested fields. Resistant cultivars have been detected within the set of released cultivars in Germany soon after the first discovery of the disease in Europe in 1978. However, at that time and in the 1980s, resistant cultivars were considerably lower yielding than susceptible ones (Table 33.2). Today, barley breeding has succeeded in combining complete resistance to BaMMV/BaYMV and high yield as most of the released cultivars are resistant and do actually outyield susceptible varieties.

However, the resistance of these cultivars is still mainly due to the *rym4/rym5* locus on chromosome 3H. Meanwhile, molecular markers have been developed for many additional BaMMV/BaYMV resistance loci (Ordon 2009; Kai et al. 2012) facilitating efficient pyramiding of resistance genes (Werner et al. 2005). Pyramiding may become of special importance in the future as many of the recessive resistance genes known are not effective against all strains of the barley yellow mosaic virus complex (Habekuß et al. 2008). By this approach, resistance genes (partly) overcome by "new" virus strains can again be used in commercial winter barley breeding.

While many BaMMV/BaYMV resistant winter barley cultivars are available today, the number of cultivars tolerant to barley yellow dwarf disease caused by different strains of *barley yellow dwarf virus*, (BYDV), *cereal yellow dwarf virus* (CYDV) is still very limited, e.g. the Italian cultivar 'Doria' (Kosova et al. 2008). However, the disease is supposed to become more important in the future due to climate change (Habekuß et al. 2009). In this respect, it has recently been shown that by a marker-based combination of *Ryd2* located on chromosome 3H (Collins et al. 1996) and *Ryd3* on chromosome 6H (Niks et al. 2004) quantitative resistance rather than tolerance can be achieved (Riedel et al. 2011, Fig. 33.3). Respective DH lines carrying positive alleles at all loci have already found their way into applied barley breeding programmes, rising hope that this resistance will be incorporated into adapted high-yielding barley cultivars in the near future.

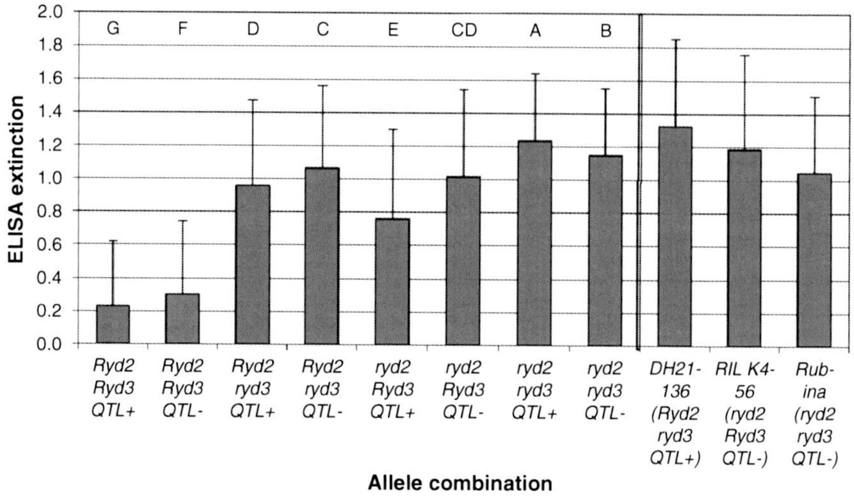

Fig. 33.3 Average ELISA extinction (405 nm) and standard deviation in the barley DH population "RIL K4-56"×"DH21-136" carrying different allele combinations at the *Ryd2* locus, *Ryd3* locus and the QTL on chromosome 2H. Results of field experiments with BYDV-PAV inoculation in four locations and 2 years. *Different letters* indicate significant differences (Tukey-test, $\alpha=0.05$). Data of parental lines and the susceptible check cv. Rubina are shown for comparison (Riedel et al. 2011)

Combinations of superior resistance and high grain yield are not limited to virus diseases but were also achieved for fungal diseases, e.g. powdery mildew, as elucidated by a comparison of winter barley cultivars released in Germany in 1985 and 2010, respectively. While in 1985 out of 35 cultivars released only two were more or less mildew resistant but low yielding, in 2010 already 29 cultivars out of 65 released were highly resistant and most of them also high yielding (Table 33.3).

Today, newly released cultivars in general combine high yield potential with good or reasonable resistances to most of the important pathogens, as shown in Table 33.4. It is striking that nearly all new cultivars are resistant to BaMMV/BaYMV, possess a good to medium resistance against all important fungal pathogens and are high to very high yielding (scores 7–9). Besides this, the first two-rowed cultivars have been released reaching the high-yield level (score 9) of six-rowed cultivars (Table 33.4, Anonymous 2011).

33.5 New Quality Traits

In addition to methodological progress, "new" traits including special qualities ("functional barley") for alternative, e.g. chemical purposes, such as high amylose and waxy barley, would potentially allow the diversification and extension of barley

Table 33.3 Progress in combining superior powdery mildew resistance (scores 1–3) and high-yield performance in winter barley cultivars released in Germany, 1985 versus 2010

	Grain yield[a]								
Mildew susceptibility[a]	1	2	3	4	5	6	7	8	9
				1985 ($n=35$)					
1									
2									
3			1	1					
				2010 ($n=65$)					
1							1		
2						2	3		
3				1	1	2	9	8	2

[a]Scales: 1–9, 1 = lowest, 9 = highest trait expression

Table 33.4 Average scores for grain yield and disease resistances of winter barley cultivars newly released in Germany in the year 2011 (Anonymous 2011)

		Susceptibility to				
Cultivar	Grain yield	Powdery mildew	Net blotch	Scald	Leaf rust	BaMMV/BaYMV
6-rowed						
Amelie	9	5	3	4	3	1
Amrai	8	3	4	5	3	1
Henriette	8	4	3	5	3	1
Meridian	9	5	4	4	3	1
Tenor	9	3	6	3	3	1
Medina	9	5	3	4	3	1
2-rowed						
Augusta	8	3	4	3	3	1
Marielle	7	2	3	4	3	1
Matros	9	3	5	3	3	9
Precosa	7	5	4	4	6	1

Scale: 1 = minimum, 9 = maximum value

cultivation and use. In Germany, the winter barley cultivar 'Waxyma' (http://www.dieckmann-seeds.de/) carrying an amylopectin content of about 95% instead of 70% normally found in barley starch has been released in 2008. The starch of this so-called waxy barley has completely different characteristics than that of ordinary barley especially concerning viscosity which is generally higher and starts rising already at a lower temperature. Besides, this starch type binds water durably and does not crystallise due to the low amylose content. Amylopectin starch is suited therefore to be used for special industrial purposes, e.g. in glues, in bonding agents but also in the food industry as a thickening agent. With respect to food, it also has to be noted that the flour of waxy barley contains on average more than 6% beta-glucan which is advantageous in human diets with respect to lowering the LDL cholesterol concentration in the blood (Ahlemeyer et al. 2009). These new qualities of barley in combinations with sufficient agronomic performance may open the way to a wider range of applications of barley in the future.

33.6 Conclusions and Perspectives

Barley breeding in Europe has been extremely successful throughout the last century. Cycles of cross-breeding which first made use of hybridisations between European landraces, later exploiting more distant and alien germplasms providing valuable disease resistances, and finally combining high-yield and high-quality varieties, have led to highly productive modern cultivars, both in spring and winter barley. In earlier time, the practical breeding techniques comprised little more than manual hybridisation, careful observation, precise testing and conscious selection. More recently, breeding programmes have been enhanced by the implementation of modern biotechnology tools, such as the "haploidy method". Today, it is obvious that haploid techniques combined with marker-assisted selection procedures–particularly PCR-based techniques combined with fast and high-throughput analysis also addressed as SMART breeding–can further enhance the process of selecting resistant varieties with superior agronomic performance. As a consequence, existing breeding schemes including "fast track" procedures are being accelerated and carried out more efficiently than in the past.

Recently, the genomic sequence of monocot species, i.e. *Brachypodium* and sorghum, has become available in addition to the rice genome, and efficient tools for exploiting the synteny between these species and barley have been developed which together with the constantly rising sequence information in barley itself (Mayer et al. 2011) will lead to an enhanced isolation of genes underlying traits of agronomic importance. The isolation of genes involved in e.g. virus resistance will transfer breeding to the allele level facilitating the identification of novel alleles and their directed use in molecular breeding strategies. Besides this, the rising number of markers and sequence information may facilitate genomic selection procedures in future barley breeding (Heffner et al. 2009).

In summary, all these advances in biotechnology will improve barley breeding and will enable breeders to react in a more directed and fast manner to the challenges arising from climatic change, new virulent strains of pathogens and the need to feed the earth's growing population. For example, a significant genotypic variation for late leaf spots due to solar radiation has been observed in a barley collection, in which specific genotypes, e.g. the cv. Primadonna, showed less spotting than others such as cv. Quench (M. Herz, pers. comm.). On this basis, it is expected that new cultivars better adapted to the conditions of climate change can be developed by systematic breeding employing modern genomic tools.

Acknowledgement We wish to sincerely thank Dr. Jutta Ahlemeyer and Deutsche Saatveredelung AG for continuous support and excellent collaboration.

References

Abeledo, L. G., Calderini, D. F., & Slafer, G. A. (2002). Physiological changes associated with genetic improvement of grain yield in barley. In G. A. Slafer, J. L. Molina-Cano, R. Savin, J. L. Araus, & I. Romagosa (Eds.), *Barley science: Recent advances from molecular biology to agronomy of yield and quality* (pp. 361–386). New York/London/Oxford: Food Products Press.

Aghnoum, R., Marcel, T. C., Johrde, A., Pecchioni, N., Schweizer, P., & Niks, R. E. (2010). Basal host resistance of barley to powdery mildew: Connecting quantitative trait loci and candidate genes. *Molecular Plant-Microbe Interactions, 23*, 91–102.

Ahlemeyer, J. (2009). Nur noch wenig Ertragszuwächse. *DLG Saatgutmagazin, 7*, 22–24.

Ahlemeyer, J., Aykut, J. F., Köhler, W., Friedt, W., & Ordon, F. (2008). Genetic gain and genetic diversity in German winter barley cultivars. In *Proceedings of the Eucarpia Cereal Science and Technology for Feeding Ten Billion People: Genomics Era and Beyond, Leida, Spain* (Options Méditerranéennes Serie A 81, pp. 43–47).

Ahlemeyer, J., Münzing, K., Jansen, G., & Dieckmann, K. (2009). Waxy barley an ancient grain fulfilling a modern health claim. *Agro Food, 20*, 62–63.

Ahokas, H. (1979). Cytoplasmic male sterility in barley. III. Maintenance of sterility and restoration of fertility in the msm1 cytoplasm. *Euphytica, 28*, 409–419.

Ahokas, H. (1982). Cytoplasmic male sterility in barley. XI. The msm2 cytoplasm. *Genetics, 102*, 285–295.

Anonymous. (1986). *Beschreibende Sortenliste 1986*. Hannover: Bundessortenamt.

Anonymous. (1995). *Beschreibende Sortenliste 1995*. Hannover: Bundessortenamt.

Anonymous. (2005). *Beschreibende Sortenliste 2005*. Hannover: Bundessortenamt.

Anonymous. (2011). *Beschreibende Sortenliste 2011*. Hannover: Bundessortenamt.

Collins, N. C., Paltridge, N. G., Ford, C. M., & Symons, R. H. (1996). The *Yd2* gene for barley yellow dwarf virus resistance maps close to the centromere on the long arm of barley chromosome 3. *Theoretical and Applied Genetics, 92*, 858–864.

Friedt, W., Werner, K., & Ordon, F. (2000). Genetic progress as reflected in highly successful and productive modern barley cultivars. In *Proceedings of the 8th International Barley Genetics Symposium* (Vol. I, pp. 271–279), Adelaide, Australia.

Grausgruber, H., Bointer, H., Tumpold, R., & Ruckenbauer, P. (2002). Genetic improvement of agronomic and qualitative traits of spring barley. *Plant Breeding, 121*, 411–416.

Habekuß, A., Kuhne, T., Kramer, I., Rabenstein, F., Ehrig, F., Ruge-Wehling, B., Huth, W., & Ordon, F. (2008). Identification of Barley mild mosaic virus isolates in Germany breaking rym5 resistance. *Journal of Phytopathology, 156*, 36–41.

Habekuß, A., Riedel, C., Schliephake, E., & Ordon, F. (2009). Breeding for resistance to insect transmitted viruses – An emerging challenge due to global warming. *Journal of Cultivated Plants, 61*, 87–91.

Heffner, E. L., Sorrells, M. E., & Jannink, J. L. (2009). Genomic selection for crop improvement. *Crop Science, 49*, 1–12.

Kai, H., Takata, K., Tsukazaki, M., Furusho, M., & Bana, T. (2012). Molecular mapping of Rym17, a dominant and rym18 a recessive barley yellow mosaic virus (BaYMV) resistance genes derived from *Hordeum vulgare* L. *Theoretical and Applied Genetics*. doi:10.1007/s00122-011-1730-5.

Kosova, K., Chrpova, J., & Sip, V. (2008). Recent advances in breeding of cereals for resistance to barley yellow dwarf virus – A review. *Czech Journal Genetics Plant Breeding, 4*, 1–10.

Mac Key, J. (1993). Demonstration of genetic gain from Swedish cereal breeding. *Sveriges Utsadesforenings Tidskrift, 103*, 33–43.

Matsui, K., Mano, Y., Taketa, S., Kawada, N., & Komatsuda, T. (2001). Molecular mapping of a fertility restoration locus (Rfm1) for cytoplasmic male sterility in barley (*Hordeum vulgare* L.). *Theoretical and Applied Genetics, 102*, 477–482.

Mayer, K. F. X., Martis, M., Hedley, P. E., Kimková, H., Liu, H., Morris, J. A., Steuernagel, B., Taudien, S., Roessner, S., Gundlach, H., Kubaláková, M., Suchánková, P., Murat, F., Felder, M., Nussbaumer, T., Graner, A., Salse, J., Endo, T., Sakai, H., Tanaka, T., Itoh, T., Sato, K., Platzer, M., Matsumoto, T., Scholz, U., Doleqel, J., Waugh, R., & Stein, N. (2011). Unlocking the barley genome by chromosomal and comparative genomics. *The Plant Cell, 23*, 1249–1263.

Munoz, P., Voltas, J., Araus, J. I., Igartua, E., & Romagosa, I. (1998). Changes over time in the adaptation of barley releases in north-eastern Spain. *Plant Breeding, 117*, 531–535.

Murakami, S., Matsui, K., Komatsuda, T., & Furuta, Y. (2005). AFLP-based STS markers closely linked to fertility restoration locus (*Rfm1*) for cytoplasmic male sterility in barley. *Plant Breeding, 124*, 133–136.

Niks, R. E., Habekuß, A., Bekele, B., & Ordon, F. (2004). A novel major gene on chromosome 6H for resistance to barley against the barley yellow dwarf virus. *Theoretical and Applied Genetics, 109*, 1536–1543.

Ordon, F. (2009). Coordinator's report: Disease and pest resistance genes. *Barley Genetics Newsletter, 39*, 58–67.

Ortiz, R., Nurminiemi, M., Madsen, S., Rognli, O. A., & Bjornstad, A. (2002). Genetic gains in Nordic spring barley over sixty years. *Euphytica, 126*, 283–289.

Peltonen-Sainio, P., & Karjalainen, R. (1991). Genetic yield improvement of cereal varieties in northern agriculture since 1920. *Acta Agriculturae Scandinavica, 41*, 267–273.

Riedel, C., Habekuss, A., Schliephake, E., Niks, R., Broer, I., & Ordon, F. (2011). Pyramiding of Ryd2 and Ryd3 conferring tolerance to a German isolate of Barley yellow dwarf virus (BYDV-PAV-ASL-1) leads to quantitative resistance against this isolate. *Theoretical and Applied Genetics, 123*, 69–76.

Rode, J., Ahlemeyer, J., Friedt, W., & Ordon, F. (2011). Identification of marker-trait associations in the German winter barley breeding gene pool (*Hordeum vulgare* L.). Molecular Breeding. doi:10.1007/s11032-011-9667-6.

Waugh, R., Marshall, D., Thomas, B., Comadran, J., Russell, J., Close, T., Stein, N., Hayes, P., Muehlbauer, G., Cockram, J., O'Sullivan, D., Mackay, I., Flavell, A., & Ramsay, L. (2010). Whole-genome association mapping in elite inbred crop varieties. *Genome, 53*, 967–972.

Werner, K., Friedt, W., & Ordon, F. (2005). Strategies for pyramiding resistance genes against the barley yellow mosaic virus complex (BaMMV, BaYMV, BaYMV-2). *Molecular Breedeeing, 16*, 45–55.

Chapter 34
Variation in Phenological Development of Winter Barley

Novo Pržulj, Vojislava Momèiloviæ, Dragan Perović, and Miloš Nožinić

Abstract Phenology of small grains as a complex trait that matches plant development with growing conditions is an important factor that influences plant adaptation to a particular environment and final yield of the plant. Scales of growth based on plant phenology and scales of development based on apical morphology are very precise, but relationships between them are not always clear. Some morphogenetic features of cereal growth can be predicted on the basis of leaf appearance, which means that shoot apex development is coordinated with leaf appearance and total number of leaves formed. The objective of this research was to study the relationship between apical development and plant phenology and the variability in phenological development of diverse winter barley cultivars under field conditions. Twelve barley cultivars differing in origin, pedigree, and agronomic traits were used in this study conducted during six growing seasons in the location of Novi Sad (45°20′N, 15°51′E, altitude 86 m) under rainfed conditions. Phyllochron approach was used as a method for determination of initial/final stages of apex development. The duration of the phases was converted to cumulated growing degree days (GDD). Spikelet initiation started at 2.5 Haun stage and finished at 7.1 Haun stage. Preanthesis phases (single ridge – SR, double ridge – DR, spikelet development – SD, heading – H, anthesis – A) and grain filling period (GFP) were

N. Pržulj (✉) • V. Momèiloviæ
Institute of Field and Vegetable Crops, Maksima Gorkog 30, 21000 Novi Sad, Serbia
e-mail: novo.przulj@ifvcns.ns.ac.rs

D. Perović
Federal Research Centre for Cultivated Plants, Julius Kuehn-Institute, Erwin-Baur-Str. 27, 06484 Quedlinburg, Germany

M. Nožinić
Agricultural Institute of the Republic of Srpska, Knjaza Miloša 19, 71000 Banja Luka, Republic of Srpska, BiH

G. Zhang et al. (eds.), *Advance in Barley Sciences: Proceedings of 11th International Barley Genetics Symposium*, DOI 10.1007/978-94-007-4682-4_34,

under significant effect of genetic (G), environmental (E), and interaction factors (G×E). Of the total variation in the DR phase, 55.7% was due to G×Y, 22.1% due to E, and 12.3% due to G. Spikelet development was mainly under control of G and E, 42% and 44%, respectively. Heritability was 0.57 and 0.95 for DR and SD, respectively. Across cultivars and environments, of the total growing period, 138 GDD belonged to sowing-first leaf period, 161 to SR, 326 to DR, 541 to SD, 254 to flag leaf-H, 142 to H-A, and 732 to GF. Although variability was found in the duration of the preanthesis phases and GF, especially in spikelet and flower development, positive genotypic correlations were found to exist between most of the phases.

Keywords Barley (*Hordeum vulgare* L.) • Spike development • Single ridge • Double ridge • Spikelet development • Grain filling

34.1 Introduction

Phenology can be defined as the phenotype expression of a plant development through successive growth stages. Phenology of small cereals is a complex trait that harmonizes plant development with growing conditions (temperature, day length, water, radiation, nutrition) and influences plant adaptation to a particular environment as well as the final yield (Richards 1991; Slafer and Whitechurch 2001). This is particularly important for the continental climate in the Pannonian zone, in which severe drought stresses occur on a regular basis (Olesen et al. 2011).

Scales of growth based on plant phenology (Feekes, Haun, Zadoks) and scales of development based on apical morphology (Banerjee and Wienhues 1965; Gardner et al. 1985; Kirby 2002) are very precise, but relationships between them are not always clear. Rickman and Klepper (1995) developed a model which connected phenological development of the apex with the external development of the plant and thermal time. They explained the initiation and development of vegetative and generative plant parts in function of sums of average temperature, using number of days, growing degree days, or photothermal units to estimate the length of growth stages. According to most models, the synchronous initiation and development of all vegetative and reproductive parts of the plant are in function of average temperature sums. The phyllochron is an alternative measure for assessing the duration of some growth stages; it is more flexible than the other approaches as it takes into account developmental processes within the plant (McMaster et al. 1992).

The objective of this research was to study (1) relationships between apical development and plant phenology and (2) the variability in phenological development of diverse winter barley cultivars grown under field conditions.

34.2 Materials and Methods

34.2.1 Cultivars and Crop Management

The 12 winter barley cultivars (Novosadski 525 – SRB, Novosadski 581 – SRB, Sonate – DEU, Boreale – FRA, Monacco – FRA, Cordoba – DEU, Kompolti-4 – HUN, Skorohod – RUS, Plaisant – FRA, Nonius – SRB, Novator – UKR, and Kredit – CZE) that differed in origin, pedigree, and agronomic traits were used in this study. All the cultivars have been grown commercially or are still in production.

A 6-year experiment was conducted in 2002/2003 to 2007/2008 growing seasons (GS) in Novi Sad location (45°20′N, 15°51′E, altitude 86 m) at the experiment field of Institute of Field and Vegetable Crops under rainfed conditions. Planting density in all GSs was 300 germinated seeds per m^2 for six-rowed barley and 350 germinated seeds per m^2 for two-rowed barley. The experiment was conducted in a randomized complete block design with 3 replications in each year. To avoid negative effects of diseases, the trials were sprayed with the fungicide Tilt 250 EC at Zadoks phase 64 and with insecticide Karate Zeon applied when needed. Weed control was performed by hand.

34.2.2 Samples Collecting and Plant Measurements

One hundred plants in each plot were tagged with colored rings 10 days after the coleoptiles emerged from the soil. After emergence, samples of three plants per plot from each replication were pooled out, every 3–4 days, put in plastic bags, and used for dissection to determine phase of apex development. Number of leaves (LN) on the main shoot was recorded according to the Haun scale (Haun 1973) prior to dissection. Determination of the apex development phase was carried out with a Carl Zeiss Stemi 2000-C stereoscopic binocular microscope allowing for magnification up to 100 ×. Emergence date was defined as the date when 50% of the plants had emerged. The time of emergence, tillering, first node detectable, flag leaf extension, heading, and anthesis for the main stem on the sampled plants were recorded following the Zadoks' scale (Zadoks et al. 1974).

Emergence is defined as the moment when first true leaf becomes visible (Zadoks stage 8–10), the onset of stem elongation when first node is 0.5 cm above the soil surface (Zadoks stage 30), flag leaf emergence when ligules of the flag leaf became visible (Zadoks stage 39), heading when heads are fully emerged from the leaf sheath on the flag leaf (Zadoks stage 55), anthesis when the central florets have shed their pollen (Zadoks stage 61), and physiological maturity when the flag leaf and the spike turned yellow (Zadoks stage 92). Final leaf number (FLN) was determined as the number of leaves on the main stem. GDD was calculated for each of the phases with base temperature of 0 °C.

34.2.3 Weather Conditions

Weather data were collected from the meteorological station located within the experiment field. Temperature and precipitation were collected daily. The average temperatures at the level of the long-term average and abundant water in October 2002 allowed a rapid crop establishment and fast progression through vegetative phase. During the generative phase in November, average temperatures were at the level of summer temperature, and there occurred a water deficit. The generative phase continued in the spring under conditions of extreme drought and low temperatures. The grain filling proceeded under drought conditions and high temperatures. Excessive rainfall, several-fold above the annual average, and average temperatures accompanied the emergence stage in the 2003/2004 growing season. Temperatures were somewhat lower during the vegetative phase in late October and early November. During the early stages of the DR stage in late November, temperatures were extremely high and precipitation was scant. The completion of the generative phase next spring and grain filling stage unfolded under normal conditions. In the 2004/2005 growing season, temperature and rainfall conditions were favorable during the vegetative phase. In the fall, favorable weather conditions lasted 3 weeks longer than the average data, which helped the generative phase to proceed normally. The remainder of the growing season and the jointing and grain filling stages continued under favorable conditions. In the 2005/2006 growing season, barley germination and the vegetative stage took place in a drought, while the generative phase in November and December unfolded under favorable conditions. The generative phase that continued next spring took place under conditions of high temperature fluctuations, while high temperatures and water deficit coincided with the jointing stage. High temperatures and water deficit in October and November 2006 obstructed winter barley emergence and the progression of the vegetative phase. Favorable rainfall and temperatures during December and the subsequent winter months enabled the plants accumulate the necessary amounts of effective temperatures and compensate for the previous delay in development. The completion of the generative phase, jointing and filling stages took place under high temperatures and excess water. In the 2007/2008 growing season, the germination, vegetative phase and the beginning of the generative phase unfolded under somewhat low temperatures and a large surplus of water. The generative phase, jointing and grain filling continued in the spring of 2008 under favorable conditions.

34.2.4 Statistical Analysis

All data were subjected to the analysis of variance using Statistica 9.0 program (StatSoft, Tulsa, OK, USA). When differences among earliness groups (early, medium early, late), duration of developmental phases and agronomic traits were tested; four cultivars from each group were considered as replication as samples for detection of developmental phases.

34.3 Results and Discussion

The life cycle of cereals is divided into two main periods, period till anthesis and grain filling period. From a practical point of view and considering the changes at the shoot apex, the period till anthesis can be divided in three phases: leaf initiation (vegetative phase), spikelet initiation (early reproductive phase), and spike growth (late reproductive phase) (Slafer and Whitechurch 2001). Miglietta (1991) found that phasic development and some morphogenetic features of wheat growth can be predicted on the basis of leaf appearance, which means that shoot apex development is coordinated with leaf appearance and total number of formed leaves. Nonlinear relationship that exists for temperature and day length effects on growth may be successfully represented by beta function (Jame et al. 1998).

The apex produces leaf and spikelet primordia in chronological order – first spikelet primordium is initiated only after last leaf primordium has been formed (Rickman and Klepper 1995; Arduini et al. 2010). Initiation of first leaf and first spikelet primordia is rather difficult to determine. We found that leaf primordium was already present when first leaf emerged in all three groups of cultivars, early (E), medium (M), and late (L) (Table 34.1). According to our result, there was no variability among the tested cultivars for the beginning of leaf primordium initiation. Arduini et al. (2010) found a strong correlation between total number of primordia (leaf and spikelet) on the main stem apex and Haun stage. He applied a logistic curve to define this relationship, and according to the selected equation, approximately three primordia were already present on the main stem apex at plant emergence and the maximum number of primordia was achieved with the emission of the 10th leaf. Our results are different in regard to the number of leaf primordia at emergence and last spikelet primordium (Table 34.1), which could be

Table 34.1 Phase variability of barley apex development of four early (E), four medium (M), and four late (L) barley cultivars evaluated by Haun stage.

Phase	Haun stage from–to	Haun average	B&W scale	Zadoks stage
Emergence	1	1	–	11
Leaf primordium initiation (SR vegetative phase)	E: 1–1 M: 1–1 L: 1–1	1	1	07
1st spikelet primordium visible (DR generative phase)	E: 2.1–2.9 M: 2.2–3.2 L: 2.2–3.4	2.51 ± 0.038	4	13
Last spikelet primordium initiation	E: 5.9–7.8 M: 6.1–7.9 L: 6.2–7.9	7.08 ± 0.067	6a, 6b, 7	31
Spike growth and differentiation	E: 11.8–13.5 M: 12.7–14.0 L: 13.2–14.5	13.50 ± 0.13	8	39

Analogous codes of Banerjee and Wienhues (1965) and Zadoks et al. (1974) scales are provided due to different barley genotypes used and different growing conditions. Single ridge or vegetative phase was under significant control of genotype, year and their interaction, with the strongest influence of year (Table 34.2).

Arduini et al. (2010) found that spikelet initiation in winter, alternative, and spring barleys started at 2.4–2.5 Haun stage, Juskiw et al. (2001) found 2.8 Haun stage for spring barley, and other authors found that beginning of this phase for barley and wheat started in the period from 2.4 to 4.0 Haun stage. During the period of spikelet initiation, leaves elongate and appear on the outside. End of spikelet initiation in barley is difficult to identify since apex initiates an indefinite number of primordia, where latest-formed ones die and do not differentiate into spikelets (Arisnabarreta and Miralles 2006). The last productive spikelet is initiated when glume initials appear and when the flag leaf primordium starts to elongate (Arduini et al. 2010). From the phenological point, spikelet initiation ended at 7.1 Haun stage or at the beginning of jointing (39 Zadoks stage), when first node is detectable (Kirby et al. 1985).

Beginning of spikelet initiation (double ridge phase, DR) is associated with the formation of double ridges (Baker and Gallagher 1983) and rapid expansion of shoot apex (Kirby 2002). In our cultivars, beginning of spike initiation across varieties and years started at 2.5 Haun stage (Table 34.1). Depending on variety and year, first double ridge appeared at 2.1 Haun stage in early and last at 3.4 Haun stage in late cultivars (Table 34.1). Haun stage of initiation of first and last spikelets and the rate of initiation depended on genotype, probably due to different final leaf numbers and not environmental conditions (Arduini et al. 2010). Variation of this phase is modest; range of variation was 1.3 leaves, and coefficient of variation was 16.8% (Table 34.1). In our previous work (Pržulj and Momčilović 2011), we found a positive correlation between apex stage and earliness, where early cultivars were first and late varieties were last to reach the DR phase. All components of variance were significant with the highest participation of interaction, 55.7%. Our earlier results (Pržulj and Momčilović 2011) with three divergent barley cultivars showed a high participation of genetic variance, 80%. Heritability for spike initiation was medium (Table 34.2) due to low genetic variability. The period of spike initiation was shortest in favorable environments and longest in the conditions of high temperature and water deficit. Spikelet initiation was completed at the beginning of jointing at 7.1 Haun stage (Table 34.1).

Spike growth and differentiation occurred from jointing to the moment when ligule of flag leaf (FL) just became visible, that is, during stem elongation or jointing (J). The number of fertile florets at anthesis is determined during this phase, which in turn determines the final number of grains per spike (Fischer 2007; Miralles and Slafer 2007). Beginning of floret differentiation and internode expansion matches with the extension of the flag leaf (Fournier et al. 2004). This is a relatively long phase during which almost 6.5 leaves appeared, including the flag leaf (from Table 34.1). Duration of this phase was 541 GDD (Table 34.3), and this duration participated with 27.7% in total thermal requirements of winter barley (data not shown). Spike growth and differentiation depended mainly on year and genotype, with 44.1% and 41.9% participation of the components of variance, respectively

Table 34.2 Mean squares and percentages of components of variance from the analysis of variance for GDD for duration of SR, DR, J, FL, H, GFP, EH, EA, and JA of 12 barley cultivars grown in 2002/2003 to 2007/2008 growing seasons at Novi Sad location

Source	df	SR	DR	J	FL	H	GFP	EH	EA	JA
Replication	2	83.2	323.9	806.0	1203.8	1.2	614.2	106.4	1458.2	827.3
Genotype (G)	11	2459.0**	3642.0**	111296.0**	16287.8**	471.1**	53015.6**	164815.1**	251563.9**	172525.1**
Year (Y)	5	18506.8**	103426.0**	236350.7**	36580.0**	471.2**	377855.5**	566792.9**	999355.8**	430434.9**
G × Y	55	345.9**	1618.3**	5579.5**	4033.5**	317.5**	9508.5**	7328.9**	10824.1**	7596.4**
Pooled error	142	42.00	90.4	156.5	658.8	7.88	658.9	441.9	603.6	617.1
% of variance components										
σ^2_G		29.55	12.30	41.89	20.21	5.85	26.94	42.38	42.65	42.24
σ^2_Y		34.38	22.13	44.10	26.86	17.95	32.83	44.36	44.56	44.19
σ^2_{GY}		25.49	55.70	12.89	33.39	70.79	32.88	11.12	10.86	10.73
σ^2_E		10.58	9.88	1.12	19.55	5.40	7.34	2.14	1.92	2.84
h^2_b		0.86	0.56	0.95	0.75	0.33	0.82	0.96	0.96	0.96

SR single ridge, *DR* double ridge, *J* jointing, *FL* flag leaf, *H* heading, *GFP* grain filling period, *EH* emergence-heading, *EA* emergence-anthesis period, *JA* jointing-anthesis period

*Significant at the 0.05 level

**Significant at the 0.01 level

Table 34.3 Mean, standard error, standard deviation, range of variation, and coefficient of variation for the duration of phenological stages of winter barley

	Separately					Cumulative GDD				
Trait	$\overline{X}\pm$	S	Min	Max	CV	$\overline{X}\pm$	S	Min	Max	CV
E	137.7±1.97	16.7	94.7	159.7	12.1	137.7±1.97	16.8	94.7	159.7	12.1
SR	161.3±2.99	25.3	127.1	224.2	15.7	299.0±269	22.8	228.8	357.9	7.6
DR	325.8±6.45	54.7	215.5	461.6	16.8	624.8±7.96	67.5	495.5	780.5	10.7
J	541.2±13.21	112.1	301.1	780.2	20.7	1166.0±18.28	155.1	843.2	1528.2	13.2
FL	254.4±5.49	46.6	170.4	387.2	18.3	1420.4±21.88	185.6	1094.2	1866.4	13.0
H	142.3±4.72	40.1	60.7	279.5	28.2	1562.7±25.37	215.3	1210.7	2145.9	13.7
A	63.7±1.27	10.8	42.0	90.0	16.9	1626.4±25.44	215.9	1276.5	2203.6	13.2
GFP	668.4±13.88	117.8	401.2	884.3	17.6	2294.8±25.13	213.2	1859.8	2784.7	9.2
FLN	13.5±0.13	1.1	10.0	15.0	8.1					
PH	75.7±0.94	8.0	63.2	94.3	10.6					

E emergence, *SR* single ridge, *DR* double ridge, *J* jointing, *FL* flag leaf, *H* heading, *A* anthesis, *GFP* grain filling period, *FLN* final leaf number, *PH* phyllochron

(Table 34.2). Significant importance of genetics in the duration of this phase confirms a high value of heritability, 0.95.

Adjustment of plant phenology to the environmental factors could be an effective method for yield potential improvement. Lengthening the duration of stem elongation phase increases the number of grains per square meter, which could increase the yield potential in small grains (Slafer et al. 2005; Fischer 2007; Miralles and Slafer 2007). Lengthening of one phase can be achieved without changing the total time to anthesis, which is the basis for adaptability improvement.

Duration of heading across 12 varieties and 6 years was 142 GDD, or about 7 days (Table 34.3). This phase was very variable, with the range of variation from 60.7 to 279.5 GDD and CV of 28.2%. As expected, the participation of genotype in total variance was very low, 5.85%, and heritability was correspondingly low (Table 34.2). Genotype × year interaction participated with 70.79% in total variation. Variation in heading duration is more due to different final leaf numbers than environment. Anthesis lasted for a few days, with average duration of 63.7 GDD or approximately 3 days (Table 34.3).

Winter barley needed in average 1,626 GDD from planting to anthesis, with a rather wide range of variation from 1,276 to 2,203 GDD (Table 34.3). Average grain filling duration was 668 GDD or approximately 33 days. Environment exhibited a stronger influence on grain filling than genotype, although contributions of all three components of variation (genotype, year, and their interaction) were approximately equal. If one analyzes the phases till anthesis, a large variation can be observed with different values of heritability, from low, for heading duration, to very high, for spike growth and differentiation, that is, jointing (Table 34.2). When more than a single phase till anthesis was observed, for example, emergence-heading, emergence-anthesis, or jointing-anthesis, participation of the genotype component increased till 42% and heritability stabilized at 0.96 (Table 34.2). In other words, the higher number of the preanthesis phases, the lower the interaction and the higher the genotype and year influences.

Testing sets of different cultivars with similar times to anthesis, Whitechurch et al. (2007) in wheat and Kernich et al. (1997) and Borràs et al. (2009) in barley found variability in the duration of preanthesis phases. Whitechurch et al. (2007) and Borràs-Gelonch et al. (2010) found that leaf and spikelet initiation were independent of stem elongation, that is, these phases were under different genetic controls. Zhou et al. (2001) found in rice some independent QTLs for the duration of vegetative and reproductive phases. According to these results, it would be possible to modify genetically the pre-heading phases without modifying the total time to heading. However, there are scant data regarding the genetic control of different preanthesis phases in either wheat or barley (Borràs-Gelonch et al. 2010).

To connect plant phenotype with spike morphology is useful for plant modeling, crop management, and yield prediction. Some authors found a linear relationship between the number of emerged leaves and the number of primordia on main culm apex (Kirby 1990; González et al. 2001) with different rates of initiation. Emphasize different responses of leaf appearance and apex stage to significant effect of environmental factors.

34.3.1 Genotypic and Phenotypic Correlations

Knowledge of relationships between pairs of quantitative traits and the degree of their correlation may facilitate interpretation of results and contribute to successful practical breeding. The existence of strong genetic correlations between low-heritability traits, for example, yield, and high-heritability traits can be effectively used to indirectly breed for the former traits by breeding for the latter ones. In that case, correlation level and type, that is, covariances of the two traits, must be taken into account.

In our investigation, phenotypic correlations were mostly nonsignificant, and genotypic correlations between the tested traits were mostly significant (Table 34.4). The single ridge phase was in significant positive genetic correlation with all phases till anthesis and with grain filling period. The double ridge phase was positively correlated with spike growth and differentiation, heading, grain filling, and total growth time. Spike growth and differentiation were positively correlated with all preanthesis phases, grain filling, and total growth time. These results are different from those obtained in our previous study where phenotypic and genotypic correlations were mostly similar and DR had negative genetic correlations with all phases of development except the SR phase.

In the additive model, phenotypic correlations consist of genotypic components, and the environmental component and their relations are determined by the relative importance of heritable and nonheritable effects. If phenotypic variation results from similar actions of genetic and environmental factors,

Table 34.4 Genotypic (above diagonal) and phenotypic (under diagonal) correlations between the growing apex stage and some other phenological phases in 12 winter barley varieties across six growing seasons ($n = 72$)

	SR	DR	J	FL	H	A	EA	GF	TGT
SR		0.715**	0.930**	0.619**	0.673**	0.959**	0.182	0.856**	0.949**
DR	0.450		0.599*	0.451	0.578*	0.019	0.698**	0.829**	0.756**
J	0.137	0.107		0.513*	0.533*	0.257	0.961**	0.787**	0.928**
FL	0.215	0.185	0.467		0.930**	0.147	0.711**	0.737**	0.736**
H	0.282	0.280	0.499	0.729**		−0.164	0.713**	0.788**	0.754
A	0.167	0.019	0.255	0.144	−0.160		0.269	0.120	0.228
EA	0.093	0.083	0.528*	0.176	0.145	0.012		0.892**	0.989**
GF	0.179	0.209	0.644**	0.353	0.322	0.011	0.806**		0.948**
TGT	0.065	0.063	0.390	0.130	0.109	0.007	0.570*	0.289	

SR single ridge, *DR* double ridge, *J* jointing, *FL* flag leaf, *H* heading, *A* anthesis, *EA* emergence-anthesis period, *GF* grain filling period, *TGT* total growing time
* Significant at the 0.05 level
** Significant at the 0.01 level

genetic and environmental correlations must necessarily be similar, given that developmental traits are in correlation. As phenotypic correlations represent a sum of genotypic and environmental components, genotypic and phenotypic correlations will be similar regardless of heritability value. Differences between genotypic and phenotypic correlations may be due to different environmental and genetic effects on the phenotype and/or due to sampling error (Cheverud 1988).

In our study, the genotypic and phenotypic correlations had the same sign, but their values differed significantly; the genotypic correlations were highly significant, while phenotypic correlations failed to confirm association between some phases (Table 34.4). Cheverud (1988) stated that high heritability increases the proportional contribution of the genetic component to phenotypic correlation, which contributes to a higher similarity between these two correlations. In our study, with high and low heritability alike, significant differences were found between the coefficients of genotypic and phenotypic correlations.

34.4 Conclusion

Leaf primordia initiation starts before first leaf appearance above soil surface, and it ceases at 2.5 Haun scale. Spikelet initiation ceases at 7.1 Haun scale and spike growth and development at 13.5 Haun scale. Preanthesis phases were longer in unfavorable environments than in favorable conditions, while duration of grain filling decreased in unfavorable conditions.

References

Arduini, I., Ercoli, L., Mariotti, M., & Masoni, A. (2010). Coordination between plant and apex development in *Hordeum vulgare* ssp. *distichum*. *Comptes Rendus Biologies, 333*, 454–460.

Arisnabarreta, S., & Miralles, D. J. (2006). Floret development and grain setting in near isogenic two- and six-rowed barley lines (*Hordeum vulgare* l.). *Field Crops Research, 96*, 466–476.

Baker, C. K., & Gallagher, J. N. (1983). The development of winter wheat in the field. 1. Relationship between apical development and plant morphology within and between seasons. *The Journal of Agricultural Science Cambridge, 101*, 327–335.

Banerjee, S., & Wienhues, F. (1965). Comparative studies on the development of the spike in wheat, barley and rye. *Zeitschrift für Pflanzenzüchtung, 54*, 130–142.

Borràs, G., Romagosa, I., van Eeuwijk, F., & Slafer, G. A. (2009). Genetic variability in duration of pre-heading phases and relationships with leaf appearance and tillering dynamics in a barley population. *Field Crops Research, 113*, 95–104.

Borràs-Gelonch, G., Slafer, G. A., Casas, A., van Eeuwijk, F., & Roagosa, I. (2010). Genetic control of pre-heading phase and other traits related to development in a double-haploid barley (*Hordeum vulgare* L.) population. *Field Crops Research, 119*, 36–47.

Cheverud, J. M. (1988). A comparison of genetic and phenotypic correlations. *Evolution, 42*, 958–968.

Fischer, R. A. (2007). Understanding the physiological basis of yield potential in wheat. *Journal of Agricultural Science, 145*, 99–113.

Fournier, C., Durand, D. J., Ljutovac, S., Schäufele, R., Gastal, F., Andrieu, B. (2004). A functional-structural model of the elongation of the grass leaf and of its relationship to the phyllochron. In C. Godin (Ed.), *Fourth international workshop on functional-structural plant models, Montpellier, France* (pp. 98–104).

Gardner, J. S., Hess, W. M., & Trione, E. J. (1985). Development of the young wheat spike: A stem study of Chinese spring wheat. *American Journal of Botany, 72*(4), 548–559.

González, F. G., Slafer, G. A., & Miralles, D. J. (2001). Vernalization and photoperiod responses in wheat pre-flowering reproductive phases. *Field Crops Research, 74*, 183–195.

Haun, J. R. (1973). Visual quantification of wheat development. *Agronomy Journal, 65*, 116–117.

Jame, Y. W., Cutforth, H. W., & Ritchie, J. T. (1998). Interaction of temperature and daylength on leaf appearance rate in wheat and barley. *Agricultural and Forest Meteorology, 92*, 241–249.

Juskiw, P. E., Jame, Y. W., & Kryzanowski, L. (2001). Phenological development of spring barley in a short-season growing area. *Agronomy Journal, 93*, 370–379.

Kernich, G. C., Halloran, G. M., & Flood, R. G. (1997). Variation in duration of pre-anthesis phases of development in barley (*Hordeum vulgare*). *Australian Journal of Agricultural Research, 48*(1), 59–66.

Kirby, E. J. M. (1990). Co-ordination of leaf emergence and leaf and spikelet primordium initiation in wheat. *Field Crops Research, 25*, 253–264.

Kirby, E. J. M. (2002). Botany of the wheat plant. In B. C. Curtis, S. Rajaram, & H. Gómez-Macpherson (Eds.), *Bread wheat: improvement and production*. Rome: FAO.

Kirby, E. J. M., Appleyard, M., & Fellowes, G. (1985). Effect of sowing date and variety on main shoot leaf emergence and number of leaves of barley and wheat. *Agronomie, 5*, 117–126.

McMaster, G. S., Wilhelm, W. W., & Morgan, J. A. (1992). Simulating winter wheat shoot apex phenology. *Journal of Agricultural Science, 119*, 1–12.

Miglietta, F. (1991). Simulation of wheat of ontogenesis. I Appearance of main stem leaves in the field. *Climate Research, 1*, 145–150.

Miralles, D. J., & Slafer, G. A. (2007). Sink limitations to yield in wheat: How could it be reduced? *Journal of Agricultural Science, 145*, 139–149.

Olesen, J. E., Trnka, M., Kersebaum, K. C., Skjelvåg, A. O., Seguin, B., Peltonen-Sainio, P., Rossi, F., Kozyra, J., & Micale, F. (2011). Impact and adaptation of European crop production systems to climate change. *European Journal of Agronomy, 34*, 96–112.

Pržulj, N., & Momčilović, V. (2011). Importance of spikelet formation phase in the yield biology of winter barley. *Field and Vegetable Crops Research, 48*, 37–48.

Richards, R. A. (1991). Crop improvement for temperature Australia: Future opportunities. *Field Crops Research, 26*, 141–169.

Rickman, R. W., & Klepper, B. L. (1995). The phyllochron: Where do we go in the future? *Crop Science, 34*, 44–49.

Slafer, G. A., & Whitechurch, E. M. (2001). Manipulating wheat development to improve adaptation and to search for alternative opportunities to increase yield potential. In M. P. Reynolds, J. I. Ortiz-Monasterio, & A. McNab (Eds.), *Application of physiology in wheat breeding* (pp. 160–170). Mexico: CIMMYT.

Slafer, G. A., Araus, J. L., Royo, C., & García del Moral, L. F. (2005). Promising ecophysiological traits for genetic improvement of cereals yields in Mediterranean environments. *Annals of Applied Biology, 46*, 61–70.

Whitechurch, E. M., Slafer, G. A., & Miralles, D. J. (2007). Variability in the duration of stem elongation in wheat and barley genotypes. *Journal of Agronomy and Crop Science, 193*, 138–145.

Zadoks, J. C., Chang, T. T., & Konzak, C. F. (1974). A decimal code for the growth stages of cereals. *Weed Research, 14*, 415–421.

Zhou, Y., Li, W., Wu, W., Chen, Q., Mao, D., & Worland, A. J. (2001). Genetic dissection of heading time and its components in rice. *Theoretical and Applied Genetics, 10*, 1236–1242.

Chapter 35
Leaf Number and Thermal Requirements for Leaf Development in Winter Barley

Vojislava Momčilović, Novo Pržulj, Miloš Nožinić, and Dragan Perović

Abstract Development and growth of cereal leaves significantly affect grain yield since dry matter accumulation depends on the leaf area that intercepts light. Phyllochron is defined as time interval between the appearance of successive leaves on the main stem. The aim of this study was to determine the effect of year and cultivars on final leaf number (FLN) and phyllochron (PHY) in winter barley. Thermal unit was used as timescale with 0 °C as base temperature. Twelve cultivars differing in origin and time to anthesis (early, medium, and late) were tested during six growing seasons (GS), 2002/2003–2007/2008. FLN across cultivars and GSs was 13.5, and PHY was 75.7 GDD leaf^{-1}. The highest FLN across GSs was in the late, six-rowed barley cultivar Kredit (14.7). The lowest FLN was in the early, two-rowed barley cultivar Novosadski 581 (11.3). In regard to earliness, lowest FLN was in the early cultivars (12.9) and highest in the late ones (13.9). The highest PHY across GSs was in the two-rowed cultivar Cordoba, 81.6 GDD leaf^{-1}, the lowest in the two-rowed cultivar Novosadski 581, 71.0 GDD leaf^{-1}. The early cultivars had fast leaf development, the medium cultivars medium, and the late cultivars slow development, 72.5 GDD leaf^{-1}, 75.6 GDD leaf^{-1}, and 78.9 GDD leaf^{-1}, respectively. The tested cultivars showed significant variability in the FLN and PHY, which can be used for selecting most adaptable genotypes for specific conditions.

Keywords Barley (*Hordeum vulgare* L.) • Leaf number • Phyllochron

---

V. Momčilović (✉) • N. Pržulj
Institute of Field and Vegetable Crops, Maksima Gorkog 30, 21000 Novi Sad, Serbia
e-mail: vojislava.momcilovic@nsseme.com

M. Nožinić
Agricultural Institute of the Republic of Srpska, Knjaza Miloša 19, 71000 Banja Luka, RS, Bosnia and Herzegovina

D. Perović
Institute of Field and Vegetable Crops, Julius Kuehn-Institute, Federal Research Centre for Cultivated Plants, Erwin-Baur-Str. 27, 06484 Quedlinburg, Germany

G. Zhang et al. (eds.), *Advance in Barley Sciences: Proceedings of 11th International Barley Genetics Symposium*, DOI 10.1007/978-94-007-4682-4_35,

35.1 Introduction

The life cycle of cereals is divided into two main periods, period till anthesis and grain filling period. From practical point of view and considering the changes at the shoot apex, the period till anthesis can be divided in three phases: leaf initiation (vegetative phase), spikelet initiation (early reproductive phase), and spike growth (late reproductive phase) (Slafer and Whitechurch 2001).

At the time of seedling emergence, the shoot apex has five to seven leaf primordia, and this is considered to be the range in minimum final leaf number (FLN) per wheat plant (Robertson et al. 1996). FLN is determined by the number of primordia initiated before the beginning of the double ridge stage. Kirby (1992) showed that variation in the number of leaves produced by winter wheat in the field in response to sowing date and location could be explained, in part, by differences in exposure to low temperature during the phase when leaf primordia were being initiated. The vernalization response is important for fitting the plant life cycle to the environment in which it is grown, so that it can make the best use of the seasonal opportunities for growth and avoid adverse climatic factors (Tottman 1977). The major effect of vernalization is to shorten the duration of the phase of leaf primordia production (Griffiths et al. 1985).

The phyllochron (PHY) or number of growing degree days (GDD) between a leaf number of n and $n+1$ is a measure of rate of development of plant leaves. It is a measure of plant development that can be used to assess how the plant has responded to environmental conditions or to predict how it is going to respond. PHY is a result of combination of genetic and environmental factors, which interact to produce plant leaves in a predictable manner. Knowledge of the PHY has been widely accepted by crop modelers for predicting plant development and by producers for determining the timing of management practices such as irrigation, fertilizer, and pesticide application. The effect of environmental changes on the rate of leaf emergence in barley must be understood in order to make accurate predictions of the cropping technologies. Kirby (1995) found that the rate of leaf appearance was set early in the life cycle.

Air temperature is the main factor affecting the PHY (Rickman and Klepper 1991). Studies conducted under constant temperature have shown that the PHY increases steadily with temperature from 7.5 to 25.8 °C in wheat (Cao and Moss 1989a) and from 12.5 8 to 27.5 8 °C in barley (Tamaki et al. 2002). Other environmental factors such as day length, water stress, carbohydrate reserves, and nutrient stress have been shown to have little effect on the PHY of grasses (Kiniry et al. 1991). Long photoperiod increases the rate of leaf appearance, i.e., decreases the PHY in wheat and barley (Cao and Moss 1991; Slafer et al. 1994). Water and nitrogen deficits decrease the PHY (Cutforth et al. 1992; Longnecker et al. 1993). Leaf emergence rate in wheat depended on the cultivar (Cao and Moss 1989b; Cutforth et al. 1992) and the sowing date (Cao and Moss 1991; Miralles et al. 2001), resulting in fewer leaves per plant in later sowing dates. In controlled environments under lower light intensities, Slafer (1995) found no effect of light intensity on PHY in wheat.

Researchers have concentrated on understanding how environmental factors, first of all temperature and photoperiod and then water and nutrition, affect the PHY. Only a few studies have evaluated cultivar effect on the PHY. In this research, we studied the effect of cultivar/genotype and year on FLN number and PHY of winter barley.

35.2 Materials and Methods

35.2.1 Cultivars and Crop Management

Twelve barley cultivars (Novosadski 525 – SRB, Novosadski 581 – SRB, Sonate – DEU, Boreale – FRA, Monacco – FRA, Cordoba – DEU, Kompolti-4 – HUN, Skorohod – RUS, Plaisant – FRA, Nonius – SRB, Novator – UKR, and Kredit – CZE) which differed in origin, pedigree, and agronomic traits were used in this study. All cultivars were or still are commercially grown.

A 6-year experiment was conducted from 2002/2003 to 2007/2008 growing seasons (GS) at the experiment field of Institute of Field and Vegetable Crops in Novi Sad (45 °20′N, 15 °51′E, altitude 86 m) on a chernozem soil and under rainfed conditions. The plots were 1.2 m wide and 10 m long, with 0.20 m spacing between rows. The experimental plots were planted on 9 October 2002, 15 October 2003, 22 October 2004, 18 October 2005, 12 October 2006, and 13 October 2007, with soybean as the preceding crop in all GSs. Planting density in all GSs was 300 viable seeds per m^2 for six-rowed barley and 350 viable seeds per m^2 for two-rowed barley. The experiment was conducted in a randomized complete block design with three replications in each year. To avoid negative effects of diseases and pests, the experiments were sprayed with the fungicide Tilt 250 EC in Zadoks phase 64 and with the insecticide Karate Zeon applied when needed. Weed control was performed by hand.

To determine the FLN on the main stem, recording was done according to the Haun scale on three tagged plants in each replication. FLN was determined as the number of leaves on the main stem, including the flag leaf. GDD was calculated for each of the phases as $T_n = [(T_7 + T_{14} + 2T_{21})]/4$, where T_7, T_{14}, and T_{21} are temperatures at 7 a.m., 2 and 9 p.m., respectively (Pržulj 2001). Base temperature was 0 °C.

35.2.2 Statistical Analysis

All data were subjected to the analysis of variance using Statistica 9.0 program (StatSoft, Tulsa, OK, USA). When differences among earliness groups (early, medium early, late), duration of developmental phases, and agronomic traits were tested, four cultivars from each group were considered as replication as samples for detection of developmental phases.

35.2.3 Weather Conditions

Weather data were obtained from the meteorological station located at the experiment field. Temperature and precipitation were collected daily. The average temperatures at the level of the long-term average and abundant rainfall in October 2002 allowed a rapid crop development and fast progression through vegetative phase. During the early reproductive phase in November, average temperatures were at the level of summer temperatures, and there occurred a water deficit. The early reproductive phase continued in the spring under conditions of extreme drought and low temperatures. Grain filling proceeded under drought conditions and high temperatures. Excessive rainfall, several-fold above the annual average, and average temperatures accompanied the emergence stage in the 2003/2004 GS. Temperatures were somewhat lower during the vegetative phase in late October and early November. During the early stages of the double ridge stage in late November, temperatures were extremely high and precipitation was limited. The completion of the generative phase next spring and grain filling stage unfolded under normal conditions. In the 2004/2005 GS, both temperature and rainfall conditions were favorable during the vegetative phase. In the fall, favorable weather conditions lasted 3 weeks longer than the average data, which helped the generative phase to proceed normally. The remainder of the GS and the jointing and grain filling stages continued under favorable conditions. In the 2005/2006 GS, barley germination and the vegetative stage took place under a drought, while the generative phase in November and December unfolded under favorable conditions. The generative phase that continued next spring took place under conditions of high temperature fluctuations, while high temperatures and water deficit coincided with the jointing stage. Drought and high temperatures in October and November 2006 obstructed winter barley emergence and the development of the vegetative phase. Favorable rainfall and temperatures during December and the subsequent winter months enabled the plants to accumulate the necessary amounts of effective temperatures and compensate for the previous delay in development. The end of the generative phase, jointing and filling stages took place under high temperatures and excess rainfall. In the 2007/2008 GS, the germination, vegetative phase, and the beginning of the generative phase unfolded under somewhat low temperatures and a large surplus of water. The generative phase, jointing and grain filling continued in the spring of 2008 under favorable conditions.

35.3 Results and Discussion

Phenological development and PHY are results of genetics and many environmental factors (McMaster 2005). FLN and PHY were controlled by all three factors, cultivar, year, and their interaction (Table 35.1). Contribution of year in FLN variation was highest, about 74%. It means that the tested cultivars were genetically similar in FLN. Low value of interaction showed stability of FLN from year to year. Considered across the GSs, the early cultivar Novosadski 581 had the

Table 35.1 Mean squares of final leaf number (FLN) per main stem and phyllochron (PHY) for winter barley

Sourcc	df	FLN	PHY
Cultivar	11	13.98**	176.78**
Year	5	10.17**	1929.41**
C ×Y	55	0.98**	41.54**
% of variance components			
Cultivar		17.20	23.09
Year		73.96	30.18
C xY		7.19	40.45
Heritability		0.933	0.77

**Significant at the 0.01 level

lowest and the late cultivar Kredit the highest FLN (Table 35.2). In Novosadski 581, early maturity was due to a reduction in FLN (Table 35.2) and a short PHY (Table 35.3). Juskiw and Helm (2003) found that, in spring barley, earliness was due to accelerated postanthesis growth rather than reduction in FLN and PHY. Even though the cultivar × year effect for FLN was significant and participated in total variation with 7.2% (Table 35.1), the FLN variability from year to year was rather small (Table 35.2).

Although differences in FLN and PHY were observed among the cultivars (Table 35.1, 35.2, 35.3), no relationship could be established in the rank of the cultivars for the FLN and PHY. For example, the early cultivar Kompolti-4 had one of the highest FLNs in 2002/2003 GS and one of the lowest in 2005/2006 GS (Table 35.2). The average FLN for the barley main stem was 13.5.

If it is accepted that PHY is shorter in the late sowing (Miglietta 1991), one can conclude that sowing was late in 2003/2004, 2004/2005, and 2005/2006 since PHY was 70.9, 71.7, and 70.9 GDD, respectively (Table 35.3). However, sowing was in regular time, October, and three GSs were different; 2003/2004 was with water excess and optimal temperature till double ridge (DR) and water deficit and optimal temperature from DR till jointing (J); 2004/2005 was with water and temperature optimum till DR but excess water and very low temperature from DR till J, and 2005/2006 was with water deficit and optimum temperature till DR and excess water and low temperature from DR till J. So, shorter PHY was a result of unfavorable environments, first of all temperature and water from sowing till flag leaf appearance. In time of emergence, elevated temperature in soil space of seed only shortened the duration of germination and seedling emergence and had no effect on either the PHY (McMaster et al. 2003) or phenological development (McMaster and Wilhelm 2003).

The early cultivars Skorohod and Novosadski 581 had lowest GDD requirements, or shortest PHY, and the late cultivars Novator, Kredit, Monaco, and Cordoba had highest GDD requirements or longest PHY (Table 35.3). Across GSs and maturity classes, the early maturity group completed the FLN fastest (Fig. 35.1a). Quadratic equation fitted best the relationship between GDD requirement and FLN per main stem, with $R^2 > 0.99$. Also, quadratic equation fitted best the relationship

Table 35.2 FLN per main stem of 12 winter barley cultivars during six GSs (2002/2003–2007/2008)

	GS						
Cultivar	2002/2003	2003/2004	2004/2005	2005/2006	2006/2007	2007/2008	Average
Kompolti-4	14	15	13	13	15	14	14.0
Skorohod	13	14	12	13	14	14	13.3
Novosadski 525	13	13	12	13	13	13	12.8
Novosadski 581	10	12	10	12	12	12	11.3
Plaisant	15	13	13	13	15	14	13.8
Gotic	15	13	14	13	15	14	14.0
Sonate	13	13	12	13	14	13	13.0
Boreale	13	13	12	13	14	13	13.0
Novator	15	15	13	14	15	14	14.3
Kredit	15	15	14	15	15	14	14.7
Monaco	13	13	13	13	14	14	13.3
Cordoba	14	14	13	14	14	15	14.0
Average	13.6	13.6	12.6	13.2	14.2	13.7	13.5
LSD		*Cultivar*	*Year*	*C ×Y*	*CV 1.9%*		
	0.05	*0.17*	*0.12*	*0.42*			
	0.01	*0.23*	*0.16*	*0.56*			

Table 35.3 Phyllochron for 12 winter barley cultivars during six growing seasons (GS) (2002/2003–2007/2008)

	GS						
Cultivar	2002/2003	2003/2004	2004/2005	2005/2006	2006/2007	2007/2008	Average
Kompolti-4	73.0	63.2	68.8	73.7	87.9	71.9	73.1
Skorohod	70.1	68.0	68.1	63.9	86.0	71.3	71.2
Novosadski 525	72.3	71.9	72.9	70.7	90.0	70.5	74.7
Novosadski 581	70.7	65.4	68.6	68.4	83.5	69.4	71.0
Plaisant	65.9	70.9	70.0	69.6	88.6	80.6	74.3
Gotic	67.7	68.9	70.4	70.5	89.6	85.1	75.4
Sonate	77.8	68.7	75.0	69.7	91.2	75.4	76.3
Boreale	77.3	69.5	72.6	70.6	90.3	76.2	76.1
Novator	71.8	73.3	70.4	70.7	90.8	85.4	77.1
Kredit	76.8	72.3	69.8	72.0	94.3	86.9	78.7
Monaco	82.1	78.0	70.7	74.2	92.8	75.6	78.9
Cordoba	81.1	81.3	83.3	77.2	94.0	72.6	81.6
Average	73.9	70.9	71.7	70.9	89.9	76.7	75.7
LSD		*Cultivar*	*Year*	*C × Y*	*CV*		
	0.05	*0.9*	*0.7*	*2.7*	*1.9%*		
	0.01	*1.2*	*0.9*	*3.0*			

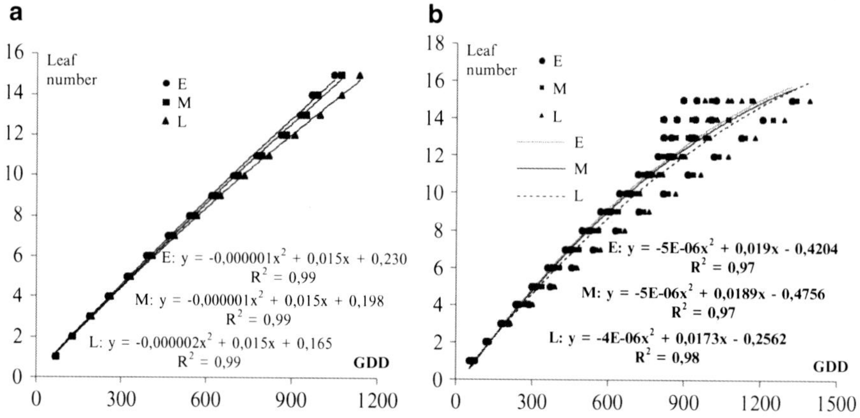

Fig. 35.1 (**a**) Leaf number development in four early (E), four medium (M), and for late (L) winter barley cultivars across six GSs. (**b**) Leaf number development of three winter barley maturity classes (E – early, M – medium, L – late) in different GSs. Each point represents the average of four cultivars belonging to a maturity group in individual GS. There are six points for each leaf per maturity group

Table 35.4 Simple correlations between final leaf number (FLN) and temperature and precipitation during some phenological growth stages (PGS)

PGS	GDD from E till DR	Precipitation from E till DR	GDD during DR	Precipitation during DR	GDD during J	Precipitation during J
FLN	0.35**	−0.25*	0.34**	−0.27*	0.79*	−0.16

E emergence; *DR* double ridge; *J* jointing

between leaf number development across cultivars in the same maturity group and GS, with $R^2>0.97$ (Fig. 35.1a, b).

The early cultivars showed no consistency in PHY to earliness relationship, while the medium early and late two-rowed barley cultivars were found to have a longer PHY. While Frank and Bauer (1995) found two-rowed spring barley cultivars to have a short PHY, Juskiw et al. (2001) found them to have a long PHY. Cao and Moss (1989c) pointed out that PHY in spring barley varied in dependence of genotype and combination of temperature and day length, but that it invariably increased as temperature increased or day length decreased.

In this study, the average PHY was 75.7 GDD, with a range of 71.0–81.6 GDD (Table 35.3). In relation to spring barley, this PHY was longer than the mean of 64.5 GDD for Alaska growing conditions (Dofing and Karlsson 1993) and shorter than the mean of 77.2 GDD for North Dakota growing conditions (Frank and Bauer 1995). McMaster et al. (1992), pooling ten winter wheat cultivars across 19 site-years in different environments of Central Great Plains, USA, and different production technologies, observed the PHY of 107 GDD. Temperature and photoperiod are the two most important factors that affect PHY in wheat (Cao and Moss 1989a, b; Kirby 1995; Slafer and Rawson 1995). In our investigation, the FLN was positively correlated with GDD accumulated till flag leaf completion (Table 35.4), while the effect of precipitation was less important.

The time from sowing to emergence was 138 GDD (data not shown), which was about two PHYs. Juskiw et al. (2001) attribute one PHY for the development of the coleoptiles and one PHY for the first true leaf.

Juskiw et al. (2001) found that PHY is prone to error because temperature and day length are known to affect the leaf appearance rate. Our results confirm this statement, where error for PHY of specific leaf was rather high, 13.63–33.47% (Table 35.4). Leaves 13th and 14th were exceptions, with the errors of 8.85 and 4.69%, respectively. There was a paradox associated with these two leaves: the lower the degree of freedom, the lower the error. It might be due to the lower variability for leaves 13th and 14th of the remaining cultivars since those that had lower values of PHY (Novosadski 581, Novosadski 525, Skorohod, Sonate, and Boreale) did not have 13th and 14th leaves (Table 35.2) and were not included in statistical calculations. The values of PHY heritability were rather high, although cultivar participation in total variation was less than 10% (Table 35.5).

When the cultivars were sorted according to earliness, each cultivar representing a replication, then maturity classes and years determined FLN and PHY (Table 35.6). The interaction maturity class × year was not significant either for FLN or PHY, i.e., the early cultivars always had lowest FLN and shortest PHY and the late ones always had highest FLN and longest PHY regardless of the year (Table 35.7, 35.8). McMaster et al. (1992) found no influence of cultivar and maturity class on the PHY in ten cultivars of winter wheat.

Across the studied GSs, the early cultivars had 12.9, medium early 13.5, and late 13.9 FLN of main stem (Table 35.7). The maturity classes differed significantly in PHY, the early group having the shortest PHY (72.5 GDD leaf^{-1}), the late one the longest (78.9 GDD leaf^{-1}) (Table 35.8).

The rate of change of leaf appearance as a function of leaf number increased throughout the growing season from about 50° GDD for the first leaf to 100 GDD for last leaves. The rate of change fitted the quadratic equation with the R^2 value >0.93 (Fig. 35.2a). The rate of leaf appearance depended greatly on GS which participated in the total variation with >30% (Table 35.1), and a summary equation for different GSs is not recommended since R^2 is <0.58 (Fig. 35.2b). Itoh and Sano (2006) found that the quartic polynomial regression fitted well the change of PHY of rice grown under controlled conditions.

Some studies indicated that the rate of leaf emergence or PHY was constant in both wheat and barley from seedling emergence to the appearance of the flag leaf (Kamali and Boyd 2000; Juskiw et al. 2001; Juskiw and Helm 2003). Other studies showed that PHY varied with plant development and that the pattern of leaf emergence was bilinear rather than linear. A change in PHY may occur between leaves 6 and 8 (Jamieson et al. 1995; Miralles et al. 2001). Flood et al. (2000) suggested that variation in PHY may be due to ontogenetic changes, the changes occurring around the double ridge stage. Using a single cultivar of day length-sensitive wheat and another of barley, Miralles and Richards (2000) found a linear relationship under long days and a bilinear relationship for leaf emergence under short days.

Table 35.5 Percentages of the components of variance and heritability for the phyllochron of individual main stem leaves across 12 winter barley cultivars and six growing seasons (2002/2003–2007/2008)

Leaf	Total df	Variance component				Percentage of variance components				h_b^2
		σ_G^2	σ_Y^2	σ_{GY}^2	Error	Cultivar	Year	C × Y	Error	
1st	215	1.55	68.46	0.00	35.22	1.47	65.06	0.00	33.47	0.44
2nd	215	3.34	84.47	0.00	28.73	2.86	72.49	0.00	24.65	0.69
3rd	215	2.70	74.87	2.85	16.10	2.80	77.57	2.96	16.68	0.66
4th	215	4.81	79.06	5.56	17.74	4.49	73.77	5.19	16.55	0.72
5th	215	5.13	57.55	9.12	13.20	6.04	67.70	10.73	15.53	0.69
6th	215	5.51	49.41	8.45	16.29	6.92	62.03	10.61	20.45	0.70
7th	215	4.93	49.48	12.62	12.00	6.23	62.61	15.97	15.19	0.64
8th	215	5.44	50.86	11.49	13.70	6.67	62.42	14.10	16.81	0.67
9th	215	3.86	53.77	13.82	12.57	4.60	63.99	16.45	14.96	0.56
10th	209	5.74	50.51	13.33	14.50	6.83	60.07	15.85	17.25	0.65
11th	209	4.75	54.41	18.32	12.23	5.29	60.65	20.42	13.63	0.56
12th	182	6.56	72.76	19.85	16.61	5.67	62.84	17.14	14.35	0.61
13th	104	15.79	97.61	36.52	14.56	9.60	59.35	22.20	8.85	0.70
14th	53	11.67	316.73	5.81	16.43	3.33	90.33	1.66	4.69	0.84
Flag leaf	215	20.77	114.95	36.30	32.95	10.13	56.08	17.71	16.08	0.72

Table 35.6 Mean squares of final leaf number (FLN) per main stem and phyllochron (PHY) for three maturity classes (E – early, M – medium, L – late) of winter barley

Source	df	FLN	PHY
Maturity class	2	8.56**	257.88**
Year	5	3.39**	643.35**
Maturity class × year	10	0.58ns	12.61ns
% of variance components			
Maturity class		14.90	28.28
Year		20.82	30.21
Maturity class × year		25.91	0.00
Heritability		0.71	0.96

ns nonsignificant
**Significant at the 0.01 level

Table 35.7 Final leaf number per main stem for three maturity classes (E – early, M – medium, L – late) across six growing seasons (GS) (2002/2003–2007/2008)

Maturity class	GS 2002/2003	2003/2004	2004/2005	2005/2006	2006/2007	2007/2008	Average
E	12.5	13.5	11.8	12.8	13.5	13.2	12.9
M	14.0	13.5	12.7	13.0	14.5	13.5	13.5
L	14.2	13.8	13.2	14.0	14.5	14.2	13.9
Average	13.6	13.6	12.6	13.2	14.2	13.7	13.5
		Maturity cl.	*Year*	*Mc × Y*	*CV*		
LSD	*0.05*	*0.54*	*0.76*	*1.31*	*6.9%*		
	0.01	*0.71*	*1.01*	*1.75*			

Table 35.8 Phyllochron for the three maturity classes (E – early, M – medium, L – late) across six growing seasons (GS) (2002/2003–2007/2008)

Maturity class	GS 2002/2003	2003/2004	2004/2005	2005/2006	2006/2007	2007/2008	Average
E	71.5	67.1	69.6	69.2	86.9	70.8	72.5
M	72.2	70.5	72.0	70.1	89.9	79.3	75.6
L	77.9	75.3	73.5	73.5	93.0	80.1	78.9
Average	73.9	70.9	71.7	70.9	89.9	76.7	75.7
		Maturity cl.	*Year*	*Mc × Y*	*CV*		
LSD	*0.05*	*2.2*	*3.1*	*5.5*	*5.1%*		
	0.01	*3.0*	*4.2*	*7.3*			

35.4 Conclusion

The average final number of leaves on the main stem of winter barley under the conditions of the Pannonian Plain was 13.5, with phyllochron average of 75.7 GDD leaf^{-1}. The early cultivars had one leaf less and the phyllochron shorter by 6.4 °C than the late cultivars. Earliness is rather the result of leaf number reduction than grain filling reduction. The rate of leaf appearance and phyllochron fitted the quadratic equation.

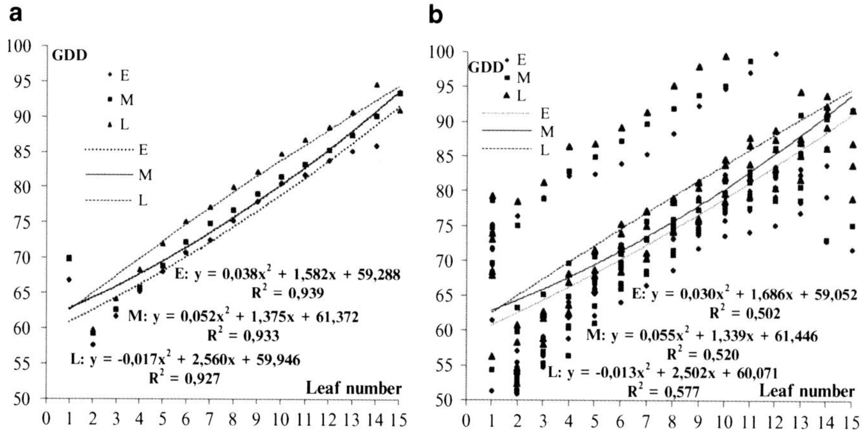

Fig. 35.2 (**a**) Phyllochron in four early (E), four medium (M), and four late (L) winter barley cultivars across six growing seasons. (**b**) Phyllochron in three winter barley maturity classes (E – early, M – medium, L – late) in the six growing seasons. Each symbol for any leaf represents the average of four cultivars that belong to each maturity class in one growing season

References

Cao, W., & Moss, D. N. (1989a). Temperature effect on leaf emergence and phyllochron in wheat and barley. *Crop Science, 29*, 1018–1021.

Cao, W., & Moss, D. N. (1989b). Day length effect on leaf emergence and phyllochron in wheat and barley. *Crop Science, 29*, 1021–1025.

Cao, W., & Moss, D. N. (1989c). Temperature and day length interaction on phyllochron in wheat and barley. *Crop Science, 29*, 1046–1048.

Cao, W., & Moss, D. N. (1991). Phyllochron change in winter wheat with planting date and environmental changes. *Agronomy Journal, 83*, 396–401.

Cutforth, H. W., Jame, Y. W., & Jefferson, P. G. (1992). Effect of temperature, vernalization and water stress on phyllochron and final main-stem leaf number of HY320 and Neepawa spring wheats. *Canadian Journal of Plant Science, 72*, 1141–1151.

Dofing, S. M., & Karlsson, M. G. (1993). Growth and development of uniculm and conventional-tillering barley lines. *Agronomy Journal, 85*, 58–61.

Flood, R. G., Moody, D. B., & Cawood, R. J. (2000). The influence of photoperiod on barley development. *Cereal Research Communications, 28*, 371–378.

Frank, A. B., & Bauer, A. (1995). Phyllochron differences in wheat, barley and forage grasses. *Crop Science, 35*, 19–23.

Griffiths, F., Lyndon, R. F., & Bennett, M. D. (1985). The effects of vernalization on the growth of the wheat shoot apex. *Annals of Botany, 56*, 501–511.

Itoh, Y., & Sano, Y. (2006). Phyllochron dynamics under controlled environments in rice. *Euphytica, 150*, 87–95.

Jamieson, P. D., Brooking, I. R., Porter, J. R., & Wilson, D. R. (1995). Prediction of leaf appearance in wheat: a question of temperature. *Field Crops Research, 41*, 35–44.

Juskiw, P. E., & Helm, J. H. (2003). Barley response to seeding date in central Alberta. *Canadian Journal of Plant Science, 83*, 275–281.

Juskiw, P. E., Jame, Y. W., & Kryzanowski, L. (2001). Phenological development of spring barley in a short-season growing area. *Agronomy Journal, 93*, 370–379.

Kamali, M. R. J., & Boyd, W. J. R. (2000). Quantifying the growth and development of commercial barley cultivars over two contrasting growing seasons in Western Australia. *Australian Journal of Agricultural Research, 51*, 487–501.

Kiniry, J. R., Rosenthal, W. D., Jackson, B. S., & Hoogenboom, G. (1991). Predicting leaf development of crop plants. In T. Hodges (Ed.), *Predicting crop phenology* (pp. 29–42). Boca Raton: CRC Press.

Kirby, E. J. M. (1992). A field study of the number of main shoot leaves in wheat in relation to vernalization and photoperiod. *Journal of Agricultural Science, 118*, 271–278.

Kirby, E. J. M. (1995). Factors affecting rate of leaf emergence in barley and wheat. *Crop Science, 35*, 11–19.

Longnecker, N., Kirby, E. J. M., & Robson, A. (1993). Leaf emergence, tiller growth, and apical development of nitrogen-deficient spring wheat. *Crop Science, 33*, 154–160.

McMaster, G. S. (2005). Phytomers, phyllochrons, phenology and temperate cereal development. *The Journal of Agricultural Science (Cambridge), 43*, 1–14.

McMaster, G. S., & Wilhelm, W. W. (2003). Phenological responses of wheat and barley to water and temperature: improving simulation models. *The Journal of Agricultural Science (Cambridge), 141*, 129–147.

McMaster, G. S., Wilhelm, W. W., & Morgan, J. A. (1992). Simulating winter wheat shoot apex phenology. *The Journal of Agricultural Science (Cambridge), 119*, 1–12.

McMaster, G. S., Wilhelm, W. W., Palic, D. B., Porter, J. R., & Jamieson, P. D. (2003). Spring wheat leaf appearance and temperature: extending the paradigm? *Annals of Botany, 91*, 697–705.

Miglietta, F. (1991). Simulation of wheat ontogenesis: I. Appearance of main stem leaves in the field. *Climate Research, 1*, 145–150.

Miralles, D. J., & Richards, R. A. (2000). Responses of leaf and tiller emergence and primordium initiation in wheat and barley to interchanged photoperiod. *Annals of Botany, 85*, 655–663.

Miralles, D. J., Ferro, B. C., & Slafer, G. A. (2001). Developmental responses to sowing date in wheat, barley and rapeseed. *Field Crops Research, 71*, 211–223.

Pržulj, N. (2001). Cultivar and year effect on grain filling of winter barley. *Plant Breeding and Seed Science, 45*(2), 45–58.

Rickman, R. W., & Klepper, B. L. (1991). Tillering in wheat. In T. Hodges (Ed.), *Predicting crop phenology* (pp. 73–83). Boston: CRC.

Robertson, M. J., Brooking, I. R., & Ritchie, J. T. (1996). Temperature response of vernalization in wheat: Modelling the effect on the final number of mainstem leaves. *Annals of Botany, 78*, 371–381.

Slafer, G. A. (1995). Wheat development as affected by radiation at two temperatures. *Journal of Agronomy and Crop Science, 175*, 249–263.

Slafer, G. A., & Rawson, H. W. (1995). Rates and cardinal temperatures for processes of development in wheat: Effect of temperature and thermal amplitude. *Australian Journal of Plant Physiology, 22*, 913–926.

Slafer, G. A., & Whitechurch, E. M. (2001). Manipulating wheat development to improve adaptation and to search for alternative opportunities to increase yield potential. In M. P. Reynolds, J. I. Ortiz-Monasterio, & A. McNab (Eds.), *Application of physiology in wheat breeding* (pp. 160–170). Mexico City: CIMMYT.

Slafer, G. A., Connor, D. J., & Halloran, G. M. (1994). Rate of leaf appearance and final number of leaves in wheat: effects of duration and rate of change of photoperiod. *Annals of Botany, 74*, 427–436.

Tamaki, M., Kondo, S., Itani, T., & Goto, Y. (2002). Temperature responses of leaf emergence and leaf growth in barley. *The Journal of Agricultural Science (Cambridge), 138*, 17–20.

Tottman, D. R. (1977). The identification of growth stages in winter wheat with reference to the application of growth regulating herbicides. *Annals of Applied Biology, 87*, 213–224.

Chapter 36
Characterization of the Barley (*Hordeum vulgare* L.) miRNome: A Computational-Based Update on MicroRNAs and Their Targets

Moreno Colaiacovo, Cristina Crosatti, Lorenzo Giusti, Renzo Alberici, Luigi Cattivelli, and Primetta Faccioli

Abstract MicroRNAs (miRNAs) are a class of endogenous noncoding small RNAs acting on gene regulation at posttranscriptional level, a phenomenon known in plants as posttranscriptional gene silencing. miRNAs are known to play key regulatory roles in plant response to stress, besides being involved in development and morphogenesis. To extend and to update information on miRNAs and their targets in barley and to identify candidate polymorphisms at miRNA target sites, the features of previously known plant miRNAs have been used to systematically search for barley miRNA homologues and targets in the publicly available EST database. Matching sequences have then been related to UniGene clusters on which most of this study was based. One hundred fifty-six microRNA mature sequences belonging to 50 miRNA families have been found to significantly match at least one EST sequence in barley. The predicted miRNA targets were ascribed to different pathways, among others the response to abiotic stress. To verify experimentally the barley miRNAs and their involvement in stress response, the barley miRNome of plants exposed to low temperatures has been characterized by a deep sequencing approach on Illumina GAIIx. Many of the known miRNAs have been found as different isomeric variants, so-called isomiRs, which might increase the target repertoire of the miRNA gene they derive from.

Keywords Barley • MicroRNA • Abiotic stress response

M. Colaiacovo • C. Crosatti • L. Giusti • R. Alberici • L. Cattivelli (✉) • P. Faccioli
CRA Genomics Research Centre, Via San Protaso 302,
29017 Fiorenzuola d'Arda (PC), Italy
e-mail: luigi.cattivelli@entecra.it

G. Zhang et al. (eds.), *Advance in Barley Sciences: Proceedings of 11th International Barley Genetics Symposium*, DOI 10.1007/978-94-007-4682-4_36,

36.1 Background

MicroRNAs (miRNAs) are a class of noncoding small RNAs with fundamental roles in key plant biological processes such as development, signal transduction, and environmental stress response (Bartel 2004). Many plant species have been investigated during recent years for miRNA identification and characterization. However, extensive studies describing the organization of miRNA families in barley are lacking, despite its economic importance and its role as model species for Triticeae (Schreiber et al. 2009).

The conservation of miRNA sequences across species provides a powerful tool for the identification of novel miRNA genes based on homology with miRNAs previously described in other species. Without genome sequence information, a powerful alternative data source comes from ESTs. The perfect or near-perfect complementarities between an miRNA and its target mRNA, which is a peculiar feature of plant miRNAs, give a powerful tool for the identification of target genes through BLAST analysis of miRNA mature sequences vs EST/genomic sequences. The correct binding of miRNA to its cognate mRNA is critical for regulating the mRNA level and protein expression. Therefore, small polymorphisms in miRNA targets can have a relevant effect on gene and protein expression and represent a type of genetic variability that can influence agronomical traits.

To extend and to update information about miRNAs and their targets in barley and to identify candidate polymorphisms at miRNA target sites, barley EST sequences have been screened and related to UniGene clusters. Mining SNPs from ESTs allows the exploitation of genetic variability based on published sequences, and the analysis of UniGene clusters can be very helpful for this purpose (Picoult-Newberg et al. 1999).

This study provides an update of the information on barley miRNAs and their targets representing a foundation for future studies. The majority of previously known plant microRNAs have homologues in barley with expected important roles during development, nutrient deprivation, biotic and abiotic stress response, and other important physiological processes. Putative polymorphisms at miRNA target sites have been identified, and they represent an interesting source for the identification of functional genetic variability (Colaiacovo et al. 2010).

36.2 Barley miRNAs

Mature miRNA sequences have been used as queries for BLAST search against *Hordeum vulgare* ESTs (Zhang et al. 2006). One hundred fifty-six microRNA mature sequences belonging to 50 miRNA families have been found to significantly match at least one EST sequence in barley. Some ESTs have matched to more than one miRNA belonging either to the same family or to different families. The first case can be due to the high level of similarity among mature sequences from different members of the same family, while ESTs matching to different miRNA families could represent examples of multi-microRNA-based control. Transcripts targeted by more than one miRNA have also been found in other plant species such as rice (Zhou et al. 2010).

To identify and annotate potential microRNA-regulated genes in barley, the 855 matching ESTs were related to UniGene clusters. Clusters annotated as protein-coding sequences were then selected for subsequent analysis. A total of 121 different UniGene clusters putatively representing the targets for 37 miRNA families have been found. In some cases, different targets for a specific miRNA are members of the same gene family (e.g., miR156-SBP family), while in other cases, there is no evident relationship among the putative targets of a given miRNA (e.g., miR1121). Although several of the candidate miRNA/target pairs here identified have the same functional annotation reported in previously studied species, some putative novel microRNA/target pairs have been discovered (Dryanova et al. 2008). Most of the novel miRNA/target pairs refer to miRNAs recently discovered and thus probably less studied (i.e., miR1120, miR1122, miR1134).

Transcription factor families comprise most of the highly conserved miRNA targets such as SBP family for miRNA 156, AP2 family for miR172, GRAS family for miR171, myb family for miR159, GRF family for miR396, and ARF family for miR160. These results confirmed what was previously observed in Triticeae and in other species (Dryanova et al. 2008). Conserved miRNAs also target genes involved in their own biogenesis and function. miRNAs regulate gene expression also by targeting enzymes of the ubiquitination pathway.

The predicted targets have been grouped into functional categories. Biological processes known to be regulated by miRNAs, such as development and response to biotic and abiotic stress, have been highlighted both in known and in novel targets. Moreover, most of the molecular functions are related to transcriptional regulation and DNA/nucleotide binding in both groups. For some UniGene clusters, the annotation was related to transcribed genes rather than protein-coding sequences. These UniGenes could represent miRNA-coding genes as shown by other authors (Guo et al. 2007; Xie et al. 2007).

36.3 Genetic Variation at miRNA Target Sites

EST-derived SNPs can provide a rich source of biologically useful genetic variation due to the redundancy of gene sequence, the diversity of genotypes present in the databases, and the fact that each putative polymorphism is associated with an expressed gene. Variations both in functional regions of putative miRNAs (mature sequence) and at miRNA target sites have been found.

Hv.5064, the candidate for miR1137 coding sequence, has been tested for modifications of pre-miRNA structure due to a base substitution in position 13 (C/G, Fig. 36.1). To evaluate the possible impact of this SNP on pre-miRNA secondary structure, Gibbs free energy (ΔG) and MFEI from each version of pre-miRNA were calculated using mfold program. Data in Fig. 36.1 show the structural

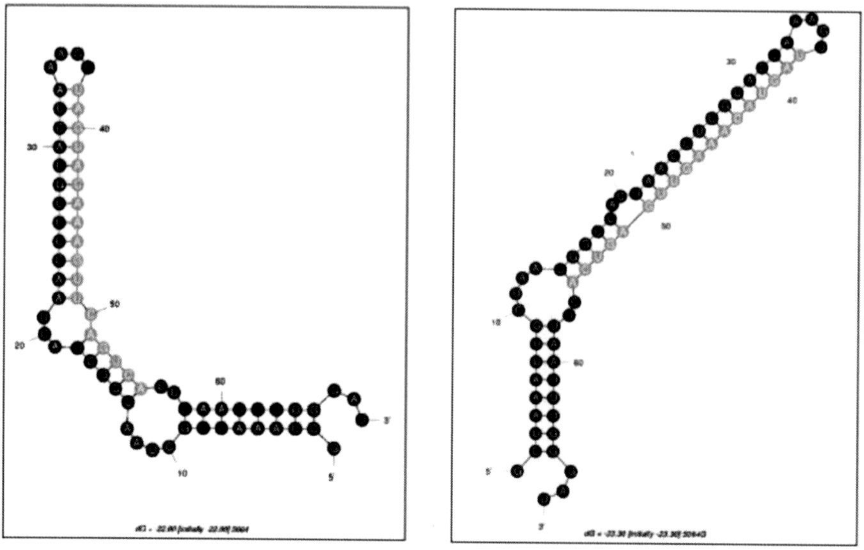

	ΔG (kcal/mol)	length (nt)	%GC	MFEI	mismatch miRNA/miRNA*	arm of the mature sequence
miR1137 (C)	-22.0	68	32.35	1.00	2	3'
miR1137 (G)	-23.3	68	32.35	1.06	2	3'

Fig. 36.1 A high number of putative polymorphisms were detected after comparison among EST sequences inside UniGene clusters, without any selection against false positives. Some of these nucleotide variation could be due to sequencing errors or related to very similar genes belonging to a specific family, nevertheless when the SNPs/indels rely on two or more copies of independent sequences it can be considered a good candidate for a true positive polymorphic target site (Batley et al. 2003). For example, a polymorphism in miRNA 408 target site detected by AutoSNP in contig 2094 (coding for a plastocyanin) is based on sequences from two different cultivars reporting the same allelic variant as part of a haplotype where a SSR (Simple Sequence Repeat) polymorphism is located upstream the target sequence

variation obtained when moving from "C variant" to "G variant" with a higher MFEI for the second one and thus a greater stability of the molecule (miRNA-miRNA* pairing enhanced in the G variant).

The Squamosa promoter-binding protein (SBP) is a known target family for miR156. miR156 performs a critical function in mediating developmental processes, and it is also related to the response to biotic stress. The screening of barley databases has identified two SBP genes targeted by miR156 for which two nucleotide variations occur in critical positions (11–12). If these SNPs will be experimentally confirmed, they could have the effect of destabilizing the interaction between the miRNA and the mRNA, which could consequently avoid cleavage and lead to phenotypical variations in developmental features or in the resistance to viral infection. SNPs have been identified also in other conserved miRNA targets and in previously not reported miRNA targets.

36.4 Conclusions

This study has thus provided an update of the information on barley miRNAs and their targets representing a foundation for future studies. Novel putative target genes have been identified, and most of them are involved in stress and hormone response. In particular, protein kinases such as protein kinase C and serine/threonine kinase, known to be important regulator on abiotic stress resistance, are largely present in novel miRNA/target pairs identified. The results have also shown that miRNA target sites can be an interesting source for the identification of functional genetic variability, representing an interesting source of candidate molecular markers for application in barley breeding.

References

Bartel, D. P. (2004). MicroRNAs: Genomics, biogenesis, mechanism and function. *Cell, 116*, 281–297.

Batley, J., Barker, G., O'Sullivan, H., Edwards, K. J., & Edwards, D. (2003). Mining for single nucleotide polymorphisms and insertions/deletions in maize expressed sequence tag data. *Plant Physiology, 132*, 84–91.

Colaiacovo, M., Subacchi, A., Bagnaresi, P., Lamontanara, A., Cattivelli, L., & Faccioli, P. (2010). A snapshot into microRNA-based gene regulation in barley (*Hordeum vulgare* L.). *BMC Genomics, 11*, 595.

Dryanova, A., Zakharov, A., & Gulick, P. J. (2008). Data mining for miRNAs and their targets in the Triticeae. *Genome, 51*, 433–443.

Guo, Q., Xiang, A., Yang, Q., & Yang, Z. (2007). Bioinformatic identification of microRNAs and their target genes from *Solanum tuberosum* expressed sequence tags. *Chinese Science Bulletin, 52*(17), 2380–2389.

Picoult-Newberg, L., Ideker, T. E., Pohl, M. G., Taylor, S. L., Donaldson, M. A., Nickerson, D. A., & Boyce-Jacino, M. (1999). Mining SNPs from EST databases. *Genome Research, 9*, 167–174.

Schreiber, A. W., Sutton, T., Caldo, R. A., Kalashyan, E., Lovell, B., Mayo, G., Muehlbauer, G. J., Druka, A., Waugh, R., Wise, R. P., Langridge, P., & Baumann, U. (2009). Comparative transcriptomics in the Triticeae. *BMC Genomics, 10*, 285.

Xie, F. L., Huang, S. Q., Guo, K., Xiang, A. L., Zhu, Y. Y., Nie, L., & Yang, Z. M. (2007). Computational identification of novel microRNAs and targets in *Brassica napus*. *FEBS Letters, 581*, 1464–1474.

Zhang, B., Pan, X., Cannon, C. H., Cobb, G. P., & Anderson, T. A. (2006). Conservation and divergence of plant microRNA genes. *The Plant Journal, 46*, 243–259.

Zhou, M., Gu, L., Li, P., Song, X., Wei, L., Chen, Z., & Cao, X. (2010). Degradome sequencing reveals endogenous small RNA targets in rice (*Oryza sativa* L., ssp. *indica*). *Frontiers of Biology, 5*(1), 67–90.

Chapter 37
The Construction of Molecular Genetic Map of Barley Using SRAP Markers

Leilei L. Guo, Xianjun J. Liu, Xinchun C. Liu, Zhimin M. Yang, Deyuan Y. Kong, Yujing J. He, and Zongyun Y. Feng

Abstract Sequence-related amplified polymorphism (SRAP) markers, a novel PCR-based molecular marker technique, were successfully applied in map construction, cultivar identification, diversity evaluation, comparative genomics, and gene location of different plant species. The molecular genetic map of SRAP markers in Steptoe/Morex DH population was constructed in this study using 216 SRAP markers and 312 SSR markers. Twenty-one of 216 SRAP markers generated 78 polymorphic loci, and 98 of 312 SSR markers produced 107 polymorphic loci. Among the 185 loci, 175 loci (70 SRAP loci, 105 SSR loci) were assigned to nine linkage groups. The map covered 1,475 cM with a mean density of 8.7 cM per locus. Thirty-three of all the loci (17.84%) showed significant segregation distortion. Twenty-three of the 33 loci (69.7%) skewed toward the parent Steptoe, whereas the remaining loci (21.3%) deviated toward the parent Morex. And some of these distorted loci tended to cluster at the end of linkage groups, others dispersed on linkage groups in a decentralized fashion. The three putative segregation distortion regions (SDRs) were detected on chromosomes 2H, 4H, and 5H, respectively. This linkage map should indicate its importance in quantitative trait loci (QTLs) mapping, marker-assisted selection (MAS), and integrative analysis for further genetic studies with other linkage maps in barley.

Keywords Barley • Sequence-related amplified polymorphism (SRAP) • Molecular genetic map • SSR • DH population

---

Presenting author, Leilei L. Guo

L.L. Guo • X.J. Liu • X.C. Liu • Z.M. Yang • D.Y. Kong • Y.J. He • Z.Y. Feng (✉)
Barley Research Centre, Department of Plant Genetics and Breeding, College of Agronomy, Sichuan Agricultural University, 211 Huimin Road, Wenjiang District, 611130 Chengdu, Sichuan, China
e-mail: leilguo@yahoo.cn; zyfeng49@yahoo.com.cn

G. Zhang et al. (eds.), *Advance in Barley Sciences: Proceedings of 11th International Barley Genetics Symposium*, DOI 10.1007/978-94-007-4682-4_37,

37.1 Introduction

Barley (*Hordeum vulgare* L.) primarily used as feed of livestock and malts for beer, whisky, etc., is a widely variable species cultivated in nearly all regions of the world and also is the fourth most important cereal crop (Zohary and Hopf 1988; Feng et al. 2006). For scientists as well as commercial breeders, barley is considered a promising crop with a broad genetic potential. It has received considerable research attention as a model crop for genetic analysis. Researchers and breeders have increasingly been adopting molecular markers to identify genomic regions influencing traits and to select for desirable phenotypes based on identified marker-trait associations (Langridge and Barr 2003). Molecular genetic maps of crop species were suggested to play an important role in breeding and genomics research. Genetic linkage map is a basis of quantitative trait loci (QTL) mapping, map-based cloning, and molecular marker-assisted selection (MAS) (Wang et al. 2008). In barley, about 71 molecular genetic maps including RAPD, RFLP, AFLP, SSR, DArT markers, etc., have been acknowledged. Sequence-related amplified polymorphism (SRAP), a novel based-PCR molecular marker technique newly developed by Li and Quiros(2001), has simplicity, reliability, moderate throughput ratio and facile sequencing of selected bands as well as richer information than RAPD, AFLP, ISSR, and SSR (Li and Quiros 2001; Budak et al. 2004, 2005; Ferriol et al. 2003; Zheng et al. 2008). The linkage maps of SRAP markers in Brassica (Li and Quiros 2001) and cotton (Lin et al. 2009) were successfully constructed. In barley, the utilization of SRAP markers in the evaluation of genetic diversity of the developed qingke (hulless barley) cultivars from the Qinghai-Tibet plateau regions of China was first reported (Yang et al. 2008, 2010). Also exhibited a high level of genetic diversity from barley germplasm developed for scald disease resistance based on SRAP markers. But the map construction of SRAP markers in barley has not been reported.

In this study, the combined map of SRAP and SSR markers in barley was constructed to evaluate the potential of SRAP markers in genetic map construction.

37.2 Materials and Methods

37.2.1 Plant Materials and DNA Extraction

One hundred fifty DH lines derived from a cross (Morex × Steptoe) and their parents, kindly offered by Dr. Sato (Okayama University, Japan), were planted at barley experiment fields in Ya'an, Sichuan, China, in 2009–2010. At tilling period, fresh young leaves of these materials were used to extract genomic DNA using CTAB protocol with some modifications (Del Sal et al. 1989). The isolated genomic DNA was stored at −20 °C for usage. DNA concentration was adjusted to 50 ng/μL for PCR amplification.

Table 37.1 List of primer sequences (forward and reverse) used for SRAP marker in this study

Primer	5′→3′	Primer	5′→3′
Me1	TGAGTCCAAACCGGATA	Em9	GACTGCGTACGAATTACG
Me2	TGAGTCCAAACCGGAGC	Em10	GACTGCGTACGAATTTAG
Me3	TGAGTCCAAACCGGAAT	Em11	GACTGCGTACGAATTTCG
Me4	TGAGTCCAAACCGGACC	Em12	GACTGCGTACGAATTGTC
Me5	TGAGTCCAAACCGGAAG	Em13	GACTGCGTACGAATTGGT
Me6	TGAGTCCAAACCGGTAG	Em14	GACTGCGTACGAATTCAG
Me7	TGAGTCCAAACCGGTTG	Em15	GACTGCGTACGAATTCTG
Me8	TGAGTCCAAACCGGTGT	Em16	GACTGCGTACGAATTCGG
Me9	TGAGTCCAAACCGGTCA	Em17	GACTGCGTACGAATTCCA
Em1	GACTGCGTACGAATTAAT	Em18	GACTGCGTACGAATTCAA
Em2	GACTGCGTACGAATTTGC	Em19	GACTGCGTACGAATTCGA
Em3	GACTGCGTACGAATTGAC	Em20	GACTGCGTACGAATTCCT
Em4	GACTGCGTACGAATTTGA	Em21	GACTGCGTACGAATTCCG
Em5	GACTGCGTACGAATTAAC	Em22	GACTGCGTACGAATTAAG
Em6	GACTGCGTACGAATTGCA	Em23	GACTGCGTACGAATTATC
Em7	GACTGCGTACGAATTATG	Em24	GACTGCGTACGAATTTGT
Em8	GACTGCGTACGAATTAGC		

Notes: Me1–Me 9 and Em1–Em 24 indicate forward primers and reverse primers, respectively

37.2.2 SSR and SRAP Profiling

Three hundred twelve SSR primer pairs covering the whole barley genome (http://wheat.pw.usda.gov/cgi-bin/graingenes/) were screened for polymorphism between parents of the DH population. And these primers were synthesized by Shanghai Invitrogen (China). The PCR reaction system and the detection of PCR products were carried out employing the methods of Liu et al. (2011). Thirty-three SRAP primers including 9 forward and 24 reverse primers (Table 37.1), designed following Li and Quiros (2001), were synthesized (Shanghai Invitrogen, China). A total of 216 primer combinations were used to search for polymorphism between the parents. Informative primer combinations were used to genotype the mapping population. The PCR amplification was conducted in an MJ Thermo Cycler in a total volume of 10 μL, including 50 ng of genomic DNA, 1×PCR buffer containing $MgCl_2$, 0.2 mM dNTPs, 0.3 μM of each primer, and 0.5 units of Taqase. The PCR reaction condition was as follows: an initial denaturing step at 94 °C for 5 min; followed by five cycles at 94 °C for 1 min, 35 °Cfor 1 min, and 72 °C for 1 min; and subsequently followed by 35 cycles at 94 °C for 1 min, 50 °C for 1 min, and 72 °C for 1 min with a final extension step at 72 °C for 10 min (Wang et al. 2008). The PCR products were separated on 6 or 8% denatured polyacrylamide gels and visualized by silver staining.

37.2.3 Data Analysis

The genetic linkage was analyzed using the mapping software Mapmaker 3.0 based on the segregation data (Lander et al. 1987), with LOD=3.0, and the Kosambi map function was used to construct genetic linkage map (Kosambi 1944). And the graphic representation of the linkage group was drawn by MapDraw software. The segregation ratio across the mapping population was tested against a 1:1 ratio using *Chi*-square test. The segregation of markers which did not fit the 1:1 ratio ($P<0.05$) was treated as distorted. SPSS 17.0 was used for *Chi*-square test.

37.3 Results

A total of 528 markers including 216 SRAP markers and 312 SSR markers were employed to screen the polymorphism between parents of the DH lines. Twenty-one of 216 SRAP markers generated 78 polymorphic loci. And 98 of 312 SSR markers produced 107 polymorphic loci. Among the 185 polymorphic loci, 175 loci (70 SRAP loci, 105 SSR loci) were assigned to nine linkage groups, which were assigned to corresponding chromosomes by SSR markers with known chromosome locations in barley. The map covered 1,475 cM with a mean density of 8.7 cM per locus (Fig. 37.1).

And the phenomenon of markers' segregation distortion was also investigated. Thirty-three of all the loci (17.84%) showed segregation distortion, of which 26 loci showed significant segregation distortion at 0.05 probability level, while seven loci significant at 0.01 level. Twenty-three of the 33 loci (69.7%) skewed toward the parent Steptoe, whereas the remaining loci (21.3%) deviated toward the parent Morex. And some of these distorted loci tended to cluster at the end of linkage groups, others dispersed on linkage groups in a decentralized fashion. The three putative segregation distortion regions (SDRs) were detected on chromosomes 2H, 4H, and 5H, respectively (Fig. 37.1).

37.4 Discussion

Since the advent of molecular marker and linkage-mapping technologies, the number of marker loci placed on genetic maps is exponentially increasing (Varshney et al. 2007). In barley, molecular genetic maps including RAPD, RFLP (Graner et al. 1991; Karakousis et al. 2003), AFLP (Heun et al. 1991; Becker et al. 1995; Karakousis et al. 2003), SSR (Karakousis et al. 2003), DArT (Wenzl et al. 2006; Hearnden et al. 2007) markers, etc., have been acknowledged. It is difficult for the QTLs detected by the linkage map constructed by the present markers, far from the linked markers, to meet the marker-assisted selection (Varshney et al. 2007). Therefore, the molecular map construction using new molecular markers is an important work in barley molecular genetics and breeding research. The SRAP markers, designed by Li and Quiros (2001), have been described as being simple and reliable to operate, with its multiplexing capacity delivering many markers from a single assay (Zhang et al. 2011). This type of markers was successfully applied in

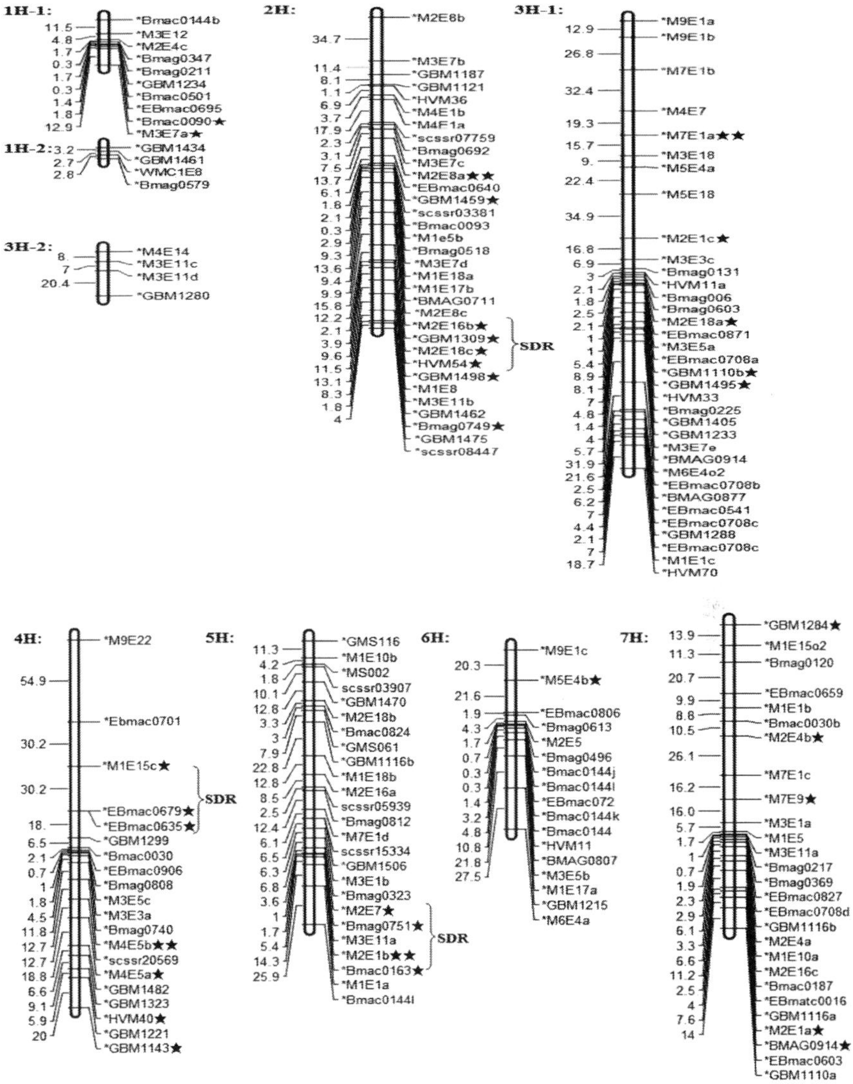

Fig. 37.1 The linkage maps constructed using SSR and SRAP markers in DH population (M-E-represent SRAP marker. ★ and ★ ★indicate significant levels of distorted segregation at 5 and 1%, respectively)

map construction (Lin et al. 2009; Zhang et al. 2011), diversity evaluation (Ferriol et al. 2003; Budak et al. 2004; Sun et al. 2006; Yang et al. 2010) comparative genomics (Li et al. 2003), gene location, and cultivar identification (Li et al. 2006). In barley, SRAP markers were only used for diversity analysis and genetic relationships (Yang et al. 2010; Dizkirici et al. 2010). Up to now, it has no report about the utility of

SRAP markers in the linkage map construction of barley. The map constructed in this study is just a framework to see whether the SRAP marker system could be used in barley. An appealing aspect of the SRAP system in this study is able to amplify more polymorphic loci. Twenty-one of 216 SRAP markers generated 78 polymorphic loci, and 98 of 312 SSR markers produced 107 polymorphic loci, indicating that SRAP is a more efficient technique. Moreover, the report from Li et al. (2008) indicated that the average number of polymorphic bands detected by each SRAP primer combination was much higher than that revealed by RAPD and ISSR primers (Liu et al. 2008). Another appealing aspect of the SRAP system is able to amplify the whole genome of barley. Seventy of 78 SRAP loci were mapped on nine linkage groups, which were assigned to corresponding chromosomes by SSR markers with known chromosome locations in barley. From what was mentioned above, the SRAP marker system is suitable for linkage and quantitative trait loci mapping in barley.

To summarize, SPAP marker-based linkage mapping in barley is first reported. The future utility of this map for barley should be further integrated into all the maps from S/M DH population to indicate its importance in quantitative trait locus mapping, marker-assisted selection (MAS), and integrative analysis for further genetic studies with other linkage maps in barley.

Acknowledgments We thank J. M., Yang X. Y., and Yuan J. E. for some technical experiments and Dr. Yang P. (IPK, Germany) for his assistance in data analysis. The present work was funded by grants from China Agriculture Research System (CARS-05) and also Major Project from Sichuan Education Department in China.

References

Becker, J., Vos, P., Kuiper, M., Salamini, F., & Heum, M. (1995). Combined mapping of AFLP and RFLP markers in barley. *Molecular & General Genetics, 249*, 65–73.

Budak, H., Shearman, R. C., Parmaksiz, I., & Dweikat, I. (2004). Comparative analysis of seeded and vegetative biotype buffalo grasses based on phylogenetic relationship using ISSRs, SSRs, RAPDs, SRAPs. *Theoretical and Applied Genetics, 109*, 280–288.

Budak, H., Shearman, R. C., Gulsen, O., & Dweikat, I. (2005). Understanding ploidy complex and geographic origin of the buchloe dactyloides genome using cytoplasmic and nuclear marker system. *Theoretical and Applied Genetics, 111*, 1545–1552.

Del Sal, G., Manfiloeti, G., & Schneider, C. (1989). The CTAB-DNA precipitation method: a common mini-scale preparation of template DNA from phagemids, phages or plasmids suitable for sequencing. *BioTechniques, 7*, 514–520.

Dizkirici, A., Zeki, K., Elif, Güren H., & Hikmet, Budak. (2010). Barley germplasms developed for scald disease resistance exhibited a high level of genetic diversity based on SRAP markers. *Turkish Journal of Biology, 34*, 271–279.

Feng, Z. Y., Liu, X. J., Zhang, Y. Z., & Ling, H. Q. (2006). Genetic diversity analysis of Tibetan wild barley using SSR markers. *Acta Genetica Sinica, 33*, 917–928.

Ferriol, M., Picó, B., & Nuez, F. (2003). Genetic diversity of a germplasm collection of Cucurbita pepo using SRAP and AFLP markers. *Theoretical and Applied Genetics, 107*, 271–282.

Graner, A., Jahoor, A., Schondelmaier, J., Siedler, H., Pillen, K., Fischbeck, G., Wenzel, G., & Herrmann, R. G. (1991). Construction of an RFLP map of barley. *Theoretical and Applied Genetics, 83*, 250–256.

Hearnden, P. R., Eckermann, P. J., McMichael, G. L., Hayden, M. J., Eglinton, J. K., & Chalmers, K. J. (2007). A genetic map of 1,000 SSR and DArT markers in a wide barley cross. *Theoretical and Applied Genetics, 115*, 383–391.

Heun, M., Kennedy, A. E., Anderson, J. A., Lapitan, N. L. V., Sorrells, M. E., & Tanksley, S. D. (1991). Construction of a restriction fragment length polymorphism map of barley (*Hordeum vulgare*). *Genome, 34*, 437–477.

Karakousis, A., Gustafson, J. P., Chalmers, K. J., Barr, A. R., & Langridge, P. (2003). A consensus map of barley integrating SSR, RFLP, and AFLP markers. *Australian Journal of Agricultural Research, 54*, 1173–1185.

Kosambi, D. D. (1944). The estimation of map distances from recombination values. *Annals of Eugenics, 12*, 172–175.

Lander, E. S., Green, P., Abrahamson, J., Barlow, A., Daly, M. J., Lincoln, S. E., & Newberg, L. A. (1987). MAPMAKER: an interactive computer package for constructing primary genetic linkage maps of experimental and natural populations. *Genomics, 1*, 174–181.

Langridge, P., & Barr, A. R. (2003). Preface. *Australian Journal of Agricultural Research, 54*, i–iv.

Li, G., & Quiros, C. F. (2001). Sequence-related amplified polymorphism (SRAP), a new marker system based on a simple PCR reaction: Its application to mapping and gene tagging in Brassica. *Theoretical and Applied Genetics, 103*, 455–461.

Li, G., Gao, M., Yang, B., & Quiros, C. F. (2003). Gene for gene alignment between the Brassica and Arabidopsis genomes by direct transcriptome mapping. *Theoretical and Applied Genetics, 107*, 168–180.

Li, L., Zheng, X. Y., & Lin, L. W. (2006). Analysis of genetic diversity and identification of cucumber varieties by SRAP. *Molecular Plant Breeding, 4*, 702–708.

Lin, Z. X., Zhang, Y. X., Zhang, X. L., & Guo, X. P. (2009). A high-density integrative linkage map for Gossypium hirsutum. *Euphytica, 166*, 35–45.

Liu, L. W., Zhao, L. P., Gong, Y. Q., Wang, M. X., Chen, L. M., Yang, J. L., Wang, Y., Yu, F. M., & Wang, L. Z. (2008). DNA fingerprinting and genetic diversity analysis of late-bolting radish cultivars with RAPD, ISSR and SRAP markers. *Scientia Horticulturae, 116*, 240–247.

Liu, X. J., You, J. M., Guo, L. L., Liu, X. C., He, Y. J., Yuan, J. E., Liu, G. X., & Feng, Z. Y. (2011). Genetic analysis of segregation distortion of SSR markers in F_2 population of barley. *Journal of Agricultural Science, 3*, 172–177.

Sun, S. J., Gao, W., Lin, S. Q., Zhu, J., Xie, B. G., & Lin, Z. B. (2006). Analysis of genetic diversity in Ganoderma population with a novel molecular marker SRAP. *Applied Microbiology and Biotechnology, 72*, 537–543.

Varshney, R. K., Marcel, T. C., Ramsay, L. R., Russell, J., Röder, M. S., Stein, N., Waugh, R., Langridge, P., Niks, R. E., & Graner, A. (2007). A high density barley microsatellite consensus map with 775 SSR loci. *Theoretical and Applied Genetics, 114*, 1091–1103.

Wang, J. S., Yao, J. C., & Li, W. (2008). Construction of a molecular map for melon (*Cucumis melo* L.) based on SRAP. *Frontiers of Agriculture in China, 2*, 451–455.

Wenzl, P., Li, H. B., Carling, J., Zhou, M. X., Raman, H., Paul, E., Hearnden, P., Maier, C., Xia, L., Caig, V., Ovesná, J., Cakir, M., Poulsen, D., Wang, J. P., Raman, R., Smith, K. P., Muehlbauer, G. J., Chalmers, K. J., Kleinhofs, A., Huttner, E., & Kilian, A. (2006). A high-density consensus map of barley linking DArT markers to SSR, RFLP and STS loci and agricultural traits. *BMC Genomics, 7*, 206.

Yang, P., Liu, X. J., Liu, X. C., Wang, X. W., He, S. P., Li, G., Yang, W. Y., & Feng, Z. Y. (2008). Genetic diversity analysis of the developed qingke (hulless barley) varieties from the plateau regions of Sichuan province in China revealed by SRAP markers. *Hereditas (Beijing), 30*, 115–122.

Yang, P., Liu, X. J., Liu, X. C., Yang, W. Y., & Feng, Z. Y. (2010). Diversity analysis of the developed qingke (hulless barley) cultivars representing different growing regions of the Qinghai-Tibet Plateau in China using sequence-related amplified polymorphism (SRAP) markers. *African Journal of Biotechnology, 9*, 8530–8538.

Zhang, F., Chen, S. M., Chen, F. D., Fang, W. M., Chen, Y., & Li, F. T. (2011). SRAP-based mapping and QTL detection for inflorescence-related traits in chrysanthemum (*Dendranthema morifolium*). *Molecular Breeding, 27*, 11–23.

Zheng, J., Zhang, Z. S., Chen, L., Wan, Q., Hu, M. C., Wang, W., Zhang, K., Liu, D. J., Chen, X., & Wei, X. Q. (2008). Intron-targeted intron-exon splice conjunction (IT-ISJ) marker and its application in construction of upland cotton linkage map. *Agricultural Sciences in China, 7*, 1172–1180.

Zohary, D., & Hopf, M. (1988). *Domestication of plants in the old world* (3rd ed.). Oxford: Clarendon.

Chapter 38
Phenotypic Evaluation of Spring Barley RIL Mapping Populations for Pre-harvest Sprouting, Fusarium Head Blight and β-Glucans

Linda Legzdina, Mara Bleidere, Guna Usele, Daiga Vilcane, Indra Beinarovica, Ieva Mezaka, Zaiga Jansone, and Nils Rostoks

Abstract The overall objective of the research is to develop molecular markers which can be used in spring barley breeding. The aim of this study was to summarise phenotyping data from recombinant inbred line (RIL) populations for mapping the QTLs for resistance to pre-harvest sprouting and Fusarium head blight (FHB) as well as content of β-glucans. The field and laboratory experiments were performed at the State Priekuli Plant Breeding Institute and at the State Stende Cereal Breeding Institute for two seasons (2010–2011). The mapping populations for pre-harvest sprouting consist of 93 (RILs produced from a cross between hulless barley (HB) breeding line 'PR 3642' (susceptible) and HB variety 'CDC Rattan' (resistant), and of 94 RILs from a cross between HB variety 'CDC Freedom' (susceptible) and hulled variety 'Samson' (resistant). Eighty-six RILs for mapping resistance to FHB were derived from a cross between 'Fontana' (susceptible) and 'ND 16461' (resistant). The content of β-glucans was evaluated in 106 RIL population developed from the spring barley cross of HB lines 'KM-1910' (low content of β-glucans) and 'KM-2084' (high) and in 117 RILs developed from cross of hulled variety 'Justina' (low content of β-glucans) and 'KM-2084' (high). The noticeable variation among the recombinant inbred lines is found for all evaluated traits.

Keywords Spring barley • Pre-harvest sprouting • FHB • β-Glucans • RIL • Phenotyping

L. Legzdina • G. Usele • D. Vilcane • I. Beinarovica • I. Mezaka
State Priekuli Plant Breeding Institute, Zinatnes Str. 1a, Priekuli 4126, Latvia

M. Bleidere (✉) • Z. Jansone
State Stende Cereal Breeding Institute, "Dizzemes", Dizstende, 3258, Latvia
e-mail: maara.bleidere@gmail.com

N. Rostoks
Faculty of Biology, University of Latvia, 4 Kronvalda Blvd., Riga 1586, Latvia

G. Zhang et al. (eds.), *Advance in Barley Sciences: Proceedings of 11th International Barley Genetics Symposium*, DOI 10.1007/978-94-007-4682-4_38,

38.1 Introduction

The genetic dissection of complex traits in spring barley (*Hordeum vulgare* L.) still presents a challenge.

Due to climatic changes caused by global warming, precipitation in Europe has increased 10–40% in the last century (McCarthy et al. 2001). Therefore, barley varieties with resistance to pre-harvest sprouting and Fusarium head blight (FHB) are increasingly needed also under conditions of Latvia.

Sprouting of grains, this is premature germination of kernels in spikes during harvest; it causes considerable economic loss due to the lower yield and inferior technological value of grains especially for hulless barley (HB). Sprouted grain cannot be used as seed material and, as far as barley is concerned, it becomes poor quality. Resistance to sprouting is genetically determined and also under a strong influence of environmental factors (Fox et al. 2003) In addition, this trait is a resultant of a complex set of morphological, anatomical, physiological and biochemical factors as well as variable environmental conditions. Among morphological and anatomical characters, such traits are the structure and chemical composition of hulls, structure and chemical composition of the seed coat. The depth and length of the post-harvest dormancy are most important physiological factors related to the development and growth of kernels (Gubler et al. 2008).

Cereal disease FHB caused by fungi of the *Fusarium* genus has gained a greater importance because food contamination by fusaria mycotoxins thus markedly affects economics, international trade and human and animal health. Therefore, establishing the testing methods and a search for the resistant varieties to FHB in barley has been one of the tasks of different barley breeding programmes (Takeda 2004; Nesvadba et al. 2006). Significant genetic variation for FHB severity and for mycotoxin contamination was also found among covered and hulless barley lines of Latvian origin (Legzdina and Buerstmayr 2004). One of the possibilities to identify genotypes resistant to FHB, and thus to promote development of new more resistant genotypes, is to use molecular markers in selection process. Identification of spring barley genotypes with different susceptibility to Fusarium head blight has already been done up to now in practice using molecular markers in the population of doubled haploid lines (Nesvadba et al. 2006).

An important aspect of creating value for customers by breeding is developing varieties with traits affecting grain quality of barley, such as β-glucan content. Amount of β-glucan in barley may differ in genotypes which vary in major morphological traits. β-Glucan content in grain is higher in genotypes with hulless grain, short awns and waxy endosperm (Fastnaught et al. 1996).

Marker-assisted selection (MAS) has been proposed as a way to increase gains from selection for quantitative traits (Berloo and Stam 1998). As pre-harvest sprouting, FHB and β-glucan content are also controlled by multiple factors, the efficiency of breeding programmes depends on knowledge of the genetic control and genomic location of quantitative trait loci (QTLs) governing these traits of interest. Choice of appropriate mapping population is very critical for the success of any QTL mapping project. Populations for QTL mapping can be broadly classified into two: experimental populations for linkage-based QTL mapping (e.g. inbred lines for autogamous or self-pollinating species) and natural or breeding populations for linkage disequilibrium-based association mapping (Ullrich et al. 2008).

Linkage-based QTL mapping depends on well defined populations developed by crossing two parents. In autogamous species, QTL mapping studies make use of F_2 or F_x derived families, backcross (BC), recombinant inbred lines (RILs), near isogenic lines (NILs) and double haploids (DH). These populations are developed by crossing two inbred parents with clear contrasting difference in phenotypic trait(s) of interest. Each mapping population developed from inbred parents has its own advantages and disadvantages, and the researchers need to decide the appropriate population depending on project objective, trait complexity, available time and whether the molecular markers to be used for genotyping are dominant or codominant (Semagn et al. 2010). The power of genetic mapping is strongly dependent upon the quality of phenotypic data. Also, environmental factors may trigger and modify gene actions, and thereby further complicate the analysis. Nevertheless, the application of MAS in autogamous crops, with the objective obtaining transgressive genotypes, can improve selection results if compared to conventional selection procedures (Berloo and Stam 1998).

The overall objective of the research is to develop molecular markers which can be used in spring barley breeding. The aim of this study was to summarise phenotyping data from recombinant inbred line (RIL) populations for mapping the QTLs for resistance to pre-harvest sprouting and FHB as well as β-glucan content.

38.2 Materials and Methods

38.2.1 Plant Material

The mapping populations of RILs for resistance to *pre-harvest sprouting* consist of 93 (RILs produced from a cross between Latvian hulless barley (HB) breeding line ‘PR 3642’ (susceptible) and HB variety ‘CDC Rattan’ (resistant), and of 94 RILs from a cross between HB variety ‘CDC Freedom’ (susceptible) and covered barley ‘Samson’ (resistant).

Eighty-six RILs for mapping resistance to *FHB* were derived from a cross between north American line ‘ND 16461’ (resistant) and susceptible variety ‘Fontana’. Content of *β-glucans* was evaluated in 106 RIL population developed from the spring barley cross of HB lines from the Czech Republic ‘KM-1910’ (low content of β-glucans) and ‘KM-2084’ (high) and in 117 RILs developed from cross of HB genotype ‘KM-2084’ (high content of β-glucans) and hulled variety ‘Justina’ (low β-glucans). Non-waxy starch parent was chosen to exclude the influence of *wax* allele.

Resistant and susceptible parents for crosses were chosen according to results of previous investigations (Legzdina et al. 2010; Legzdina and Buerstmayr 2004;

Legzdina 2001). Chosen parent genotypes were crossed in 2007 and hybrids of generations F_5 and F_6 were phenotyped. Seeds from one spike per F_4 plant were sown to obtain F_5 lines; DNA was extracted from one plant per F_5 line, and seeds were collected from the same plant to sew F_6 generation trials.

38.2.2 Phenotyping

The field and laboratory experiments for pre-harvest sprouting and FHB were performed at the State Priekuli Plant Breeding Institute (Latvia) for two seasons (2010–2011). In 2010, the first-year field trial and laboratory testing for β-glucans were carried out at the State Stende Cereal Breeding Institute (Latvia).

To assess the germination in laboratory conditions, three spikes per genotype were collected in grain ripening (growth stage 92 according to decimal code of growth stages of cereals); each spike was presumed as one replication. Spikes were placed in Petri dishes between moist filter paper layers and kept at 20 °C with 16/8 h photoperiod. The testing method was adapted with modifications from Derycke et al. (2002). The amount of seeds with visible germination symptoms was determined in the 7th day after initiation of the test (Legzdina et al. 2010).

To assess resistance to FHB, two methods were applied: cut spike inoculation method adapted from K. Takeda (2004) in 2010 and artificial inoculation with *Fusarium culmorum* under field conditions in 2011. Plants were grown in 1-m long one- or two-row plots. In 2010, three spikes per line were collected at anthesis (growth stage 65 according to decimal code of growth stages of cereals); each spike has been assumed as one replicate. Spikes were detached from the plants at the second internode from the top, and put in humidity chamber with air humidifier. Air humidifier was filled two times a day and it kept working 5 h after filling. The suspension of conidia was made from dry *Fusarium culmorum* infected grains (produced by Prophyta, Germany). Five grams of infection material was incubated in 1 L of water for 1 h and then filtered. The final spore concentration for inoculation was 1×10^5 conidiospores per litre. Inoculations were carried out in the evening. Temperature and light in the chamber were not controlled and depended from the outdoor conditions. In each spike, the percentage of visibly infected spikelets was scored according to a linear 0–100% scale 8 days after inoculation to measure FHB severity. In 2011, inoculation method in field conditions was used. Every genotype was inoculated individually at its susceptible stage (50% of spikes in anthesis) and inoculation was repeated after 3 days. Single-spore isolate of local of *Fusarium culmorum* strain Nr.52 (27–2) was used for inoculation (macro-conidia suspension 5×10^4 spores per mL^{-1} water). Inoculation was performed in late afternoon, when air temperature was less than 20 °C. Mist irrigation was not applied and air humidity depended of whether conditions. The percentage of spikelets with FHB symptoms was visually estimated three times at days 18, 22 and 26 after inoculation on a whole plot basis. The area under the disease progress curve (AUDPC) was calculated for each plot and used as the measure of disease severity as described by Buerstmayr et al. (2000).

β-Glucan content (g kg^{-1}) was measured using chemical kits from Megazyme (Megazyme International Ireland Ltd).

38.2.3 Statistical Analysis

The obtained results were statistically processed by MS Excel program package using the methods of descriptive statistics. Comparisons were made using analysis of variance and *t*-test: two samples assuming equal variance. Expected values for skewness and kurtosis are zero for a normal phenotype distribution. Broad-sense heritability for pre-harvest sprouting and FHB was estimated as an intergeneration correlation (r) calculated between results in 2 years of investigation.

38.3 Results

38.3.1 Pre-harvest Sprouting

There was significant ($p<0.01$) difference between susceptible female parent and male parent genotypes in both RIL populations for pre-harvest sprouting (Table 38.1).

In the RIL population PR 3642/CDC Rattan in the year of 2010, the average amount of germinated grains varied between 35 and 100% (mean value 71%), but in 2011, it varied from 0 to 68% (mean value 30%) (Table 38.1). As indicated by the partitioning of sum of squares in 2011, the effect of genotype on variation of germination

Table 38.1 Phenotypic data for pre-harvest sprouting RIL mapping populations, % (2010–2011)

	PR 3642/CDC Rattan		CDC Freedom/Samson	
Indices	2010	2011	2010	2011
Female parent	x	74^{a*}	x	74^{a}
Male parent	x	26^{b}	x	9^{b}
RIL mean	71	30	28	21
RIL range	35–100	0–68	0–80	0–71
SD	13.4	20.3	25.6	19.9
Skewness	0.20	0.06	0.53	0.63
Kurtosis	0.21	1.07	−1.08	−0.86
$Rs_{0.05}$	27.3	28.7	26.8	24.7
η^2_G, %	45	83	84	80
$r_{10/11}$	0.26**		0.479**	

*Trait means followed by different letters within a rows between the parents are significant at the level of $p<0.01$; **significant at $p<0.01$

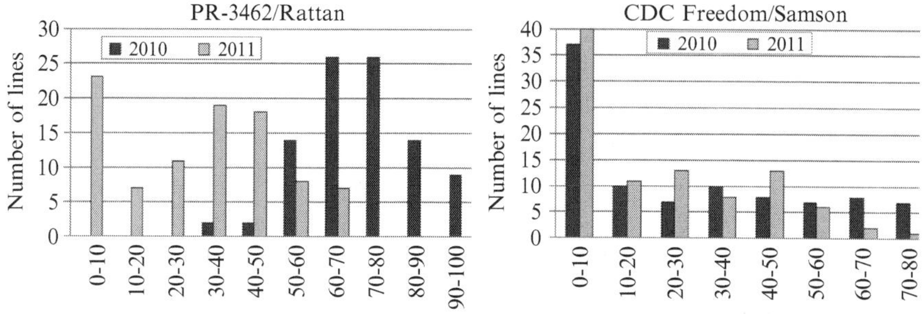

Fig. 38.1 Frequency distribution of pre-harvest sprouting (%) of the RIL populations PR 3642/CDC Rattan and CDC Freedom/Samson (2010–2011)

Table 38.2 Phenotypic data for Fusarium head blight RIL mapping population Fontana/ND 1646, (2010–2011)

	Parents		RIL population				
Year	Fontana	ND 1646	Mean	Range	SD	Skewness	Kurtosis
FHB severity, %							
2010	x	x	55.6	13.6–89.6	19.1	−0.38	−0.58
FHB severity 26 days post-inoculation, %							
2011	17.8 [a]	2.8 [b]	14.2	5–30	6.0	0.77	0.14
FHB progress (AUDPC)							
2011	151.7[a]	18.7[b]	73.1	16.5–210.0	40.9	1.19	1.26

*Trait means followed by different letters between the parents are significant at the level of $p<0.01$

of grains was noticeably higher than it was in 2010 ($\eta^2_G=83$ and 45%, respectively), but average value was lower (71 and 30%, respectively). High effect of genotype on variation of germination of grains in both years of investigation was within RIL population Samson/CDC Freedom ($\eta^2_G=80$–84%). For this RIL population in both years of investigation considerable variation of trait was observed, the highest amount of lines had 0–10% of germinated grains (Fig. 38.1). Higher difference in distribution of lines according to pre-harvest sprouting between years of investigation was detected for RIL population PR 3642/CDC Rattan than it was for Samson/CDC Freedom.

In this population intergeneration, correlation coefficient estimated between years was significantly positive, although this value was low ($r=0.261$, $p<0.01$). In RIL population Samson/CDC Freedom, this correlation was significantly positive ($r=0.479$, $p<0.01$) as well.

38.3.2 Fusarium Head Blight (FHB)

In the mapping population for FHB Fontana/ND 1646 in the year of 2010, high variation of trait was observed (Table 38.2). The average amount of infected spikelets

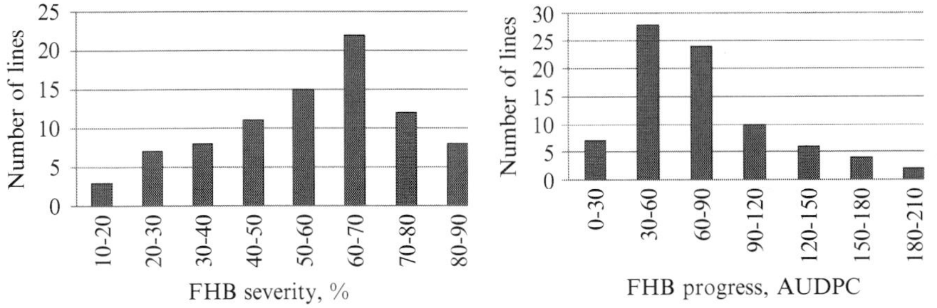

Fig. 38.2 Frequency distribution of FHB severity, % (2010) and FHB progress, AUDPC (2011) of the RIL population Fontana/ND 1646

within RIL population varied between 13.6 and 89.6% with average value 55.6%. The effect of genotype as a factor to trait variation was significant ($p<0.01$, $\eta^2_G=54.7\%$). According to data of 2010, the highest proportion of evaluated lines (22 lines) were characterised with FHB severity between 60 and 70% (Fig. 38.2).

In 2011, significant difference ($p<0.01$) in FHB severity among parent varieties at 26 days post-inoculation was observed –17.8% for susceptible parents 'Fontana' and 2.8% for resistant genotype 'ND 1646'. This difference was significant also for AUDPC values (AUDPC = 151.7 and 18.7, respectively).

Amount of diseased spikelets at 26 days post-inoculation within RIL population varied between 5 and 30%, but variation of FHB progress in AUDPC values was from 16.5 to 210.0 with average value 73.1 (SD = 40.9) in this RIL mapping population. In 2011, the highest amount of lines was with FHB progress in AUDPC values between 30 and 60% (28 lines). Obtained data showed that intergeneration correlation coefficient estimated between years and at the same time between both used evaluation methods was not significant.

38.3.3 Content of β-Glucans

In RIL population derived from the spring barley cross of HB lines 'KM-1910' and 'KM-2084', the difference between both parents was 11.7 $g\,kg^{-1}$ (40.7 and 52.4 $g\,kg^{-1}$, respectively) (Table 38.3).

The content of β-glucans in RIL mapping population ranged from 35.8 to 76.8 $g\,kg^{-1}$ with coefficient of variation 14.5%. The highest proportion of evaluated lines (totally 54 lines) was with β-glucan content between 50 and 60 $g\,kg^{-1}$.

Also in RIL population KM-2084/Justina where hulless and hulled parent barley genotypes were used, considerable variation in β-glucan content was detected (Table 38.3).

This RIL population consists from 95 hulled lines and 22 hulless RILs. Average β-glucan content for hulless barley RILs was significantly ($p<0.01$) higher than for hulled, but trait variation within hulled RILs population was higher than hulless ones (CV = 12.3 and 7.3%, respectively). The highest proportion of evaluated lines

Table 38.3 Phenotypic data for β-glucans (g kg^{-1}) RIL mapping populations (2010)

	KM-1910/KM-2084	KM-2084/Justina	
Indices	Hulless RILs	Hulled RILs	Hulless RILs
Female parent	52.4	x	52.4
Male parent	40.7	41.0	x
RIL mean	51.9	45.9[a]	49.9[b]
RIL range	35.8–76.8	28.7–59.1	44.9–57.6
SD	7.5	5.6	0.78
CV%	14.5	12.3	7.3
Skewness	0.25	−0.34	0.60
Kurtosis	0.24	0.65	−0.49

*Trait means followed by different letters between the hulled and hulless genotypes are significant at the level of $p<0.01$

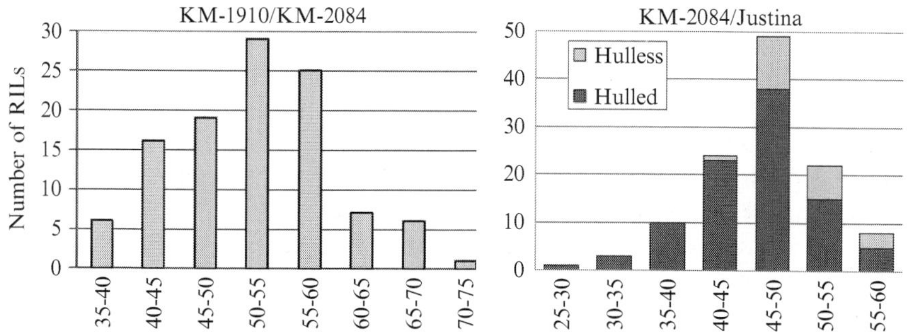

Fig. 38.3 Frequency distribution of β-glucans (g kg^{-1}) of the RIL population KM-1910/KM-2084 and KM-2084/Justina

for both barley types (totally 49 lines) was characterised with β-glucan content between 45 and 50 g kg^{-1} (Fig. 38.3).

Overall results of this investigation showed that the values of skewness and kurtosis of the distributions of traits were mainly less than 1.0 or had a value of these indices slightly more than 1.0 in absolute values, suggesting that traits within RIL populations approximately fit normal distributions and the data are suitable for QTL mapping.

38.4 Discussion

This paper analysed the preliminary phenotypic data for resistance to pre-harvest sprouting, FHB and β-glucan content with the objective of research to develop recombinant inbred line (RIL) populations for mapping the QTL for these traits suitable for Latvian conditions. Recombinant inbred lines (RILs) obtained from divergent parental accessions have already led to the molecular identification of

QTL for a number of important complex traits, also for traits evaluated in the present study. Previously, several quantitative trait loci (QTL) also for pre-harvest sprouting (Ullrich et al. 2008), FHB resistance (Ma et al. 2000) and β-glucan content (Li et al. 2008) have been mapped in bi-parental populations. Understanding the genetic networks underlying quantitative trait variation will provide new targets for plant breeders. However, as they are generally under the control of many genes, those characters are quantitatively variable and their study requires specific strategies and techniques. As the basic phenotypic data required for QTL mapping are the estimates of phenotypic performance of individuals across environments, the accuracy and precision of phenotyping determines how realistic the QTL mapping results are. Increased precision of phenotyping increases heritability, which, in turn, increases the statistical power of QTL detection (Xu et al. 2005). The heritability of a trait depends in part on whether the phenotyping is repeatable across different seasons, locations and environments. Therefore, environmental comparison of phenotyping is needed in order to determine how the marker-trait association identified under one environment can be used for selection under another (Xu and Crouch 2008).

There is usually a high cost associated with genotyping (generation of molecular marker data) and phenotyping (field, greenhouse or laboratory evaluation for the phenotypic trait) of large population size, particularly for traits requiring extensive field trials or complex analysis. Consequently, the size of the mapping population and the number of replications and sites (environments) for phenotyping are often limited (Semagn et al. 2010). Overall, the QTL mapping literature has shown that if a breeder can develop a mapping population of 100–150 progenies, it could be possible to obtain reasonably good phenotypic data for the traits of interest (Bernando 2008). In our study, amount of samples which were included in the RILs populations was rather high – 86–117 RILs per population. The cost and logistics of phenotyping are the main reasons imposing limits on number of RILs. This is especially true of phenotypes involving complex traits and traits where scoring of phenotyping data in the field conditions are inconvenient. Pre-harvest sprouting in field conditions can be scored rarely. In order to be able to perform selection in breeding programme and provide the possibility to obtain information every season, the indirect testing method was adapted and used for evaluation and phenotyping of RILs. The research that is done up to now in Latvian spring barley breeding programme approved existence of genetic diversity and selection possibilities for pre-harvest sprouting resistance (Legzdina et al. 2010). Parent varieties used in present study to obtain RIL populations for pre-harvest sprouting were chosen according to results in the field conditions as well as in laboratory conditions. Evaluation results showed that parents were characterised with high difference in amount of germinated seeds. It was good precondition for obtaining RILs populations PR 3642/CDC Rattan and CDC Freedom/Samson with significant difference among genotypes and high variability (Table 38.1). Although the intergeneration correlation coefficients among years of investigation for both RIL populations were significant, nevertheless, there was reasonable difference between years of investigation in the average amount of germinated grains for both RIL populations. It can be explained mainly by the difference in meteorological conditions around the time of maturity when the spikes were collected. In 2010, it was rather rainy and part of the spikes was already moist before starting the test, but in 2011, there was no rain during the time of collection of spike samples. Previous evaluation results and experience in application of laboratory methods were used to choose the parents for hybridisations and to obtain the RIL population described in this study for detecting QTLs for FHB (Legzdina and Buerstmayr 2004). In RIL population Fontana/ND 16461 using two different FHB evaluation methods, the high difference between susceptible and resistant parent, and high variability among RILs were found (Table 38.2). Nevertheless, results of 2 years showed that there was low intergeneration correlation among two methods which were applied in 2010 and 2011. It can be explained by the two different methods used for phenotyping. Dill-Macky (2003) concluded that one of the main problems in testing Fusarium resistance is to reproduce experimental results. Low correlation between greenhouse and field data was mentioned by Rudd et al. (2001). The weather conditions around inoculation play a crucial role in the infection process. Varying atmospheric conditions may lead to a significant bias in the resistance evaluations and to sometimes large G×E interactions and poor correlations between experiments (Mesfin et al. 2003), especially when the genotypes under investigation differ significantly in flowering date (Buerstmayr et al. 2004). Although in 2011 the meteorological conditions after inoculation were generally favourable for the disease development (it was rainy and warm), the observed disease severity was comparatively low. It can be caused by insufficient air moisture because no mist irrigation was used.

The first-year data according to RILs populations for β-glucan content showed significant variation in both populations. Population KM-2084/Justina consisted from hulled and hulless RILs. In the cross combinations between covered and hulless parents, there is a possibility to obtain covered lines with increased β-glucan content and hulless lines with decreased β-glucan content (Bleidere and Belicka 2009). The results of current investigation also showed that in population KM-2084/Justina were hulled lines with increased β-glucan content (Fig. 38.3). It gives the possibility to increase the diversity of hulled RILs. As the β-glucan content usually was significantly higher in hulless, barley genotypes compared to hulled barley genotypes *nud* locus itself could influence the β-glucan content in barley; this fact needs to be taken into account in process of QTLs detection for this trait (Mezaka et al. 2011).

In general, this study showed that all traits in this study showed notable difference among the RILs and among the parents, revealing that they are polymorphic among the parents. Within all populations, we observed an important variation in traits investigated and a transgression in both directions when in several cases the range of variation found in the RIL population extended beyond the variation in phenotypes of the parental genotypes. Phenotypic data showed that also evaluated trait values within the RIL populations approximately fit normal distributions and therefore the data are suitable for QTL mapping.

Acknowledgements This study is performed with financial support of European Social Fund co-financed project 2009/0218/1DP/1.1.1.2.0/09/APIA/VIAA/099. Thanks to Dr. K. Vaculova, Dr. B. Rossnagel and Dr. H. Buerstmayr for providing genotypes used as parents in crosses.

References

Berloo, R., & Stam, P. (1998). Marker assisted selection in autogamous RIL populations: A simulation study. *Theoretical and Applied Genetics, 96*, 147–154.

Bernando, R. (2008). Molecular markers and selection for complex traits in plants: Learning from the last 20 years. *Crop Science, 48*, 1649–1664.

Bleidere, M., & Belicka, I. (2009). Characteristic of grain quality for early generation lines in the crossings between covered and hulless barley. In *Research for Rural Development 2009. International Scientific Conference Proceedings* (pp. 14–20). Jelgava: LLU.

Buerstmayr, H., Steiner, B., Lemmens, M., & Ruckenbauer, P. (2000). Resistance to *fusarium* head blight in winter wheat: Heritability and trait associations. *Crop Science, 40*, 1012–1018.

Buerstmayr, H., Legzdina, L., Steiner, B., & Lemmens, M. (2004). Variation for resistance to *Fusarium* head blight in spring barley. *Euphytica, 137*, 279–290.

Derycke, V., Haesaert, G., Latre, J., Struik, P. C. (2002, June/July). *Relation between laboratory sprouting resistance tests and field observations in triticale genotypes*. Proceedings of the 5th Triticale Symposium, Radzikow, Poland, pp. 123–133.

Dill-Macky, R. (2003). Inoculation methods and evaluation of *Fusarium* head blight resistance in wheat. In P. E. Nelson, T. A. Toussoun, & R. J. Cook (Eds.), *Fusarium: Diseases, biology and taxonomy* (pp. 184–210). St Paul: APS Press.

Fastnaught, C. E., Berglund, P. T., & Holm, E. T. (1996). Genetic and environmental variation in β-glucan content and quality parameters of barley for food. *Crop Science, 36*, 941–946.

Fox, G. P., Panozzo, J. F., Li, C. D., Lance, R. C. M., Inkerman, P. A., & Henry, R. J. (2003). Molecular basis or barley quality traits. *Australian Journal of Agricultural Research, 54*, 1081–1101.

Gubler, F., Hughes, T., Waterhouse, P., & Jacobsen, J. (2008). Regulation of dormancy in barley by blue light and after-ripening: Effects on abscisic acid and gibberellin metabolism. *Plant Physiology, 147*, 886–896.

Legzdina, L. (2001). *Yield and grain quality of diverse origin hulless barley in Latvian growing conditions*. Youth Seeks Progress'2001. Paper collection of scientific conference of Ph.D. students, pp. 74–78.

Legzdina, L., & Buerstmayr, H. (2004). Comparison of infection with *Fusarium* head blight and accumulation of mycotoxins in grain of hulless and covered barley. *Journal of Cereal Science, 40*, 61–67.

Legzdina, L., Mezaka, I., & Beinarovica, I. (2010). Hulless barley (*Hordeum vulgare* L.) resistance to pre-harvest sprouting: Diversity and development of method for testing of breeding material. *Agronomy Research, 8*, 645–652.

Li, J., Baga, M., Rossnagel, B. G., Legge, W. G., & Chibbar, R. N. (2008). Identification of quantitative trait loci for β-glucan concentration in barley grain. *Journal of Cereal Science, 48*, 647–655.

Ma, Z., Steffenson, B. J., Prom, L. K., & Lapitan, N. L. (2000). Mapping of quantitative trait loci for *Fusarium* head blight resistance in barley. *Phytopathology, 90*, 1079–1088.

McCarthy, J. J., Canziani, O. F., Leary, N. A., & Dokken, D. J. (2001). *Climate change 2001: Impacts, adaptation, and vulnerability. intergovernmental panel on Climate Change (IPCC) Working Group II* (p. 1005). Cambridge: Cambridge University Press.

Mesfin, A., Smith, K. P., Dill-Macky, R., Evans, C. K., Waugh, R., Gustus, C. D., & Muehlbauer, G. J. (2003). Quantitative trait loci for *Fusarium* head blight resistance in barley detected in a two-rowed by six-rowed population. *Crop Science, 43*, 307–318.

Mezaka, I., Bleidere, M., Legzdina, L., & Rostoks, N. (2011). Whole genome association mapping identifies naked grain locus NUD as determinant of β-glucan content in barley. *Žemdirbystė=Agriculture* (in press).

Nesvadba, Z., Vyhnanek, T., Jeziskova, I., Tvaruzek, L., Spunarova, M., & Spunar, J. (2006). Evaluation of spring barley genotypes with different susceptibility to Fusarium head blight using molecular markers. *Plant, Soil and Environment, 52*(11), 485–491.

Rudd, J. C., Horsleyb, R. D., McKendryc, A. L., & Eliasb, E. M. (2001). Host plant resistance genes for Fusarium head blight sources, mechanisms, and utility in conventional breeding systems. *Crop Science, 41*, 620–627.

Semagn, K., Bjornstad, A., & Xu, Y. (2010). The genetic dissection of quantitative traits in crops. *Electronic Journal of Biotechnology, 13*(5), 1–14.

Takeda, K. (2004, June). *Inheritance of the Fusarium head blight resistance in barley*. Proceedings of the 9th International Barley Genetics Symposium, Brno, Czech Republic, pp. 302–307.

Ullrich, S. E., Clancy, J. A., del Blanco, I., Lee, H., Jitkov, V. A., Han, F., Keinhofs, A., & Matsui, K. (2008). Genetic analysis of pre-harvest sprouting in a six-row barley cross. *Molecular Breeding, 21*, 249–259.

Xu, Y., & Crouch, J. H. (2008). Marker-assisted selection in plant breeding: from publications to practice. *Crop Science, 48*(2), 391–407.

Xu, Z., Zou, F., & Vision, T. (2005). Improving quantitative trait loci mapping resolution in experimental crosses by the use of genotypically selected samples. *Genetics, 170*, 401–410.

Chapter 39
Molecular Mechanisms for Covered vs. Naked Caryopsis in Barley

Shin Taketa, Takahisa Yuo, Yuko Yamashita, Mika Ozeki, Naoto Haruyama, Maejima Hidekazu, Hiroyuki Kanamori, Takashi Matsumoto, Katsuyuki Kakeda, and Kazuhiro Sato

Abstract Typical barley cultivars have caryopses with adhering hulls at maturity, known as covered (hulled) barley. However, a few barley cultivars are a free-threshing variant called naked (hulless) barley. The covered vs. naked caryopsis is controlled by a single locus (*nud*) on chromosome arm 7HL. Positional cloning identified that an ERF (ethylene response factor) family transcription factor gene controls the covered vs. naked caryopsis phenotype. This conclusion was further supported by (1) fixation of the 17-kb deletion, harboring the ERF gene, among all 100 naked cultivars studied; (2) five induced *nud* alleles with a DNA lesion at a different site, each affecting the putative functional motif; and (3) gene expression strictly localized

S. Taketa (✉) • T. Yuo • Y. Yamashita
Group of Genetic Resources and Functions, Institute of Plant Science and Resources, Okayama University, 2-20-1 Chuo, Kurashiki, 710-0046, Japan
e-mail: staketa@rib.okayama-u.ac.jp

M. Ozeki • N. Haruyama
Tochigi Prefectural Agriculture Experiment Station, 2920 Otsuka-cho, Tochigi-City, Tochigi, 328-0007, Japan

M. Hidekazu
Nagano Agricultural Experiment Station, Suzuka Nagano 383-0051, Japan

H. Kanamori
Institute of Society for Techno-Innovation of Agriculture, Forestry, and Fisheries, Tsukuba, 305-8602, Japan

T. Matsumoto
National Institute for Agrobiological Sciences, Tsukuba, 305-8602, Japan

K. Kakeda
Graduate School of Bioresources, Mie University, Tsukuba, 514-8507, Japan

K. Sato
Group of Barley Resources, Institute of Plant Science and Resources, Okayama University, 2-20-1 Chuo, Kurashiki, 710-0046, Japan

G. Zhang et al. (eds.), *Advance in Barley Sciences: Proceedings of 11th International Barley Genetics Symposium*, DOI 10.1007/978-94-007-4682-4_39,

to the testa. Survey of natural variation at the *nud* locus indicates that naked barley has monophyletic origin but that covered barley is classified into some clusters, suggesting plural lineages. The *Nud* gene has homology to the *Arabidopsis WIN1/SHN1* transcription factor gene, whose deduced function is control of a lipid biosynthesis pathway. Staining with a lipophilic dye (Sudan Black B) detected a lipid layer on the pericarp epidermis only in covered barley. This observation indicates that in covered barley, lipids on the surface of caryopses act as a glue for their tight adhesion with hulls. Separation of hulls in naked barley is due to the absence of surface lipids on caryopses. Genetic complementation experiment is in progress toward functional validation of the *Nud* gene.

Keywords ERF transcription factor • Lipids • Positional cloning

39.1 Introduction

Barley typically has covered (hulled) caryopses in which the hull (outer lemma and inner palea) is firmly glued to the pericarp epidermis at maturity. But there also exist naked (hulless) variants whose caryopses are free threshing (Fig. 39.1). Both caryopsis types have agronomic values and are utilized in different purposes. Covered barley is mainly used for animal feed and brewing. The hull of covered barley is nutritionally low but serves as filtration medium during brewing. The hull also protects embryos from mechanical damage and ensures uniform germination when preparing malts. In contrast, naked barley is preferable for human food because extensive pearling to remove the hull is unnecessary; this minimizes loss of nutrients and minerals in the grain. Although barley consumption as human food is still limited, current consumers' interest in nutrition and health may boost the status of barley as the human diet rich in dietary fiber (Liu et al. 1996; Taketa et al. 2011).

The covered or naked caryopsis is controlled by a single locus (*Nud/nud*, for *nud*um) located on the long arm of chromosome 7H (Franckowiack 1996), and the covered caryopsis gene (*Nud*) is dominant over the naked one (*nud*). Harlan (1920) reported that in covered barley, a sticky adhesive substance appears 10 days after flowering on the surface of the caryopsis. Transmission electron microscope observation showed that a cementing substance is started to be secreted by the pericarp epidermis 2 days after flowering and increased its thickness during kernel development, but its chemical composition remains unknown (Gaines et al. 1985). To understand the mechanism underlying this character, we recently isolated the *Nud/nud* gene by means of positional cloning (Taketa et al. 2008). This chapter summarizes our *nud* gene cloning work. We have identified three additional naked caryopsis alleles of artificial origin since then, which provide excellent materials for functional analysis of the barley *Nud/nud* gene. Finally, on the basis of the molecular variation of the *Nud/nud* gene, the issue of the origin of naked barley is revisited.

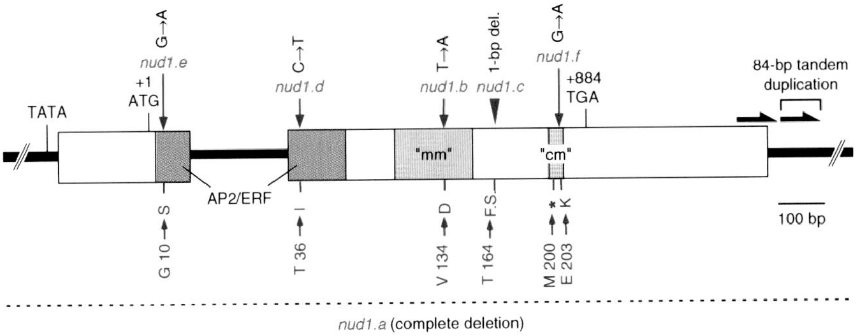

Fig. 39.1 Structure of *Nud* gene and mutation sites of five allelic-induced naked mutants (AP2/ERF domain, middle motif "mm," and C-terminal motif "cm" are shaded)

39.2 Materials and Methods

39.2.1 Plant Materials and Positional Cloning

Genetic mapping was performed in a total of 2,828 plants segregating for *nud* from two populations: Kobinkatagi (*nud*)×Triumph (*Nud*) and Karafuto Zairai (*Nud*)×Aizu Hadaka 3 (*nud*)

Chromosome walking from tightly linked markers was performed using the Haruna Nijo barley artificial chromosome (BAC) library (Saisho et al. 2007). Clones were selected eith er by PCR screening of DNA pools or hybridizing the arrayed BAC colony filters with digoxigenin-labeled probes. BAC DNA was extracted by a standard alkali-SDS procedure and analyzed by pulsed-field gel electrophoresis after digestion with *Not*I, *Asc*I, or *Pme*I. Overlapping BACs were identified either by Southern hybridization or by PCR marker analysis. A minimum tiling path of BAC clones encompassing the *nud* locus was shotgun sequenced. Annotation analyses were carried out according to Taketa et al. (2008).

39.2.2 Mutant Analysis

For validation of gene isolation, *nud* alleles of five induced naked mutants were analyzed together with their wild types. Mut.4129 and Mut.3041/6a are X-ray-induced mutants in the genetic background of Haisa's and Ackermann's Donaria, respectively (Scholz 1955). Three sodium azide-induced naked mutants are identified: AM6 from Fiber Snow and KM13 and KM14 both from Sachiho Golden. We confirmed allelism of these mutants to *nud* by crossing experiments. For analysis of natural variation in the *Nud* gene, 33 covered barley accessions of diverse origin were used.

39.2.3 Histochemical and Gene Expression Analyses

For these analyses, cv. Bowman and its near-isogenic naked line (abbreviated *nud*-Bowman) were used (a kind gift from Dr. J. Franckowiack). Plants were grown in a controlled environment room at a constant temperature of 15 °C under natural light. Dehulled caryopses were stained with a lipophilic dye Sudan Black B. RT-PCR and RNA *in situ* hybridization were performed as reported by Taketa et al. (2008).

39.2.4 Phylogenetic Analysis

The deduced amino acid sequences of the *Nud* gene and the ERF family transcription factor genes from Arabidopsis (AtWIN1/SHN1: AAR20494, AtSHN2: AAT44932, AtSHN3: AAO63881) and rice (OsSHN1: Os02g0202000, OsSHN2: Os06g0604000) were aligned using the computer program ClustalW (Thompson et al. 1994). A phylogenetic tree was constructed using the neighbor-joining method provided by the same program. Genomic sequences of the *Nud* gene from 33 covered accessions and two induced naked mutants were used for construction of the phylogenetic tree as described above.

39.3 Results

39.3.1 Positional Cloning of Nud

Fine mapping using flanking markers reported previously (Taketa et al. 2006) mapped *nud* to a 0.64-cM interval between markers sKT3 and sKT9. We then exploited barley/rice microsynteny for further localization. Integration of the flanking markers into a high-density barley expressed sequence tag (EST) map (Sato et al. 2009) selected two barley ESTs flanking the *nud* locus (accession no. BJ462032 for marker 3G12 and AV935407 for 82C6). BLASTN analysis identified their respective rice homologs (AK068856 and AK070667) 370 kilobase (kb) apart on the long arm of rice chromosome 6. Two rice genes (AK061163 and AK121264) within the collinear region were successfully used as vehicles to develop closer barley markers (ABRS3 and ABRS9). Starting with markers sKT9 and ABRS3, which are respectively 0.04 cM distal to and cosegregating with *nud*, we screened the barley BAC library. A 500-kb contig spanning the *Nud* locus was constructed (Taketa et al. 2008).

Four BAC clones that form the minimum tiling path were shotgun sequenced. The contig sequence is 244,108 base pairs (bp). Annotation showed two predicted genes, and about 50% of the contig sequence was classified as repetitive elements. One gene encodes putative iron-deficiency specific 4 (*ids-4*)-like protein, and the other is a transcription factor belonging to the ethylene response factor (ERF) family. The ERF transcription factor is the only gene that lies in the region delimited by mapping and therefore is considered a strong candidate of *Nud*.

To isolate the candidate gene from naked cultivars, PCR amplification was attempted. However, no fragment was amplified from naked cultivars. Detailed analysis using many PCR primer pairs revealed a deletion of 16,680 bp compared with the corresponding region of covered cultivar. The 16,680-bp deletion (hereafter called the 17-kb deletion) included the entire ERF gene. The 17-kb deletion is fixed in all 100 naked cultivars studies.

39.3.2 Sequence Analysis of Induced Naked Mutants

As shown in Fig. 39.1, each mutant carries a lesion that causes amino acid substitutions within functionally important domain/motif of the ERF gene. We named induced naked mutant alleles *nud1.b* (Mut.4129), *nud1.c* (Mut.3041/6a), *nud1.d* (AM6), *nud1.e* (KM13), and *nud1.f* (KM14), respectively, according to the proposal of Franckowiack and Konishi (1997). The sequence analysis of the induced mutants validated that the ERF gene comprises the *nud* locus, and therefore, we will refer to the gene as *Nud*.

39.3.3 Structure of *Nud*

Comparisons between the genomic and cDNA sequences revealed that *Nud* consists of two exons and one intron, and the ORF encodes a deduced protein of 227 amino acids (Fig. 39.1). The deduced amino acid sequence of *Nud* is 59 and 74% identical to *Arabidopsis* WIN1/SHN1 protein (AAR20494) and rice putative ERF protein (Os06g0604000) on chromosome 6, respectively. The functions of *WIN1/SHN1* are implicated in control of cutin and wax production (Aharoni et al. 2004; Broun et al. 2004; Kannangera et al. 2007).

39.3.4 Expression Analysis of *Nud*

RT-PCR showed that in Bowman, *Nud* is expressed only in the caryopses with a peak at 2 weeks after anthesis; no expression was detected in hulls or leaves. RNA *in situ* hybridization using the antisense probe showed that in Bowman, *Nud* is expressed strongly in the ventral side of the testa and very weakly expressed in the dorsal side of the testa (data not shown). Thus, these experiments revealed that *Nud* is expressed in the tissues where adherence occurs.

39.3.5 Histochemical Analyses

To determine whether developing barley caryopses accumulate lipids on its surface, we stained dehulled caryopses of the isogenic lines of Bowman and *nud*-Bowman with a lipophilic dye Sudan Black B. In Bowman, dehulled caryopses of 2 and 3 weeks old showed strong staining of the caryopsis surface except for areas over the embryo. In *nud*-Bowman, no staining of the caryopsis surface was observed at any stage studied. Staining of longitudinal grain section with Sudan Black B detected a clear lipid layer on the pericarp epidermis of Bowman only. Thus, the presence or absence of the lipid layer on the pericarp epidermis is a critical difference between covered and naked barleys.

39.3.6 Natural Allelic Variation in Nud

Natural allelic variation of *Nud* sequences was studied in 33 covered lines of diverse origin [22 domesticated (12 two-rowed and 10 six-rowed) and 9 *H. vulgare* subsp. *spontaneum* and 2 subsp. *agriocrithon* lines] and two induced naked mutants. Sequencing of about 1.7-kb region of the *Nud* gene including 5′ and 3′ noncoding region detected various types of nucleotide polymorphisms at 16 sites relative to the standard sequence of Haruna Nijo. Nucleotide changes were also observed in the first intron and 5′ and 3′ noncoding region. In the 3′ noncoding region, an 84-bp tandem duplication was detected in most domesticated lines (Fig. 39.1), but all subsp. *spontaneum* and *agriocrithon* lines sequenced did not carry it. Figure 39.2 shows a phylogenetic tree based on the *Nud* sequence. Sequence analysis revealed that covered barley has two lineages.

39.4 Discussion

By means of positional cloning, we revealed that *Nud* encodes an ERF family transcription factor gene in barley. This conclusion was validated by high-resolution genetic and physical mapping, annotation of the candidate region, fixation of the 17-kb deletion harboring the ERF gene among all 100 naked cultivars studied, five induced *nud* mutant alleles with an amino acid substitution(s) that affect the putative functional motif, and testa-specific gene expression. Fixation of the 17-kb deletion in all naked cultivars unequivocally shows its monophyletic origin, which was also suggested from our previous study using a closely linked marker (Taketa et al. 2004). For functional validation of the gene isolation, complementation experiments to produce transgenic barley are under the way.

The *Nud* gene is the barley ortholog of the *Arabidopsis WIN1/SHN1* transcription factor whose presumed function is regulation of a lipid biosynthesis pathway.

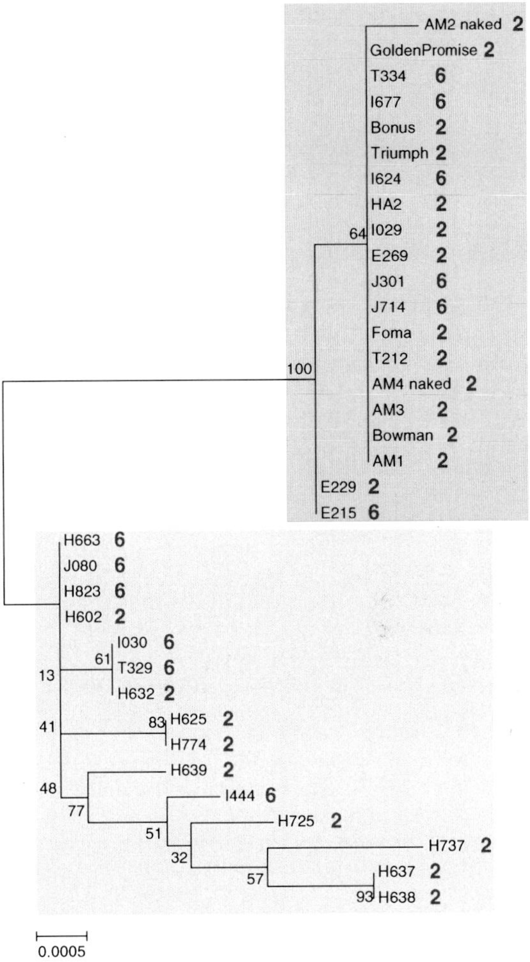

Fig. 39.2 Phylogenetic tree based on genomic sequences of the *Nud* gene from 33 covered accessions and two induced naked accessions (Numbers indicate bootstrap values. The *upper clade* contains accessions with an 84-bp duplication in the 3′ UTR, while the *lower clade* includes accessions without it. Numbers after the accession names, 2 and 6, indicate two-rowed and six-rowed, respectively)

The present results demonstrated that lipids play important roles in adhesion of caryopses to the hull. Chemical compositions of lipids involved in the caryopsis-hull adhesion needs to be identified.

The present study isolated naked caryopsis gene, which is one of important domestication genes in barley. The cloning of *Nud* in barley may help confer the naked caryopsis trait to wild *Hordeum* species with agronomic potential.

Acknowledgments We thank former students for technical assistance. Dr. M. Ichii is acknowledged for encouragement. This research was supported by grants from the Ministry of Agriculture, Forestry and Fisheries of Japan (TRC1007 and TRG1007) and Grant-in-Aid for Scientific Research (C) (18580006) from the Ministry of Education, Culture, Sports, Science and Technology, Japan.

References

Aharoni, A., Dixit, S., Jetter, R., Thoenes, E., van Arkel, G., & Pereira, A. (2004). The SHINE clade of AP2 domain transcription factors activates wax biosynthesis, alters cuticle properties, and confers drought tolerance when overexpressed in Arabidopsis. *The Plant Cell, 16*, 2463–2480.

Broun, P., Poindexter, P., Osborne, E., Jiang, C. Z., & Riechmann, J. L. (2004). WIN1, a transcriptional activator of epidermal wax accumulation in *Arabidopsis*. *Proceedings of the National Academy of Sciences of the United States of America, 101*, 4706–4711.

Franckowiack, J. D. (1996). Revised linkage maps for morphological markers in barley, *Hordeum vulgare*. Naked caryopsis. *Barley Genetics, 26*, 51–52. Special Issue.

Franckowiack, J. D., & Konishi, T. (1997). Naked caryopsis. *Barley Genetics Newsletter, 26*, 51–52.

Gaines, R. L., Bechtel, D. B., & Pomeranz, Y. (1985). A microscopic study on the development of a layer in barley that causes hull-caryopsis adherence. *Cereal Chemistry, 62*, 35–40.

Harlan, H. V. (1920). Daily development of kernels of Hannchan barley from flowering to maturity at Aberdeen, Idaho. *Journal of Agricultural Research, 19*, 393–429.

Kannangera, R., Branigan, C., Liu, Y., Penfield, T., Rao, V., Mouille, G., Hofte, H., Pauly, M., Riechmann, J. L., & Broun, P. (2007). The transcription factor WIN1/SHN1 regulates cutin biosynthesis in *Arabidopsis thaliana*. *The Plant Cell, 19*, 1278–1294.

Liu, C. T., Wesenberg, D. M., Hunt, C. W., Branen, A. L., Robertson, L. D., Burrup, D. E., Dempster, K. L., & Haggerty, R. J. (1996). *Hulless barley: A new look for barley in Idaho*. Resources for Idaho. http://info.ag.uidaho.edu/

Saisho, D., Myoraku, E., Kawasaki, S., Sato, K., & Takeda, K. (2007). Construction and characterization of a bacterial artificial chromosome (BAC) library from the Japanese malting barley 'Haruna Nijo'. *Breeding Science, 57*, 29–38.

Sato, K., Nankaku, N., & Takeda, K. (2009). A high-density transcript linkage map of barley derived from a single population. *Heredity, 103*, 110–117.

Scholz, F. (1955). Mutationsversuche an Kulturpflanzen. IV. *Kulturpflanze, 3*, 69–89.

Taketa, S., Kikuchi, S., Awayama, T., Yamamoto, S., Ichii, M., & Kawasaki, S. (2004). Monophyletic origin of naked barley inferred from molecular analyses of a marker closely linked to the naked caryopsis gene (*nud*). *Theoretical and Applied Genetics, 108*, 1236–1242.

Taketa, S., Awayama, T., Amano, S., Sakurai, Y., & Ichii, M. (2006). High-resolution mapping of the *nud* locus controlling the naked caryopsis in barley. *Plant Breed, 125*, 337–342.

Taketa, S., Amano, S., Tsujino, Y., Sato, T., Saisho, D., Kakeda, K., Nomura, M., Suzuki, T., Matsumoto, T., Sato, K., Kanamori, H., Kawasaki, S., & Takeda, K. (2008). Barley grain with adhering hulls is controlled by an ERF family transcription factor gene regulating a lipid biosynthesis pathway. *Proceedings of the National Academy of Sciences of the United States of America, 105*, 4062–4067.

Taketa, S., Yuo, T., Tonooka, T., Tsumuraya, Y., Inagaki, Y., Haruyama, N., Larroque, O., & Jobling, S. A. (2011). Functional characterisation of barley betaglucanless mutants demonstrates a unique role for CslF6 in (1,3;1,4)-β-D-glucan biosynthesis. *Journal of Experimental Botany*. doi:10.1093/jxb/err285.

Thompson, J. D., Higgins, D. G., & Gibson, T. J. (1994). CLUSTAL W: improving the sensitivity of progressive multiple sequence alignment through sequence weighting, position-specific gap penalties and weight matrix choice. *Nucleic Acids Research, 22*, 4673–4680.

Printed by Publishers' Graphics LLC
AMZ20130104.19.20.20